Inorganic Biochemistry

Volume 3

A Specialist Periodical Report

Inorganic Biochemistry
Volume 3

A Review of the Literature Published up to 1980

Senior Reporter
H. A. O. Hill *Inorganic Chemistry Laboratory, University of Oxford*

Reporters
J. R. Arthur *Rowett Research Institute, Aberdeen*
N. J. Birch *Wolverhampton Polytechnic*
I. Bremner *Rowett Research Institute, Aberdeen*
M. Brunori *University of Rome*
A. E. G. Cass *University of Oxford*
J. K. Chesters *Rowett Research Institute, Aberdeen*
R. R. Crichton *Catholic University of Louvain, Belgium*
A. Galdes *Harvard Medical School, Boston, Massachusetts, USA*
B. Giardina *University of Rome*
M. N. Hughes *Queen Elizabeth College, London*
H. A. Kuiper *University of Rome*
C. A. McAuliffe *UMIST, Manchester*
A. R. McEuen *University of Liverpool*
J.-C. Mareschal *Catholic University of Louvain, Belgium*
P. J. Sadler *Birkbeck College, London*

The Royal Society of Chemistry
Burlington House, London W1V 0BN

British Library Cataloguing in Publication Data

Inorganic biochemistry.–Vol. 3.–(Specialist periodical report/
    Royal Society of Chemistry)
    1. Biological chemistry – Periodicals
    2. Chemistry, Inorganic – Periodicals
    I. Royal Society of Chemistry
    547.19′ 214′05   QP531

    ISBN 0-85186-565-8
    ISSN 0142-9698

Printed in Great Britain by Adlard and Son Ltd
Bartholomew Press, Dorking

Foreword

The interest in the multifarious roles of the inorganic elements in biology continues to grow unabated. Consequently the task undertaken by the Reporters increases proportionately. I am very grateful to them for their efforts; I would like very much to know if readers appreciate their labours. It might be thought that such appreciation would naturally be reflected in sales of the Report, but, in these difficult times, this need not necessarily be so. Though such comments that have come my way are favourable, the main criticism is concerned with the lengthy interval between the final date of the period covered in the Report and that of publication. It is obvious that, if further volumes are to be successful, or even appear, different methods of preparation and production must be used. I would greatly appreciate receiving the views of readers on the Report; its content, format, and the use to which they put it.

Volume 3 follows the pattern set by those that preceded it. Unfortunately, it was not possible to include a chapter on redox proteins. It is my intention that extra space will be made available for an extended coverage in Volume 4.

H. A. O. HILL

Contents

Chapter 4   Oxygen-Transport Proteins   126
By M. Brunori, B. Giardina, and H. A. Kuiper

1
Inorganic Analogues of Biological Molecules

C. A. McAULIFFE

1 Complexes of Amino-Acids and Peptides

Simple Amino-Acids and Peptides.—The structural characteristics of the binding of calcium ion to the aminocarboxylates EDTA and NTA (nitrilotriacetate) are shown in Ca[Ca(EDTA)]·7H$_2$O and Na[Ca(NTA)]. In the former, the Ca^{2+} has eight-fold co-ordination with a sexidentate EDTA ligand and in the latter the Ca^{2+} has seven-fold co-ordination with a quadridentate NTA ligand.[1] Infrared band assignments have been made for cadmium glycinate mono-hydrate.[2]

The compound [Mo$_2$(gly)$_4$Cl$_2$]·xH$_2$O (gly = glycine) has been obtained in two different crystalline forms; form 1 (x = 3) contains two equivalent [Mo$_2$(gly)$_4$]$^{4+}$ ions per unit cell, with d(Mo—Mo) = 2.112 Å and weak axial co-ordination of Cl$^-$ [d(Mo $\cdots$ Cl) = 2.882 Å]. For form 2 (x = 2.67) there are three crystallo-graphically independent molecules, each residing on a crystallographic inversion centre.[3] When solutions of [Mo$_2${O$_2$CCH(NH$_3$)R}$_4$]$^{4+}$ and KNCS are mixed, red to red-purple complexes [Mo$_2${O$_2$CCH(NH$_3$)R}$_2$(NCS)$_4$]·nH$_2$O are obtained. For the glycinate (R = H, n = 1) and L-isoleucinate (R = CHMeEt, n = 4.5) there are cisoid arrangements of the amino-acid groups, and the four N-bonded NCS ions, together with the molybdenum atoms, form a sawhorse arrange-ment; the Mo–Mo distances are in the range 2.132—2.154 Å.[4] The solution structure and equilibria of vanadium(V), molybdenum(VI), and tungsten(VI) complexes formed by H$_4$EDTA, ethylenediamine-NN-diacetic (H$_2$EDDA), H$_3$NTA, and iminodiacetic (H$_2$IDA) acids show that formation constants of the 1:1 complexes increase with the number of chelate rings, as expected, and de-crease from V^V to Mo^{VI} and W^{VI} for a given ligand.[5]

First-row transition-metal(II) complexes of N-acetyl-DL-valine of the type [M(AcValO$^-$)]·xH$_2$O (M = Co or Ni, x = 2; M = Zn, x = 0) (AcValO$^-$ = N-acetyl-DL-valinate) and their amine adducts of the type [M(AcValO$^-$)$_2$]·B$_2$ (B = pyridine, 3- or 4-methylpyridine, or 1,10-phenanthroline) have been isolated; the Co^{II} and Ni^{II} compounds have MO$_6$ and MO$_4$N$_2$ chromophores.[6] In [Co(dien){(S)-glutamato-2}]$^+$ the (S)-glutamic acid behaves as a bidentate

[1] L. Barnett and V. A. Uchtman, *Inorg. Chem.*, 1979, **18**, 2674.
[2] M. L. Niven and D. A. Thornton, *Inorg. Chim. Acta*, 1979, **32**, 205
[3] A. Bino, F. A. Cotton, and P. E. Fanwick, *Inorg. Chem.*, 1979, **18**, 1719.
[4] A. Bino and F. A. Cotton, *Inorg. Chem.*, 1979, **18**, 1381.
[5] K. Zare, P. Lagrange, and J. Lagrange, *J. Chem. Soc., Dalton Trans.*, 1979, 1372.
[6] G. Marcotrigiano, L. Menaube, G. C. Pellacani, and M. Saladini, *Inorg. Chim. Acta*, 1979, **32**, 149.

ligand with a dangling $-(CH_2)_2COO^-$ group.[7] 1:1 Complexes between bivalent Co, Ni, and Cu and tetrapeptides containing tyrosine and glycine residues are formed in solution over a wide pH range; protons are ionized from terminal groups as well as from peptidic nitrogen atoms.[8] Resulting from the intramolecular addition of cobalt(III)-bound H_2O and OH^- to glycinamide, glycylglycine isopropyl ester and glycylglycine also co-ordinated to cobalt(III) in the *cis*-$[Co(en)_2(OH_2/OH)(GlyNHR)]^{3+/2+}$ ions ($R=H$, $CH_2CO_2C_3H_7$, or CH_2CO^-). Both the aquo- and the hydroxo-species form $[Co(en)_2(GlyO)]^{2+}$ for the dipeptide complex ($R = CH_2CO_2C_3H_7$), but loss of hydroxide also occurs, resulting in the chelated amide $[Co(en)_2(GlyNHR)]^{3+}$.[9] The c.d. spectra of complexes of the type $[Co(amOH)(N)_2(O)_2]$, where amOH = 2-aminoethanol or (*S*)-2-aminopropan-1-ol and $(N)_2(O)_2 = (gly)_2$, (β-ala)$_2$, (ox)(en), or (ox)$(NH_3)_2$, have been measured.[10] The kinetics and steric course of basic hydrolysis (displacement of X^-) of the species *trans*-$[Co(en)_2X\{O_2CCH(R)NH_2\}]^+$ show that, for the DL-alanine and DL-aminobutyric acid complexes, $65 \pm 5\%$ of *trans*- and $35 \pm 5\%$ of *cis*-$[Co(en)_2$-$OH\{O_2CCH(R)NH_2\}]^+$ are formed.[11] The quadridentate ligand (2*S*,2'*S*)-1,1'-(ethane-1,2-diyl)bis(pyrrolidine-2-carboxylic acid) dihydrochloride (H_2pren·2·HCl) has been derived from (*S*)-proline, and *cis*-α geometrical isomers have been obtained, *viz* Na$[Co(pren)(CO_3)]$·3H_2O, $[Co(pren)(H_2O)_2]ClO_4$·2H_2O, and $[Co(pren)(H_2O)Cl]$·1$\frac{1}{2}H_2O$.[12] The differential reactivity of the α-methylene protons of bis(pyridoxylideneglycinato)cobalt(III) may have inferences for reactions that are catalysed by vitamin B_6.[13] The X-ray crystal structure of *fac*(N)-Δ-tris-(L-asparaginato)cobalt(III) trihydrate shows that the amide groups of the two side-chains approach the central metal atom, and the oxygen atoms are connected through intramolecular hydrogen-bonds.[14] The separation of the meridional isomers of $[Co(\alpha_1\text{-}\alpha_2)_2]^-$ ($H\alpha_1$-$H\alpha_2$ is a dipeptide, H_2NCHR^1-$CONHCHR^2CO_2H$, ranging from Gly-Gly to L-Phe-L-Phe) has been achieved by Gillard and co-workers. Detailed comparison of the ^1H n.m.r. spectrum of the pairs of diastereoisomers enabled determination of their absolute configuration unambiguously for $[Co(L-Phe-GlyO)_2]^-$ and $[Co(Gly-L-PheO)_2]^-$.[15] The Hg^{2+}-catalysed removal of Br^- from *cis*-$[Co(en)_2Br(GlyNHR)]^{2+}$ results in the immediate formation of $[Co(en)_2(GlyNHR)]^{3+}$ containing the chelated amide or dipeptide; full retention of configuration about the Co^{III} centre obtains, and no intermediate aqua-complex is formed.[16] An effective t.l.c. technique has been developed for the separation of cobalt(III) geometrical isomers and diastereoisomers and various derivatized amino-acid and peptide ligands.[17]

[7] F. Jursik, B. Hájek, and M. S. Abdel-Moez, *Inorg. Chim. Acta*, 1979, **33**, L123.
[8] M. S. El-Eazby, J. M. Al-Hassan, N. F. Eweiss, and F. Al-Massaad, *Can. J. Chem.*, 1979, **57**, 104.
[9] C. J. Boreham, D. A. Buckingham, and F. R. Keene, *J. Am. Chem. Soc.*, 1979, **101**, 1409.
[10] K. Okazaki and M. Shibata, *Bull. Chem. Soc. Jpn.*, 1979, **52**, 1391.
[11] K. B. Nolan and A. A. Soudi, *J. Chem. Soc., Dalton Trans.*, 1979, 1419.
[12] T. C. Woon and M. J. O'Connor, *Aust. J. Chem.*, 1979, **32**, 1661.
[13] J. R. Fisher and E. H. Abbott, *J. Am. Chem. Soc.*, 1979, **101**, 2781.
[14] M. Sekizaki, *Bull. Chem. Soc. Jpn.*, 1979, **52**, 403.
[15] L. V. Boas, C. A. Evans, R. D. Gillard, P. R. Mitchell, and D. A. Phipps, *J. Chem. Soc., Dalton Trans.*, 1979, 582.
[16] C. J. Boreham, D. A. Buckingham, and F. R. Keene, *Inorg. Chem.*, 1979, **18**, 28.
[17] B. D. Warner and J. I. Legg, *Inorg. Chem.*, 1979, **18**, 1839.

Single-crystal X-ray structures of the hippurates $[M(hipp)_2(H_2O)_3] \cdot 2H_2O$ (M = Ni or Cu) show the compounds to be isostructural and to crystallize as linear chains with canted metal octrahedra.[18] Mixed ligand complexes [MHAL] and [MAL] (A = histamine, 1,10-phenanthroline, or α,α'-bipyridyl; L = O-phospho-DL-serine; M = Cu, Ni, Co, or Zn) contain terdentate O-phosphoserine in MAL for Ni, Co, or Zn but bidentate for Cu. Owing to the planar configuration in the Cu complex, the phosphate moiety is not bound.[19]

Farago and Mullen have examined the copper complex of a ninhydrin-positive ligand in the water extract from the roots of the copper-tolerant strain of *Armeria maritima*; this may be a copper–proline species.[20] An increasing number of Schiff-base amino-acid ligands are being complexed to copper(II). Thus, the glycine residue in N-salicylideneglycyl-L-valinatocopper(II) reacts with formaldehyde in aqueous solution at pH 8.5, and the resulting complex gives seryl-L-valine that contains optically active serine.[21] The copper complexes obtained from the reactions of Cu^{2+} with Schiff bases derived from histamine and salicylaldehyde or pyridine-2-carbaldehyde in acidic solutions are mono-nuclear, whilst tetranuclear complexes are obtained in basic solutions.[22] In μ-(N-salicylidene-L-valinato-O)-N-salicylidene-L-valinatodiaquadicopper(II) there is one square-pyramidal and one planar co-ordination of copper. The N-sali-cylidene-L-valinato-group occupies three positions of the basal plane around Cu-1 and a co-ordinated water molecule occupies the fourth position. An oxygen atom of the adjacent carboxylate group occupies the axial position. The basal plane of Cu-2 contains three atoms of the ligand and a water mole-cule.[23] Copper(II) complexes with terdentate Schiff-bases derived from the condensation of (+)-(hydroxymethylene)camphor or (+)-(hydroxymethylene)-menthone with a series of (S)- and (R)-amino-acids have been isolated; little interaction between the various chiral centres has been found, and the conforma-tion of the chelate rings depends mainly on the configuration of the α-carbon atom of the amino-acid.[24] The crystalline *trans*-isomer of bis(glycinato)copper-(II) is, contrary to many reports, a monohydrate. The dehydration of the solid *cis*-monohydrate at sufficiently high temperatures leads to the mainly anhydrous *trans*-complex, which readily re-hydrates to give the *trans*-monohydrate; this is a useful preparative technique.[25] The reaction of a methanolic suspension of CuCOCl with histamine (hm) at 0 °C gives a solution from which [Cu(hm)(CO)] BPh₄ is recovered, whilst a methanolic suspension of CuI in the presence of histamine reversibly absorbs carbon monoxide (1 molecule of CO per Cu atom), giving $[Cu_2(hm)_3(CO)_2]^{2+}$. An X-ray analysis of the BPh_4^- salt shows the presence of a dimeric cation, with one histamine molecule chelated to each copper and the other

[18] M. M. Morelock, M. L. Good, L. M. Trefonas, D. Karraker, L. Maleki, H. R. Eichel-berger, R. Majeste, and J. Dodge, *J. Am. Chem. Soc.*, 1979, **101**, 4858.
[19] M. S. Mohan, D. Bancroft, and E. H. Abbott, *Inorg. Chem.*, 1979, **18**, 2468.
[20] M. E. Farago and W. A. Mullen, *Inorg. Chim. Acta*, 1979, **32**, L93.
[21] S. Suzuki, H. Narita, and K. Harada, *J. Chem. Soc., Chem. Commun.*, 1979, 29.
[22] Y. Nakao, W. Mori, N. Okuda, and A. Nakahara, *Inorg. Chim. Acta*, 1979, **35**, 1.
[23] K. Korhonen and R. Hamalainen, *Acta. Chem. Scand., Ser. A*, 1979, **33**, 569.
[24] L. Casella, M. Gullotti, A. Pasini, and A. Rockenbauer, *Inorg. Chem.*, 1979, **18**, 2825.
[25] B. W. Delf, R. D. Gillard, and P. O'Brien, *J. Chem. Soc., Dalton Trans.*, 1979, 1301.

one bridging the two metal atoms.[26] Crystal structures of bis-(L-leucinato)-
and bis-(D,L-2-aminobutyrato)-copper(II) show both to contain tetragonally
co-ordinated copper ions, arranged in isolated sheets. Equatorial N_2O_2 ligation is
provided by the *trans* co-ordination of two amino-acids, whilst axialCu—O ligation
by two neighbouring amino-acids completes the co-ordination of the metal and
links the CuL_2 units to form carboxylate-bridged sheets of Cu^{II} ions.[27] The lines
in the n.m.r. spectra of a single crystal of *trans*-[Cu(DL-Ala)$_2$]·H$_2$O are shifted
by both the electron–nuclei dipole–dipole and the Fermi-contact interactions;
analysis of the spectra can yield both Cu–proton distances and isotropic coupling
constants of protons.[28] A dependence of the paramagnetic line-broadening of n.m.r.
spectra on pH has been observed for histidine- (*e.g.* Gly-Gly-His) and glycine-
containing (*e.g.* Gly, Gly-Gly, Gly-Gly-Gly, and Gly-Gly-Gly-Gly) peptide–cop-
per(II) systems.[29] A cyanogen complex $CuC_3N_3H_3O$ gives glycine on hydrolysis.[30]
Potentiometry of the system of copper(II) and L-3-(3,4-dihydroxyphenyl)-2-methyl-
alanine (methyldopa) in aqueous solution, using a glass electrode, shows that
both mononuclear and oligonuclear complexes occur, the former including a
series of successively deprotonated species at higher ligand:metal ratios.[31] An
equilibrium study of the mixed-ligand complexes of copper(II) and amino-acids
and 2,2′-bipyridyl with thiodicarboxylic and pyridinedicarboxylic acids has been
reported.[32] Margerum and co-workers have continued their work on copper(III)–
peptide complexes, and report electron-transfer reactions of these species with
hexachloroiridate(III),[33] the oxidative decarboxylation of glyoxalate ion by a
deprotonated-amine copper(III)–peptide complex,[34] and the electron-transfer
reactions of copper(III)–peptide complexes with [Co(phen)$_3$]$^{2+}$.[35]

A series of trimethylplatinum(IV) complexes [PtMe$_3$(AA)$_2$]$^-$ and [PtMe$_3$(AA)L]
(HAA = alanine, valine, phenylalanine, or α-aminoisobutyric acid) have been
isolated; mixtures of diastereoisomers are formed when the amino-acid also
contains an asymmetric carbon atom.[36] The solution and solid-state c.d. spectra
of the amino-acid [(S)-serine, (S)-valine, and (S)-proline] complexes of bivalent
Pd and Pt show a fairly consistent pattern, which is opposite to that shown by
the dipeptide [Gly-(S)-Ala and (S)-Ala-Gly] complexes.[37] The c.d. spectra of
bivalent Pd, Ni, and Cu complexes of NN'-bis-(L-alanyl)-1,3-propanediaminate
anion, ML, are discussed in comparison with those of *cis*- and *trans*-bis(amino-
carboximidato)-complexes.[38] The nature of diastereoisomeric discrimination in

[26] M. Pasquali, C. Floriani, A. Gaetani-Manfredotti, and C. Guastini, *J. Chem. Soc.,
Chem. Commun.*, 1979, 197.
[27] T. G. Fawcett, M. Ushay, J. P. Rose, R. A. Lalancette, J. A. Potenza, and H. J. Schugar,
Inorg. Chem., 1979, **18**, 327.
[28] T. Sandreczki, D. Ondercin, and R. W. Kreilick, *J. Am. Chem. Soc.*, 1979, **101**, 2880.
[29] Y. Kuroda and H. Aiba, *J. Am. Chem. Soc.*, 1979, **101**, 6837.
[30] M. T. Beck, V. Gaspar, and J. Ling, *Inorg. Chim. Acta*, 1979, **33**, L177
[31] C. V. Fazakerley, P. W. Linder, R. G. Torrington, and M. R. W. Wright, *J. Chem. Soc.,
Dalton Trans.*, 1979, 1872.
[32] D. N. Shelke, *Inorg. Chim. Acta*, 1979, **32**, L45.
[33] G. D. Owens, K. L. Chellappa, and D. W. Margerum, *Inorg. Chem.*, 1979, **18**, 960.
[34] S. T. Kirksey and D. W. Margerum, *Inorg. Chem.*, 1979, **18**, 966.
[35] J. M. Dekorte, G. D. Owens, and D. W. Margerum, *Inorg. Chem.*, 1979, **18**, 1538.
[36] T. G. Appleton, J. R. Hall, and T. G. Jones, *Inorg. Chim. Acta*, 1979, **32**, 127.
[37] E. A. Sullivan, *Can. J. Chem.*, 1979, **57**, 62.
[38] T. Komorita and Y. Shimura, *Bull. Chem. Soc. Jpn.*, 1979, **52**, 1832.

the series of platinum(II) complexes *trans*(*N,olefin*)-chloro[*N*-methyl-(*S*)-prolinato](olefin)platinum has been investigated by ^{1}H, ^{13}C, and ^{195}Pt n.m.r. methods.[39]

Dimethylgold(III) complexes with the anions of glycine, alanine, valine, cysteine, histidine, and imidazole, *i.e.* [Me$_2$AuL], have been the subject of an infrared spectroscopic study.[40] The kinetics and mechanism of formation of pentammine(glycine)rhodium(III) ion from pentammineaquarhodium(III) ion and glycine in weakly acidic media have been reported.[41]

Sulphur-containing Amino-Acids.—The three possible donor sites in methionine have attracted interest for some time. To the [Cu(met)$_2$] complex that was originally prepared[42] has recently been added the copper(III) complex of L-methionine methyl ester.[43] The L-methionylglycinato-group acts as a quinquedentate ligand (involving amino-nitrogen, ionized amide-nitrogen, one carboxylato-, and one bridged carboxylato-group) towards copper(II).[44] *cyclo*-L-Methionyl-L-methionine has been shown to co-ordinate as a neutral species to palladium(II), gold-(III), and copper(II).[45]

Organotin chlorides have been shown to bind to L-cysteine, DL-penicillamine, L-cysteine ethyl ester, *N*-acetyl-L-cysteine, L-cysteic acid, and glutathione.[46] The mononuclear molybdenum(V) complexes of cysteinyl peptide are suggested to resemble those of the molybdenum centres in nitrate reductase and other molybdo-enzymes.[47] Some reactions of triorganotin(IV) compounds with L-cysteine, L-cysteine ethyl ester, *N*-acetyl-L-cysteine, and glutathione have also been reported.[48] The *X*-ray crystal structure of (2-aminothiazoline)cobalt(III) shows N$_5$O co-ordination around the cobalt, with the cysteinyl sulphur unbound.[49] Four kinds of [Co(N)$_2$(O)$_3$(S)] mixed complexes (L- or D-aspartato, or L-methioninato) have been prepared.[50]

D-Penicillamine forms species MA and MA$_2$ (M=Ni) that contain *N,S* co-ordination in aqueous solution.[51] Copper and zinc complexes of Schiff-base ligands that contain penicillamine have been prepared.[52]

The fourth *N*-terminal amino-acid in the peptide chain of corticotropin is methionine. The electron-pair-donor ability of the thioether sulphur atom in

[39] S. Shinoda, Y. Yamaguchi, and Y. Saito, *Inorg. Chem.*, 1979, **18**, 673.
[40] R. S. Tobias, C. E. Rice, W. Beck, B. Purucker, and K. Bartel, *Inorg. Chim. Acta*, 1979, **35**, 11.
[41] C. Chatterjee and A. K. Basak, *Bull. Chem. Soc. Jpn.*, 1979, **52**, 2710.
[42] C. A. McAuliffe, J. V. Quagliano, and L. M. Vallarino, *Inorg. Chem.*, 1966, **5**, 1996.
[43] H. Kozlowski and T. Kowalik, *Inorg. Chim. Acta*, 1979, **34**, L231.
[44] J. Dehand, J. Jordanov, F. Keck, A. Mosset, J. J. Bonnet, and J. Galy, *Inorg. Chem.*, 1979, **18**, 1543.
[45] R. Ettorre, V. Guantieri, A. M. Tamburro, and L. De Nardo, *Inorg. Chim. Acta*, 1979, **32**, L39.
[46] G. Domazetis, M. F. Mackay, R. J. Magee, and B. D. James, *Inorg. Chim. Acta*, 1979, **34**, L247.
[47] C. D. Garner, F. E. Mabbs, and D. T. Richens, *J. Chem. Soc., Chem. Commun.*, 1979, 415.
[48] G. Domazetis, R. J. Magee, and B. D. James, *Inorg. Chim. Acta*, 1979, **32**, L48.
[49] G. J. Gainsford, W. G. Jackson, and A. M. Sargeson, *J. Am. Chem. Soc.*, 1979, **101**, 3966.
[50] T. Isago, K. Igi, and J. Midaka, *Bull. Chem. Soc. Jpn.*, 1979, **52**, 407.
[51] S. H. Laurie, D. H. Prime, and B. Sarkar, *Can. J. Chem.*, 1979, **57**, 1411.
[52] L. Macdonald, D. H. Brown, and W. E. Smith, *Inorg. Chim. Acta*, 1979, **33**, L183.

the macromolecule has been elucidated on the basis of its ability to co-ordinate to silver(I) ion.[53]

Amino-Acids that contain Heterocyclic *N*-Donors. The six isomers of the [Co(L- or D-Asp)(L-His)] complex have been prepared and the isomerization in the absence of any catalyst has been studied. The equilibrium mole fractions of these isomers in water have been found for L-*trans*O_5cisN_5, L-*cis*$O_5transN_5$, and L-*fac*, to be 0.53, 0.06, and 0.41, respectively.[54] The *X*-ray crystal structure of *trans*-amine-bis-(L-histidinato)cobalt(III) perchlorate dihydrate shows octahedral co-ordination with the amino-nitrogen, the imidazole nitrogen, and a carboxylate oxygen atom of each histidinate. The two imidazole groups, as well as the two carboxylate groups, are ligated *cis* to one another; the amines are *trans*.[55] Some linear and circular dichroism spectra in the visible range for two isomers of the bis(histidinato)cobalt(III) complex in the crystal phase, *i.e.* all-*cis*- and *trans*-amine, have been reported.[56] The anion D-$H_2NCH(CH_2C_5H_4N)$-CO_2^-, a terdentate analogue of histidinate, complexes with cobalt(III) to form [Co(D-PyAla)$_2$]NO$_3\cdot\frac{1}{2}H_2O$; the most stable form of this is that in which the carboxylate groups of the two ligands are mutually *trans*.[57] The hydrolysis of *cis*-[CoCl(en)$_2$(imH)]$^{2+}$ (imH = imidazole) by base has been studied over the pH range 8.26—9.74.[58]

N-Acetyl-DL-tryptophanate forms the complexes [M(AcTrp)$_2$]·xH$_2$O (M = Co or Zn, x = 2; M = Ni, x = 3) and amine adducts [M(AcTrp)$_2$]·B$_2$ (B = py, 3-pic, or 4-pic); in all cases, carboxylate co-ordination occurs.[59] Stability constants of heterobinuclear complexes of Cu(L-His) with bivalent nickel, zinc, and cadmium in water have been reported.[60] It has been observed that a terdentate Schiff-base complex of nickel(II), *i.e.* triaquotribenzo[*b,f,j*]triazacyclodecanenickel(II), shows substantial stereoselectivity on complexing with several amino-acids; it can be resolved, using histidine.[61] The kinetics of complexing of this species by histidine, 3-methylhistidine, histamine, histidine methyl ester, and glycine have been reported.[62] Three-component equilibria between nickel(II), imidazole, and OH$^-$ have been measured, using various concentrations of these reagents.[63]

Ternary copper(II) complexes [Cu(his)(AA)] (his = L- or D-histidinate; AA = L- or D-asparaginate, L-glutaminate, L-serinate, L-homoserinate, or L-citrullinate) have been isolated as crystals. Facile isolation of these ternary complexes and successful resolution of DL-histidine have been interpreted in terms of the solubility differences and the intramolecular hydrogen-bond between COO$^-$ of the ligand (his) and the amide or hydroxyl group in the side-chain of AA.[64]

[53] B. Noszal and K. Burger, *Inorg. Chim. Acta*, 1979, **35**, L387.
[54] M. Watabe, H. Yano, and S. Yoshikawa, *Bull. Chem. Soc. Jpn.*, 1979, **52**, 61.
[55] N. Thorup, *Acta Chem. Scand., Ser. A*, 1979, **33**, 759.
[56] H. P. Jensen, *Acta Chem. Scand., Ser. A*, 1979, **33**, 563.
[57] S. R. Ebner, R. A. Jacobson, and R. J. Angelici, *Inorg. Chem.*, 1979, **18**, 765.
[58] R. W. Hay, M. Tajik, and P. R. Norman, *J. Chem. Soc., Dalton Trans.*, 1979, 636.
[59] G. Marcotrigiano, L. Antolini, L. Menaube, and G. C. Pellacani, *Inorg. Chim. Acta*, 1979, **35**, 177.
[60] P. Amico, P. G. Daniele, G. Arena, G. Ostacoli, E. Rizzarelli, and S. Sammartano, *Inorg. Chim. Acta*, 1979, **35**, L383.
[61] B. Erno and R. B. Jordan, *Can. J. Chem.*, 1979, **57**, 883.
[62] R. B. Jordan and B. Erno, *Inorg. Chem.*, 1979, **18**, 2893.
[63] W. Forsling, *Acta Chem. Scand., Ser. A*, 1979, **33**, 641.
[64] O. Yamauchi, T. Sakurai, and A. Nakahara, *J. Am. Chem. Soc.*, 1979, **101**, 4164.

Solution equilibria in the ternary copper(II)–L-histidine–diglycl-L-histidine system have also been reported.[65] L-Mimosine [α-amino-β-(3-hydroxy-4-oxo-1,4-dihydropyridin-1-yl)propanoic acid] binds bivalent copper and zinc ions more strongly than do more simple amino-acids.[66]

The reactions of OsO_4 with model side-chains of tissue protein (the ligands used include histidine and imidazole derivatives) suggest a possible role for OsO_4 in cross-linking proteins and lipids during fixation of biological tissues.[67] Some complexes of imidazole with dioxouranium(VI) nitrate and acetate have been isolated.[68]

2 The Binding of Small Molecules by Transition-Metal Complexes

Dioxygen.—This is an area of considerable growth, not only in investigations of the binding of dioxygen by simple transition-metal complexes, but also, as will be seen later, by macrocyclic complexes. Not surprisingly, interest is still strong in cobalt-based systems. Thus, the observation of the first O–O stretching frequency (1137 cm^{-1}) in a mononuclear cobalt(II) complex in pyridine–DMF solution has been reported.[69] E.p.r. measurements on the dioxygen adducts $[Co^{II}L]\cdot O_2$ of a number of quadridentate Schiff-base chelates in pyridine solution show some variation in the hyperfine parameters of cobalt, and the Co–O–O angle lies in the range 115—120°.[70] An X-ray crystal structure of the dioxygen adduct of aquo[NN'-(1,1,2,2-tetramethylethylene)bis-(3-methoxysalicylideniminato)]cobalt(II) shows planar co-ordination of the ligand, and the dioxygen molecule is disordered between two positions, with an average $d(Co-O)$ of 1.88Å and $d(O-O)$ of 1.25Å.[71] A crystal structure of the reversible dioxygen carrier $[Co(pyDPT)]_2\cdot O_2\cdot I_4\cdot 3H_2O$ shows that the pyridyl nitrogens are *cis* to one another and to the O_2 bridge in the disturbed octahedral geometry around the cobalt atoms, so that the imine nitrogen is *trans* to the O_2 bridge. The $d(O-O)$ of 1.456Å is consistent with a peroxide group, and thus the cobalt is in a tripositive oxidation state. The differences between complexes of pyDPT (1) and pyDIEN (2) have been compared.[72] The reactions of [(tren)Co(O_2,tren)Co-(tren)]$^{4+}$, a new complex, have been compared with those of [(tren) (NH$_3$) (CoO$_2$-Co)(NH$_3$)(tren)]$^{4+}$. In acidic solution, the former mainly decomposes to Co^{II}

(1)

[65] T. Sakurai and A. Nakahara, *Inorg. Chim. Acta*, 1979, **34**, L245.

[66] H. Stunzi, D. D. Perrin, T. Teitei, and R. L. N. Harris, *Aust. J. Chem.*, 1979, **32**, 21.

[67] A. J. Nielson and W. P. Griffiths, *J. Chem. Soc., Dalton Trans.*, 1979, 1084.

[68] A. Marzotto, M. Nicolini, F. Braga, and G. Pinto, *Inorg. Chim. Acta*, 1979, **34**, L295.

[69] T. Szymanski, T. W. Cape, R. P. Van Duyne, and F. Basolo, *J. Chem. Soc., Chem. Commun.*, 1979, 5.

[70] R. L. Lancashire, T. D. Smith, and J. R. Pilbrow, *J. Chem. Soc., Dalton Trans.*, 1979, 66.

[71] B. T. Huie, R. M. Leyden, and W. P. Schaeffer, *Inorg. Chem.*, 1979, **18**, 125.

[72] J. H. Timmons, A. Clearfield, A. E. Martell, and R. H. Niswander, *Inorg. Chem.*, 1979, **18**, 1042.

(2)

and O_2.[73] Potentiometric, visible—u.v., and n.m.r. measurements of the oxygenation of aqueous solutions in which there are 5:2 and 3:1 molar ratios of ethylenediamine to cobalt(II) indicate that the equilibrium product is tetrakis(ethylenediamine)-μ-(ethylenediamine)-μ-peroxo-dicobalt(III); $[Co_2(trien)(en)(O_2)]$[44] has also been identified in solution.[74] Dioxygen(pyridine)-NN'-ethylenebis(acetylacetoniminato)cobalt(II) reacts[75] with acids in organic solvents that contain excess pyridine to yield 0.5 mole of O_2, 0.5 mole of H_2O_2, and 1 mole of [Co(py)$_2$(ligand)]$^+$. Co-condensation reactions of cobalt atoms with CO/O_2 mixtures at 10—12 K give rise to a plethora of reaction products, including $[Co(CO)_4(O_2)]$, that involve O_2^-.[76]

The 8-quinolinato ($=Q$) complex $[VO(Q)_2(py)]$ reacts very easily at room temperature and atmospheric pressure with O_2 to produce the diamagnetic dinuclear μ-oxo-complex $[(Q)_2OV\text{-}O\text{-}VO(Q)_2]$.[77] Amongst the manganese complexes reported is the large range of complexes $[Mn(L)X_2]$ (L = tertiary phosphine, but not PPh$_3$), which rapidly and reversibly react with O_2 to form $[MnL(O_2)X_2]$. These interesting complexes $[Mn(L)X_2]$ also reversibly bind other small molecules, *e.g.* CO.[78] Taylor's group have continued their investigations of salen-type ligand complexes of manganese(II), *i.e.* Mn^{II} (ligand dianion), which react with O_2 to form manganese(IV) complexes that contain μ-dioxobridges.[79] The reactions of $[Mn^{III}(EDTA)]^-$ and related species with superoxide have been followed by stopped-flow methods.[80] The reaction of [Ni(Gly-Gly-L-His)] with O_2 results in a yellow species, which may be decarboxylated.[81] The disproportionation of H_2O_2 (*i.e.* catalase-like activity) that is caused by Cu^{2+} in aqueous solution is promoted by the co-ordination of 2,2'-bipyridyl to the copper.[82]

Direct transfer of oxygen in the intermediate rhodium–dioxygen complex is the preferred mechanism for the [(Ph$_3$E)$_3$RhCl]-catalysed co-oxidation of terminal olefins and the Ph$_3$E (E = P or As) ligands.[83] Some new binuclear palladium complexes, involving chloride bridges and both metal–carbon σ- and

[73] M. Zehnder, U. Thewalt, and S. Fallab, *Helv. Chim. Acta*, 1979, **62**, 2099.
[74] M. Crawford, S. A. Bedell, R. I. Patel, L. W. Young, and R. Nakon, *Inorg. Chem.*, 1979, **18**, 2075.
[75] J. J. Pignatello and F. R. Jensen, *J. Am. Chem. Soc.*, 1979, **101**, 5929.
[76] G. A. Ozin, A. J. L. Hanlon, and W. J. Power, *Inorg. Chem.*, 1979, **18**, 2390.
[77] M. Pasquali, A. Landi, and C. Floriani, *Inorg. Chem.*, 1979, **18**, 2397.
[78] C. A. McAuliffe, H. Al-Khateeb, M. H. Jones, W. Levason, K. Minten, and F. P. McCullough, *J. Chem. Soc., Chem. Commun.*, 1979, 736.
[79] S. J. E. Titus, W. M. Barr, and L. T. Taylor, *Inorg. Chim. Acta*, 1979, **32**, 103.
[80] J. Stein, J. P. Fackler, G. J. McClune, J. A. Fee, and L. T. Chan, *Inorg. Chem.*, 1979, **18**, 3511.
[81] T. Sakurai and A. Nakahara, *Inorg. Chim. Acta*, 1979, **34**, L243.
[82] H. Sigel, K. Wyss, B. E. Fischer, and B. Prijs, *Inorg. Chem.*, 1979, **18**, 1354.
[83] R. Tang, F. Mares, N. Neary, and D. E. Smith, *J. Chem. Soc., Chem. Commun.*, 1979, 274.

π-bonds, react with superoxide, leading to the formation of new dioxygen complexes that have olefinic ligands.[84] A ^{1}H n.m.r. double-resonance technique has been used to detect pure n.q.r. resonances of $^{17}O-^{17}O$ that is reversibly bonded to $[Ir(O_2)Cl(CO)(PPh_3)_2]$. Two sets of lines are observed, corresponding to $e^2qQ = 16.9$ MHz with $v = 0.7$ and to $e^2qQ = 15.6$ MHz with $v = 0.9$. This reveals the inequivalence of the charge distributions about each oxygen atom; this is a phenomenon that was not previously revealed by internuclear distances that were determined by using X-ray diffraction.[85]

Though not of direct interest here, mention is made of the unlikelihood of an electron-transfer (Haber–Weiss) reaction between superoxide and peroxides;[86] the entropy-controlled reactivity of singlet O_2 towards furans and indoles in toluene;[87] the detection of O_2^- in aqueous media by N-halogeno-pyridines;[88] the transfer of O_2 from 4a-hydroperoxyflavin anion to a phenolate ion, which is a flavin-catalysed dioxygenation reaction;[89] and the fact that no evidence for ozone–olefin complexes has been obtained in ozone–olefin matrices at low temperatures.[90]

Carbon Monoxide and Carbon Dioxide.—The vibrational frequency of carbon monoxide has been measured when it interacts with uni-, bi-, and ter-valent metal fluorides in argon matrices. It has been found to shift to higher values relative to free CO in all cases. Within an isovalent group, the shifts vary with the reciprocal of the ionic radius of the metal.[91] The reaction of CO with $[Ta(\eta^5\text{-}C_5H_5)Me_4]$, involving intramolecular coupling of CO, proceeds *via* an 'η^2-acetone' intermediate.[92] The catalytic reaction between CO and N_2O over Cr_2O_3 involves a rate-controlling step which includes a carbonate-like species during the consecutive decomposition of two N_2O molecules.[93]

The NN'-bis-(3-thioxo-1-methylbutylidene)ethylenediaminatocobalt(II) complex adds CO to become pentaco-ordinate.[94] It has been shown that co-ordinated carbon monoxide can be converted into ethene with $> 95\%$ selectivity.[95] Trifluorophosphine and CO react with $CuAsF_6$ in SO_2 solution to form 1:1 addition compounds, the vibrational spectra of which are consistent with the formulations $[CuPF_3]^+ [AsF_6]^-$ and $[CuCO]^+ [AsF_6]^-$ {this has the highest value for a $v(CO)$ stretch, of 2180 cm^{-1}, yet observed for a monocarbonyl}.[96]

The reaction of the rhodium(I)–hydride complexes $[RhH(PPr^i_3)_3]$, $[RhH(N_2)(PPhBu^t_2)_2]$, and $[Rh_2H_2(\mu\text{-}N_2)\{P(c\text{-}C_6H_{11})\}_4]$ with CO_2 in the presence of H_2O

[84] H. Suzuki, K. Mizutani, Y. Moro-oka, and T. Ikawa, *J. Am. Chem. Soc.*, 1979, **101**, 748.
[85] O. Lumpkin, W. T. Dixon, and J. Poser, *Inorg. Chem.*, 1979, **18**, 982.
[86] M. J. Gibian and T. Ungermann, *J. Am. Chem. Soc.*, 1979, **101**, 1291.
[87] A. A. Gorman, G. Lovering, and M. A. Rodgers, *J. Am. Chem. Soc.*, 1979, **101**, 3050.
[88] A. Rigo, E. Argese, E. F. Orsega, and P. Viglino, *Inorg. Chim. Acta*, 1979, **35**, 161.
[89] C. Kemal and T. C. Bruice, *J. Am. Chem. Soc.*, 1979, **101**, 1635.
[90] B. Nelander and L. Nord, *J. Am. Chem. Soc.*, 1979, **101**, 3769.
[91] R. H. Mange, S. E. Gransden, and J. L. Margrave, *J. Chem. Soc., Dalton Trans.*, 1979, 745.
[92] C. D. Wood and R. R. Schrock, *J. Am. Chem. Soc.*, 1979, **101**, 5421.
[93] B. W. Krupay and R. A. Ross, *Can. J. Chem.*, 1979, **57**, 718.
[94] M. Sakadura, Y. Sasaki, M. Matsui, and T. Shigematsu, *Bull. Chem. Soc. Jpn.*, 1979, **52**, 1861.
[95] C. D. Desjardins, D. B. Edwards, and J. Passmore, *Can. J. Chem.*, 1979, **57**, 2714.
[96] J. Passmore, personal communication.

affords[97] the novel dihydro-bicarbonate complexes $[RhH_2(O_2COH)L_2]$ (L= phosphine). Carbon dioxide is quantitatively deoxygenated to carbon monoxide by $[Cp_2TiCl]_2$, which is converted into $[Cp_2TiCl]_2O$, while $[Cp_2Ti(CO)_2]$ promotes the disproportionation of CO_2 to CO and carbonate.[98] Reduction of [Co(salen)] with Li or Na in THF affords the bimetallic systems [Co(salen)NaTHF] and $[Co(salen)Li(THF)_{1.5}]$,[99] which are effective in fixing carbon dioxide.

Nitrogen Oxides.—The reaction of Ag_2O with NO, studied in the range 333–473 K, produces $AgNO_2$, $AgNO_3$, Ag, and NO_2.[100] The preparation of the nucleophile trans-$[RuCl(NO)L]$ $(L=Ph_2PCH_2C_{18}H_{10}CH_2PPh_2)$ and those of the pentaco-ordinate $[RuCl(CO)(NO)L]$ and $[RuCl(CO)(NO)(Ph_2PCH_2Ph)_2]$ have been reported.[101] The complex $[Co(saloph)(py)(NO_2)]$ (saloph = NN'-bis-salicylidene-o-phenylenediamino) can be used in the stoicheiometric and catalytic oxidation of Ph_3P to Ph_3PO by an oxygen-transfer mechanism.[102] The reaction of trans-$[Ni(NO_2)_2(PEt_3)]$ with CO under mild conditions has been shown to afford CO_2 in addition to $[Ni(NO_2)(NO)(PEt_3)_2]$. An ^{18}O-labelling study, using trans-$[Ni(N^{18}O_2)_2(PEt_3)_2]$ and CO, has unambiguously established NO_2^- as the source of oxygen in the conversion of CO into CO_2.[103]

Sulphur Dioxide.—An important study of the diagnostic features of co-ordination geometries of transition metal–SO_2 complexes has been reported by Kubas; the co-ordination geometries of SO_2 were coplanar (MSO_2), pyramidal (MSO_2), and bridging (MSO_2M), with O,S-bonded SO_2 or ligand–SO_2 interaction.[104] The co-ordination about the cobalt in $[Co(NO)(SO_2)(PPh_3)_2]$ is approximately tetrahedral, there being coplanar, sulphur-bonded M–SO_2 geometry and a linear nitrosyl group.[105] In contrast, the isoelectronic complex $[Rh(NO)(SO_2)-(PPh_3)_2]$ exhibits a bent nitrosyl group and an η^2-SO_2 co-ordination in which the sulphur and one oxygen atom are approximately equidistant from the rhodium atom.[106] The structures of $[Ni(SO_2)(PPh_3)_3]$ and $[Ni(SO_2)_2(PPh_3)_2]$ show that the SO_2 groups are S-bonded in both complexes, and are coplanar with their respective Ni—S bonds.[107] Sulphur dioxide inserts into the metal–metal bond of $[Pd_2(dpm)_2X_2]$ (dpm = $Ph_2PCH_2PPh_2$; X = Cl or Br), $[Pd_2-(dam)_2Cl_2]$ (dam = $Ph_2AsCH_2AsPh_2$), and $[Pt_2(dpm)_2Cl_2]$. The structure of $[Pd_2(dpm)_2(\mu$-$SO_2)Cl_2]$ has been determined by X-ray methods.[108] In liquid SO_2, very weak N- and O-donors form transition-metal complexes.[109]

97 T. Yoshida, D. L. Thorn, T. Okano, J. A. Ibers, and S. Otsuka, *J. Am. Chem. Soc.*, 1979,
    101, 4212.
98 G. Fachinetti, C. Floriani, A. Chiesi-Villa, and C. Guastini, *J. Am. Chem. Soc.*, 1979,
    101, 1767.
99 G. Fachinetti, C. Floriani, P. F. Zanazzi, and A. R. Zanzari, *J. Am. Chem. Soc.*, 1979, **101**,
    3469.
100 S. Kagawa, H. Furukawa, and M. Iwamoto, *J. Chem. Soc., Dalton Trans.*, 1979, 74.
101 R. Holderegger, L. M. Venanzi, F. Bachechi, P. Mura, and L. Zambonelli, *Helv. Chim.
    Acta*, 1979, **62**, 2159.
102 B. S. Tovrog, S. E. Diamond, and F. Mares, *J. Am. Chem. Soc.*, 1979, **101**, 270.
103 D. T. Toughty, G. Gordon, and R. P. Stewart, *J. Am. Chem. Soc.*, 1979, **101**, 2645.
104 G. J. Kubas, *Inorg. Chem.*, 1979, **18**, 182.
105 D. A. Johnson and V. C. Dew, *Inorg. Chem.*, 1979, **18**, 3273.
106 D. C. Moody, R. R. Ryan, and A. C. Larson, *Inorg. Chem.*, 1979, **18**, 227.
107 D. C. Moody and R. R. Ryan, *Inorg. Chem.*, 1979, **18**, 223.
108 A. L. Balch, L. S. Benner, and M. M. Olmstead, *Inorg. Chem.*, 1979, **18**, 2996.
109 R. Mews, *J. Chem. Soc., Chem. Commun.*, 1979, 278.

The ^{13}C n.m.r. spectrum of the non-fluxional π-CS_2 complex [Pt(PPh$_3$)$_2$-(η^2-CS_2)] has been reported.[110] The addition of the 'carbenoid' [V(Cp)$_2$] to CS_2 gives the known [V(Cp)$_2$(CS_2)] complex, containing the CS-side-bonded CS_2 moiety.[111]

Dinitrogen.—Fewer dinitrogen complexes appear to have been synthesized than last year. Thermochemical reactions of [Ru(NH$_3$)$_5$N$_2$]X$_2$ (X = Cl, Br, or I) have been reported.[112] The conversion of ligating dinitrogen into hydrazine with hydrazido(1 −) complexes as intermediates is possible with, for example, *cis*-[Mo(N$_2$)$_2$(PPhMe$_2$)$_4$].[113] The influence of ancillary ligands in dinitrogen complexes of iridium has been studied, and the electronic structures of tetrahedral dinitrogen complexes, *e.g.* [Ni(N$_2$)$_4$], [Co(N$_2$)$_4$]$^-$, and [Fe(N$_2$)$_4$]$^{2-}$, have been calculated by multiple-scattering methods.

3 Non-Haem Iron

Iron–Sulphur Compounds.—Holm and co-workers have shown that a reaction system consisting of [(R$_4$N)$_2$MoS$_4$], FeCl$_3$, NaOMe, and ethanethiol in methanol affords at least three anionic Mo–Fe–S–SEt clusters. Iron-57 Mössbauer and EXAFS studies of two of these clusters, *i.e.* [Mo$_2$Fe$_6$S$_8$(SEt)$_9$]$^{3-}$ and [Mo$_2$Fe$_6$S$_9$-(SEt)$_8$]$^{3-}$, are discussed in terms of a 'double-cubane' structure.[114] Using anaerobic conditions, in methanol, a mixture of iron chlorides, elemental sulphur, and the Li or Na salts of a thiol gives [Fe$_4$S$_4$(SR)$_4$]$^{2-}$ in good yield.[115] The X-ray crystal structure of [Et$_4$N]$_3$[Fe$_4$S$_4$(SCH$_2$Ph)$_4$] shows it to contain six short (2.302 Å) and six long (2.331 Å) Fe—S* bonds, affording D_{2d} symmetry. This contrasts with the D_{2d} elongated geometry of [Fe$_4$S$_4$(SPh)$_4$]$^{3-}$, which is an analogue of the Fd$_{red}$ site.[116] This phenomenon has been discussed in terms of a general change in electron-transfer reactions in analogues of 4-Fe sites.[117]

Sykes and co-workers have employed an extended range of oxidants in studying the reaction of reduced 2-Fe ferredoxins of parsley; their results indicate that there are three different reaction sites on the protein.[118] The reaction of (3-mercaptopropyl)trimethoxysilane with silica gel produces a gel surface with SiCH$_2$CH$_2$CH$_2$SH groups, and the reaction of this with [Fe$_4$S$_4$(SBu^t)$_4$]$^{2-}$ produces a Fe$_4$S$_4$ unit bound to the surface. No hydrogenation or oxidation activity was seen in attempted catalysis by these materials.[119] EXAFS studies of proteins and model compounds containing dimeric and tetrameric Fe–S clusters provide strong evidence that structural changes at the active site(s) are energetically

[110] P. J. Vergamini and P. G. Eller, *Inorg. Chim. Acta*, 1979, **34**, L291.
[111] G. Fachinetti, C. Floriani, A. Chiesi-Villa and C. Guastini, *J. Chem. Soc., Dalton Trans.*, 1979, 1612.
[112] S. Kohata, N. Itoh, H. Kawaguchi, and A. Ohyoshi, *Bull. Chem. Soc. Jpn.*, 1979, **52**, 2264.
[113] T. Takahashi, Y. Mizobe, M. Sato, Y. Uchida, and M. Midai, *J. Am. Chem. Soc.*, 1979, **101**, 3405.
[114] T. E. Wolff, J. M. Berg, K. O. Hodgson, R. B. Frankel, and R. H. Holm, *J. Am. Chem. Soc.*, 1979, **101**, 4104.
[115] G. Christou, and C. D. Garner, *J. Chem. Soc., Dalton Trans.*, 1979, 1093.
[116] J. M. Berg, K. O. Hodgson, and R. H. Holm, *J. Am. Chem. Soc.*, 1979, **101**, 4586.
[117] E. J. Laskewski, J. G. Reynolds, R. B. Frankel, G. C. Papaefthymiou, and R. H. Holm, *J. Am. Chem. Soc.*, 1979, **101**, 6562.
[118] F. A. Armstrong, R. A. Henderson, and A. G. Sykes, *J. Am. Chem. Soc.*, 1979, **101**, 6912.
[119] R. G. Bowman and R. L. Bunvell, *J. Am. Chem. Soc.*, 1979, **101**, 2877.

insignificant, as judged by comparing the structures of the proteins with those of the model compounds.[120] A ^{19}F n.m.r. method has been developed for the identification of Fe–S cores that have been extruded from active centres of proteins, and it has application to structural studies of milk xanthine oxidase and of nitrogenase.[121] The ^{1}H n.m.r. spectra of derivatives show that the Fe–S frameworks of $[(C_5H_4R)_4Fe_4S_6]$ and $[Fe_2(CO)_6SR]_2S$ are chiral in solution.[122]

Tris-(NN-dibenzyldithiocarbamato)iron(III) has a temperature-dependent magnetic moment; a change in μ_{eff} (*i.e.* a change in T) results in a change in Fe–S and also in a small change in the geometry of the FeS$_6$ core.[123] The crystal structure of chlorobis-(NN-di-isopropyldithiocarbamato)iron(III) is similar to those of other $[FeX(S_2CNR_2)_2]$ derivatives, the iron atom lying in a pseudo-square-pyramidal environment with apical chloride; Fe–Cl = 2.281 Å, Fe–S = 2.287 Å (av.).[124]

Transport of Iron.—Raymond and co-workers have produced two types of sequestering agent for ferric ion. Both are tris-(2,3-dihydroxybenzoyl) (DHB) derivatives of triamines, patterned after the microbial iron-chelating agent enterobactin.[125] The apparent first-order rate constants for removal of iron from transferrin by the synthetic sequestering agents 1,5,10-$NN'N''$-tris-(5-sulpho-2,3-dihydroxybenzoyl)triazadecane (3,4-LICAMS), 1,3,5-$NN'N''$-tris-(2,3-di-hydroxybenzoyl)aminomethylbenzene (MECAM), and the natural siderophore enterobactin have been determined, at 25 °C. The results suggest that these catecholate ligands are both kinetically and thermodynamically capable of removal of iron from transferrin at physiologically accessible concentrations. In contrast, the hydroxamate-based sequestering agents are kinetically hindered in this reaction.[126] Rhodotorulic acid, H$_2$RA, is a dihydroxamic acid that is produced in high yields by yeast as an iron-transport agent (siderophore). At 25 °C, the predominant ferric species at neutral pH is $[Fe_2RA_3]$, in which each iron atom is bound to three hydroxamate groups. Below pH 3, this complex dissociates into the monomer $[FeRA]^+$.[127] A model for the microbial transport of iron may be *trans*-tris(benzohydroxamato)chromium(III)-2-(2-propanol).[128, 129] Aerobactin, a dihydroxamate derivative of citric acid, is a siderophore that is produced by *Aerobacter aerogenes*. The iron complex has been isolated as the trisodium salt.[130] The complexation of six bidentate hydroxamic acids R^{1}C(O)-N(OH)R^2 (R^1 = Me or Ph; R^2 = H, Me, or Ph) with high-spin iron(III) has been

[120] B.-K. Teo, R. G. Shulman, G. S. Brown, and A. E. Meixner, *J. Am. Chem. Soc.*, 1979, **101**, 5624.

[121] S. B. Wong, D. M. Kurtz, R. H. Holm, L. E. Mortenson, and R. G. Upchurch, *J. Am. Chem. Soc.*, 1979, **101**, 3078.

[122] R. T. Edidim, L. A. Zyzyck, and J. R. Norton, *J. Chem. Soc., Chem. Commun.*, 1979, 580.

[123] J. Albertsson, I. Elding, and A. Oskarsson, *Acta Chem. Scand., Ser. A*, 1979, **33**, 703.

[124] S. Mitra, B. N. Figgis, C. L. Raston, B. W. Skelton, and A. H. White, *J. Chem. Soc., Dalton Trans.*, 1979, 753.

[125] F. L. Weitl and K. N. Raymond, *J. Am. Chem. Soc.*, 1979, **101**, 2728.

[126] C. J. Carrano and K. N. Raymond, *J. Am. Chem. Soc.*, 1979, **101**, 5401.

[127] C. J. Carrano, S. R. Cooper, and K. N. Raymond, *J. Am. Chem. Soc.*, 1979, **101**, 599.

[128] K. Abu-Dan, J. D. Ekstrand, D. P. Freyberg, and K. N. Raymond, *Inorg. Chem.*, 1979, **18**, 108.

[129] D. P. Freyberg, K. Abu-Dan, and K. N. Raymond, *Inorg. Chem.*, 1979, **18**, 3037.

[130] W. R. Harris, C. J. Carrano, and K. N. Raymond, *J. Am. Chem. Soc.*, 1979, **101**, 2722.

reported.[131] Spectral and magnetic properties of Fe^{III}, Co^{II}, and Ni^{II} complexes of monohydroxamic acids indicate that the complexes are octahedral, the cobalt and nickel species being polymeric.[132] EXAFS measurements on the iron core of ferritin and on an Fe–glycine model show the iron atoms to be surrounded by 6.4 ± 0.6 oxygen atoms at 1.95Å; each iron also has 7 ± 1 iron atoms as neighbours at an average distance of $3.29 \pm 0.05 \text{Å}$.[133] The spectrophotometric determination of the proton-dependent stability constant ($\log K = 52$; $pM = 35.5$) of ferric enterobactin has been reported.[134] Iron(III) complexes with the two triketones 1,5-diphenylpentane-1,3,5-trione and 1-phenyl-5-methylpentane-1,3,5-trione have been prepared.[135]

4 Copper Proteins

A number of interesting spectroscopic and synthetic reports have appeared. Low-temperature absorption and room-temperature c.d. and m.c.d. measurements have been made on two multi-copper oxidases, *i.e.* laccase from *Rhus vernicifera* and laccase from *Polyporus versicolor*. Near-infrared bands have been assigned to *d–d* transitions of the type 1 (blue) copper(II).[136] Azide- and thiocyanate-bound caeruloplasmins have also been examined. The combined spectroscopic evidence suggests that both blue copper sites in caeruloplasmin are related to the $Cu(N)_2(S)(S^*)$ (N = his, S = cys, S* = met) unit in azurin and in plastocyanin.[137] Manganese(II) and nickel(II) derivatives of azurin have been prepared.[138] The compound obtained from zinc-free thermolysin and copper(II) sulphate, which is biologically inactive, has been characterized, together with some inhibitor derivatives, by e.s.r. and n.m.r. methods.[139] Kinetically determined pK_a values have provided evidence for binding sites in some blue copper proteins.[140] The oxidation of parsley plastocyanin with $[Co(4,7\text{-DPSphen})_3]^{3-}$, $[Fe(CN)_6]^{3-}$, and $[Co(phen)_3]^{3+}$ has shown that there is strong protein–complex association prior to transfer of electron(s).[141, 142] X-Ray PES studies of plastocyanin may not, on their own, be used to infer Cu–S co-ordination in proteins.[143]

The crystal structure of potassium *p*-nitrobenzenethiolato[hydrotris-(3,5-dimethyl-1-pyrazolyl)borato]cuprate(I) diacetone shows trigonally distorted

[131] B. Monzyk and A. L. Crumbliss, *J. Am. Chem. Soc.*, 1979, **101**, 6203.
[132] P. A. Brown, D. McKeith, and W. K. Glass, *Inorg. Chim. Acta*, 1979, **35**, 5.
[133] S. M. Heald, W. R. Harris, C. J. Carrano, and K. N. Raymond, *J. Am. Chem. Soc.*, 1979, **101**, 2213.
[134] W. R. Harris, C. J. Carrano, and K. N. Raymond, *J. Am. Chem. Soc.*, 1979, **101**, 2213.
[135] L. L. Borer and W. Vanderbout, *J. Am. Chem. Soc.*, 1979, **101**, 526.
[136] D. M. Dooley, J. Rawlings, J. H. Dawson, P. J. Stephens, L. E. Andreasson, B. G. Malmström, and H. B. Gray, *J. Am. Chem. Soc.*, 1979, **101**, 5038.
[137] J. H. Dawson, D. M. Dooley, R. Clark, P. J. Stephens, and H. B. Gray, *J. Am. Chem. Soc.*, 1979, **101**, 5046.
[138] D. L. Tennent and D. R. McMillin, *J. Am. Chem. Soc.*, 1979, **101**, 2307.
[139] I. Bertini, G. Canti, H. Kozlowski, and A. Scozzafava, *J. Chem. Soc., Dalton Trans.*, 1979, 1270.
[140] A. G. Lappin, M. G. Segal, D. C. Weatherburn, and A. G. Sykes, *J. Chem. Soc., Chem. Commun.*, 1979, 38.
[141] A. G. Lappin, M. G. Segal, D. C. Weatherburn, and A. G. Sykes, *J. Am. Chem. Soc.*, 1979, **101**, 2297.
[142] A. G. Lappin, M. G. Segal, D. C. Weatherburn, and A. G. Sykes, *J. Am. Chem. Soc.*, 1979, **101**, 2302.
[143] M. Thompson, J. Whelan, D. J. Zemon, and B. Bosnich, *J. Am. Chem. Soc.*, 1979, **101**, 2482.

tetrahedral CuN_3S geometry,[144, 145] whereas CuN_2S_2 co-ordination is found in [2-(2-pyridyl)ethyl]bis-[2-(ethylthio)ethyl]aminecopper(I) tetraphenylborate.[146] A new type of bis(benzimidazole)thioether (N_2S) ligand forms [CuL]BF_4 complexes; these react with RS^- to form [CuL(SR)].[147] The structure of *cis*-dichloro-(2,2'-*o*-phenylenebisbenzothiazole)copper(II) is square-planar, with the copper atom bound to two *cis* chlorines and the nitrogen atoms of the chelating benzo-thiazole ligand.[148]

5 Complexes of Constituents of Nucleic Acids

The interaction between Cu^{II} or Mn^{II} and co-ordination sites on some purine bases in aqueous solution shows that Cu^{II} forms a strong chelate with adenines through N-7 and the amino (or imine) group. When the substituent at C-6 is an oxygen donor, only unidentate co-ordination, to N-7 exists. Manganese(II) interacts with hypoxanthines and inosines only through chelation *via* N-7 and the substituent at C-6; Mn^{II} is predominantly phosphate-bound with cyclic IMP.[149] Perturbation of the u.v. absorption spectrum of the adenine ring has been used to detect co-ordination by metal ions.[150] The structure of the bis(acetylacetonato)(nitro)(1,9-dimethyladeninium)cobalt(III) cation contains an *N*-7-bonded adeninium nucleus in an axial position.[151] The kinetics and equi-libria of interactions between methylmercury and adenine have been investi-gated by ^1H n.m.r. three-site lineshape analysis.[152] The structure of the complex [(MeHg)$_2$(9-methyladenine)]$^+$ shows that there is binding of Hg to N-1 and to the deprotonated amino-group of 9-methyladenine, whereas co-ordination to N-9 and N-7 occurs in [(MeHg)$_2$(adenine)]$^+$, even though N-1 is free.[153]

The crystal structure of (glycylglycinato)(7,9-dimethylhypoxanthine)copper(II) tetrahydrate shows planar copper, with terdentate Gly-Gly and N-1 of the 7,9-dimethylhypoxanthine molecule bonded to copper; $Cu-N = 1.977 Å$.[154] The 1:1 complex of silver(I) with 1-methylthymine (HMT) contains one-half of the silver atoms with linear co-ordination, bound strongly to N-3 of two deprotona-ted ligands; $Ag-N = 2.081 Å$. The resulting planar $Ag(MT)_2$ units are connected by oxygen atoms to the remaining silver atoms.[155] The electronic properties of the 1:1 complex of nickel(II) with inosine 5'-phosphate have been described,[156]

[144] J. S. Thompson, T. J. Marks, and J. A. Ibers, *J. Am. Chem. Soc.*, 1979, **101**, 4180.
[145] J. S. Thompson, T. Sorrell, T. J. Marks, and J. A. Ibers, *J. Am. Chem. Soc.*, 1979, **101**, 4193.
[146] K. D. Karlin, P. L. Dahlstrom, M. L. Stamford, and J. Zubieta, *J. Chem. Soc., Chem. Commun.*, 1979, 465.
[147] J. V. Dagdigian and C. A. Reed, *Inorg. Chem.*, 1979, **18**, 2623.
[148] R. G. Ball and J. Trotter, *Can. J. Chem.*, 1979, **57**, 1368.
[149] G. V. Fazakerley, G. E. Jackson, M. A. Phillips, and J. C. van Wiekerk, *Inorg. Chim. Acta*, 1979, **35**, 151.
[150] Y. H. Mariam and R. B. Martin, *Inorg. Chim. Acta*, 1979, **35**, 23.
[151] C. C. Chiang, L. A. Epps, L. G. Marzilli, and T. J. Kistenmacher, *Inorg. Chem.*, 1979, **18**, 791.
[152] D-L. Hoo and B. McConnell, *J. Am. Chem. Soc.*, 1979, **101**, 7470.
[153] L. Prizant, M. J. Olivier, R. Rivest, and A. L. Beauchamp, *J. Am. Chem. Soc.*, 1979, **101**, 2765.
[154] L. G. Marzilli, K. Wilkowski, C. C. Chiang, and T. J. Kistenmacher, *J. Am. Chem. Soc.*, 1979, **101**, 7504.
[155] F. Guay and A. L. Beauchamps, *J. Am. Chem. Soc.*, 1979, **101**, 6260.
[156] H. G. Nelson, *Inorg. Chim. Acta*, 1979, **32**, L51.

as have the copper(II) complexes of 2-thiocytosine and 2,4-dithiouracil.[157]
The crystal structure of [CdCl$_2$(1-methylcytosine)] shows cadmium to have
four strong bonds (two Cd–Cl, of 2.497 and 2.485 Å; two Cd–N, of 2.281 and
2.296 Å) that define a tetrahedron and two weak Cd–O (2.677, 2.677 Å) bonds.[158]
The packing motif in [Ag(1-methylcytosine)(NO$_3$)]$_2$ has each ligand residue
doubly cross-linked by two Ag$^+$ ions through the binding sites N-3 and O-2
of the base. Propagation of this dimeric unit along the crystallographic *c*-axis
via a second type of Ag–O(2) bond and base–base overlap yields a two-stranded,
cross-linked polymer, with the base residues slightly out of register. An attempt
has been made to relate these observed properties to the binding of Ag to regions
of high G-C content in duplex DNA.[159] Monomeric units of [Ru(NH$_3$)$_5$(1-
MeCyt)] (PF$_6$)$_2$ contain the base anion bound to the metal by the deprotonated
exocyclic amine nitrogen, this being the first example of such co-ordination by
a purine or pyrimidine.[160]

Pyrimidin-2-one (Hpymo) forms complexes with the bivalent metals Mn to
Zn and Cd. The exocyclic oxygen in Hpymo has much less tendency to bind to
these metal ions than does the sulphur atom in pyrimidine-2-thione.[161, 162]

The increased lability of C(8)–H bonds of purine nucleosides upon co-
ordination of a metal ion to the adjacent N-7 position has been used to explain
the observed formation of C(8)-bonded inosine and guanosine methyl-
mercurials.[163] Reactions of the anti-tumour complex [Rh$_2$(O$_2$CMe)$_4$] with
adenine nucleosides and with nucleotides give 1:1 and 1:2 adducts, the former
being polymeric bridged adducts with N-1 and N-7 sites bonded to Rh.[164] The
polymeric complex [{Cu$_3$(Guo-P)$_3$(OH)·8.5H$_2$O}$_n$] (Guo-P = guanosine monophosphate) shows bonding of the PO$_4$ unit to copper.[165] Complexes of stoicheiometry [RHg(GuaH$_{-1}$)], [RHg(Gua)]NO$_3$, and [(RHg)$_2$(GuaH$_{-1}$)]NO$_3$
result from the reaction of guanine (Gua) and RHg^{II} in water (R = Me) or
aqueous ethanol (R = Et).[166] In [Cu$_3$(2'-deoxyguanosine)$_2$(OH)$_4$]·4H$_2$O the
sugar-containing nucleic acid component is part of a bent trinuclear complex.[167]
In [Pt(diethylenetriamine)(guanosine)](ClO$_4$)$_2$ the four-co-ordinate platinum
is bonded to guanosine by N-7.[168]

The co-ordination of ATP to [Pd^{II}(glycyl-L-histidine)] in aqueous solution
shows that the purine ring influences the chemical shifts of the imidazole proton
(H-2) of the histidine residue.[169] The crystal structures of the ternary complexes
[Cd(5'-UMP)(dpa)(H$_2$O)$_2$]·5H$_2$O (5'-UMP = uridine 5'-phosphate; dpa = 2,2'-
dipyridylamine) and [Cu(5'-CMP)(dpa)(H$_2$O)]$_2$·5H$_2$O (5'-CMP = cytidine 5'-

[157] H. G. Nelson and J. F. Villa, *Inorg. Chim. Acta*, 1979, **34**, L235.
[158] C. Gagnon, A. L. Beauchamp, and D. Tranqui, *Can. J. Chem.*, 1979, **57**, 1372.
[159] T. J. Kistenmacher, M. Rossi, and L. G. Marzilli, *Inorg. Chem.*, 1979, **18**, 240.
[160] B. J. Craves and D. J. Hodgson, *J. Am. Chem. Soc.*, 1979, **101**, 5608.
[161] D. M. L. Goodgame and I. Jeeves, *Inorg. Chim. Acta*, 1979, **32**, 157.
[162] D. M. L. Goodgame and G. A. Leach, *Inorg. Chim. Acta*, 1979, **32**, 69.
[163] E. Buncel, A. R. Norris, W. J. Racz, and S. E. Taylor, *J. Chem. Soc., Chem. Commun.*, 1979, 562.
[164] G. Pneumatikakis and N. Hadjiliadis, *J. Chem. Soc., Dalton Trans.*, 1979, 596.
[165] J. F. Villa, F. J. Rudd, and H. C. Nelson, *J. Chem. Soc., Dalton Trans.*, 1979, 110.
[166] A. J. Canty and R. S. Tobias, *Inorg. Chem.*, 1979, **18**, 413.
[167] H. C. Nelson and J. F. Villa, *Inorg. Chem.*, 1979, **18**, 1725.
[168] R. Melanson and F. D. Rochon, *Can. J. Chem.*, 1979, **57**, 57.
[169] H. Kozłowski and E. Matczak-Jon, *Inorg. Chim. Acta*, 1979, **32**, 143.

phosphate) show the former to be polymeric, with a stacked structure between the uracil base and the dpa ligand *via* a metal–phosphate bridge, and the latter to be dimeric, with an unstacked structure involving metal bonding to the phosphate group, and *not* to the cytosine base of the nucleotide.[170] The binding of copper(II) to the bases of DNA is influenced by the nature of the other cations present.[171] The intramolecular and dimensionless equilibrium constant, K', for the equilibrium between open and aromatic-ring-stacked isomers of ternary complexes formed between adenosine 5'-triphosphate, inosine 5'-triphosphate, or uridine 5'-triphosphate, Mg^{2+} or Zn^{2+}, and L-tryptophanate, 2,2'-bipyridyl, or 1,10-phenanthroline have been estimated by ^1H n.m.r. shift measurements in D_2O.[172] Similar studies have been performed with Mg^{2+} and a number of mono- and di-nucleotides (AMP, TMP, GMP, CMP, d-ApA, d-TpT, d-GpG, d-CpC, d-pTpT).[173]

6 Molybdenum and Tungsten Complexes

The structures of *cis*-$[MoO_2Br_2(PPh_3O)_2]$ and *cis*-$[MoO_2Cl_2(PPh_3)_2]$ are of distorted octahedral species with *cis* PPh_3O ligands.[174] The known molybdenum-(V) complex $[OMoCl(acac)_2]$ has been prepared by a new method, involving an overall oxygen–chlorine exchange between $[MoO_2(acac)_2]$ and $[MoCl_2(acac)_2]$. The $[OMoCl_2(acac)_2]$ undergoes substitution with the acid forms of a number of bi- and quadri-dentate ligands to produce $[OMoCl(L)_2]$ (L = substituted quino-lines) and $[OMoCl_2(L^1)]$ (L^1 = quadridentate Schiff-base anion).[175] A large number of bi-, ter-, and quadri-dentate Schiff-base ligands, containing *O*-, *N*-, and *S*-donors, have been complexed to the *cis*-MoO_2 moiety. In the case of the complexes containing quadridentate ligands, the nature of the backbone between the imine groups determines which *cis*-isomer is formed.[176] Sawyer and co-workers have isolated a binuclear bis(catechol) complex of oxomolyb-denum(V) and two related catechol complexes of molybdenum(VI) that result from the reaction of molybdenum(VI) with 3,5-di-t-butylcatechol.[177] The in-fluence of the nucleophilicity of the Mo–O bond in a number of oxomolyb-denum(VI) complexes on the ^{17}O n.m.r. spectra has been investigated.[178] The chemical evolution of a nitrogenase model has been reported in which the simula-tions of steric and of inhibiting effects of the enzymic entire site, with acetylenes and nitriles as the substrates, and molybdenum catalysts are discussed.[179] Structural results that are relevant to the molybdenum sites in xanthine oxidase[180]

[170] K. Aoki, *J. Chem. Soc., Chem. Commun.*, 1979, 589.
[171] P. M. Lansant and R. P. Ferrari, *Inorg. Chim. Acta*, 1979, **33**, 145.
[172] P. R. Mitchell, B. Prijs, and H. Sigel, *Helv. Chim. Acta*, 1979, **62**, 1723.
[173] H. H. Trimm and R. C. Patel, *Inorg. Chim. Acta*, 1979, **35**, 15.
[174] R. J. Butcher, B. R. Penfold, and E. Simm, *J. Chem. Soc., Dalton Trans.*, 1979, 668.
[175] G. J.-J. Chen, J. W. McDonald, and W. E. Newton, *Inorg. Chim. Acta*, 1979, **35**, 93.
[176] W. E. Hill, N. Atabay, C. A. McAuliffe, F. P. McCullough, and S. M. Razzoki, *Inorg. Chim. Acta*, 1979, **35**, 35.
[177] J. P. Wilshire, L. Leon, P. Basserman, and D. T. Sawyer, *J. Am. Chem. Soc.*, 1979, **101**, 3379.
[178] K. F. Miller and R. A. D. Wentworth, *Inorg. Chem.*, 1979, **18**, 984.
[179] B. J. Weathers, J. H. Grate, and G. N. Schrauzer, *J. Am. Chem. Soc.*, 1979, **101**, 917.
[180] J. M. Berg, K. O. Hodgson, S. P. Cramer, J. L. Corbin, A. Elsberry, N. Pariyadath, and E. I. Stiefel, *J. Am. Chem. Soc.*, 1979, **101**, 2774.

and sulphite oxidase[181] have been reported with the X-ray structures of $[Mo(O_2)L]$ {L = $(SCH_2CH_2)_2NCH_2CH_2X$, where X is SMe or NMe_3}, together with evidence from X-ray absorption spectroscopy.[182]

Measurements of $[NEt_4]_3[Fe_6W_2S_8(SPh)_6(OMe)_3]$, *via* X-ray crystallography and Mössbauer spectroscopy, have confirmed that the anion in this complex contains two {Fe_3WS_4} cubane-like clusters, with terminal benzenethiolato-ligands co-ordinated to the iron atoms and three μ_2-methoxo-groups bridging the tungsten atoms.[183] The isolation and crystal structure of $[Et_4N]_3[Fe_6Mo_2S_8$-$(SEt)_9]$ have been reported,[184] and the structure of $[Et_4N]_3[Fe_6Mo_2S_8(SCH_2$-$CH_2OH)]$ and its thiol-exchange reactions,[185] a novel bimetallic crystal structure of bis[(methyldiphenylphosphine)silver]tetrathiotungsten {$[(Ph_2PMe)_4Ag_4$-$W_2S_8]$},[186] the crystal structure of the binuclear complex with the $FeMoS_2$ core, *i.e.* bis(tetraethylammonium) di-μ-thio-[bis(phenylthio)ferrate(III)[dithiomolyb-date(V)] {$[Et_4N]_2[(PhS)_2FeMoS_4]$},[187] and the preparation, crystal structure, and molecular structure of $[MoO(SCH_2CH_2PPh_2)_2]$[188] have been reported.

The oxo-complexes $[Mo(O_2)L_2]$, where L is oximate, S_2CNMe_2, S_2CNEt_2, $S_2CN(CH_2)_5$, or S_2CNPh_2, react with the hydrazines R_2NNH [R_2 = Me_2, $(CH_2)_5$, or Ph_2] to give the hydrazido(2–) complexes $[Mo(NNR_2)OL_2]$. The crystal and molecular structures of $[\{Mo(NNMe_2)O\}\{S_2(NMe_2)_2\}]$ have been determined by single-crystal X-ray diffraction methods.[189] The crystal and molecular structures of tris-(NN-diethyldithiocarbamato)nitrosomolybdenum and tris-(NN-dimethyldithiocarbamato)thionitrosylmolybdenum have been determined,[190] together with those for diaquahydrogen(I) μ-oxo-[bis(diethyldithio-carbamato)oxomolybdate(IV) [bis(diethyldithiocarbamato)oxomolybdenum(V)] bis-{μ-oxo-bis[bis(diethyldithiocarbamato)oxomolybdenum(V)]} and μ-oxo-bis[bis(diethyldithiocarbamato)oxomolybdenum(V)].[191] The effects of neighbouring groups on redox potentials for organodiazenido-complexes of molybdenum[192] and the thionitrosyl complex of molybdenum have also attracted attention.[193] A series of *cis*-dioxomolybdenum(VI) complexes of the type [Mo-$(O_2)(L-L)_2$] was prepared with a variety of chelating monoanions (L-L), to

[181] S. P. Cramer, H. B. Gray, and K. V. Rajagopalan, *J. Am. Chem. Soc.*, 1979, **101**, 2772.
[182] T. D. Tullius, D. M. Kurtz, jr., S. D. Conradson, and K. O. Hodgson, *J. Am. Chem. Soc.*, 1979, **101**, 2776.
[183] G. Christou, C. D. Garner, T. J. King, C. E. Johnson, and J. D. Rush, *J. Chem. Soc., Chem. Commun.*, 1979, 503.
[184] S. R. Acott, G. Christou, C. D. Garner, T. J. King, F. E. Mabbs, and R. M. Miller, *Inorg. Chim. Acta*, 1979, **35**, L337.
[185] G. Christou, C. D. Garner, F. E. Mabbs, and M. G. B. Drew, *J. Chem. Soc., Chem. Commun.*, 1979, 91.
[186] J. K. Stalick, A. R. Siedle, A. D. Mighell, and C. R. Hubbard, *J. Am. Chem. Soc.*, 1979, **101**, 2903.
[187] D. Coucouvanis, E. D. Simhon, D. Swenson and N. C. Baenziger, *J. Chem. Soc., Chem. Commun.*, 1979, 361.
[188] J. Chatt, J. R. Dilworth, J. A. Schmutz, and J. A. Zubieta, *J. Chem. Soc., Dalton Trans.*, 1979, 1595.
[189] M. W. Bishop, J. Chatt, J. R. Dilworth, M. B. Hursthouse, and M. Motevalli, *J. Chem. Soc., Dalton Trans.*, 1979, 1600.
[190] M. B. Hursthouse and M. Motevalli, *J. Chem. Soc., Dalton Trans.*, 1979, 1362.
[191] C. D. Garner, N. C. Howlader, F. E. Mabbs, A. T. McPhail, and K. D. Onan, *J. Chem. Soc., Dalton Trans.*, 1979, 962.
[192] G. Butler, J. Chatt, G. J. Leigh, and C. J. Pickett, *J. Chem. Soc., Dalton Trans.*, 1979, 113.
[193] M. W. Bishop, J. Chatt, and J. R. Dilworth, *J. Chem. Soc., Dalton Trans.*, 1979, 1.

examine their relative stabilities. Complexes $[Mo(O_2)(L-L)_2]$ with L-L = dialkyl-dithiocarbamato (R_2dtc) or similar sulphur-containing ligands were found to be effective in the dehydrogenation of hydrazobenzene to azobenzene and in the oxygenation of PPh_3, while those with O,O- and O,N-chelate ligands were inactive. These reactivities have been discussed in terms of the electronic effects of the ligands.[194, 195]

Addition of acids HX (X = BF_4, FSO_3, HSO_4, $ClSO_3$, Cl, or Br) to the complexes *trans*-$[M(CNR)_2(dppe)_2]$ (M = Mo or W; R = Me, Bu^t, Ph, or C_6H_4Me-p; dppe = $Ph_2PCH_2CH_2PPh_2$) under carefully controlled conditions has given the complexes $[MH(CNR)_2(dppe)_2][X]$ and the hydrido-carbyne complexes $[MH\{CN(H)R\}(CNR)(dppe)_2][X]_2$.[196] The acylhydrazines $RCONHNH_2$ (R = Ph, p-ClC_6H_4, p-$NO_2C_6H_4$, or 1-$C_{10}H_7$) react with $[MoOCl_2(PR_3)_3]$ (PR$_3$ = PMe$_2$Ph, PEt$_2$Ph, PPr$^n_2$Ph, PMePh$_2$, or PEt$_3$) to give red complexes $[MoCl(N_2COR)$-$(NHNCOR)(PR_3)_2]$.[197] Chatt *et al.* have also reported how the complexes $[MoCl(O)(P-P)_2][MoCl_3O(RCON_2Ph)]$ (R = Ph, p-ClC_6H_4, p-$MeOC_6H_4$, or Me; P-P = $Ph_2PCH_2CH_2PPh_2$ or $Ph_2PCH=CHPPh_2$) have been prepared by the reactions of the hydrazines $RCONHNHPh$ with $[\{MoCl_2O(P-P)\}_2]$ or $[MoCl_3O(P-P)]$ in refluxing methanol.[198] Connor and Riley have described how the oxidation of *cis*-$[Mo(CO)_2(dmpe)_2]$ with silver(I) salts of co-ordinating anions Ag$_n$X (n = 1, X = [NCS]$^-$, [NO$_3$]$^-$, or [NO$_2$]$^-$; n = 2, X = [CO$_3$]$^{2-}$ or [SO$_4$]$^{2-}$) produces the seven-co-ordinate molybdenum(II) complexes *cis*-$[MoX(CO)_2(dmpe)_2]X$ (X = NCS or NO$_3$), $[MoX(CO)(NO)(dmpe)_2]$ (X = NO$_2$), and the zwitterions *cis*-$[Mo(OZO_m)(CO)_2(dmpe)_2]$ (Z = C, m = 2; Z = S, m = 3).[199] George and Archer have discussed the displacement of co-ordinated dinitrogen by dihydrogen in low-valent molybdenum complexes.[200]

The oxygenation reactions of aldehydes, tertiary phosphines, and phosphites that are catalysed by the *cis*-dioxomolybdenum(VI)-ethyl-L-cystinate complexes have been studied by Speier.[201] Some attempts were made to find a correlation between reactivity and the structure of the substrate. Chatt *et al.* have reported a series of complexes $[Mo(N_2Q)(S_2CNMe_2)_3]$ (Q = alkyl, aryl, or alkoxycarbonyl) and $[Mo(N_2Q)_2(S_2CNMe_2)_2]$, prepared from $[MoO_2(S_2CNMe_2)_2]$ and the appropriate hydrazine. The electrochemical properties of the new complexes are discussed.[202]

The crystal structure and the molecular structure of (N-benzoyl-N'-p-tolyl-di-azene-$N'O$)dichloro(dimethylphenylphosphine)(p-tolylimido)molybdenum have been described by Chatt *et al.*[203] As a model system for the molybdenum enzyme

194 A. Nakamura, M. Nakayama, K. Sugihashi, and S. Otsuka, *Inorg. Chem.*, 1979, **18**, 394.
195 E. A. Maatta and R. A. D. Wentworth, *Inorg. Chem.*, 1979, **18**, 2409.
196 J. Chatt, A. J. L. Pombeiro, and R. L. Richards, *J. Chem. Soc., Dalton Trans.*, 1979, 1585.
197 A. V. Butcher, J. Chatt, J. R. Dilworth, G. J. Leigh, M. B. Hursthouse, S. A. A. Jaya-weera, and A. Quick, *J. Chem. Soc., Dalton Trans.*, 1979, 921.
198 M. W. Bishop, J. Chatt, J. R. Dilworth, M. B. Hursthouse, and M. Motevalli, *J. Chem. Soc., Dalton Trans.*, 1979, 1603.
199 J. A. Connor and P. I. Riley, *J. Chem. Soc., Dalton Trans.*, 1979, 1231.
200 L. J. Archer and T. A. George, *Inorg. Chem.*, 1979, **18**, 2079.
201 G. Speier, *Inorg. Chim. Acta*, 1979, **32**, 139.
202 M. W. Bishop, G. Butler, J. Chatt, J. R. Dilworth, and G. J. Leigh, *J. Chem. Soc., Dalton Trans.*, 1979, 1843.
203 M. W. Bishop, J. Chatt, J. R. Dilworth, M. B. Hursthouse, S. A. A. Jayaweera, and A. Quick, *J. Chem. Soc., Dalton Trans.*, 1979, 914.

nitrate reductase, the reduction of NO_3^- by monomeric Mo^V complexes of the formulae $[MoOCl_3L]$ ($L = \alpha\alpha'$-bipyridyl or 1,10-phenanthroline), $[MoOClL_2]$ ($L =$ 8-hydroxyquinoline or 8-mercaptoquinoline), $[MoOCl(MeOH)L]$ $\{L = o$-(salicylideneimino)phenyl$\}$, and $[MoOClL]$ $\{L = o$-bis-(salicylideneimido)benzene$\}$ has been investigated by Spence *et al.*[204] Three types of mixed-ligand tungsten(II) chelates have been isolated, with simultaneous co-ordination of the hard ligands phenolato and oxygen, heterocyclic nitrogen (borderline), and the soft ligands carbonyl and phosphine. The physical and chemical properties of the diamagnetic chelates are consistent with seven-co-ordinate geometries.[205] Tungsten carbonyl complexes of 2,3-O-isopropylideneguanosine and 6-mercaptopurine have been shown by Beck to be useful model compounds for studying metal-binding sites of nucleic acid components.[206] McAuliffe *et al.*[207] have reported the synthesis of a series of oxomolybdenum(v) complexes of substituted hydroxyquinolines and quinolines by the reaction of $[MoOCl_3$-$(THF)_2]$ ($THF =$ tetrahydrofuran) with 8-hydroxyquinoline, isoquinoline, 3-methylisoquinoline, 6-methylquinoline, 7-methylquinoline, 5,6-benzoquinoline, 8-mercaptoquinoline, and 2,2'-biquinolyl.

7 Photolysis of Water

Work in this area continues to grow, and a review article has appeared on the subject.[208] Many of the investigations involve the excitation of ruthenium(II) complexes, particularly of $[Ru(bipy)_3]^{2+}$. The rate constants for the reaction of titanium(III) with the excited states of $[RuL_3]^{2+}$ complexes have been found to be insensitive to the reduction potential of the ruthenium(II) complexes. Quenching reactions proceed predominantly by energy-transfer mechanisms.[209] Quenching of $[Ir(bipy)_2(H_2O)(bipy)]^{3+}$ and of $[Ir(bipy)_2(H_2O)(bipy)]^{2+}$ by a variety of charged metal complexes and neutral biacetyl is similar to that of $[Ru(bipy)_3]^{2+}$ with the same species.[210] Flash photolysis and luminescence techniques have been used to investigate the properties and the behaviour of the ^2E excited state of the chromium(III) complexes of bipy, phen, terpy, and some of their methyl, phenyl, and chloro-derivatives.[211] The photo-excited triplet-state properties of chlorophyll aggregate systems *in vitro* have been examined by zero-field optically detected magnetic resonance spectroscopy at 2 K. Measurement of the triplet-state zero-field splittings, overall triplet lifetimes, and individual rate constants for sublevel intersystem crossing were obtained for solutions of chlorophyll *a* and zinc-substituted chlorophyll-*a* and for the covalently linked dimeric derivative of pyrochlorophyllide-*a*.[212] Electron-transfer reactions involving porphyrins (P) and quinones (Q), studied by pulse radiolysis, indicate

[204] R. D. Taylor, P. G. Todd, N. D. Chasteen, and J. T. Spence, *Inorg. Chem.*, 1979, **18**, 44.
[205] W. H. Batschelet, R. D. Archer, and D. R. Whitcomb, *Inorg. Chem.*, 1979, **18**, 48.
[206] N. Kottmair and W. Beck, *Inorg. Chim. Acta*, 1979, **34**, 137.
[207] C. A. McAuliffe, A. Hosseiny, and F. P. McCullough, *Inorg. Chim. Acta*, 1979, **33**, 5.
[208] G. Porter and M. D. Archer, *Interdisc. Sci. Rev.*, 1976, **1**, 119.
[209] B. S. Brunschwig and N. Sutin, *Inorg. Chem.*, 1979, **18**, P1731.
[210] S. F. Bergeron and R. J. Watts, *J. Am. Chem. Soc.*, 1979, **101**, 3151.
[211] N. Serpone, M. A. Jamieson, M. S. Henry, M. Z. Hoffman, F. Bolletta, and M. Maestri, *J. Am. Chem. Soc.*, 1979, **101**, 2907.
[212] R. H. Clarke, D. R. Hobart, and W. R. Leenstra, *J. Am. Chem. Soc.*, 1979, **101**, 2416.

that the former are rapidly reduced by electron transfer from $(CH_3)_2CO^-$. In some reactions there is evidence for the intermediate complex $[P \cdots Q.^-]$.[213] Yields for intersystem crossing in ruthenium(II) and osmium(II) photosensitizers have been confirmed as unity.[214]

Three new ruthenium bipyridyl surfactant complexes, *i.e.* [Ru(bipy){bipy-$(C_{19}H_{39})_2$}]$^{2+}$, [Ru(bipy)(CN)$_2${bipy($C_{19}H_{39})_2$}], and [Ru{bipy(COO)$_2$}$_2${bipy-$(C_{19}H_{39})_2$}]$^{2-}$, have been prepared and their properties as monolayers described.[215] The covalent attachment of [Ru(bipy)$_3$]$^{2+}$ complexes to silanised platinum oxide electrodes produces species which show an impressive stability towards cycling between the Ru^{II} and Ru^{III} oxidation states.[216] The alkaline hydrolysis of [Ru(bipy)$_2${bipy(COOR)$_2$}]$^{2+}$ in monolayer films (R = $C_{18}H_{37}$) is found to proceed more rapidly than in homogeneous, aqueous solution (R = Et) at the same bulk pH.[217] The [Ru(bipy)$_2$(CN)$_2$] and [Ru(phen)$_2$(CN)$_2$] complexes appear to be unique in that inversion of the lowest charge-transfer and ligand-localized triplet, $^3\pi-\pi^*$, states occurs on protonation, with a complete change in the emission character of the systems.[218] The complex [Ru(py)$_6$](BF$_4$)$_2$ has been produced by extended refluxing of a solution of [RuH(S)$_3$(PPh$_3$)$_2$]BF$_4$ (S = MeOH or H$_2$O) and pyridine in methanol.[219] Some photo-reactions of bis(bipyridyl)-(poly-4-vinylpyridine)ruthenium(II) complexes have been briefly described.[220] A simple photochemical synthesis of [Ru(bipy)$_2$X$_2$] complexes has been discovered which makes use of the low solubility of these complexes in ethanol or water.[221] Ruthenium(II)–alcohol complexes can be oxidized by O$_2$ to H$_2$O$_2$ and the corresponding ruthenium(IV)–alcohol complexes, which then undergo metal–ligand redox, leading to a ruthenium(II)–ketone complex.[222]

The redox properties of a series of manganese(III) complexes incorporating dianionic quinquedentate Schiff-base ligands have been examined, using cyclic voltammetry. Substituents on the aromatic portion of the ligand and charges on the central donor atom were shown to cause changes in the reduction potentials of the Mn^{III} ion. Electron-withdrawing substituents resulted in lower reduction potentials relative to complexes possessing ligands with electron-donating groups. The electrochemical data have been correlated with previously observed reactivity patterns of these complexes towards O$_2$ and NO.[223] Richens and Sawyer have isolated (Me$_4$N)$_2$[Mn^{IV}(C$_6$H$_{12}$O$_6$)$_3$], where C$_6$H$_{14}$O$_6$ is sorbital (= D-glucitol); this is the first authentic example of a high-spin tris(chelate) monomeric complex of manganese(IV). The complex exhibits an e.s.r. spectrum which is consistent with a $\frac{1}{2} \to -\frac{1}{2}$ transition of the spin quartet ground state in an axially distorted octahedral ligand field.[224] Sorbitol, which is a straight-chain

213 P. Neta, A. Scherz, and H. Levanon, *J. Am. Chem. Soc.*, 1979, **101**, 3624.
214 J. N. Demas and D. G. Taylor, *Inorg. Chem.*, 1979, **18**, 3177.
215 S. J. Valenty, D. E. Behnken, and G. L. Gaines, *Inorg. Chem.*, 1979, **18**, 2160.
216 H. D. Abruna, T. J. Meyer, and R. W. Murry, *Inorg. Chem.*, 1979, **18**, 3233.
217 S. J. Valenty, *J. Am. Chem. Soc.*, 1979, **101**, 1.
218 S. H. Peterson and J. N. Dernas, *J. Am. Chem. Soc.*, 1979, **101**, 6571.
219 J. L. Templeton, *J. Am. Chem. Soc.*, 1979, **101**, 4906.
220 J. M. Clear, J. M. Kelly, D. C. Pepper, and J. G. Vos, *Inorg. Chim. Acta*, 1979, **33**, L139
221 W. M. Wallace and P. E. Hoggard, *Inorg. Chem.*, 1979, **18**, 2934.
222 B. S. Tovrog, S. E. Diamond, and F. Mares, *J. Am. Chem. Soc.*, 1979, **101**, 5067.
223 W. M. Coleman and L. T. Taylor, *Inorg. Chim. Acta*, 1979, **34**, L291.
224 D. T. Richens and D. T. Sawyer, *J. Am. Chem. Soc.*, 1979, **101**, 3681.

hexitol, complexes with and stabilizes the $+2$, $+3$, and $+4$ oxidation states of manganese in aqueous alkaline solution. Molecular oxygen oxidizes the Mn^{II} and Mn^{III} complexes to the Mn^{III} and Mn^{IV} complexes, respectively, and is reduced to the peroxide anion in each case.[225] Sawyer's group has investigated the oxidation–reduction chemistry of hydrogen peroxide in aprotic and aqueous solutions.[226]

Grätzel has applied his Pt sol photolysis catalyst to systems where other reducing radicals appear as intermediates in photochemical reactions; thus molecular hydrogen has been evolved from the photolysis of alcoholic solutions of benzophenone.[227] When the activated catalyst Eu^{III}–Y zeolite is exposed to degassed water at 25–50 °C, the emission band for Eu_2^+ disappears and dihydrogen is evolved.[228] A photoelectrochemical cell, based on electron-transfer quenching of the excited state of $[Ru(bipy)_3]^{2+}$, has been described by Meyer in which visible photolysis gives both H_2 and an appreciable photocurrent.[229]

Ultraviolet excitations of aerated, acidic aqueous solutions of $[Rh(NH_3)_4-(OH_2)H]^{2+}$ result in the efficient formation of $[Rh(NH_3)_4(OH_2)O_2H]^{2+}$. The photoinitiated thermal reaction appears to be a chain reaction, with $[Rh(NH_3)_4]^{2+}$ (aq.) acting as a chain carrier which 'activates' dioxygen by co-ordination. The hydroperoxo-complex decomposes slowly to form a coloured, paramagnetic superoxo-complex.[230] An X-ray structural analysis of $[Rh_2(bridge)_4Cl_2]Cl_2\cdot8H_2O$ (bridge = 1,3-di-isocyanopropane) shows the cation to have approximately D_{4h} symmetry. Each Rh is hexaco-ordinate, with Cl and Rh atoms in axial positions of a $Cl(RhC_4)Rh$ unit.[231]

Lehn *et al.* have described fully their approach to photochemical conversion and storage of solar energy. The components of this system for the generation of dihydrogen by reduction of water under irradiation by visible light or sunlight are (i) a photosensitizer, *e.g.* the $[Rh(bipy)_3]^{2+}$ complex, for absorption of visible light, (ii) a relay species, *e.g.* the $[Rh(bipy)_3]^{3+}$ complex, which mediates the reduction of water by intermediate storage of electrons *via* a reduced state, (iii) an electron donor, *e.g.* triethanolamine, which provides the electrons for the reduction process, and (iv) a redox catalyst, *e.g.* colloidal platinum, which facilitates the formation of dihydrogen.[232] Molecular oxygen is generated when aqueous $[Fe(bipy)_3]^{3+}$ is brought into contact with catalytic amounts of colloidal RuO_2, as shown in reaction (1). The surprisingly low energy losses (160 mV)

$$4[Fe(bipy)_3]^{3+} + 2H_2O \xrightarrow{RuO_2} 4[Fe(bipy)_3]^{2+} + 4H^+ + O_2 \qquad (1)$$

in this reaction make the construction of a system that is capable of four-quanta splitting of water feasible.[233] Sutin and co-workers have described the mediation

225 D. T. Richens, C. G. Smith, and D. T. Sawyer, *Inorg. Chem.*, 1979, **18**, 706.
226 M. M. Morrison, J. L. Roberts, and D. T. Sawyer, *Inorg. Chem.*, 1979, **18**, 1971.
227 C. K. Grätzel and M. Grätzel, *Inorg. Chem.*, 1979, **18**, 7743.
228 T. S. Arakana, T. Takata, G. Adachi, and J. Shiokawa, *J. Chem. Soc., Chem. Commun.*, 1979, 453.
229 B. Durham, W. J. Dressick, and T. J. Meyer, *J. Chem. Soc., Chem. Commun.*, 1979, 381.
230 J. F. Endicott, C.-L. Wong, T. Inoue, and P. Natarajan, *Inorg. Chem.*, 1979, **18**, 450.
231 K. R. Mann, R. A. Bell, and H. B. Gray, *Inorg. Chem.*, 1979, **18**, 2671.
232 M. Kirch, J.-M. Lehn, and J.-P. Sauvage, *Helv. Chim. Acta*, 1979, **62**, 1345.
233 K. Kalyanasundaram, O. Micic, E. Pramauro and M. Grätzel, *Helv. Chim. Acta*, 1979, **62**, 2432.

of a $[Ru(bipy)_3]^{2+}$–cobalt(II) macrocycle in the photoreduction of water by visible light.[234, 235] Fong and co-workers have described experiments that delineate the effects of heat and visible light on Pt in deoxygenated water; evolution of dihydrogen occurs when Pt and water are heated together in the relatively low temperature range of 130—210 °C.[236] Reduced polypyridyl complexes of ruthenium can be generated by electrochemical reduction in acetonitrile, and the reduction of water can subsequently be achieved.[237] Completely synthetic dihexadecylphosphate vesicles undergo photoionization and electron-transfer processes.[238]

Also of interest in this section are the reports of the preparation of chlorophyll-*a* and -*b* from spinach leaves by means of column chromatography with Sephasorb HP Ultrafine.[239] New water-soluble complexes of tertiary phosphines with rhodium(I) catalyse the water-gas shift.[240] Some e.s.r. measurements of manganese(II) complexes should prove useful in structural analyses.[241]

8 Complexes of Metals with Cyclic Ligands

Porphyrins and Metalloporphyrins.—*X*-Ray photoelectron spectroscopic data for *N*-methyltetraphenylporphyrin, its dianionic salt, and a number of metal complexes have been reported, with special attention being given to the 1*s* binding energies of nitrogen.[242] Smith and co-workers[243] have synthesized a series of novel meso-substituted metalloporphyrins *via* the reaction of octaethylmetalloporphyrin π-cation radicals with a variety of nucleophiles. A convenient method has been developed by Linn and Schreiner[244] for the preparation of protoporphyrin-IX dimethyl ester from haemin, and Ullman and co-workers[245] have prepared tetraphenylporphyrins in which the two NH groups are replaced by the heteroatoms S, Se, and Te (of Group 6A). The reaction of the μ-oxo-dimer of iron(III) tetraphenylporphyrin with aqueous tetrafluoroboric acid, in the presence of ethanol, yields the bis(ethanol)tetraphenylporphyrinatoiron(III) tetrafluoroborate complex.[246] The compound has been characterized by *X*-ray crystallography and the iron(III) atom has been shown to have an $S = \frac{5}{2}$ ground

234 G. M. Brown, B. S. Brunschwig, C. Creutz, J. F. Endicott, and N. Sutin, *J. Am. Chem. Soc.*, 1979, **101**, 296.
235 G. M. Brown, S-F. Chan, C. Creutz, H. A. Schwarz, and N. Sutin, *J. Am. Chem. Soc.*, 1979, **101**, 7638.
236 L. Salloway, D. R. Fruge, G. M. Haley, A. B. Caddington, and F. K. Fong, *J. Am. Chem. Soc.*, 1979, **101**, 229.
237 H. D. Abruna, A. Y. Teng, G. J. Samuels, and T. J. Meyer, *J. Am. Chem. Soc.*, 1979, **101**, 6945.
238 J. R. Escabi-Perez, A. Romero, S. Lukac, and J. H. Fendler, *J. Am. Chem. Soc.*, 1979, **101**, 2231.
239 K. Iriyama, M. Yoshiura, and M. Shiraki, *J. Chem. Soc., Chem. Commun.*, 1979, 406.
240 R. G. Nuzzo, D. Feitler, and G. M. Whitesides, *J. Am. Chem. Soc.*, 1979, **101**, 3683.
241 R. D. Dowsing, J. F. Gibson, M. Goodgame, and P. J. Hayward, *J. Chem. Soc. A*, 1970, 1133.
242 D. K. Lavallee, J. B. Brace, and N. Winograde, *Inorg. Chem.*, 1979, **18**, 1776.
243 K. M. Smith, G. H. Barnett, B. Evans, and Z. Martynenko, *J. Am. Chem. Soc.*, 1979, **101**, 5953.
244 J. A. Linn and A. F. Schreiner, *Inorg. Chim. Acta*, 1979, **35**, L339.
245 A. Ulman, J. Manassen, F. Frolow, and D. Rabinovich, *J. Am. Chem. Soc.*, 1979, **101**, 7055.
246 P. Gans, G. Buisson, E. Duee, J. R. Regnard, and J. C. Marchon, *J. Chem. Soc., Chem. Commun.*, 1979, 393.

state. A number of other crystallographic studies of iron porphyrins have also appeared.[247-249] $\alpha\beta\gamma\delta$-Tetraphenylporphyrin derivatives with acrylic (*cis* and *trans* isomers) and propionic side-chains that have terminal imidazole groups have been synthesized, together with the iron(III) and iron(II) complexes of these ligands.[250] A series of n.m.r. studies of both naturally occurring and synthetic metalloporphyrins have been reported.[251-254] From a study of the [1]H n.m.r. spectra of the histidine residues in ferrous carbon monoxide sperm-whale myoglobin, Bradbury *et al.*[251] have estimated the pK' value for the distal histidine to be 5.2 at 40 °C.

Kozuka and Nakamoto[255] have studied the adducts of [Co(TPP)] (TPP = tetraphenylporphyrin) with a range of axial ligands, including CO, NO, and O_2, in argon and/or krypton matrices, by infrared spectroscopy.

The temperature dependence of magnetic moments of haemichrome salts in the solid state and in solution has been interpreted in terms of a spin equilibrium between an $S' = \frac{1}{2}$ and an $S' = \frac{5}{2}$ state.[256] Measurements of average magnetic susceptibility and magnetic anisotropy on single crystals of [Fe(TPP)] over a range of temperatures have been reported. The results, together with other evidence, establish an $S = 1$ spin state for the iron(II) in [Fe(TPP)].[257] The inter-action of low-spin [(TPP)Fe(2-MeIm)$_2$] with [Cu(acac)$_2$] has been shown to produce the high-spin dimeric species [(TPP)Fe-(2-MeIm)-Cu(acac)$_2$].[258] The temperature dependence of the magnetic susceptibility of an azide complex of an octapeptide in ferric haem has been investigated by n.m.r. methods.[259] Equilibrium data on the binding of axial bases by iron(III) porphyrins has recently appeared.[260, 261] The effect of substituents on the electrochemical properties of porphyrins and metalloporphyrins has been studied,[262, 263] as have the super-oxide dismutase activities of a series of iron complexes.[264]

Quantum-mechanical calculations on iron–porphyrin model complexes have been reported.[265] Analyses of both ground and excited states of dioxygen and

[247] G. B. Jameson and J. A. Ibers, *Inorg. Chem.*, 1979, **18**, 200.
[248] K. M. Adams, P. G. Rasmussen, W. R. Scheidt, and K. Hatano, *Inorg. Chem.*, 1979, **18**, 1892.
[249] K. Hatano and W. R. Scheidt, *Inorg. Chem.*, 1979, **18**, 887.
[250] M. Momenteau, B. Loock, E. Bisagni, and M. Rougee, *Can. J. Chem.*, 1979, **57**, 1804.
[251] J. H. Bradbury, S. L. M. Deacon, and M. D. Ridgway, *J. Chem. Soc., Chem. Commun.*, 1979, 997.
[252] I. Morishima, T. Inubushi, T. Hayashi, and T. Yonezawa, *J. Chem. Soc., Chem. Commun.*, 1979, 483.
[253] Y. Ozaki and Y. Kyogoku, *J. Chem. Soc., Chem. Commun.*, 1979, 76.
[254] J. Mispelter, M. Momenteau, and J. M. Lhoste, *J. Chem. Soc., Chem. Commun.*, 1979, 808.
[255] M. Kozuka and K. Nakamoto, *J. Am. Chem. Soc.*, 1979, **103**, 2162.
[256] H. A. O. Hill, P. D. Skyte, J. W. Buchler, H. Lueken, M. Tonn, A. K. Gregson, and G. Pellizer, *J. Chem. Soc., Chem. Commun.*, 1979, 151.
[257] P. D. W. Boyd, D. A. Buckingham, R. F. McMeeking, and S. Mitra, *Inorg. Chem.*, 1979, **18**, 2585.
[258] T. Prosperi and A. A. G. Tomlinson, *J. Chem. Soc., Chem. Commun.*, 1979, 196.
[259] Y-P. Huang, and R. J. Kassner, *J. Am. Chem. Soc.*, 1979, **101**, 5707.
[260] T. Yoshimura and T. Ozaki, *Bull. Chem. Soc. Jpn.*, 1979, **52**, 2268.
[261] A. N. Thompson and M. Krishnamurthy, *Inorg. Chim. Acta*, 1979, **34**, 145.
[262] A. Giraudeau, H. J. Callot, and M. Gross, *Inorg. Chem.*, 1979, **18**, 201.
[263] A. Giraudeau, H. J. Callot, J. Jordan, I. Ezhar, and M. Gross, *J. Am. Chem. Soc.*, 1979, **101**, 3857.
[264] R. F. Pasternack and B. Halliwell, *J. Am. Chem. Soc.*, 1979, **101**, 1026.
[265] D. A. Case, B. H. Huynh, and M. Karplus, *J. Am. Chem. Soc.*, 1979, **101**, 4433.

carbon monoxide adducts have been performed. Calculated Mössbauer splittings and infrared stretching frequencies are in good agreement with experimental values, and lend support to the treatment. The FeO_2 unit has been shown to be well represented as an equal mixture of $Fe^{2+}(S=0)$, $O_2(S=0)$ and $Fe^{2+}(S=1)$, $O_2(S=1)$, whereas the FeCO unit has been shown to correspond closely to an idealized $Fe^{2+}(S=0)$, $CO(S=0)$ species. Spiro and co-workers[266-268] have employed resonance Raman techniques to investigate expansion of the core and transfer of charge in iron(III) porphyrins. Spectra are reported for high-spin bis(dimethyl sulphoxide)iron(III) complexes of protoporphyrin-IX dimethyl ester, octaethylporphyrin, and tetraphenylporphyrin, together with aquo- and fluoro-methaemoglobin and myoglobin, particular attention being paid to three bands that are known to be sensitive to the spin state of iron. Spiro *et al.* concluded that both expansion of the core and tilting of pyrrole rings contribute to lowering the frequency of these sensitive bands, with core expansion being the dominant factor for moderate tilt angles. Laser photolysis has provided some useful data on the kinetics of binding of nitrogenous bases to four- and five-co-ordinate iron(II) tetraphenylporphin.[269] Baldwin, Basolo, and co-workers[270, 271] have studied the binding of dioxygen by iron(II) and cobalt(II) 'cap' and 'homologous cap' porphyrin complexes. They have shown that the five-co-ordinate, 'homologous cap' complex [Fe(HmCap)(1-MeIm)] is capable of weakly binding a second molecule of 1-MeIm, and that the complex that is so formed can reversibly bind dioxygen without displacing the weakly bound 1-MeIm ligand. The 'homologous cap' complexes are worse carriers of dioxygen than their 'cap' analogues, and this has been explained in terms of an increase in the conformational strain energy of the 'cap' porphyrins upon oxygenation. A systematic study of the affinity of myoglobins and of their cobalt-substituted analogues for dioxygen has recently been described,[272] and kinetic studies on the binding of carbon monoxide by iron porphyrins have provided some useful mechanistic information.[273-276] A fairly comprehensive review by Traylor of compounds that are models of haemoproteins has recently appeared.[277]

[266] T. G. Spiro, J. D. Strong, and P. Stein, *J. Am. Chem. Soc.*, 1979, **101**, 2648.

[267] J. R. Kincade, T. G. Spiro, J. S. Valentine, D. D. Saperstein, and A. J. Rein, *Inorg. Chim. Acta*, 1979, **33**, L181.

[268] P. G. Wright, P. Stein, J. M. Burke, and T. G. Spiro, *J. Am. Chem. Soc.*, 1979, **101**, 3531.

[269] D. Lavalette, C. Tetreau, and M. Momenteau, *J. Am. Chem. Soc.*, 1979, **101**, 5395.

[270] J. R. Budge, P. E. Ellis, R. D. Jones, J. E. Lenard, F. Basolo, J. E. Baldwin, and R. L. Dyer, *J. Am. Chem. Soc.*, 1979, **101**, 4760.

[271] J. R. Budge, P. E. Ellis, R. D. Jones, J. E. Lenard, T. Szymanski, F. Basolo, J. E. Baldwin, and R. L. Dyer, *J. Am. Chem. Soc.*, 1979, **101**, 4763.

[272] M-Y. R. Wang, B. M. Hoffman, S. J. Shine, and F. R. N. Gurd, *J. Am. Chem. Soc.*, 1979, **101**, 7394.

[273] T. G. Traylor, D. Campbell, V. Sharma, and J. Gerbel, *J. Am. Chem. Soc.*, 1979, **101**, 5376.

[274] T. G. Traylor, D. Campbell, and S. Tsuchiya, *J. Am. Chem. Soc.*, 1979, **101**, 4748.

[275] T. Mencey and T. G. Traylor, *J. Am. Chem. Soc.*, 1979, **101**, 765.

[276] J. C. Sirantz, M. A. Stanford, J. N. Moy, B. M. Hoffman, and J. S. Valentine, *J. Am. Chem. Soc.*, 1979, **101**, 3397.

[277] T. G. Traylor, *Acc. Chem. Res.*, 1981, **14**, 102.

Interest in cytochrome *P*-450 and related model systems continues.[278-285] From results on model systems for cytochrome *c*, Reed, Scheidt, and co-workers[279] have suggested that (*a*) Fe–S bond lengths in the haemoprotein should be close to 2.34Å and insensitive to the valency of iron if the protein is unconstrained, (*b*) thioethers are better ligands to Fe[III] than had previously been thought, and (*c*) thioether–imidazole ligation is responsible for a shift of about 160 mV in the iron(II/III) redox potential relative to ligation by two molecules of imidazole, independent of environment. The work of Tabushi and Koga[285] on manganese porphyrins as models for cytochrome *P*-450 is a particularly interesting development, as is the work of Groves and Kruper[286] on oxoporphyrinatochromium(v). Reed, Scheidt, and co-workers[287] have structurally characterized the low-spin bis(pyridine)(meso-tetra-*p*-tolylporphyrinato)-chromium(II) complexes, whilst Scheidt *et al.*[288] have also characterized the tetra-*p*-tolylporphyrinatomanganese(II) nitrosyl complexes [Mn(TTP)(NO)] and [Mn(TTP)(NO)(py)] by *X*-ray diffraction methods. Some unusual cobalt(II) and copper(II) porphyrin complexes have also been reported.[289,290] The kinetics of complexation of metal ions with *N*-methyltetraphenylporphyrin have been studied and a general mechanism of metallation of porphyrins has been postulated.[291] A study of the Soret red shift for zinc tetraphenylporphyrin in the presence of uncharged Lewis bases has appeared.[292] Reversible binding of dioxygen by a titanium(III) porphyrin complex has been reported.[293]

Irradiation of degassed solutions of [Pd(TPP)] in the presence of 0.5—5M-dimethylacetamide in a number of solvents (*e.g.* benzene, pyridine, acetone, 2-propanol, or isobutyronitrile) with light that is of such a wavelength that it is absorbed only by the porphyrin leads to the formation of either (3) or (4).[294,295]

Metalloporphyrins in which the metal is bonded to fewer than four nitrogen atoms, sitting atop complexes, can be considered as models for the initial steps of the metallation of the macrocycle. Weiss and co-workers[296] have published

[278] C. A. Reed, T. Mashiko, S. P. Bentley, M. E Kastner, W. R. Scheidt, K. Spartalian, and G. Lang, *J. Am. Chem. Soc.*, 1979, **101**, 2948.

[279] T. Mashiko, J-C. Marchon, D. T. Musser, C. A. Reed, M. E. Kastner, and W. R. Scheidt, *J. Am. Chem. Soc.*, 1979, **101**, 3654.

[280] M. A. Phillippi and H. M. Goff, *J. Am. Chem. Soc.*, 1979, **101**, 7642.

[281] D. Mansuy, J.-P. Battioni, J-C. Chottard, and V. Ullrich, *J. Am. Chem. Soc.*, 1979, **101**, 3971.

[282] J. T. Groves, T. E. Nemo, and R. S. Myers, *J. Am. Chem. Soc.*, 1979, **101**, 1032.

[283] C. K. Chang and M-S. Kuo, *J. Am. Chem. Soc.*, 1979, **101**, 3413.

[284] C. Cason, L. Ricard, and M. Schappach, *J. Am. Chem. Soc.*, 1979, **101**, 7401.

[285] I. Tabushi and N. Koga, *J. Am. Chem. Soc.*, 1979, **101**, 6456.

[286] J. T. Groves and W. J. Kruper, *J. Am. Chem. Soc.*, 1979, **101**, 7614.

[287] W. R. Scheidt, A. C. Brinegar, J. F. Kirner, and C. A. Reed, *Inorg. Chem.*, 1979, **18**, 3610.

[288] W. R. Scheidt, K. Hatano, G. A. Rupprecht, and P. L. Piciulo, *Inorg. Chem.*, 1979, **18**, 292.

[289] V. Favaudon, M. Momenteau, and J.-M. Lhoste, *Inorg. Chem.*, 1979, **18**, 2355.

[290] G. Paquette and M. Zador, *Can. J. Chem.*, 1979, **57**, 2916.

[291] M. J. Bain-Ackerman, and D. R. Lavallee, *Inorg. Chem.*, 1979, **18**, 3358.

[292] O. W. Kolling, *Inorg. Chem.*, 1979, **18**, 1175.

[293] J-M. Latour, J. C. Marchon, and M. Nakap *J. Am. Chem. Soc.*, 1979, **101**, 3975.

[294] J. A. Mercer-Smith, C. R. Sutcliffe, R. H. Schmehl, and D. G. Whitten, *J. Am. Chem. Soc.*, 1979, **101**, 3995.

[295] J. A. Mercer-Smith and D. G. Whitten, *J. Am. Chem. Soc.*, 1979, **101**, 6620.

[296] H. J. Callot, B. Chevrier, and R. Weiss, *J. Am. Chem. Soc.*, 1979, **101**, 7729.

(3)   (4)

R = CH₂NMePh

rhe crystal structure of the bischloromercury(II) complex of *N*-tosylamino-octaethylporphyrin; this represents the first structural proof for the existence of mercury(II)–porphyrin complexes of this type. Their results show the complex to be monomeric, with one mercury being tetraco-ordinate (three porphyrin N atoms and one Cl) and the other Hg being linearly co-ordinated to the remaining N atom of the porphyrin. Reduction of dichloromolybdenum meso-tetra-*p*-tolylporphyrin by zinc analgam under NO leads to the formation of two low-valent nitrosyl molybdenum porphyrins, the structures of which have been established by *X*-ray diffraction methods.[297] The structures of the molybdenum-(IV) oxo-complexes [MoO(TTP)] and [Mo(Cl)₂(TTP)] have also been determined by *X*-ray diffraction.[298]

The crystal structures of tri-*μ*-oxo-bis[tetraphenylporphyrinatoniobium(IV)] and acetato-oxo-(tetraphenylporphyrinato)niobium(V)–acetic acid have also recently appeared,[299] as has that of a tetraphenylporphyrinatodihydroxytin(IV)-bis(chloroform)bis(carbon tetrachloride) solvate.[300]

A number of new tungsten porphyrin complexes have been reported.[301, 302] An e.s.r. study of dimeric copper(II) complexes of octaethylporphyrin and meso-substituted octaethylporphyrins has appeared,[303] as has a kinetic study of electrophilic exchange reactions between cadmium porphyrin and zinc ion.[304]

Cobalamins and Cobaloximes.—Schrauzer and Hashimoto[305] have shown that straight-chain aliphatic carboxylic acids and certain dicarboxylic as well as branched carboxylic acids can be oxidized to organic radicals in the presence of vitamin B_{12r} and converted into organocobalamins. Owing to the instability of secondary and tertiary alkylcobalamins, and other steric effects, vitamin B_{12r} acts as a selective scavenger of primary alkyl and α-carboxyalkyl radicals. This new method may provide a useful synthetic pathway to otherwise difficultly

[297] T. Diebold, M. Schappacher, B. Chevrier, and R. Weiss, *J. Chem. Soc., Chem. Commun.*, 1979, 693.
[298] T. Diebold, B. Chevrier, and R. Weiss, *Inorg. Chem.*, 1979, **18**, 1193.
[299] C. Lecomte, J. Protas, R. Guilard, B. Fliniaux, and P. Fournari, *J. Chem. Soc., Dalton Trans.*, 1979, 1306.
[300] P. G. Harrison, K. Molloy, and E. W. Thornton, *Inorg. Chim. Acta*, 1979, **33**, 137.
[301] M. Krishnamurthy, *Inorg. Chim. Acta*, 1979, **32**, L32.
[302] B. Fleischer, R. D. Chapman, and M. Krishnamurthy, *Inorg. Chem.*, 1979, **18**, 2156.
[303] M. M. Chikira, H. Kan, R. A. Hawley, and K. M. Smith, *J. Chem. Soc., Dalton Trans.*, 1979, 245.
[304] J. Reid and P. Hambright, *Inorg. Chim. Acta*, 1979, **33**, L135.
[305] G. N. Schrauzer and M. Hashimoto, *J. Am. Chem. Soc.*, 1979, **101**, 4593.

accessible compounds. The Co^I cobaloxime $[(PBu^n_3)Co(dmgH)_2]$ reacts with various compounds R_2AsCl to produce air-sensitive species $[R^1R^2AsCo(dmgH)_2-(PBu^n_3)]$.[306] The synthesis of various previously inaccessible secondary alkyl- and cycloalkyl-cobalamins *via* the reaction between olefins, alkyl iodides or bromides, and hydridocobalamin has been described by Grate and Schrauzer.[307] Thayer[308] has investigated the kinetics of the reaction of methylcobalamin with a variety of methyl derivatives of heavier metals, *e.g.* Me_4PI, Me_4AsI, and Me_4SbI, amongst others. The methylation of platinum complexes by methylcobalamin has been studied by Wood and co-workers.[309] They have presented kinetic equilibrium and ^1H n.m.r. evidence to show that an 'outer-sphere' complex is formed between charged Pt^{II} salts and the corrin macrocycle. This formation of a complex catalytically labilizes the Co—C bond to electrophilic attack. The copper-containing analogue of vitamin B_{12}, copper cobalamin, has been studied by c.d., m.c.d., and frozen-solution e.p.r. and ENDOR spectroscopy.[310] The bonding scheme of the metal, with reference to the separate roles of σ- and π-bonding, has been derived from the spectroscopic parameters. The synthesis and characterization of neutral thiocyanate- and selenocyanate-bridged cobaloximes has been described by Ciskowski and Crumbliss.[311]

Phthalocyanines.—The reactions of iron phthalocyanines with molecular oxygen have been studied by various workers, using a variety of physical techniques.[312–314] When dissolved in concentrated sulphuric acid (96%), iron(II) phthalocyanine reacts with O_2 *via* a two-stage process, the first step being reversible and the second irreversible.[312] Core- and valence-level X-ray photoemission spectra of monomeric, dimeric, and polymeric iron phthalocyanines have been related to the catalytic activity of these compounds in the electrochemical reduction of oxygen.[313] Collamati has reported the isolation of a solid dioxygen derivative of phthalocyaninatoiron(II).[314] The preparation and properties of some phthalocyanine carbonyl derivatives $[Fe(pc)(CO)L]$, where L is DMF, THF, DMSO, NH_3, or Pr^nNH_2, have been reported.[315] The crystal and molecular structures of phthalocyaninato-manganese(II) and -cobalt(II) have been determined by X-ray diffraction methods.[316] Lever *et al.*[317] have reported that manganese(II) phthalocyanine reacts reversibly with dioxygen in pure NN'-dimethylacetamide solution; the adduct has been formulated as $[(pc)Mn^{III}]\cdot O_2^-$. Phthalocyanines have also been employed as semiconductor films and in conductive polymers.[318, 319]

[306] W. R. Cullen, D. Dolphin, F. W. B. Einstein, L. M. Mihichuk, and A. C. Willis, *J. Am. Chem. Soc.*, 1979, **101**, 6898.
[307] J. H. Grate and G. N. Schrauzer, *J. Am. Chem. Soc.*, 1979, **101**, 4601.
[308] J. S. Thayer, *Inorg. Chem.*, 1979, **18**, 1171.
[309] Y-T. Fanchiang, W. P. Ridley, and J. M. Wood, *J. Am. Chem. Soc.*, 1979, **101**, 1442.
[310] K. A. Rubinson, *J. Am. Chem. Soc.*, 1979, **101**, 6105.
[311] J. M. Ciskowski and A. L. Crumbliss, *Inorg. Chem.*, 1979, **18**, 638.
[312] C. Ercolani, F. Monacelli, and G. Rossi, *Inorg. Chem.*, 1979, **18**, 712.
[313] S. Maroie, M. Savy, and J. J. Verbist, *Inorg. Chem.*, 1979, **18**, 2560.
[314] I. Collamati, *Inorg. Chim. Acta*, 1979, **35**, L303.
[315] B. S. Holla and S. Y. Ambekar, *J. Chem. Soc., Chem. Commun.*, 1979, 221.
[316] R. Mason, G. A. Williams, and P. E. Fielding, *J. Chem. Soc., Dalton Trans.*, 1979, 676.
[317] A. B. P. Lever, C. P. Wilshire, and S. K. Quan, *J. Am. Chem. Soc.*, 1979, **101**, 3668.
[318] F-R. Fan and L. R. Faulkner, *J. Am. Chem. Soc.*, 1979, **101**, 4779.
[319] K. F. Schoch, jr., B. R. Kundalkar, and T. J. Marks, *J. Am. Chem. Soc.*, 1979, **101**, 7071.

Crown Ether Complexes and Cryptates.—Cram *et al.*[320-323] have performed some elaborate syntheses to obtain a large number of new macrocyclic poly-ethers, and have studied extensively the binding characteristics of these ligands. The photochemical condensation between dibenzo-18-crown-6 and ethyl *N*-chloroacetylglycinate has been found to be inhibited by the presence of K^+ ions.[324] The ionization potential of the crown ether increases upon complexa-tion, and this is considered to be the reason for the inhibition of the reaction. Isolable crown-ether complexes of Br_2 have been shown to be stereoselective brominating agents.[325] A calorimetric study of the reaction of several organic ammonium cations with 18-crown-6 has appeared.[326] Electronic factors had no significant effects on log K, ΔH, or $T\Delta S$ values, nor did variation of the length of diammonium cations. The crystal structures of a number of crown-ether and cryptate complexes have been reported.[326-328] Dicyclohexyl-18-crown-6 has been employed by Grimaldi and Lehn[329] as a selective carrier for alkali-metal cations in a multi-carrier transport system. Shaw and Shepherd[330] have reported the synthesis of palladium and platinum complexes with substituted benzo-15-crown-5 ethers that contain nitrogen donor ligand atoms. A number of uranium complexes of 18-crown-6 ethers have been reported,[331,332] as have Pr^{III} com-plexes of both 18-crown-6 and 15-crown-5 ethers.[333] Lipophilic [2,2,2] cryptands have been employed as phase-transfer catalysts in nucleophilic substitution reactions in aqueous/organic two-phase solvent systems.[334]

Other Macrocyclic Complexes.—Of particular interest amongst the very large number of studies reported are the Ni^{II} and Fe^{III} complexes prepared by Busch and co-workers.[335,336] These compounds are amongst the first totally synthetic models for haem proteins that are based on ligands with an N_4 donor

[320] A. H. Alberts and D. J. Cram, *J. Am. Chem. Soc.*, 1979, **101**, 3545.
[321] K. E. Koenig, G. M. Lein, P. Stuckler, T. Kaneda, and D. J. Cram, *J. Am. Chem. Soc.*, 1979, **101**, 3553.
[322] R. C. Helgeson, G. R. Weisman, J. L. Toner, T. L. Tarnowski, Y. Chao, J. M. Mayer, and D. J. Cram, *J. Am. Chem. Soc.*, 1979, **101**, 4928.
[323] M. Newcomb, J. L. Toner, R. C. Helgeson, and D. J. Cram, *J. Am. Chem. Soc.*, 1979, **101**, 4941.
[324] M. Tada, A. Suzuki, and H. Hirano, *J. Chem. Soc.*, *Chem. Commun.*, 1979, 1004.
[325] K. H. Pannell and A. Mayr, *J. Chem. Soc.*, *Chem. Commun.*, 1979, 132.
[326] R. M. Izatt, J. D. Lamb, N. E. Izatt, B. E. Rossiter, J. J. Christenson, and B. L. Haymore, *J. Am. Chem. Soc.*, 1979, **101**, 6273.
[327] P. K. Coughlin, J. C. Dewan, S. J. Lippard, E. Watanabe, and J.-M. Lehn, *J. Am. Chem. Soc.*, 1979, **101**, 265.
[328] T. E. Jones, L. S. W. L. Sokol, D. B. Rorabacher, and M. D. Glick, *J. Chem. Soc.*, *Chem. Commun.*, 1979, 140.
[329] J. J. Grimaldi and J. M. Lehn, *J. Am. Chem. Soc.*, 1979, **101**, 1333.
[330] B. L. Shaw and I. Shepherd, *J. Chem. Soc.*, *Dalton Trans.*, 1979, 1634.
[331] G. Folcher, P. Charpin, R.-M. Costes, N. Keller, and G. C. de Villardi, *Inorg. Chim. Acta*, 1979, **34**, 87.
[332] D. C. Moody, R. A. Penneman, and K. V. Salazar, *Inorg. Chem.*, 1979, **18**, 208.
[333] J-C. G. Bünzli, D. Wessner, and H. T. T. Oanh, *Inorg. Chim. Acta*, 1979, **32**, L33.
[334] D. Landini, A. Maid, E. Montanan, and P. Tundo, *J. Am. Chem. Soc.*, 1979, **101**, 2526.
[335] W. P. Schammel, K. S. B. Mertes, G. G. Christoph, and D. H. Busch, *J. Am. Chem. Soc.*, 1979, **101**, 1622.
[336] A. C. Melryk, N. R. Kildahl, A. R. Rendina, and D. H. Busch, *J. Am. Chem. Soc.*, 1979, **101**, 3232.

set. Also of interest is the work of both Fenton *et al.*[337-339] and Nelson *et al.*[340-344] on mononuclear and dinuclear complexes of macrocyclic Schiff-bases.

9 Platinum Complexes

Because of the proven anti-tumour properties of a number of platinum–amine complexes, interest in the properties of existing drugs and the intriguing platinum blues continues to grow. Thus, the ^{15}N n.m.r. spectra of a number of platinum amines and their solutions containing ^{15}N-labelled 1-methylimidazole[345] and an n.m.r. study of the $[Pt(bipy)_2]^{2+}$ cation have been reported.[346]

Farago continues her interesting work on the uptake of metals by plants, and has examined the uptake of platinum, palladium, and rhodium by tomato, corn, and bean plants, grown hydroponically. Platinum (and, to a lesser extent, palladium) is transported to the tops of the plants and extensively taken up by the roots. Growth of the plants and translocation of calcium are affected. Rhodium appears to have little effect on the growth of tomato and corn plants.[347]

Tetrahydrothiamine (THTH), prepared by the action of $NaBH_4$ on thiamine, reacts with K_2PtX_4 (X = Cl or Br) to produce $[Pt(THTH)_2X_2]$.[348] Theophylline reacts with $[PtCl_4]^{2-}$ to yield bis(theophyllinium) tetrachloroplatinate(II) and also theophyllinium trichlorotheophyllineplatinate(II); in the latter, Pt is bound to N-9.[349] In order to spin-label a drug, 4-amino-2,2,6,6-tetramethylpiperidin-1-oxyl was added to K_2PtCl_4 to produce the e.s.r.-active *cis*-$[PtL_2Cl_2]$ complex.[350]

Diastereoisomers of the $[platinum(NNN'N'-tetramethylethylenediamine)-(guanosine)_2]^{2+}$ cation have been produced.[351] A series of complexes *trans*-$[Pd(PBu^n_3)_2L_2]$ and $[Pd(PBu^n_3)LCl]_n$ (L = anion of adenine, cystosine, guanosine, inosine, theophylline, thymine, uracil, or uridine) as well as adenosine-bridged $[Cl_2(PBu^n_3)M-adenosine-M(PBu^n_3)Cl_2]$ (M = Pd or Pt) have been produced. Crystals of *trans*-$[Pd(PBu^n_3)_2(adeninate)_2]\cdot4MeOH$ are triclinic; adeninate is co-ordinated *via* N-9 and the purine rings are virtually orthogonal to the co-ordination plane PdP_2N_2. The crystals are stabilized by hydrogen-bonds between the oxygen atoms of the methanol molecules and N-3 and N-7 of adenine. Intermolecular hydrogen-bonding is observed between the amino-group N(6)H and N''-1. Thus, with the exception of the metallated N-9 atom, all of the N

[337] D. H. Cook, D. E. Fenton, M. G. B. Drew, A. Rodgers, M. McCann, and S. M. Nelson, *J. Chem. Soc., Dalton Trans.*, 1979, 414.
[338] D. H. Cook and D. E. Fenton, *J. Chem. Soc., Dalton Trans.*, 1979, 810.
[339] D. H. Cook and D. E. Fenton, *J. Chem. Soc., Dalton Trans.*, 1979, 266.
[340] M. G. B. Drew, M. McCann, and S. M. Nelson, *J. Chem. Soc., Chem. Commun.*, 1979, 481.
[341] M. G. B. Drew, S. G. McFall, and S. M. Nelson, *J. Chem. Soc., Dalton Trans.*, 1979, 575.
[342] S. M. Nelson, M. McCann, C. Stevenson, and M. G. B. Drew, *J. Chem. Soc., Dalton Trans.*, 1979, 1477.
[343] M. G. B. Drew, J. Cabral, M. F. Cabral, F. S. Esho, and S. M. Nelson, *J. Chem. Soc., Chem. Commun.*, 1979, 1033.
[344] S. M. Nelson, F. S. Esho, M. G. B. Drew, and P. Bird, *J. Chem. Soc., Chem. Commun.*, 1979, 1035.
[345] M. Alei, P. J. Vergamini, and W. E. Wazeman, *J. Am. Chem. Soc.*, 1979, **101**, 5415.
[346] O. Farver, O. Monsted, and G. Nov, *J. Am. Chem. Soc.*, 1979, **101**, 6118.
[347] M. E. Farago, W. A. Mullen, and J. B. Payne, *Inorg. Chim. Acta*, 1979, **34**, 151.
[348] J. Markopoulos, O. Markopoulos, and N. Hadjiliadis, *Inorg. Chim. Acta*, 1979, **34**, L299.
[349] E. H. Griffith and E. L. Archer, *J. Chem. Soc., Chem. Commun.*, 1979, 322.
[350] A. Mathew, B. Bergquist, and J. Zimbrick, *J. Chem. Soc., Chem. Commun.*, 1979, 222.
[351] R. E. Cramer and P. L. Dahlstrom, *J. Am. Chem. Soc.*, 1979, **101**, 3681.

atoms of adeninate are involved in hydrogen-bonding.[352] A non-stoicheiometric compound [Na$_2${(5'-IMPH)$_2$}·16H$_2$O]$_{0.14}$ [Na$_2${(NH$_3$)$_2$Pt(5'-IMP)$_2$}·15–16H$_2$O]$_{0.86}$ (5'-IMP = dianion of inosine 5'-monophosphate) is isomorphous with the monosodium salt of 5'-IMP and with several Pt^{II}–5'-IMP complexes that had previously been reported to be non-stoicheiometric. The co-ordination geometry around platinum is essentially planar, with N-7 of the hypoxanthine base and the ammonia ligand and their two-fold related symmetry mates occupying the four co-ordination sites.[353] The binding of aquated *cis*-[Pt^{II}(NH$_3$)$_2$] to poly(ribo-adenylic acid), poly(ribocytidylic acid), and poly(riboguanylic acid) has been studied at pH 7. The total number of binding sites per nucleotide in each polynucleotide is one; the magnitude of the intrinsic binding constant increases in the order poly(C) < poly(A) < poly(G).[354] The crystal structure and the magnetic and e.s.r. properties of *cis*-[Pt(NH$_3$)$_2$(α-pyridone)] blue have been reported; the compound is a mixed-valence tetramer with the formula [(NH$_3$)$_4$Pt$_2$(C$_5$H$_4$NO)$_2$] (NO$_3$)$_5$·H$_2$O. There is one cation per unit cell, consisting of two *cis*-Pt(NH$_3$)$_2$ units bridged by two α-pyridonate ligands, and two of these are further linked across a crystallographic inversion centre to form a tetranuclear chain. The Pt atoms at the ends of the chain are bonded to two amine ligands in a *cis* configuration and to two deprotonated nitrogen atoms of the α-pyridonate rings. There is a Pt–Pt separation of 2.87Å.[355] The extinction coefficient of the cationic chromophore is a sensitive function of the anions that are present and of the temperature. The approximately axial spin-resonance spectrum is characterized by *g* values of 2.38 and 1.976.[356] A number of paramagnetic species have been seen in the reaction of some quinones with [Pt(PPh$_3$)$_4$].[357]

10 Miscellaneous Items

Most of the peripheral work concerns copper and mercury systems. Thus, it has been seen that MeHg^{II} transfers from two heterocyclic thiones to a wide variety of other ligands.[358] The co-ordination of the ambidentate anion NCS$^-$ has been investigated in [Hg(PPh$_3$)(SCN)$_2$], which contains two infinite chains with backbones of thiocyanate-bridged mercury atoms. In chain 1, mercury is bound to P and two S atoms in the equatorial plane of a distorted trigonal bipyramid. The axial positions are occupied by N atoms of bridging anions. Chain 2 contains tetrahedral mercury with PS$_2$N co-ordination.[359] Binuclear mixed-metal mixed-ligand monomeric bridged complexes of the type [(XCN)$_2$L$_2$M(NCX)$_2$Hg(PPh$_3$)$_2$] (M^{II} = Co, Ni, Cu, or Zn; L = pyridine or nicotinamide; X = S or Se) have been isolated.[360] Some relevant chemical shifts have been obtained of

352 W. M. Beck, J. C. Calabrese, and N. D. Kottmair, *Inorg. Chem.*, 1979, **18**, 176.
353 T. J. Kistenmacher, C. C. Ching, P. Chalilpoyil, and L. G. Marzilli, *J. Am. Chem. Soc.*, 1979, **101**, 1143.
354 W. M. Scovell and R. S. Reaoch, *J. Am. Chem. Soc.*, 1979, **101**, 174.
355 J. K. Barton, D. J. Szalda, H. N. Rabinowitz, J. V. Waszcak, and S. J. Lippard, *J. Am. Chem. Soc.*, 1979, **101**, 1434.
356 J. K. Barton, C. Caravana, and S. J. Lippard, *J. Am. Chem. Soc.*, 1979, **101**, 7269.
357 G. A. Abakumov, I. A. Teplova, V. K. Cherkasov, and K. G. Shalnova, *Inorg. Chim. Acta*, 1979, **32**, L57.
358 I. Erni and G. Geier, *Helv. Chim. Acta*, 1979, **62**, 1007.
359 R. C. Makhija, R. Rivest, and A. L. Beauchamps, *Can. J. Chem.*, 1979, **57**, 2555.
360 P. P. Singh and S. P. Yadav, *Can. J. Chem.*, 1979, **57**, 394.

Hg^{II}-co-ordinated thioether complexes which might serve as potential models for the ^{13}C n.m.r. shifts that are expected in metalloproteins.[361] An analysis of the oscillations in 'beating mercury heart' systems[362] and the reactions of (monohalogenoalkyl)mercury(II) with aromatic amines to give bis-(4-amino-phenyl)alkanes in high yields[363] have been reported. Solutions in organic solvents of 1:1 complexes between organomercury(II) cations and dithizone undergo reversible photo-isomerization (yellow$\rightleftharpoons$blue) at measurable rates.[364]

The crystal structure of a heterobinuclear copper(II)–magnesium complex derived from [Cu(salen)] and [Mg(hfac}$_2$]·2H$_2$O has been obtained, and it shows CuN$_2$O$_2$ and MgO$_6$ co-ordination.[365] Benzil bisthiosemicarbazonato-copper(II) contains non-planar copper and CuN$_2$S$_2$ co-ordination.[366] Chloro-bis(thiabendazole)copper(II) chloride dihydrate contains thiabendazole as a bidentate chelate that is bonded to copper through the nitrogen atoms of the thiazolyl and benzimidazole rings in a *cis* configuration. Only one of the two chlorides is co-ordinated, and the geometry about copper is distorted trigonal-bipyramidal.[367] The reaction of *NNN'N'*-tetramethylpropane-1,3-diamine (L) with CuCl, CO$_2$, and O$_2$ in CH$_2$Cl$_2$ gives a quantitative yield of the binuclear μ-carbonato-dicopper(II) species[LCuCl(CO$_3$)ClCuL], which acts as an initiator for the oxidative coupling of phenols by dioxygen.[368] Rapid hydrolysis of amides, mediated by copper and zinc, that is reminiscent of the metallohydrol-ases has been observed.[369] Electrochemical techniques have been employed to study electron-transfer properties of a binuclear copper–macrocyclic ligand complex and to prepare several reduced derivatives,[370] and binuclear copper(II) complexes with *N*-[2-(alkylthio)ethyl]-3-aminopropanol have been prepared.[371] Riboflavin-2',3',4',5'-tetra-acetate forms complexes with bivalent Co, Ni, and Cu,[372] and human milk lactoferrin forms M$_2$L complexes (M = tervalent Cr, Mn, or Co).[373] In order to characterize the interactions of the two interconvert-ing species CO$_2$ and HCO$_3^-$ with metals in carbonic anhydrase derivatives better, the copper-substituted bovine carbonic anhydrase B (CuBCAB) has been investigated.[374]

Complexes of β-cyclodextrin with different inorganic species are known,[375] and the effect of simple metal salts on the antimicrobial activity of tetracyline

[361] R. G. Khalifah, *Inorg. Chim. Acta*, 1979, **32**, L53.

[362] J. Keizer, P. A. Rock, and S-W. Lin, *J. Am. Chem. Soc.*, 1979, **101**, 5637.

[363] J. Barluenga, P. J. Campos, M. A. Roy, and G. Asensio, *J. Chem. Soc., Chem. Commun.*, 1979, 339.

[364] A. T. Hutton and H. M. N. H. Irving, *J. Chem. Soc., Chem. Commun.*, 1979, 1113.

[365] D. E. Fenton, *J. Chem. Soc., Chem. Commun.*, 1979, 39.

[366] G. W. Bushnell and A. Y. M. Tsang, *Can. J. Chem.*, 1979, **57**, 603.

[367] M. R. Udupa and B. Krebs, *Inorg. Chim. Acta*, 1979, **32**, 1.

[368] M. R. Churchill, G. Davis, M. A. El-Sayed, M. F. El-Shazly, J. P. Hutchinson, M. W. Rupich, and K. O. Watkins, *Inorg. Chem.*, 1979, **18**, 2296.

[369] J. T. Groves and R. M. Dias, *J. Am. Chem. Soc.*, 1979, **101**, 1033.

[370] R. R. Gagné, C. A. Koval, T. J. Smith, and M. C. Cimolino, *J. Am. Chem. Soc.*, 1979, **101**, 4571.

[371] M. Mikuriya, H. Okawa, and S. Kida, *Inorg. Chim. Acta*, 1979, **34**, 13.

[372] M. Goodgame and K. W. Johns, *Inorg. Chim. Acta*, 1979, **34**, 1.

[373] E. W. Ainscough, A. M. Brodie, and J. E. Plowman, *Inorg. Chim. Acta*, 1979, **33**, 149.

[374] I. Bertini, E. Borghi, and C. Luchinat, *J. Am. Chem. Soc.*, 1979, **101**, 7069.

[375] A. Buran and L. Barcza, *Inorg. Chim. Acta*, 1979, **33**, L179.

hydrochloride shows marked variations with metal ions.[376] Ionomycin, a novel diacidic polyether antibiotic, has a high affinity for calcium ions.[377] The reduction of (E)-2-, (E)-3-, and (E)-4-cinnamoylpyridines by 1,4-dihydropyridine derivatives, to form the corresponding dihydro-ketones, is catalysed by Zn^{2+} and Mg^{2+} ions.[378] Controlled reduction of N-methyl-3-ethylpyridinium iodide affords the corresponding 1,2-dihydropyridine, which was stabilized by complexation to give 1,2- and 1,6-dihydro-3-ethyl-N-methylpyridinetricarbonyl-chromium(0).[379] Horrocks and Burlone have reported e.s.r. studies on some [Co^{II}N$_2$O$_2$] models of metalloproteins.[380] Stability constants of cadmium that is bound to fulvic acids range from 1.4 to 43×10^3.[381] Carbon-13 n.m.r. spectra confirm that co-ordination of metal ions by pyridoxine is through the oxygens attached to C-3 and C-4′ in aqueous solution.[382] Cobalt(II) complexes of chiral vic-dioximate ligands derived from D-camphor and L-β-pinene have been reported,[383] as have K(quin)·Hquin and K(quin)·2Hquin (Hquin = 8-quinolinol).[384] Copper(II) acetate reacts with methyl 3,4,6-tri-O-acetyl-2-deoxy-2-salicyl-ideneamino-β-D-glucopyranoside to form a Schiff-base complex.[385]

Measurements of the intensity of sum peaks (in the spectra of γ-rays emitted from radioactive nuclides) have been applied to biological molecules, e.g. albumin,[386] and radiolabelled technetium has been obtained in dichloro[hydrotris-(1-pyrazolyl)borato]oxotechnetium(V).[387] Spin–lattice relaxation times have been determined for several classes of selenium compounds that are biologically important.[388]

[376] R. P. Gupta, B. N. Yadav, O. P. Tiwari, and A. K. Srivastava, *Inorg. Chim. Acta*, 1979, **32**, L95.

[377] B. K. Toeplitz, A. I. Cohen, P. T. Funke, W. L. Parker, and J. Z. Gougoutas, *J. Am. Chem. Soc.*, 1979, **101**, 3344.

[378] R. A. Gase and U. K. Pandit, *J. Am. Chem. Soc.*, 1979, **101**, 7059.

[379] J. P. Kutney, R. A. Badger, W. R. Cullen, R. Greenhouse, M. Noda, V. E. Ridaura-Sanz, Y. H. So, A. Zanarotti, and B. R. Worth, *Can. J. Chem.*, 1979, **57**, 300.

[380] W. De W. Horrocks and D. A. Burlone, *Inorg. Chim. Acta*, 1979, **35**, 165.

[381] R. A. Sasr and J. H. Weber, *Can. J. Chem.*, 1979, **57**, 1263.

[382] J. S. Hartman and E. C. Kelusky, *Can. J. Chem.*, 1979, **57**, 2118.

[383] A. Nakamura, A. Konishi, and S. Otsuka, *J. Chem. Soc., Dalton Trans.*, 1979, 488.

[384] D. L. Hughes and M. R. Truter, *J. Chem. Soc., Dalton Trans.*, 1979, 520.

[385] M. J. Adam and L. D. Hall, *J. Chem. Soc., Chem. Commun.*, 1979, 234.

[386] K. Yoshihara, H. Kaji, and T. Shiokawa, *Inorg. Chim. Acta*, 1979, **32**, L43.

[387] R. W. Thomas, G. W. Estes, C. Elder, and E. Deutch, *J. Am. Chem. Soc.*, 1979, **101**, 4581.

[388] J. D. Odom, W. H. Dawson, and P. D. Ellis, *J. Am. Chem. Soc.*, 1979, **101**, 5815.

2

Storage, Transport, and Function of the Cations of Groups IA and IIA

M. N. HUGHES

Introduction.—This chapter differs from those in earlier volumes in several ways. The discussion of model ionophoric ligands has been reduced, while an attempt has been made to present a more rational discussion of membrane-bound transport ATPases. The coverage of Mg^{2+}-dependent processes has been increased.

1 Interaction of Synthetic Ionophoric and other Ligands with Cations of Groups IA and IIA

A comprehensive review of the co-ordination chemistry of these cations has been published.[1] An increasing number of publications have appeared in which the interaction of these cations with small ligands and macro-ligands has been studied by n.m.r. techniques (*i.e.* sodium-23,[2-4] magnesium-25,[5,6] calcium-43,[6] and potassium-39[2]).

The Cyclic Polyethers (Crown Ethers).—In Volume 2 it was noted that the caesium salts of aromatic dicarboxylic acids react with activated poly(ethylene glycols) to form crown ethers. It is now known[7] that the caesium salts of various phenolic compounds will also react readily with dibromo-poly(ethylene glycols) to give crown ethers. Furthermore a simple, high-yield synthesis of macrocyclic lactones has been carried out by ring-closure of caesium salts of ω-halogeno-aliphatic acids.[8] These provide useful additional synthetic methods. The synthesis of macrocyclic di- and tetra-ester compounds has been reviewed,[9] while the preparation of cyclic oligo-esters having alternating tetrahydropyran and ester moieties has been reported.[10]

The role of metal ions as templates is illustrated by the synthesis[11] of furyl-

[1] N. S. Poonia and A. V. Bajaj, *Chem. Rev.*, 1979, **79**, 389.
[2] M. M. Civan and M. Shporer, *Biol. Magn. Reson.*, 1978, **1**, 1.
[3] A. Delville, C. Detellier, and P. Laszlo, *J. Magn. Reson.*, 1979, **34**, 301.
[4] J. E. Norne, H. Gustavsson, S. Forsén, E. Chiancone, H. A. Kuiper, and E. Antonini, *Eur. J. Biochem.*, 1979, **98**, 591.
[5] E. O. Bishop, S. J. Kimber, B. E. Smith, and P. J. Beynon, *FEBS Lett.*, 1979, **101**, 31.
[6] P. Reimarsson, J. Parello, T. Drakenberg, H. Gustavsson, and B. Lindman, *FEBS Lett.*, 1979, **108**, 439.
[7] B. J. van Keulen, R. M. Kellog, and O. Peipers, *J. Chem. Soc., Chem. Commun.*, 1979, 285.
[8] W. H. Kruizinga and R. M. Kellog, *J. Chem. Soc., Chem. Commun.*, 1979, 286.
[9] J. S. Bradshaw, G. E. Maas, R. M. Izatt, and J. J. Christensen, *Chem. Rev.*, 1979, **79**, 37
[10] M. Okada, H. Sumitomo, and I. Tajima, *J. Am. Chem. Soc.*, 1979, **101**, 4013.
[11] D. H. Cook and D. E. Fenton, *J. Chem. Soc., Dalton Trans.*, 1979, 810.

containing Schiff-base macrocycles that is induced by ions of alkaline-earth metals, in which furan-2,5-dicarbaldehyde is condensed with 3,6,9-trioxaundec-ane-1,11-diamine (1) or 1,5-bis-(2-aminophenoxy)-3-oxapentane in alcoholic media. In the absence of metal cation, the macrocycles are formed in little or no yield, although Mg^{2+} is ineffective as a catalyst. These macrocycles, *e.g.* (2), are of interest as model ligands in that the furyl unit occurs in antibiotics such as the actins and X-537A (an open-chain carboxylic acid). A similar template effect for Ca^{2+} has been found in the synthesis of analogous pyridyl-containing macrocyclic Schiff-bases.[12]

(1)

(2)

Several new clasess of crown ethers have been prepared. Amongst these are 'multi-loop' crown ethers,[13] *e.g.* (3) and (4), in which several rings of various cavity sizes and donor properties are coupled *via* spiro carbon atoms. These offer the prospect of the simultaneous and selective complexation of several types of cations. Another group[14] is the 'breathing' crown ethers, with fluctuating ring size, *e.g.* (5), although their metal-binding properties await a full study.

Much effort continues to be expended on the synthesis of crown ethers that are specifically designed for molecular recognition. The series of papers on host–guest complexation by Cram and co-workers continues.[15–22] Of particular interest are studies on the binding of cations to macrocycles containing acetylace-tone groups, for which formation constants[16] vary by up to 10^6, depending upon the macrocycle in which the β-diketone is incorporated, and the synthesis of macrocyclic polyethers containing convergent methoxyaryl groups.[17] In the latter case the resulting organization of the oxygen donor atoms results in effective binding of cations, despite the presence of the aryl groups. Thus these compounds

12 D. H. Cook and D. E. Fenton, *J. Chem. Soc., Dalton Trans.*, 1979, 4013.
13 E. Weber, *Angew. Chem., Int. Ed. Engl.*, 1979, **18**, 219.
14 G. Schröder and W. Witt, *Angew. Chem., Int. Ed. Engl.*, 1979, **18**, 311.
15 G. D. Y. Sogah and D. J. Cram, *J. Am. Chem. Soc.*, 1979, **101**, 3035.
16 A. H. Alberts and D. J. Cram, *J. Am. Chem. Soc.*, 1979, **101**, 3545.
17 K. E. Koenig, G. M. Lein, P. Stukler, T. Kaneda, and D. J. Cram, *J. Am. Chem. Soc.*, 1979, **101**, 3553.
18 L. J. Kaplan, G. R. Weisman, and D. J. Cram, *J. Org. Chem.*, 1979, **44**, 2226.
19 R. C. Helgeson, G. R. Weisman, J. L. Toner, T. L. Tarnowski, Y. Chao, J. M. Mayer, and D. J. Cram, *J. Am. Chem. Soc.*, 1979, **101**, 4928.
20 R. C. Helgeson, T. L. Tarnowski, and D. J. Cram, *J. Org. Chem.*, 1979, **44**, 2538.
21 M. Newcomb, T. L. Toner, R. C. Helgeson, and D. J. Cram, *J. Am. Chem. Soc.*, 1979, **101**, 4941.
22 Y. Chao, G. R. Weisman, G. D. Y. Sogah, and D. J. Cram, *J. Am. Chem. Soc.*, 1979, **101**, 4948.

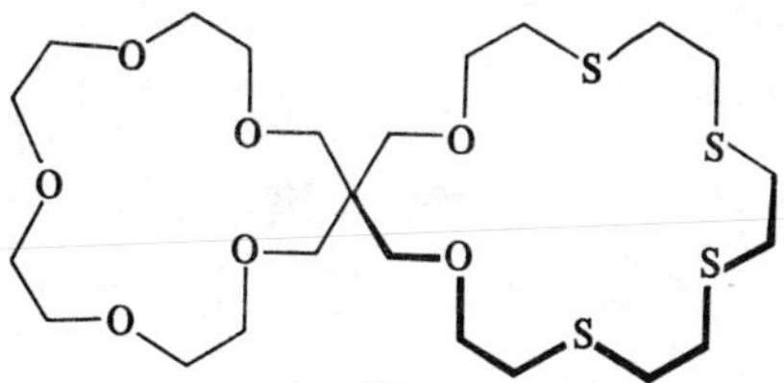

(3)

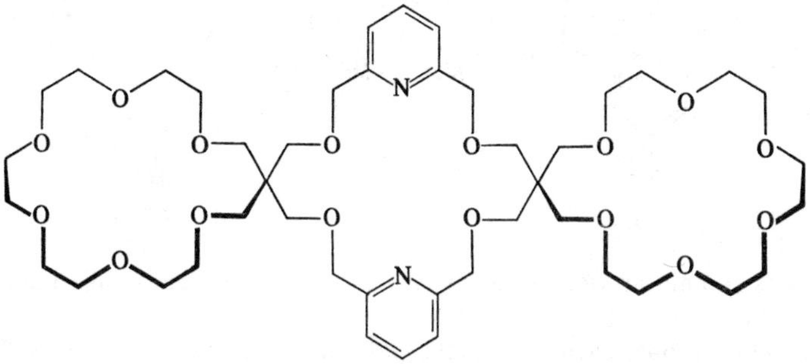

(4)

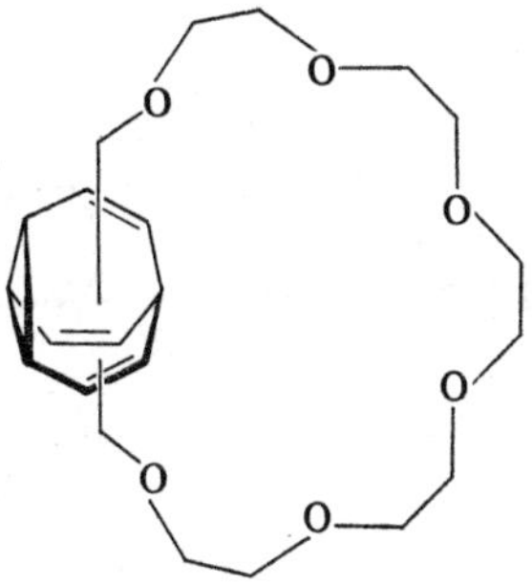

(5)

complex more effectively than the crown ethers. This particular work has been extended[23] to define a new class of ligand, the spherands, whose cation-binding potential is attributed to a relief of electron–electron repulsion on binding of cation. A typical spherand (6), a cyclohexametaphenylene system, provides a framework which holds the oxygens of the six methoxyl groups in a perfect octahedral array by their attachment to the convergent positions of the aryl groups. The cavity can only be occupied by spherical entities such as univalent cations, and not by solvent molecules. This spherand strongly binds Li$^+$ and Na$^+$, and scavenges these ions from other cations of Groups IA and IIA. The synthesis and complexing properties of spherand (7) have also been reported.[24] These

[23] D. J. Cram, T. Kaneda, R. C. Helgeson, and G. M. Lein, *J. Am. Chem. Soc.*, 1979, **101**, 6753.
[24] D. J. Cram, T. Kaneda, G. M. Lein, and R. C. Helgeson, *J. Chem. Soc., Chem. Commun.* 1979, 948.

(6)   (7)

spherands contain an organized cavity *before* complexation, unlike the crowns and cryptands.

Three-dimensional cavities in cyclic polyethers are also provided by the use of multi-strand polyethers,[20] as in (8), and the bridged polyethers[25] (9) and (10). The structure of the complex of (10) with KCl has been reported.[25]

(8)

(9)

(10) R = H or But

25 I. R. Hanson, D. G. Parsons, and M. R. Truter, *J. Chem. Soc., Chem. Commun.*, 1979, 486.

Lehn has reviewed[26] the use of macropolycyclic receptor molecules (including cryptands) in the context of recognition, catalytic, and transport processes. Other examples include the use of a chiral 1,4-dihydropyridine crown ether in asymmetric reductions,[27] nicotinic acid crown ethers,[28] and a chiral 18-crown-6 derivative incorporating a 2,5-anhydro-D-mannitol residue.[29] Complexes of γ-cyclodextrin with crown ethers have been prepared, and should be useful in model studies for the transport of ions through membranes.[30]

Solution Studies. Solvent-extraction studies of picrates of cations of Groups IA and IIA with crown ethers continue to be published (dibenzo-24-crown-8,[31] and 15-crown-5 and 18-crown-6[32]). Dibenzo-24-crown-8 is not selective, in accord with its known non-selectivity in binding studies. Studies have been reported for two t-butyl derivatives of dibenzo-18-crown-6, in the hope that their solubility in organic phases, and hence their practical uses as reagents for solvent extraction, would be enhanced.[33] This work involves measurement of formation complexes between bis-(4-t-butylbenzo)-18-crown-6 and bis-(3,5-di-t-butylbenzo)-18-crown-6 and alkali-metal cations in methanol, dimethyl sulphoxide, and acetonitrile.

The binding of K^+ and of Na^+ by crown ethers that contain a diphenyl ether group has been studied by ^1H and ^{13}C n.m.r. techniques, which have allowed the establishment of the stoicheiometry of complex formation and conformational changes.[34] The interaction of Na^+, K^+, and Cs^+ with dibenzo-30-crown-10 in a range of non-aqueous solvents has also been examined[35] by ^{13}C, ^{23}Na, and ^{133}Cs n.m.r. The ions K^+ and Na^+ form 1:1 complexes, but Na^+ gives Na_2L, Na_3L_2, and NaL in solution. The Cs complex involves a wrap-around structure. From the temperature dependence of the chemical shift of the ^{133}Cs resonance, it was concluded that the complex is enthalpy-stabilized and entropy-destabilized, but no satisfactory explanation is available at present.

Formation constants and the partial molal volume and isentropic partial molal compressibility changes of complex formation between the cations Na^+, K^+, Rb^+, Cs^+, Ca^{2+}, and Ba^{2+} and the crown ethers 12-crown-4, 15-crown-5, and 18-crown-6 have been reported[36] for water at 25 °C. Values of ΔV of complex formation are discussed in terms of two models (scaled particle theory and the Drude–Nernst continuum model), and indicate that the charge on K^+ in the 18-crown-6 complex is especially well screened from the water. Hydration numbers of complexed cations have been calculated. There is an interesting report[37] on the behaviour of the heat capacity of aqueous solutions of 18-crown-6 and KCl.

[26] J. M. Lehn, *Pure Appl. Chem.*, 1979, **51**, 979.
[27] J. G. de Vries and R. M. Kellog, *J. Am. Chem. Soc.*, 1979, **101**, 2759.
[28] G. R. Newkome and T. Kawato, *J. Org. Chem.*, 1979, **44**, 2692.
[29] J. A. Haslegrove, J. F. Stoddart, and D. J. Thompson, *Tetrahedron Lett.*, 1979, 2279.
[30] F. Vögtle and W. M. Müller, *Angew. Chem., Int. Ed. Engl.*, 1979, **18**, 623.
[31] Y. Takeda, *Bull. Chem. Soc. Jpn.*, 1979, **52**, 2501.
[32] Y. Takeda and H. Goto, *Bull. Chem. Soc. Jpn.*, 1979, **52**, 1920.
[33] L. J. Tušek-Bosić and P. R. Danesi, *J. Inorg. Nucl. Chem.*, 1979, **41**, 833.
[34] K. Torizuka and T. Sato, *Org. Magn. Reson.*, 1979, **12**, 190.
[35] M. Shansipur and A. I. Popov, *J. Am. Chem. Soc.*, 1979, **101**, 4050.
[36] H. Høiland, J. A. Ringseth, and T. S. Brun, *J. Solution Chem.*, 1979, **11**, 779. See also H. Høiland, J. A. Ringseth, and E. Vikingstad, *ibid.*, 1978, **7**, 515.
[37] C. Jolicoeur, L-L. Lemelin, and R. Lapalme, *J. Phys. Chem.*, 1979, **83**, 2806.

This shows a maximum in the heat capacity which originates in the perturbation of the complexation equilibrium that follows the temperature rise that occurs during the measurement of the heat capacity.

The Cryptands.—The protonation and deprotonation of small cryptands are slow compared to proton-transfer reactions of other systems with nitrogen and oxygen donor and acceptor atoms. The rate of protonation of the dibasic form of cryptand 2.1.1 by D_2O is some four times slower[38] than the reaction in H_2O. This substantial effect strongly suggests that the proton is transferred directly from water to the cryptand, and that the previous suggestion that the low rate of reaction results from conformational changes in the ligand is incorrect.

Thallium-205 n.m.r. studies on Tl^I complexes with the cryptands 2.2.1, 2.2.2, and 2.2.2B in water and several non-aqueous solvents lead to the conclusion that the thallium(I) ion is completely shielded by the solvent, as the chemical shifts of the complexes were independent of the solvent.[39] However, studies on the thermodynamics of 2.2.2 cryptates with a range of cations from Group IA and with Tl^+ and Ag^+ suggest that cryptand 2.2.2 does not shield the various cations from the influence of the solvent.[40]

Solvent-extraction studies, using Na^+, K^+, and Cs^+ and the cryptands 2.2.1, 2.2.2, and 2.2.2B, have been reported,[41] while the use of lipophilic cryptands that are related to 2.2.2 and which have a C_{14} linear alkyl chain attached to one strand as phase-transfer catalysts[42] has been explored. A range of new cryptands that have lipophilic structural elements, such as *o*-, *m*-, or *p*-phenylene, biphenyl, and pyridine nuclei, have been synthesized.[43] In certain of these cryptands, *e.g.* (11) and (12), the 2.2 cryptand strands are held apart at fixed and different distances, thus providing, it is suggested, model compounds for channel-forming ionophores.

Non-cyclic Ethers.—These have been reviewed.[44] Several new series of non-cyclic neutral polyethers have been synthesized.[45,46] These include linear polyethers with aromatic end-groups,[45] prepared by linking various aromatic and hetero-cyclic residues with tetra(ethylene glycol) and ethylene glycol analogues, and open-chain tripod-like polyethers prepared from triethanolamine.[45] The inter-action of these polyethers with alkali metals in methanol has been well studied, and, not surprisingly, they do not distinguish well between cations. These long, linear polyethers can form complexes with cations of at least two structural types, (*a*) in which the ligand wraps spherically around the cation with more than one turn, so that the cation is shielded from the anion, and (*b*) in which one cation is co-ordinated to each of the two loops provided by the polyether in an

[38] A. M. Kjaer, P. E. Sørensen, and J. Ulstrup, *J. Chem. Soc., Chem. Commun.*, 1979, 965.
[39] D. Gudlin and H. Schneider, *Inorg. Chim. Acta*, 1979, **33**, 205.
[40] M. H. Abraham, E. C. Viguria, and A. F. D. de Namov, *J. Chem. Soc., Chem. Commun.*, 1979, 374.
[41] M. J. Reyes, A. G. Maddock, G. Duplatre, and J. J. Schleiffer, *J. Inorg. Nucl. Chem.*, 1979, **41**, 1365.
[42] D. Landini, A. Maia, F. Montanari, and P. Tundo, *J. Am. Chem. Soc.*, 1979, **101**, 2526.
[43] E. Buhleier, W. Wehner, and F. Vögtle, *Chem. Ber.*, 1979, **112**, 546.
[44] F. Vögtle and E. Weber, *Angew. Chem., Int. Ed. Engl.*, 1979, **18**, 753.
[45] B. Tümmler, G. Maass, F. Vögtle, H. Seiger, V. Heimann, and E. Weber, *J. Am. Chem. Soc.*, 1979, **101**, 2588.
[46] W. Rasshofer, W. M. Müller, and F. Vögtle, *Chem. Ber.*, 1979, **112**, 2095.

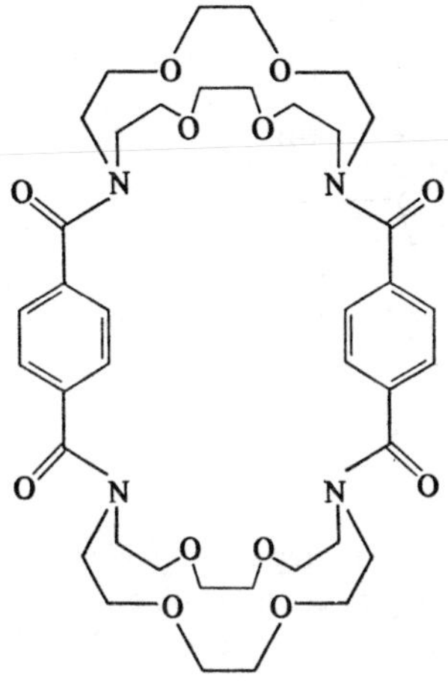

(11)

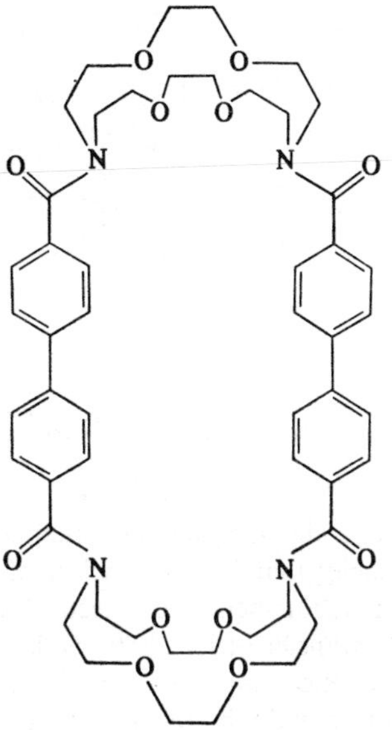

(12)

S-shaped arrangement, giving a binuclear complex, with each cation also bound to an anion. Evidence for these two structural types has been established for the complex[47] of RbI with the ligand (13) and that of KSCN[48] with the ligand (14), respectively. In the former case, the Rb⁺ is probably ten-co-ordinate and enclosed by the ligand, while the polyether (14) is S-shaped in the complex and binds a K⁺ SCN⁻ pair in each of the two S-loops.

The transport capability of the newly synthesized ether (15) for Ca^{2+} compares favourably with those of the well-known ionophores A-23187 and X-573A.[49]

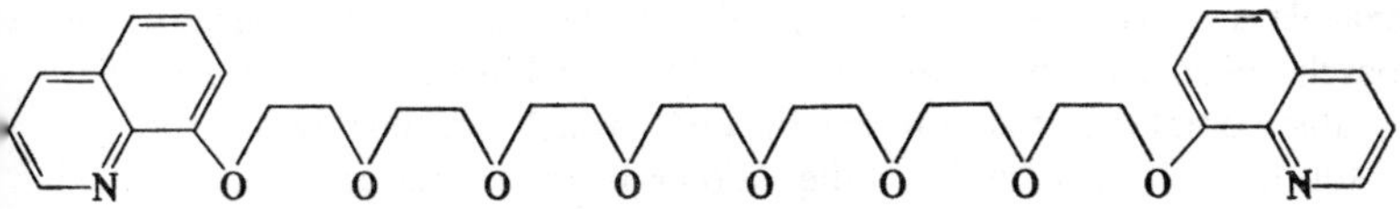

(13)

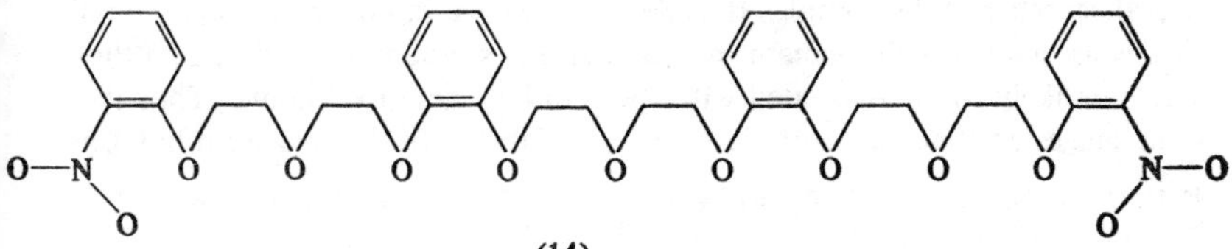

(14)

[47] G. Weber, W. Saenger, F. Vögtle, and H. Sieger, *Angew. Chem., Int. Ed. Engl.*, 1979, **18**, 226.
[48] G. Weber and W. Saenger, *Angew. Chem., Int. Ed. Engl.*, 1979, **18**, 227.
[49] W. Wierenga, B. R. Evans, and J. A. Woetersom, *J. Am. Chem. Soc.*, 1979, **101**, 1334.

(15)

Lipophilic diether diamide ionophores have been prepared[50] which have a lipophilicity that is seven orders of magnitude greater than some ion-carriers used in liquid membrane electrodes. For this type of ion-carrier, the carrier-induced transfer of ions between aqueous and membrane phases is usually subject to kinetic limitations. This is overcome by the use of carriers whose lipophilic segments remain in the membrane phase, and whose segments with the co-ordination sites are exposed to the aqueous phase during the transfer process.

2 The Ionophore Antibiotics

This section describes studies on the interaction of cations with various iono-phorous antibiotics, a subject that has been reviewed.[51] Complexes of A-23187 (whose absolute configuration has been established[52]) with Mg^{2+}, Ca^{2+}, Sr^{2+}, Ba^{2+}, Mn^{2+}, Co^{2+}, Ni^{2+}, and Zn^{2+} have been examined[53] by n.m.r. and c.d. techniques in an endeavour to relate the cation selectivity of this ionophore to its conformational response to complexation. This work suggests that the known structure for the Ca^{2+} complex (of stoicheiometry A_2Ca) is maintained for all cations. In this structure,[54] Ca^{2+} is complexed by the oxygen atoms from the benzoxazole carboxylate and ketopyrrole groups, and by the benzoxazole heterocyclic nitrogen atom from each of the two ionophore molecules. The most stable complex is that formed by Zn^{2+}, which presumably provides the best fit for the cavity. As the size of the cation is increased, the cavity is expanded by minor rotations about the C-8–C-9 and C-19–C-20 bonds. For cations smaller than Zn^{2+}, the cavity size cannot be reduced through such rotations, and the overall stability of the complex thus decreases with size. The reduction in stability that is associated with increase in cation size is substantial, so much so that it seems unlikely to be associated with the rotations described above. Thus it has been suggested that these rotations are associated with the hydrogen-bond net-

[50] U. Oesch, D. Ammann, E. Pretsch, and W. Simon. *Helv. Chim. Acta*, 1979, **62**, 2073.
[51] Yu. A. Ovchinnikov, *Eur. J. Biochem.*, 1979, **94**, 321.
[52] D. A. Evans, C. E. Sacks, W. A. Kleschick, and T. R. Taber, *J. Am. Chem. Soc.*, 1979, **101**, 6788.
[53] D. R. Pfeiffer and C. M. Deber, *FEBS Lett.*, 1979, **105**, 360.
[54] M. O. Chaney, N. D. Jones, and M. Debono, *J. Antibiot.*, 1976 **29**, 424; G. D. Smith and W. L. Duax, *J. Am. Chem. Soc.*, 1976, **98**, 1578.

work in the dimeric complex, which can therefore increase the energetic barriers to such rotations. It is noteworthy that the same structure is found for strongly and weakly interacting cations. This situation is not found for other ionophores.

The structures of the polyether antibiotics carriomycin[55] (as the thallium salt) and ionomycin[56] have been reported. The latter ionophore is a diacidic polyether, which has high affinity for calcium ions (and cadmium ions) and which gives a dimeric globular structure in which the two ionophore molecules are linked *via* hydrogen bonds. The structure of ionomycin bears distinct resemblances to the well-known antibiotic lasalocid-A (X-573A), which also only forms a dimer in non-polar solvents. The results of a spectroscopic and *X*-ray crystallographic study on free lasalocid are available,[57] and show that two different crystalline modifications of the 1 : 1 methanolate of lasalocid can be obtained from methanol solutions, and that the vibrational spectrum of the ketone carbonyl group is a sensitive monitor of the structure. The competitive binding of sodium ions and biogenic amines to lasalocid has been characterized by the use of ^{23}Na n.m.r.[58]

A modified form of nigericin has been prepared[59] in which the terminal hemiacetal group is opened. This results in a dramatic decrease in the ionophore properties of the antibiotic, thus testifying to the importance of this structural feature. The formation of the complex Mon$^-$ Na$^+$ (MonH = monensin) in methanol is enthalpy- and entropy-stabilized, and that of MonH $\cdot$ NaClO$_4$ is enthalpy-stabilized and entropy-destabilized.[60] The selectivity of the ionophore salinomycin towards cations[61] appears to be a function of both the initial conformation of the free anionic ionophore and its ability to reorientate about the cation. An analogy has been drawn with the well-known 'induced fit' model of enzyme–substrate interaction.

The ^{205}Tl chemical shifts[62] for the complexes of Tl^I with the macrotetrolides nonactin, monactin, and dinactin are intermediate between those found for the Tl^I complexes of valinomycin and the crown ethers. It appears that Tl^I interacts more favourably with ether than carbonyl oxygen atoms. The temperature dependence of the ^{205}Tl chemical shift indicates the presence of two conformational species. Other reports on the actins include the enthalpies of complex formation for tetranactin by microcalorimetry techniques[63] and the synthesis[64] of useful models for the actin antibiotics based on the 32-crown-8-tetraester family, having three-carbon bridges between the oxygen atoms.

The interaction of cations of Group IA and Tl^I with the macrotetrolides and

[55] H. Nakayama, N. Otake, H. Miyamae, S. Sato, and Y. Saito, *J. Chem. Soc., Perkin Trans. 2*, 1979, 293.

[56] B. K. Toeplitz, A. I. Cohen, P. T. Funke, W. L. Parker, and J. Z. Gougoutas, *J. Am. Chem. Soc.*, 1979, **101**, 3344.

[57] J. M. Friedman, D. L. Rousseau, C. Shen, C. C. Chiang, E. N. Duesler, and I. C. Paul, *J. Chem. Soc., Perkin Trans. 2*, 1979, 835.

[58] J. Grandjean and P. Laszlo, *Angew. Chem., Int. Ed. Engl.*, 1979, **18**, 153.

[59] L. David, M. Chapel, J. Gandreuil, G. Jeminet, and R. Durand, *Experientia*, 1979, **35**, 1562.

[60] J. G. Hoogerheide and A. I. Popov, *J. Solution Chem.*, 1979, **8**, 83.

[61] G. Painter and B. C. Pressman, *Biochem. Biophys. Res. Commun.*, 1979, **91**, 117.

[62] R. W. Briggs and J. F. Hinton, *J. Magn. Reson.*, 1979, **33**, 363.

[63] M. Ueno and H. Kishimoto, *Chem. Lett.*, 1979, 1487.

[64] A. Samat, J. Elguero, and J. Metzger, *J. Chem. Soc., Chem. Commun.*, 1979, 1182.

valinomycin has been studied[65] by the use of polarographic techniques. Stability constants and rate-constants for dissociation have been measured for valinomycin. Further work has been reported on cyclic peptide analogues for valinomycin. The peptide cyclo(-L-Val-Gly-Gly-L-Pro-)$_3$ has a conformation[66] in solvents such as CDCl$_2$ and CDCl$_3$ which lacks a cavity, and so it binds cations only weakly or not at all. In contrast, it forms strong 1:1 complexes with K$^+$ and Ba^{2+} in acetonitrile which are structurally analogous to the 'bracelet' structure of valinomycin. The poor efficiency of cyclo(-L-Pro-L-Val-D-Ala-D-Val-)$_3$ as a membrane-bound ion carrier has been attributed[67] to its slow dissociation kinetics (determined by a novel approach involving deuterium exchange between amide and solvent in mixtures of K$^+$PVAV and PVAV).

3 Cations, Ionophores, and Model Membrane Systems

Some useful reports have been published on the diffusion of ions through membranes,[68] the effect of dielectric constant of the phospholipid bilayer on permeability,[69] the role of lipids in membranes,[70] and the use of vibrational spectroscopy in the analysis of membrane structure.[71]

Channel-forming Ionophores.[51]—The bulk of published work still concentrates on gramicidin, which is an artificial channel-former that shows high transport rates (*e.g.* 10^7 Na$^+$ ions per second, which is close to fluxes in the Na$^+$ channel of nerve fibres). The well-known hypothesis that it functions *via* the formation of dimers receives further support from the demonstration, by light-scattering techniques,[72] that gramicidin A exists as dimers in methanol and dioxan and as monomers in dimethyl sulphoxide. X-Ray studies indicate[73] that the binding of K$^+$ (like that[74] of Cs$^+$) leads to conformational changes in which the length of the channel decreases from 32 to 26 Å and the diameter increases from about 5 to 6.8 Å. Each structural unit contains two of these cylindrical channels, each of which contains two bound cations. Studies on analogues of gramicidin A show that the two strands in the dimeric channel are antiparallel.[75] The widening of the channel is probably necessary to permit peptide carbonyl–metal co-ordination.

Some effort has been made to apply n.m.r. techniques to the problems of binding of cations to gramicidin (and other channel-formers). Thus the binding of Na$^+$ to gramicidin D in an emulsion that is dispersed in sodium chloride solutions has been demonstrated[76] by ^{23}Na n.m.r. Similar measurements[77] in ethanol–water mixtures (90:10) showed that Na$^+$ binds to the gramicidin A dimer at well-

[65] A. Hofmanová, J. Koryta, M. Brézina, T. H. Ryan, and K. Angelis, *Inorg. Chim. Acta*, 1979, **37**, 135.
[66] K. R. K. Easwaran, L. G. Pease, and E. R. Blout, *Biochemistry*, 1979, **18**, 61.
[67] D. G. Davis and B. F. Gisin, *J. Am. Chem. Soc.*, 1979, **101**, 3755.
[68] S. L. Hardt, *J. Membr. Biol.*, 1979, **48**, 299.
[69] J. P. Dilger, S. G. A. McLaughlin, T. McIntosh, and S. A. Simon, *Science*, 1979, **206**, 1196.
[70] P. R. Cullis and B. De Kruijff, *Biochim. Biophys. Acta*, 1979, **559**, 399.
[71] D. F. H. Wallach, S. P. Verma, and J. Fookson, *Biochim. Biophys. Acta*, 1979, **559**, 153.
[72] F. Rondelez and J. D. Litster, *Biophys. J.*, 1979, **27**, 455.
[73] R. E. Koeppe, J. M. Berg, K. O. Hodgson, and L. Stryer, *Nature (London)*, 1979, **297**, 724.
[74] R. E. Koeppe, K. O. Hodgson, and L. Stryer, *J. Mol. Biol.*, 1978, **121**, 41.
[75] J. S. Morrow, W. R. Veatch, and L. Stryer, *J. Mol. Biol.*, 1979, **132**, 733.
[76] H. Monoi and H. Uedaira, *Biophys. J.*, 1979, **25**, 535.
[77] A. Cornelis and P. Laszlo, *Biochemistry*, 1979, **18**, 2004.

defined sites, with binding constant $K = 4$ dm^4 mol^{-1}. These have been suggested to be outer sites, associated with the entrance to the channel, binding to which involves partial dehydration of Na$^+$. Rate constants for the binding and release of Na$^+$ at these sites are $\leqslant 2.2 \times 10^9$ dm^3 mol^{-1} s^{-1} and $\leqslant 5.5 \times 10^8$ s^{-1}, respectively.

The incorporation of gramicidin into membranes involves an interesting example of lipid–protein interaction. It is assumed to involve the formation of large aggregates at the interface of the lipid bilayer and the subsequent release of the aggregate that penetrates the lipid layer to form active channels. Such incorporation of gramicidin A in lysolecithin micelles is induced by elevated temperatures, as evidenced by the use[78] of ^{13}C n.m.r., which shows that there is a decrease in mobility of the micellar lipid on heating as gramicidin is incorporated. Sodium-23 n.m.r. shows the interaction of sodium to be dependent on the heat-incorporation of the channel.

The Urry head-to-head linked-dimer hypothesis is supported by studies on chemically modified gramicidin A,[79] and from ^{13}C and ^{19}F n.m.r. studies, which indicate that the CO$_2$H terminus is located near the surface of the membrane and the NH$_2$ terminus deep inside the bilayer. This evidence seems to show beyond reasonable doubt that the NH$_2$–NH$_2$ dimer is the major conformation of the channel.[80]

The electrical conductance of the gramicidin channel at low ionic concentrations in negatively charged phosphatidylserine membranes is larger[81] than in a neutral membrane, although the channel conductance approaches similar saturation values in both cases at higher ion concentrations. This increase was greater than that produced in a neutral lipid by using a negatively charged gramicidin analogue. These results are in qualitative agreement with the predictions of the Gouy–Chapman theory of the diffuse double layer.

Models for conduction through gramicidin channels depend upon the symmetry of the channel–membrane system, even though the two halves of many biological membranes show different lipid composition. As gramicidin shows different behaviour in different membranes, it may well be that it behaves asymmetrically in asymmetric bilayers. This has now been confirmed[82] for gramicidin-doped asymmetric bilayers, and it also appears that the dimerization reaction may be influenced by the variation in the structural properties of the two halves of the bilayer. A kinetic scheme has been developed to describe single-file transport through pores containing up to two ions.[83] The uniform nature of the mean potential in the gramicidin channel has been demonstrated;[84] this is an observation that is difficult to reconcile with a simple electrostatic picture. Central to

[78] A. Spisni, M. A. Khaled, and D. W. Urry, *FEBS Lett.*, 1979, **102**, 321; D. W. Urry, A. Spisni, and M. A. Khaled, *Biochem. Biophys. Res. Commun.*, 1979, **88**, 940.
[79] E. Bamberg, H. Alpes, H.-J. Apell, R. Bradley, B. Härter, M.-J. Quelle, and D. W. Urry, *J. Membr. Biol.*, 1979, **50**, 257; H.-J. Apell, E. Bamberg and H. Alpes, *ibid.*, 1979, **50**, 271.
[80] S. Weinstein, B. A. Wallace, E. R. Blout, J. S. Morrow, and W. Veatch, *Proc. Natl. Acad. Sci. USA*, 1979, **76**, 4230.
[81] H.-J. Apell, E. Bamberg, and P. Läuger, *Biochim. Biophys. Acta*, 1979, **552**, 369.
[82] O. Fröhlich, *J. Membr. Biol.*, 1979, **48**, 365, 385.
[83] B. W. Urban and S. B. Hladky, *Biochim. Biophys. Acta*, 1979, **554**, 410.
[84] R. H. Tredgold, *Biophys. J.*, 1979, **25**, 373.

this problem is the question of whether or not the ions are transported in the hydrated form (although it is clear that the channel contains water). This question has been explored by measuring single-channel conductances in H_2O and D_2O for several cations. Significant differences in conductivity were observed[85] in all cases except for lithium, the ratios of conductances (D_2O/H_2O) varying from 0.90 to 0.86 for Na^+ to Cs^+. It is noteworthy that the ratio of the viscosity of H_2O to that of D_2O at the temperature of these measurements is 0.82. This suggests that the passage of the larger ions through the channel is largely controlled by the viscosity of the medium (*e.g.*, if the water in the channel moves with the ion).

The pore systems that are produced in membranes by alamethicin,[86–88] suzukacillin,[86] and other antibiotics[89] have been examined further. The toxin from the sea anemone *Stoichactis helianthus* has been suggested to function by the formation of trans-membrane channels.[90]

Advances continue to be made in the characterization of the pores that are formed in lipid bilayers by the matrix protein (porin) of *Escherichia coli*. Thus they are more permeable to alkali-metal ions than to Cl^- ions, due probably to the presence of negative charges in the pore.[91] A porin oligomer has been purified (mol. wt 102 900) and dissociated into monomers (mol. wt $\approx 32\ 600$) by heating it, suggesting the presence of three subunits.[92] The permeability of the outer membrane of bacteria has been reviewed.[93] Channel-forming material from the outer membrane of rat liver mitochondria has been isolated, but this has greater permeability for anions over cations.[94]

Carrier Ionophores.—The standard carrier model for ion transport has been extended to predict the time-dependent currents for systems that are symmetrical at zero applied potential.[95] A kinetic study on the distribution of neutral ionophores between a solvent polymeric membrane and an aqueous phase is available.[96] Certain lipophilic β-diketones preferentially transport Mg^{2+} and Ca^{2+} through PVC-based liquid membranes when a pH gradient is applied.[97] Other carrier ionophores that have been studied include glycoproteins isolated from preparations of mitochondrial ATPase, which selectively carried Ca^{2+} through bilayer membranes,[98] and the anionic detergents SDS, cholate, and DOC, which transported bivalent cations.[99] The herbicide 2,4-dichlorophenoxyacetic acid (2.4-D)

[85] R. H. Tredgold and R. Jones. *Biochim. Biophys. Acta*, 1979, **550**, 543.
[86] H. Schindler, *FEBS Lett.*, 1979, **104**, 157.
[87] B. Sakmann and G. Boheim, *Nature (London)*, 1979, **282**, 336.
[88] J. J. Donovan and R. Latorre, *Biophys. J.*, 1979, **25**, 549.
[89] Kh. M. Kasumov, M. P. Borisova, L. N. Ermishkin, V. M. Potseluyer, A. Ya. Silberstein, and V. A. Vainshtein, *Biochim. Biophys. Acta*, 1979, **551**, 229.
[90] D. W. Michaels, *Biochim. Biophys. Acta*, 1979, **555**, 67.
[91] R. Benz, K. Janko, and P. Läuger, *Biochim. Biophys. Acta*, 1979, **551**, 238.
[92] T. Nakae, J. Ishii, and M. Tokunaga, *J. Biol. Chem.*, 1979, **254**, 1457.
[93] H. Nikaido, *Angew. Chem., Int. Ed. Engl.*, 1979, **18**, 337.
[94] M. Colombini, *Nature (London)*, 1979, **279**, 642.
[95] S. B. Hladky, *J. Membr. Biol.*, 1979, **46**, 213.
[96] U. Oesch and W. Simon, *Helv. Chim. Acta*, 1979, **62**, 754.
[97] D. Erne, W. E. Morf, S. Arvanitis, Z. Cimerman, D. Ammann, and W. Simon, *Helv. Chim. Acta*, 1979, **62**, 994.
[98] T. S. Azarashvili, A. I. Luk'yaneko, and Yu. V. Evtodienko, *Biochemistry (Engl. Transl.)*, 1979, **43**, 895.
[99] J. J. Abramson and A. E. Shamoo, *J. Membr. Biol.*, 1979, **50**, 241.

enhanced[100] the carrier-dependent transport of several ions across bilayer membranes, in accord with a number of physiological observations from which the toxicity of 2.4-D has been associated with the redistribution of ions. It has been suggested that a layer of 2.4-D is absorbed within the interfacial region, with a dipole moment directed towards the aqueous medium. Studies on the kinetics of transport of nonactin·K^+ indicate that this layer is located on the hydrocarbon side of the surface.

The effect of surface charge density on the formation of valinomycin–K^+ complexes in model membranes has been investigated[101] by the use of ionizable valinomycin analogues at various pH conditions and by changing the nature of the lipid in mixed monolayer films of valinomycin and lipid. Good agreement was obtained between the experimental results and the theoretical predictions based on the Gouy–Chapman theory. Conductance measurements of asymmetric bilayers containing the neutral egg lecithin lipid on one side and a negatively charged lipid (phosphatidylserine) on the other confirm the role of the surface charge. The rate of valinomycin-mediated uptake of rubidium into erythrocytes is decreased by increasing the amount of cholesterol in the cell membrane.[102] A similar effect was observed for large (up to 200 nm) vesicles and the reverse effect on small (less than 200 nm) vesicles. Thus the role of cholesterol in the transport properties of membranes appears to be a function of vesicle size. Ionophore A-23187 disrupts membrane structure by modifying protein–lipid interaction,[103] while the conditions under which it can be used to probe the involvement of Mg^{2+} in biological reactions have been described, with specific reference to the reactions of intact spinach chloroplasts.[104]

Interaction of Metal Ions with Membranes and Model Systems.—Models for membranes are usually lipid bilayers, but it should be noted that a very simple model system may be prepared by the sonication of di(octadecyl)dimethylammonium chloride.[105]

Measurements of relaxation rates, using ^{23}Na n.m.r.,[106] have shown that Na^+ binds specifically to phosphatidylserine (PS) vesicles, although it may be displaced partially from the binding site by K^+ and Ca^{2+}. The association constant for the Na^+–PS complex probably lies in the range 0.4—1.2 dm^3 mol^{-1}, so that in the absence of other cations an appreciable fraction of the PS sites would be associated with Na^+ at $[Na^+]=0.1$ mol dm^{-3}. This naturally has considerable implications for the numerical value of the surface potential for such bilayers. Association constants have also been assessed independently (by other techniques) for M^+–PS; *viz* Li^+ (0.8), Na^+ (0.6), NH_4^+ (0.17), K^+ (0.15), Rb^+ (0.08), Cs^+ (0.05), and Et_4N^+ (0.03 dm^3 mol^{-1}).[107]

[100] P. Smejtek and M. Paulis-Illangasekare, *Biophys. J.*, 1979, **26**, 441, 467.
[101] J. Caspers, M. Landuyt-Caufriez, M. Deleers, and J. M. Ruysschaert, *Biochim. Biophys. Acta*, 1979, **554**, 23.
[102] F. E. Labelle, *Biochim. Biophys. Acta*, 1979, **555**, 258.
[103] R. D. Klausner, M. C. Fishman, and M. J. Karnovsky, *Nature* (London), 1979, **281**, 82.
[104] P. M. Sokolove, *Biochim. Biophys. Acta*, 1979, **545**, 155.
[105] Y. Y. Lim and J. H. Fendler, *J. Am. Chem. Soc.*, 1979, **101**, 4023; K. Kano, A. Romero, B. Djermouni, H. J. Ache, and J. H. Fendler, *ibid.*, p. 4030.
[106] R. Kurland, C. Newton, S. Nir, and D. Papahadjopoulos, *Biochim. Biophys. Acta*, 1979, **551**, 137.
[107] M. Eisenberg, T. Gresalfi, T. Riccio, and S. McLaughlin, *Biochemistry*, 1979, **18**, 5213.

It has been shown[108] that small, unilamellar vesicles of cardiolipin, which is a major lipid of the mitochondrial inner membrane, fuse in the presence of either Ca^{2+} or Mg^{2+} to form the hexagonal phase. Removal of the cation by EDTA causes the hexagonal phase to revert to large unilamellar vesicles. This result allows the interesting speculation that the formation of the hexagonal phase in a membrane in this way could result in the formation of membrane pores and so modify the membrane permeability. The phase change in a model membrane made up of 20% PS and 80% egg-yolk phosphatidylethanolamine has been followed, using ^{31}P n.m.r. techniques.[109] The addition of Ca^{2+} triggers the transition to the hexagonal phase, an effect that can be reversed by the addition of the local anaesthetic dibucaine. This latter phenomenon probably reflects the displacement of Ca^{2+} from the lipid–water interface, so causing dispersal of the PS and the re-stabilization of the bilayer phase. In addition, the dibucaine could compete well for negative sites on the PS groups. The disappearance of Ca^{2+}-induced separation in mixed PS–phosphatidylcholine (PC) membranes is also caused by protonation, again reflecting competition for anionic sites.[110] The effects of Ca^{2+} on the phase behaviour of dipalmitoyl phosphatidylcholine have been reported.[111]

The cations Ca^{2+} and Mg^{2+} are well known to differ markedly in their ability to trigger fusion of vesicles. This has prompted a number of comparisons. Thus the binding of these cations to PS vesicles has been monitored by ^{31}P n.m.r. lineshifts and relaxation times.[112] Signals from inner and outer surfaces of sonicated PS vesicles can be distinguished by the effects of low concentrations of added Mg^{2+} and Ca^{2+}. The data show that Ca^{2+} binds directly to the PS phosphate, restricting the motion of the group; Mg^{2+} binds less strongly to the head-group. Raman spectra indicate that Mg^{2+} does not significantly alter the conformation of the acyl chain and the lateral packing of the bilayers in the PS vesicles, although the transition temperature is shifted. In contrast, the binding of Ca^{2+} results in highly rigid acyl chains.[113] X-Ray diffraction studies[114] confirm the existence of striking structural differences in the PS bilayers following the addition of Ca^{2+} or Mg^{2+}. Both cause aggregation, giving complexes with similar stoicheiometry ($1:2$ $M^{2+}:PS$), but the details of the interactions are quite different. Addition of Ca^{2+} gives an 'anhydrous' complex of closely apposed membranes with highly ordered acyl chains and a very high transition temperature of $T >$ 100 °C. Complex formation is associated with the release of the contents of the vesicle and fusion of the vesicles. It has been suggested that the unique PS–Ca^{2+} complex is an inter-membrane 'trans' complex, the formation of which is essential for fusion. In contrast, the PS–Mg^{2+} complex, which is more hydrated and in which there is no crystallization of the acyl chains, is suggested to be a 'cis' complex with respect to each membrane.

[108] W. J. Vail and J. G. Stollery, *Biochim. Biophys. Acta*, 1979, **551**, 74.
[109] P. R. Cullis and A. J. Verkleij, *Biochim. Biophys. Acta*, 1979, **552**, 546.
[110] S. Tokutomi, G. Eguchi, and S.-I. Ohnishi, *Biochim. Biophys. Acta*, 1979, **552**, 78.
[111] C. D'Ambrosio and L. Powers, *Biophys. J.*, 1979, **27**, 15.
[112] R. J. Kurland, M. Hammoudah, S. Nir, and D. Papahadjopoulos, *Biochem. Biophys. Res. Commun.*, 1979, **88**, 927.
[113] S. K. Hark and J. T. Ho, *Biochem. Biophys. Res. Commun.*, 1979, **91**, 665.
[114] A. Portis, C. Newton, W. Pangborn, and D. Papahadjopoulos, *Biochemistry*, 1979, **18**, 780.

The binding of Mn^{2+} to vesicles prepared from several different anionic phospholipids has been determined as a function of temperature by the use of e.p.r. spectroscopy.[115] For PS vesicles there appeared to be an enhancement of cation affinity with increased lipid fluidity, which is unexpected in the light of considerations of the surface potential.

Cation-induced interactions between phospholipid vesicles and monolayer membranes have been studied[116] as a function of $[M^{2+}]$ and time by the use of measurements of the surface tension. For PS monolayer and 1:1 PS/PC vesicles, the critical concentrations of bivalent cations that are necessary to produce a large reduction in surface tension of the monolayer (*i.e.* incorporation of the vesicle) are 16 (Mg^{2+}), 7 (Sr^{2+}), 6 (Ca^{2+}), 3.5 (Ba^{2+}), and 1.8 mmol dm^{-3} (Mn^{2+}). In contrast, for PC vesicles and PC monolayer there was no change in surface tension up to $[M^{2+}] = 25$ mmol dm^{-3}. An intermediate situation was observed for PS monolayer and PC vesicles, with the effectiveness $Mn^{2+} > Ca^{2+} > Sr^{2+}$. The interaction of lanthanide ions with PS vesicles[117] and with PC vesicles[118] has been studied.

Membrane Fusion. It is widely accepted that Ca^{2+} plays a role in the fusion of certain biological membranes, a phenomenon that has been modelled by the role of Ca^{2+} in the fusion of phospholipid vesicles. However, this latter process has now been questioned, and the suggestion has been made that the role of Ca^{2+} in vesicle fusion is actually one of causing lysis of vesicles, which subsequently reassemble to form large non-vesicular structures.[119] This suggestion has already been challenged,[120,121] but in this context it is appropriate to cite a number of techniques which appear to show that the contents of vesicles mix, under the influence of Ca^{2+}, without contamination by the surrounding medium, although there may be some 'leakage' from such vesicles. A luminescence assay for the fusion of vesicles, similar to that discussed in last years Report, has been developed.[122] Purified firefly luciferase and Mg^{2+} are incorporated into one set of phospholipid vesicles and ATP is incorporated into another set. Thus the fusion of the vesicles can be followed by the measurement of luminescence. In this case it appears, however, that less than 1% of the vesicles actually fused in the Ca^{2+}-dependent fusion reaction. Addition of EDTA after luminescence had been induced (to chelate the Ca^{2+}) resulted in a two- to three-fold increase in light emission, which rapidly decayed. This unexpected result suggests that the sudden removal of Ca^{2+} caused a transient increase in fusion, after which subsequent fusion was inhibited. Another technique for following the interaction of vesicles[121] has involved the interaction between Tb^{3+} and dipicolinic acid (dpa), $TbCl_3$ in citrate solution being present in one set of vesicles and dipicolinic acid in the other. Addition of Ca^{2+} and fusion of the vesicles results in the formation of the

[115] J. S. Puskin and T. Martin, *Biochim. Biophys. Acta*, 1979, **552**, 53.
[116] S. Ohki, and N. Düzgünes, *Biochim. Biophys. Acta*, 1979, **552**, 438.
[117] M. M. Hammoudah, S. Nir, T. Isac, R. Kornhauser, T. P. Stewart, S. W. Hui, and W. L. C. Vaz, *Biochim. Biophys. Acta*, 1979, **558**, 338.
[118] J. Westman and L. E. Goran-Eriksson, *Biochim. Biophys. Acta*, 1979, **557**, 62.
[119] L. Ginsberg, *Nature (London)*, 1978, **275**, 758.
[120] S. Nir and W. Pangborn, *Nature (London)*, 1979, **279**, 820.
[121] J. Wilschut and D. Papahadjopoulos, *Nature (London)*, 1979, **281**, 690.
[122] R. W. Holz and C. A. Stratford, *J. Membr. Biol.*, 1979, **46**, 331.

$[Tb(dpa)_3]^{3-}$ complex, which produces a 10^4-fold enhancement of the fluorescence of Tb. Leakage was shown in this case, but this was slower than the fusion reaction. Clearly it would be helpful if experiments such as these could involve phospholipids that are more resistant to leakage than PS. The Ca^{2+}-induced leakage from PS vesicles is approximately second-order with respect to vesicle concentration,[114] suggesting strongly that vesicle–vesicle interaction is a prerequisite for the observed release, and that leakage only therefore occurs from the fused vesicles, and not from separate vesicles.

The fusion of PS and PS–PC mixed vesicles depends on Ca^{2+} and temperature. Results for mixed vesicles with varying PS content suggest that fusion takes place between pure PS domains.[123] Cardiolipin is efficient in promoting the Ca^{2+}-induced fusion of proteoliposomes to an extent that is dependent on the phosphatidylethanolamine content of the vesicles.[124]

The role of Ca^{2+} in promoting fusion of vesicles is usually suggested to be related to the asymmetric distribution of the cation on either side of the membrane; Ca^{2+} will reduce electrostatic repulsions between negatively charged groups on the membrane surface, resulting in a contraction of the side that is exposed to the cation. Clearly, if both sides are exposed equally to Ca^{2+}, then there will be no overall effect. Thus PS vesicles are more stable with respect to fusion in the presence of an ionophore for Ca^{2+}. An attempt has been made[125] to assess the requirement for an asymmetrical distribution of $[Ca^{2+}]$, by observing the rates of fusion in the presence and absence of Ca^{2+} ionophores and also in the presence of internal $[Ca^{2+}]$. The extent of fusion in the former case will then reflect the relative rates of fusion and transfer. Reduction in the extent of fusion was attributed to the lowering of the degree of asymmetry of $[Ca^{2+}]$. However, another report claims that charge asymmetry does not affect the rate of the Ca^{2+}-induced aggregation of phospholipid vesicles.[126]

The role of fusogens in initiating the fusion of erythrocytes has been attributed to an increase in the permeability of the membrane to Ca^{2+}, *i.e.* the fusogen is behaving as an ionophore for Ca^{2+}.[127] Synaptic vesicles from the ray *Torpedo* aggregate[128] in the presence of Ca^{2+} (4 mmol dm^{-3}) and K^+ (50 mmol dm^{-3}). The Ca^{2+}-induced reaction involves the binding of Ca^{2+} to specific anionic groups and results in the formation of higher-order aggregates. The K^+-induced reaction leads only to dimerization, and is due to screening interactions.

4 Transport ATPases and Cation Fluxes

Proceedings of conferences on cation fluxes across biomembranes[129] and Na^+-K^+-ATPase[130] have been published. An excellent short account of ion transport

123 S. T. Sun, C. C. Hsang, E. P. Day, and J. T. Ho, *Biochim. Biophys. Acta*, 1979, **557**, 45.
124 A. E. Gad, R. Broza, and G. D. Eytan, *Biochim. Biophys. Acta*, 1979, **556**, 181.
125 M-J. Liao and J. H. Prestegard, *Biochem. Biophys. Res. Commun.*, 1979, **90**, 1274.
126 J. B. Lansman and D. H. Haynes, *Biophys. J.*, 1979, **26**, 335.
127 A. M. J. Blow, G. M. Botham, and J. A. Lucy, *Biochem. J.*, 1979, **182**, 555.
128 D. H. Haynes, J. Lansman, A. L. Cahill, and S. J. Morris, *Biochim. Biophys. Acta*, 1979, **557**, 340.
129 'Cation Flux across Biomembranes', ed. Y. Mukohata and L. Packer Academic Press, London, 1979.
130 'Na,K-ATPase – Structure and Kinetics', ed. J. C. Skou and J. G. Nørby Academic Press, London, 1979.

is available,[131] while other reviews have dealt with the use of membrane vesicles in transport studies,[132] the permeation of ions through membrane channels,[133] biomembranes,[134] and chemiosmosis,[135] while useful chapters have appeared in review series.[136]

The following discussion deals primarily with 'Class II' ATPases, which depend upon Mg^{2+} and other cations for activity. These ATPases are involved in ion translocation, the maintenance of ion gradients, and the 'pumping of cations' (Na^+/K^+ and Mg^{2+}). 'Class I' ATPases will be considered briefly: these also depend upon Mg^{2+}, and are best known for their role in oxidative phosphorylation and the pumping of protons.

Na^+-K^+-ATPases.—Several aspects of this enzyme have been reviewed.[129,130,] [137-139] The solubilization and, in some cases, the molecular weights have been reported for Na^+-K^+-ATPase isolated from the gills of *Anguilla anguilla*[140] and from the rectal glands of *Squalus acanthias*[141] and shark.[142] A new micromethod allows the determination of this enzyme in discrete segments of nephrons from rabbit, rat, and mouse kidneys,[143] while the use of [³H]ouabain has allowed its location in the epithelium of the trachea and rectal gland of the dog[144] and the dogfish[145] respectively. Biochemical evidence is available for its localization at the lateral region of the surface membrane of hepatocytes.[146]

The Na^+-K^+-ATPases that were prepared from various organs of the rat have been classified into two groups,[147] while the brain (from a range of animals) contains two distinct molecular forms of the enzyme,[148] although only one form could be found in other organs. These two forms differ in their affinity for the specific inhibitor strophanthidin, in their sensitivity to digestion by trypsin, and in the number and reactivity of SH groups.

The Role of the Lipid. Lipid-free Na^+-K^+-ATPase is inactive, and much effort is now being devoted to determining the role of the protein–lipid interaction in the overall mechanism. Modifications in such interactions have been suggested to account for the temperature dependence of transport of K^+ ion.[149] The delipidated

[131] E. Racker, *Acc. Chem. Res.*, 1979, **12**, 338.
[132] J. E. Lever, *CRC Crit. Rev. Biochem.*, 1979, **7**, 187.
[133] 'Membrane Transport Processes', ed. C. F. Stevens and R. W. Tsien, Raven Press, New York, 1979, Vol. 3.
[134] 'Methods in Enzymology', ed. S. Fleischer and L. Packer Academic Press, London, 1979.
[135] R. J. P. Williams, *FEBS Lett.*, 1979, **102**, 126; P. Mitchell, *Eur. J. Biochem.*, 1979, **95**, 1.
[136] *Current Top. Bioenerg.*, 1979, Vol. 7; A. S. Mildvan, *Adv. Enzymol. Related Areas Mol. Biol.*, 1979, **49**, 103; H. S. Penefsky, *ibid.*, p. 223.
[137] J. D. Robinson and M. S. Flashner, *Biochim. Biophys. Acta*, 1979, **549**, 145.
[138] C. M. Grisham, *Adv. Inorg. Biochem.* 1979, **1**, 193.
[139] E. T. Wallick, L. K. Lane, and A. Schwartz, *Annu. Rev. Physiol.*, 1979, **41**, 397.
[140] M. V. Bell and J. R. Sargent, *Biochem. J.*, 1979, **179**, 431.
[141] M. Esmann, J. C. Skou, and C. Christiansen, *Biochim. Biophys. Acta*, 1979, **567**, 140.
[142] D. F. Hastings and J. A. Reynolds, *Biochemistry*, 1979, **18**, 817.
[143] A. Doucet, A. I. Katz, and F. Morel, *Am. J. Physiol.*, 1979, **237**, F105.
[144] J. H. Widdicombe, C. B. Basbaum, and J. Y. Yee, *J. Cell Biol.*, 1979, **82**, 380.
[145] J. Eveloff, K. J. Karnaky, P. Silva, F. H. Eptein, and W. B. Kinter, *J. Cell Biol.*, 1979, **83**, 16.
[146] R. E. Poupon and W. H. Evans, *FEBS Lett.*, 1979, **108**, 374.
[147] O. Urayama and M. Nakao, *J. Biochem.* (*Tokyo*), 1979, **86**, 1371.
[148] K. J. Sweadner, *J. Biol. Chem.*, 1979, **254**, 6060.
[149] C. H. Joiner and P. K. Lauf, *Biochim. Biophys. Acta*, 1979, **552**, 540.

enzyme has been re-lipidated[150] with dioleoyl phosphatidylcholine and dioleoyl phosphatidylethanolamine. Regeneration of the phosphatase activity occurs before that of the ATPase activity, due (it is suggested) to one and then two sub-units of the enzyme being saturated with the phospholipid.

The activity of Na^+-K^+-ATPase decreases with increasing pressure.[151] A plot of log (activity) *vs* P shows a change in slope at a well-defined point (P_b). Values of P_b increase linearly with increasing temperature in a manner that is similar to the pressure shift for the melting transitions in phospholipids and aliphatic chains. This indicates that this process is also involved in the pressure dependence of the enzyme. The presence of fluidity-enhancing drugs had no effect on the inflection point in the Arrhenius plot. The Na^+-K^+-ATPase from a mutant line of the ovary cell of the chinese hamster which is defective in the regulation of cholesterol biosynthesis showed an activity that varied with the order parameter of the acyl chain of the membrane.[152]

The effect of the long-chain fatty acyl ester palmitylcarnitine on the structure and function of cardiac membrane has been clarified through the use of a lipid-bound fluorescent probe in Na^+-K^+-ATPase sarcolemmal membranes (and for sarcoplasmic reticulum) at concentrations of palmitylcarnitine that are high enough to inhibit the enzyme. The effect of the palmitylcarnitine is generally similar to that of detergents on these membrane systems, suggesting that delipidation occurs.[153] The use of lipid-bound fluorescent probes has allowed the study of the effect of bivalent cations on lamb kidney Na^+-K^+-ATPase, showing that the cations Mg^{2+}, Ca^{2+}, Na^+, and K^+ interact primarily with the polar head-groups of the phospholipids. Interaction of the fluorescent label with the tryptophan residues of the enzyme occurs, thus allowing the monitoring of protein–lipid interactions and the future testing of the hypothesis that conformational changes in the protein are transmitted to adjacent phospholipids, so causing changes in the affinity of the lipids for different ions.[154]

The delipidation of the Na^+-K^+-ATPase of the electric eel by a range of deter-gents has been described carefully,[155] and this study provides useful guidelines for those using detergents with this enzyme. The insertion of the ATPase subunits into eel electroplax membranes has also been described.[156]

Subunit Structure. The large (catalytic) subunit of eel electroplax Na^+-K^+-ATPase is a glycoprotein,[157] a conclusion obtained through the use of more sensitive assays than those used earlier. The association of a small proteolipid (mol. wt

150 P. Ottolenghi, *Eur. J. Biochem.*, 1979, **99**, 113.
151 H. De Smedt, R. Borghgraef, F. Ceuterick, and K. Heremans, *Biochim. Biophys. Acta*, 1979, **556**, 479.
152 M. Sinensky, F. Pinkerton, E. Sutherland, and F. R. Simon, *Proc. Natl. Acad. Sci. USA*, 1979, **76**, 4893.
153 R. J. Adams, D. W. Cohen, S. S. Gupte, J. D. Johnson, E. T. Wallick, T. Wong, and A. Schwartz, *J. Biol. Chem.*, 1979, **254**, 12 404.
154 S. S. Gupte, L. K. Lane, J. D. Johnson, E. T. Wallick, and A. Schwartz, *J. Biol. Chem.*, 1979, **254**, 5099.
155 J. R. Brotherus, P. C. Jost, O. H. Griffith, and L. E. Hokin, *Biochemistry*, 1979, **18**, 5043.
156 L. Churchill and L. E. Hokin, *J. Biol. Chem.*, 1979, **254**, 7388.
157 L. Churchill, G. L. Peterson, and L. E. Hokin, *Biochem. Biophys. Res. Commun.*, 1979, **90**, 481.

$\approx 12\,000$) with this subunit has been confirmed by the use of a photoaffinity label.[158]

The need for mobility of subunits has been demonstrated once more by the reversible inactivation of Na^+-K^+-ATPase that arises from the use of a cleavable bifunctional reagent that cross-links the subunits.[159] Cross-linking to give α,α and α,β dimers also arises[160] from the use of Cu^{2+} or Cu^{2+}–phenanthroline, although only the α,β dimer is obtained if the enzyme is pre-treated with detergents. The formation of the α,α dimer is also inhibited by a combination of K^+ and ATP, but not by Na^+ and ATP. This suggests that the various cation-induced conformational states have different reactivities to the cross-linking reaction.

The presence of different conformations in membrane-bound Na^+-K^+-ATPase is reflected in different degrees of inhibition by antibodies. Four different degrees of inhibition were obtained for varying combinations of Na^+, K^+, Mg^{2+}, and ATP,[161] which combinations led also to different patterns for the digestion of the enzyme by trypsin.[162] More work has been reported on the trypsin-catalysed hydrolysis of Na^+-K^+-ATPase under different conditions, with the objective of locating fragments containing specific sites.[163] The incorporation of such a trypsin-modified enzyme into phospholipid vesicles[164] gave a preparation with abnormal transport properties, showing a transport ratio of 1–$2Na^+$:$2K^+$ instead of $3Na^+$:$2K^+$. This may result from interference either with the binding of Na^+ or with the conformational changes associated with its translocation.

The currently accepted view of the dimeric nature of Na^+-K^+-ATPase is supported by further work on the binding of groups to the purified enzyme.[165] However, a comparison of the binding sites available for nucleotide, ouabain, and vanadate for crude mammalian kidney microsomes and detergent-treated enzyme preparations has led to an important new proposal for the subunit composition of N^+-K^+-ATPase.[166] The results are more complex in the former case, and lead to the suggestion that the enzyme involves eight large catalytic peptides, having four sites available for each of ouabain and nucleotide, although only two of the nucleotide sites are available at any time. These sites have been suggested to be distributed on four identical functional subunits, containing two large peptides each. The quadrupling of p-nitrophenylphosphatase activity that occurs when the enzyme is treated with SDS is thus accounted for by the conversion of one molecule into four independent subunits that would accompany this treatment.

Binding and Kinetic Studies. The Na^+-K^+-ATPase-catalysed hydrolysis of p-nitrophenylphosphatase is associated with high (catalytic) and low (regulatory) binding sites for K^+. The former sites are probably the extracellular binding

[158] T. B. Rogers and M. Lazdunski, *FEBS Lett.*, 1979, **98**, 373.
[159] J. J. H. H. M. de Pont, *Biochim. Biophys. Acta*, 1979, **567**, 247.
[160] W-H. Huang and A. Askari, *Biochim. Biophys. Acta*, 1979, **578**, 547; *FEBS Lett.*, 1979, **101**, 66.
[161] H. Koepsell, *J. Membr. Biol.*, 1979, **45**, 1.
[162] H. Koepsell, *J. Membr. Biol.*, 1979, **48**, 69.
[163] J. Castro and R. A. Farley, *J. Biol. Chem.*, 1979, **254**, 2221.
[164] P. L. Jørgensen, B. M. Anner, and J. Petersen, *Biochim. Biophys. Acta*, 1979, **555**, 485.
[165] F. Kudoh, S. Nakamura, M. Yamaguchi, and Y. Tonomura, *J. Biochem. (Tokyo)*, 1979, **86**, 1023.
[166] O. Hansen, J. Jensen, J. G. Nørby, and P. Ottolenghi, *Nature (London)*, 1979, **280**, 410.

sites that are associated with transport of K^+. The regulatory sites appear to have an allosteric function to unmask the catalytic sites, a function also fulfilled by the binding of Na^+ at other regulatory sites under phosphorylating conditions. While Rb^+ interacts with both types of K^+ site, Li^+ is unable to activate *p*-nitrophenylphosphate, but is able to do so when the catalytic sites are exposed by the use of Na^+ and ATP. This suggests that the two classes of sites differ in their selectivity for cations.[167]

Several problems relating to the binding of substrates to Na^+-K^+-ATPase have been resolved. Experiments[168] with a ^{3}H-labelled derivative of ouabain and ^{32}P-labelled P_i show that ouabain binds to the same peptide that is phosphorylated from $^{32}P_i$. Rapid ion-exchange techniques have been used[169] to show that K^+ ions are occluded in unphosphorylated Na^+-K^+-ATPase, thus confirming an earlier suggestion. The use of rapid filtration techniques has confirmed the simultaneous binding of $3Na^+$ and $2K^+$ per mole, and the appropriate changes in affinity that are induced by the formation of the phospho-enzyme.[170]

The inactivation of Na^+-K^+-ATPase by butane-2,3-dione is suggestive of the involvement of arginine residues in the active site. However, it appears[171] that this inactivation process is biphasic, indicating either that two arginine groups are involved, or that a single group is exhibiting different reactivities in the different conformations shown by the enzyme. The role of sulfhydryl groups has been explored further by the use of the fluorescent compound 2-(4′-maleimidylanilino)-naphthalene-6-sulphonic acid, whose reaction with —SH groups can be monitored by the increase in the fluorescence intensity of the covalently bound probe.[172] The rate of reaction with —SH groups is increased by Mg^{2+} and Ca^{2+}, owing to their interaction with the membrane lipids. Approximately two —SH groups per $\alpha_2\beta_2$ unit of the enzyme react without loss of activity, and the enzyme is completely inhibited when sixteen —SH groups have reacted. Inactivation of the enzyme by acetic anhydride and trinitrobenzenesulphonate[173] has been suggested to involve interaction with amino-groups of the enzyme.

The decrease in N^+-K^+-ATPase activity that is brought about by the exposure of erythrocytes or their membranes to ozone has been attributed either to direct inhibition of the ATPase or to an effect[174] on the membrane which inhibits cation transport. Ozone-treated[175] aqueous suspensions of natural phospholipids give at least two types of inhibitor of human erythrocyte membrane Na^+-K^+-ATPase. Inhibition by bilirubin (a physiological substance that causes brain damage) has also been studied.[176]

The most widely studied inhibitor of Na^+-K^+-ATPase at present is undoubtedly the vanadate ion. Interest in this inhibitor as a mechanistic tool has increased dramatically since its activity was reported in 1977. Vanadate also inhibits other

167 A. C. Swann and R. W. Albers, *J. Biol. Chem.*, 1979, **254**, 4540.
168 B. Forbush and J. F. Hoffman, *Biochemistry*, 1979, **18**, 2308.
169 L. A. Beaugé and I. M. Glynn, *Nature (London)*, 1979, **280**, 510.
170 M. Yamaguchi and Y. Tonomura, *J. Biochem. (Tokyo)*, 1979, **86**, 509.
171 C. M. Grisham, *Biochem. Biophys. Res. Commun.*, 1979, **88**, 229.
172 S. S. Gupte and L. K. Lane, *J. Biol. Chem.*, 1979, **254**, 10 362.
173 J. D. Robinson and M. S. Flashner, *Arch. Biochem. Biophys.*, 1979, **196**, 350.
174 A. E. Koontz and R. L. Heath, *Arch. Biochem. Biophys.*, 1979, **198**, 493.
175 L. Kesner, R. J. Kindya, and P. C. Chan, *J. Biol. Chem.*, 1979, **254**, 2705.
176 S. Kashiwamata, S. Goto, R. K. Semba, and F. N. Suzuki, *J. Biol. Chem.*, 1979, **254**, 4577.

ATPases,[177] acid and alkaline phosphatases, phosphofructokinase,[178] and adenylate kinase, although it stimulates adenylate cyclase.[179]

The inhibition of the Na^+-K^+-ATPase from the gills of *Anguilla anguilla* by vanadate has been reported[180] and a claim[181] made that it transiently stimulates Na^+-K^+-ATPase from a cat heart cell-membrane preparation. This was deduced from an initial rapid loss of NADH in the ATPase assay, but it was subsequently reported[181] to be due to the oxidation of NADH as a result of a NADH-dependent reduction of vanadate. Another study showed that much of the vanadate taken up by the red cell is converted into V^{IV} and that it binds to haemoglobin.[182]

The potentiation of inhibition by vanadate by an increase in the concentration of Mg^{2+} and K^+ has been confirmed,[181,183,184] although other cations of Group IA will also promote this inhibition.[184,185] It appears that at least two K^+ ions are required for this inhibition,[184] which is reversed by Na^+.

Vanadate binds at two sites on Na^+-K^+-ATPase, these sites being of high and of low affinity. Phosphate competes for the high-affinity site. Several groups have shown, in accord with this, that vanadate facilitates the binding of ouabain to the enzyme in much the same way as P_i does.[186–188] In the presence of Mg^{2+}, ATP, Na^+, and K^+, vanadate[187] decreased the rate of binding of ouabain by increasing the apparent affinity of the enzyme for K^+ at a site that modulates the binding of ouabin.

Vanadate modifies[189] the effect of extracellular K^+ in a similar way to lowered ATP concentration, and it was suggested that vanadate retarded the relatively slow, ATP-sensitive conformational change that is thought to follow the hydrolysis of phospho-enzyme when that hydrolysis is catalysed by extracellular K^+. This has now been confirmed by using a fluorescein-labelled enzyme.[190]

Treatment of plasma membrane with EDTA resulted in a 300-fold increase in the sensitivity of Na^+-K^+-ATPase to ouabain, due to loss of Ca^{2+} and protein from the cytoplasmic face of the membrane. Recovery of resistance to ouabain required Ca^{2+} and the binding of the protein.[191] Binding of ouabain was inhibited by antibody against the catalytic subunit, but not by antibody against the glycoprotein subunit.[192] A Na^+-K^+-ATPase from rat kidney was markedly insensitive

[177] B. J. Bowman and C. W. Slayman, *J. Biol. Chem.*, 1979, **254**, 2928.
[178] G. Choate and T. E. Mansour, *J. Biol. Chem.*, 1979, **254**, 11 457.
[179] U. Schwabe, C. Puchstein, H. Hannemann, and E. Sochtig, *Nature (London)*, 1979, **277**, 143; G. Grupp, I. Grupp, C. L. Johnson, E. T. Wallick, and A. Schwartz, *Biochem. Biophys. Res. Commun.*, 1979, **88**, 440.
[180] M. V. Bell and J. R. Sargent, *Biochem. J.*, 1979, **179**, 431.
[181] E. Erdmann, W. Krawietz, G. Philipp, I. Hackbarth, W. Schmitz, and H. Scholz, *Nature (London)*, 1979, **278**, 459; E. Erdmann, W. Krawietz, G. Philipp, I. Hackbarth, H. Scholz, and F. L. Crane, *ibid.*, 1979, **282**, 336.
[182] L. C. Cantley and P. Aisen, *J. Biol. Chem.*, 1979, **254**, 1781.
[183] P. H. Wu and J. W. Phillis, *Int. J. Biochem.*, 1979, **10**, 629.
[184] G. H. Bond and P. M. Hudgins, *Biochemistry*, 1979, **18**, 325.
[185] J. G. Grantham and I. M. Glynn, *Am. J. Physiol.*, 1979, **236**, F530.
[186] O. Hansen, *Biochim. Biophys. Acta*, 1979, **568**, 265.
[187] E. T. Wallick, L. K. Lane, and A. Schwartz, *J. Biol. Chem.*, 1979, **254**, 8107.
[188] T. D. Myers, R. C. Boerth, and R. L. Post, *Biochim. Biophys. Acta*, 1979, **558**, 99.
[189] L. Beaugé and R. DiPolo, *Biochim. Biophys. Acta*, 1979, **551**, 220.
[190] S. J. D. Karlish, L. A. Beaugé, and I. M. Glynn, *Nature (London)*, 1979, **282**, 333.
[191] L. Lelievre, A. Zachowski, D. Charlemagne, P. Laget, and A. Paraf, *Biochim. Biophys. Acta*, 1979, **557**, 399.
[192] H. M. Rhee and L. E. Hokin, *Biochim. Biophys. Acta*, 1979, **558**, 108.

to inhibition by cardiac glycosides, due, it has been suggested, to the primary structure of the enzyme.[193] Tritiated ouabain, digitoxin, and digitoxigenin will photolabel the catalytic protein, and not the glycoprotein, when they are bound and the conjugate is exposed to light.[194]

The effect of ATP, K^+, and Na^+ on the rate of inactivation of Na^+-K^+-ATPase by the thiol reagent 7-chloro-4-nitrobenzo-2,1,3-oxadiazole (NBD-Cl) correlates well with a model which suggests the presence of two binding sites for ATP and two separate binding sites for K^+ and Na^+ on the enzyme–ATP complex. The two sites for ATP are of high and low affinity, the existence of which has been attributed to the induction of antico-operative interaction by the binding of ATP, the two sites initially being equivalent. These conclusions[195] are in accord with the flip-flop mechanism for the enzyme. Various stages in the overall enzymatic reaction have been studied. The effect of Mg^{2+} on the activation of three of these stages by Na^+ lends support to the presence of high-affinity and low-affinity sites for MgATP;[196] Na^+ that is acting at high-affinity extracellular sites inhibits Na^+-ATPase activity by slowing dephosphorylation.[197] Studies[198] of the effect of ATP on $K_{0.5}$ for Na^+ and of the rate of hydrolysis at different values of pH and temperatures have led to the conclusions (a) that ATP, at a given pH, increases the affinity for Na^+ relative to that for K^+ at the internal site, and (b) that binding of the cations to external and internal sites leads to changes in pK_a which are different for Na^+ and K^+; *i.e.* the selectivity for Na^+ relative to K^+ depends upon ATP and the degree of protonation of certain groups. The intriguing suggestion has been made that the system pumps out protons in addition to sodium ions.

The phospho-enzyme complex of N^+-K^+-ATPase is usually assayed by causing its precipitation by acidification; this is a process that may give a product that differs from that formed at the neutral pH of the original reaction mixture. It is of value therefore to compare[199] this product with the heat-precipitated complex. The heat-precipitated phospho-enzyme has been shown not be be an acyl–, seryl–, threonyl–, or histidyl–phosphate complex, and thus to be a unique species, which should probably be included in the reaction sequence of Na^+-K^+-ATPase.

More detailed studies are now available on the activation of sheep kidney Na^+-K^+-ATPase by Mn^{2+} (as examined by e.p.r. techniques).[200] The enzyme binds Mn^{2+} at one tight site, other binding sites being associated with the phospholipids in the preparation. E.p.r. spectra under different conditions suggest that the true substrate for the enzyme is ATP, and not Mn^{2+}–ATP, and that the sites for bivalent metal and substrate are preserved in the delipidated enzyme, although their conformations are substantially altered. Side-specific effects have been

[193] S. M. Periyasamy, L. K. Lane, and A. Askari, *Biochem. Biophys. Res. Commun.*, 1979, **86**, 472.

[194] B. Forbush and J. F. Hoffman, *Biochim. Biophys. Acta*, 1979, **555**, 299.

[195] R. Grosse, T. Rapoport, J. Malur, J. Fischer, and K. R. H. Repke, *Biochim. Biophys, Acta*, 1979, **550**, 500.

[196] M. S. Flashner and J. D. Robinson, *Arch. Biochem. Biophys.*, 1979, **192**, 584.

[197] L. A. Beaugé and I. M. Glynn, *J. Physiol.*, 1979, **289**, 17.

[198] J. C. Skou, *Biochim. Biophys. Acta*, 1979, **567**, 421.

[199] D. A. Gewirtz, R. J. Sohn, and W. A. Brodsky, *Arch. Biochem. Biophys.*, 1979, **195**, 336.

[200] S. E. O'Connor and C. M. Grishnam, *Biochemistry*, 1979, **18**, 2315.

monitored by the use of inside-out membrane vesicles of red cells[201] and dialysed squid axons.[202]

Ca^{2+}-Mg^{2+}-ATPases.—Analytical methods for Ca^{2+} continue to attract attention. The metallochromic indicator dye Arsenazo III forms 2:1 complexes with Ca^{2+} and 1:1 complexes with Mg^{2+} under biological conditions.[203] Several problems with the use of metallochromic indicators have been discussed,[203,204] noteworthy being the fact that certain of them interact with Ca^{2+}-sequestering organelles.[205] The use of lanthanides as probes for calcium has been reviewed,[206] while manganese[207] and ruthenium red[208] have been considered further. Ruthenium red appears to be a useful resonance Raman probe of Ca^{2+}-binding sites in biological materials. Thus the low-affinity sites for Ca^{2+} in mitochondria have been shown to be phospholipid in character.

Ca^{2+}-Mg^{2+}-ATPase from Sarcoplasmic Reticulum (S.R.). Aspects of this enzyme have been reviewed,[209,210] reference being made also to calsequestrin and the high-affinity calcium-binding proteins, on which a number of papers have been published.[211,212] The biosynthesis of Ca^{2+}-Mg^{2+}-ATPase of S.R. has been studied in cell cultures of embryonic chick heart,[213] together with the effect of [Ca^{2+}] in the medium on the rate of synthesis in cultures of chick pectoralis muscle.[214]

The rotational movement of the Ca^{2+}-Mg^{2+}-ATPase of S.R. has been investigated by measuring the decay of laser flash-induced dichroism with the covalently attached triplet probe eosin thiocyanate.[215] The two discontinuities that are observed in the Arrhenius plot for rotational mobility (at ~ 15 and 30 °C) have been rationalized in terms of a conformational change in the ATPase at 15 °C and a temperature-dependent equilibrium between the conformationally altered ATPase and oligomeric forms of it in the temperature range 15–35 °C. Enzymatic activity shows a discontinuity in the Arrhenius plot for the rate of hydrolysis of ATP which corresponds to the discontinuity at 15 °C. Furthermore, there is good agreement between the activation energies for rotational motion below 15 °C and enzymatic activity. An independent report[216] confirms that the Arrhenius discontinuity of the ATPase activity is unrelated to changes in fluidity of the membrane lipid of the sarcoplasmic reticulum.

[201] R. Blostein, *J. Biol. Chem.*, 1979, **254**, 6673.
[202] L. Beaugé and R. DiPolo, *Biochim Biophys. Acta*, 1979, **553**, 495.
[203] M. V. Thomas, *Biophys. J.*, 1979, **25**, 541.
[204] S. T. Ohnishi, *Biochim. Biophys. Acta*, 1979, **586**, 217.
[205] S. T. Ohnishi, *Biochim. Biophys. Acta*, 1979, **585**, 315.
[206] R. B. Marlin and F. S. Richardon, *Q. Rev. Biophys.*, 1979, **12**, 181.
[207] D. Getz, J. F. Gibson, R. N. Sheppard, K. J. Micklem, and C. A. Pasternak, *J. Membr. Biol.*, 1979, **50**, 311.
[208] J. M. Friedman, D. L. Rousseau, G. Navon, S. Rosenfeld, P. Glynn, and K. B. Lyons, *Arch. Biochem. Biophys.*, 1979, **193**, 14.
[209] L. de Meis and A. L. Vianna, *Annu. Rev. Biochem.*, 1979, **48**, 275.
[210] D. H. MacLennan and K. P. Cambell, *Trends Biochem. Sci.*, 1979, 148.
[211] M. Varsángi and L. M. G. Heilmeyer, *FEBS Lett.*, 1979, **103**, 85; R. K. Tume, *Aust. J. Biol. Sci.*, 1979, **32**, 177.
[212] W. Wnuk, J. A. Cox, L. G. Kohler, and E. A. Stein, *J. Biol. Chem.*, 1979, **254**, 5284.
[213] P. C. Holland, *J. Biol. Chem.*, 1979, **254**, 7604.
[214] D. B. Ha, R. Boland, and A. Martonosi, *Biochim. Biophys. Acta*, 1979, **585**, 165.
[215] W. Hoffmann, M. G. Sarzala, and D. Chapman, *Proc. Natl. Acad. Sci. USA*, 1979, **76**, 3860.
[216] T. D. Madden and P. J. Quinn, *FEBS Lett.*, 1979, **107**, 110.

The possible existence of an oligomeric structure for Ca^{2+}-Mg^{2+}-ATPase has been explored by cross-linking experiments, using I_2 or 1,10-phenanthroline complexes of Cu^{II} to form S—S bridges.[217,218] Exposure to these reagents results in the formation of a series of oligomers containing either four or five molecules plus other oligomers that are too large to enter the gel. Complete oxidation at $-10\,^{\circ}C$ left most of the ATPase as monomer[218] (*i.e.* with intramolecular S—S bonds), leading to the conclusion that most of the cross-linked species arose from randomly colliding ATPase molecules which are present in the membrane at very high concentration. Such collisions would be decreased at $-10\,^{\circ}C$, in view of the increased viscosity of the lipid. Furthermore, it appears unlikely that conclusions relating to the form of the enzyme that is involved in transport can be drawn from cross-linking experiments which gave such a wide range of oligomers. However, it has been suggested that the minimal functional unit of the enzyme is a tetramer, on the basis of a study of the inhibition that is produced by dicyclohexylcarbodi-imide.[219] Furthermore, it has been claimed that Ca^{2+} is located on a site in a hydrophobic region which is not accessible to EGTA.

The lipid annulus model for the Ca^{2+}-Mg^{2+}-ATPase of S.R. postulates the presence of a shell of some 30 lipid molecules bound to the protein, which excludes cholesterol. However, the use[220] of cholesterol-enriched liposomes to vary the amount of cholesterol in the S.R. membranes leads to variation in the activity of Ca^{2+}-Mg^{2+}-ATPase. Either the annulus is unable to exclude cholesterol, or the model is incorrect. Indeed, the existence of the annulus seems to be incompatible with recent n.m.r. and e.p.r. evidence.[221] Other aspects of protein–lipid interaction have also been studied,[222] including the reconstitution of Ca^{2+}-Mg^{2+}-ATPase with synthetic phospholipids.[223] It has been suggested that the protein assembly of the enzyme within the membrane of dystrophic muscle differs from that of normal muscle. Thus, when the membrane structures of normal and of dystrophic preparations are perturbed by detergents, they show similar ATPase activity.[224]

The detailed mechanism of catalysis of the hydrolysis of ATP and transport of calcium is still not established. The enzyme is dependent upon a univalent cation, so there are at least three types of metal-ion-binding site, plus two types of ATP-binding sites, as well as both acid-labile and acid-stable sites for inorganic phosphate. The S.R. vesicles from rabbit skeletal muscle show two types of site for Ca^{2+} outside and four inside. One of the outside sites is the calcium-translocating site of the pump protein, while the inner sites include those of the

217 R. J. Baskin and S. Hanna, *Biochim. Biophys. Acta*, 1979, **576**, 61.
218 G. M. Hebdon, L. W. Cunningham, and N. M. Green, *Biochem. J.*, 1979, **179**, 135.
219 U. Pick and E. Racker, *Biochemistry*, 1979, **18**, 108.
220 T. D. Madden, D. Chapman, and P. J. Quinn, *Nature (London)*, 1979, **279**, 539.
221 D. Chapman, J. C. Gomez-Fernandez, and F. M. Goñi, *FEBS Lett.*, 1979, **98**, 211.
222 Y. H. Lau, A. H. Caswell, J. P. Brunschwig, R. J. Baerwald, and M. Garcia, *J. Biol. Chem.*, 1979, **254**, 540; G. Swoboda, J. Fritzsche, and W. Hasselbach, *Eur. J. Biochem.*, 1979, **95**, 77; D. M. Rice, M. D. Meadows, A. O. Scheinman, F. M. Goñi, J. C. Gómez-Fernández, M. A. Moscarello, D. Chapman, and E. Oldfield, *Biochemistry*, 1979, **18**, 5893.
223 J. C. Gómez-Fernández, F. M. Goñi, D. Bach, C. Restall, and D. Chapman, *FEBS Lett.*, 1979, **98**, 224.
224 S. Verjovski-Almeida and G. Inesi, *Biochim. Biophys. Acta*, 1979, **558**, 119.

high-affinity calcium-binding protein and calsequestrin.[225] Addition of Ca^{2+} in the presence of a nucleotide caused a change in the e.p.r. spectrum of spin-labelled S.R. ATPase, showing conformational changes due to the binding of Ca^{2+} to sites which undergo a reduction of calcium-binding affinity upon phosphorylation with ATP.[226]

The interactions of Gd^{3+}, Li^+, and the substrate analogues $\beta\gamma$-imido-ATP (AMP-PNP) and CrATP with the Ca^{2+}-Mg^{2+}-ATPase of S.R. have been examined[227] by ^7Li n.m.r., ^1H n.m.r. (of protons of water), and Gd^{3+} e.p.r. techniques. Steady-state phosphorylation studies indicate that Gd^{3+} binds to the Ca^{2+} activator sites on the enzyme with an affinity that is some ten times greater than that of Ca^{2+}. The Li^+ ion has been used to probe the site that binds univalent cations. Relaxation n.m.r. studies have shown that the binding of Gd^{3+} to the two Ca^{2+} sites increases the longitudinal relaxation time of enzyme-bound Li^+. Relaxation studies of the protons of water have confirmed that the ATPase binds Gd^{3+} at two tight-binding sites, and furthermore that these Ca^{2+} sites experience reduced accessibility of solvent water, due probably to the fact that they are bound within a cleft in the enzyme (see ref. 219). From the effect of Gd^{3+} on the relaxation rate of enzyme-bound Li^+, Gd^{3+}-Li^+ separations of 7.0 and 9.1 Å have been calculated, and hence an upper limit of 16.1 Å has been set on the separation between the Ca^{2+} sites on the enzyme. Kinetic evidence is also available to show that Ca^{2+}-Mg^{2+}-ATPase has distinct sites for univalent cations which interact allosterically with other regulatory sites on the enzyme.[228]

Optical probes are currently being used for measurements of membrane potentials in subcellular membranes. These probes include the cyanines (positively charged) and the oxanols (negatively charged). These have been used with S.R. membranes and vesicles to see if uptake and efflux of Ca^{2+} is associated with a change in membrane potential. It has been claimed that uptake of Ca^{2+} by S.R. vesicles is associated with a build-up of a membrane potential with a positive polarity inside the vesicles,[229] but the use of cyanine dyes in vesicles provided no evidence that uptake and release of Ca^{2+} is electrogenic.[230] Electroneutrality is possibly maintained by anion fluxes or counter-transport of Mg^{2+}. Artificial changes in membrane potentials appeared to have no effect on the release of Ca^{2+}. The same technique has been used on sarcoplasmic reticulum[231] (for both merocyanine and oxanol dyes). The changes in absorption and fluorescence that are induced by a range of cations have been ascribed to the binding of the cations to the low-affinity cation-binding sites. No clear evidence was found for electrogenic processes either during uptake or release of Ca^{2+}. Studies with the pH indicator bromothymol blue incorporated into S.R. vesicles showed that a transient alkalinization occurs inside the vesicles during uptake of Ca^{2+}, which

[225] H. Miyamoto and M. Kasai, *J. Biochem. (Tokyo)*, 1979, **85**, 765.
[226] C. Coan, S. Verjovski-Almeida, and G. Inesi, *J. Biol. Chem.*, 1979, **254**, 2968.
[227] E. M. Stephens and C. M. Grisham, *Biochemistry*, 1979, **18**, 4876.
[228] L. R. Jones, *Biochim. Biophys. Acta*, 1979, **557**, 230.
[229] K. E. O. Åkerman and C. H. J. Wolff, *FEBS Lett.*, 1979, **100**, 291.
[230] T. Beeler and A. Martonosi, *FEBS Lett.*, 1979, **98**, 173.
[231] J. T. Russell, T. Beeler, and A. Martonosi, *J. Biol. Chem.*, 1979, **254**, 2047.

has been suggested[232] to reflect the formation of a trans-membrane proton gradient, which then sustains the transport of Ca^{2+}.

The proposal that the trans-sarcolemmal influx of Ca^{2+} that occurs during the plateau of the mammalian action potential triggers the release of Ca^{2+} from the sarcoplasmic reticulum has been further confirmed.[233] Release of Ca^{2+} has also been demonstrated for a water-soluble derivative of prostaglandin B_1[234] and for ionomycin[235] (a new ionophore for Ca^{2+} which appears to be a new reagent of some promise). The effect of an increase in pH on the release of Ca^{2+} has been examined,[236] while the phenomenon of spontaneous release has been shown to be a genuine one.[237]

The effect of the following reagents on the reactions or partial reactions of Ca^{2+}-Mg^{2+}-ATPase has been examined: thymol,[238] DMSO, propranolol, and chlorpromazine,[239] Tris buffer,[240] and vanadate.[241] Vanadate is a potent inhibitor, and doubtless will receive further attention.

Fragmented S.R. that has been incorporated into a phospholipid bilayer shows the presence of a voltage-gated pathway for univalent cations, transport of which is blocked by certain bivalent cations.[242]

ATP serves as both substrate and activator for Ca^{2+}-Mg^{2+}-ATPase, although the mechanism of its regulatory role is still controversial.[243] The dependence of transport of Ca^{2+} on [ATP] is biphasic,[244] which has been attributed to dimeric enzyme–enzyme interaction, the occupancy of one site by nucleotide accelerating the turnover of the first. A further class of tight nucleotide-binding sites have been characterized which are associated[245] with translocation of Ca^{2+}. Regulation of the transport ATPase is in part due to a Ca^{2+}-dependent protein kinase and a protein phosphatase.[246] Uptake of Ca^{2+} is regulated by both a cAMP-dependent phosphorylation and a Ca^{2+}-calmodulin-dependent phosphorylation. The catalytic subunit of cAMP-dependent protein kinase catalyses the incorporation of phosphate into phospholamban (a microsomal protein of mol. wt 22 000, which serves as a modulator of the ATPase).[247] Phosphorylation of cardiac microsomal

232 V. M. C. Madeira, *Arch. Biochem. Biophys.*, 1979, **193**, 22.
233 A. Fabiato and F. Fabiato, *Nature (London)*, 1979, **281**, 146; S. T. Ohnishi, *J. Biochem. (Tokyo)*, 1979, **86**, 1147.
234 S. T. Ohnishi and T. M. Devlin, *Biochem. Biophys. Res. Commun.*, 1979, **89**, 240.
235 T. Beeler, I. Iona, and A. Martonosi, *J. Biol. Chem.*, 1979, **254**, 6229.
236 J. Dunnett and W. G. Naylor, *Arch. Biochem. Biophys.*, 1979, **198**, 434.
237 M. L. Entman, W. B. Van Winkle, E. Bornet, and C. Tate, *Biochim. Biophys. Acta*, 1979, **551**, 382.
238 K. Takishima, M. Setaka, and H. Shimizu, *J. Biochem. (Tokyo)*, 1979, **86**, 347.
239 M. Shigekawa, A. A. Akovitz, and A. M. Katz, *Biochim. Biophys. Acta*, 1979, **548**, 433.
240 G. L. Alonso, D. M. Arrigo, and S. Terradas de Fermani, *Arch. Biochem. Biophys.*, 1979, **198**, 131.
241 T. Wang, L.-I. Tsai, R. J. Solaro, A. O. Grassi de Gende, and A. Schwarz, *Biochem. Biophys. Res. Commun.*, 1979, **91**, 356; S. G. O'Neal, D. B. Rhoads, and E. Racker, *ibid.*, 1979, **89**, 845.
242 C. Miller and R. L. Rosenberg, *Biochemistry*, 1979, **18**, 1138.
243 H. M. Scofano, A. Vieyra, and L. de Meis, *J. Biol. Chem.*, 1979, **254**, 10 227.
244 S. Verjovski-Almeida and G. Inesi, *J. Biol. Chem.*, 1979, **254**, 18.
245 A. A. Aderem, D. B. McIntosh, and M. C. Berman, *Proc. Natl. Acad. Sci. USA*, 1979, **76**, 3622.
246 M. Varsanyi and L. M. G. Heilmeyer, *Biochemistry*, 1979, **18**, 4869.
247 M. Tada, F. Ohmori, M. Yamada, and H. Abe, *J. Biol. Chem.*, 1979, **254**, 319; C. J. Le Peuch, J. Haiech, and J. G. Demaille, *Biochemistry*, 1979, **18**, 5150.

proteins does not appear to affect the permeability to Ca^{2+} of the microsomal membrane.[248]

A range of reagents have been used to quench the reactions of Ca^{2+}-Mg^{2+}-ATPase. While denaturation by acid simultaneously stops all partial reactions of the catalytic cycle, other reagents (particularly La^{3+}) inhibit specific steps, permitting the remaining sequential reactions to proceed for one cycle.[249] Other mechanistic studies have been concerned with the reactions of the phospho-enzyme intermediates,[250] the phosphorylation reaction,[251] and various other aspects.[252]

Other Ca^{2+}-Mg^{2+}-ATPases. The problem of separating vesicles of cardiac sarcolemma from vesicles of cardiac sarcoplasmic reticulum has been studied further.[253] Binding of Ca^{2+} to sarcolemmal vesicles is stimulated by taurine[254] and by ATP.[255] The Ca^{2+}-Mg^{2+}-ATPase is stimulated by a cAMP-dependent protein kinase.[256] A Na^+–Ca^{2+} exchange has also been characterized[257] for this system.

The binding and transport of calcium in several erythrocyte systems have been studied,[258] and the Ca^{2+}-Mg^{2+}-ATPase from human erythrocyte membranes has been purified by means of a calmodulin affinity column.[259] The enzyme from plasma membranes of pig erythrocytes has been reconstituted in vesicles,[260] while some attention has been paid to the activator protein associated with the enzyme. Modulator protein from brain and the activator for pig erythrocyte membrane

[248] M. Weller and W. Laing, *Biochim. Biophys. Acta*, 1979, **511**, 406.

[249] M. Chiesi and G. Inesi, *J. Biol. Chem.*, 1979, **254**, 10 370.

[250] Y. Takakuwa and T. Kanazawa, *Biochem. Biophys. Res Commun.*, 1979, **88**, 1209; M. Shigekawa and A. A. Akowitz, *J. Biol. Chem.*, 1979, **254**, 4726; R. M. Chaloub, H. Guimaraes-Motta, S. Verjovski-Almeida, L. de Meis, and G. Inesi, *ibid.*, p. 9464; N. Kolassa, C. Punzengruber, J. Suko, and M. Makinose, *FEBS Lett.*, 1979, **108**, 495; H. Takisawa and Y. Tonomura, *J. Biochem. (Tokyo)*, 1979, **86**, 425.

[251] B. Plank, G. Hellman, C. Punzengruber, and J. Suko, *Biochim. Biophys. Acta*, 1979, **550**, 259; A. Vieyra, H. M. Scofano, H. Guimaraes-Motta, R. K. Tume, and L. de Meis, *Biochim. Biophys. Acta*, 1979, **568**, 437; R. Prager, C. Punzengruber, N. Kolassa, F. Winkler, and J. Suko, *Eur. J. Biochem.*, 1979, **97**, 239.

[252] R. P. Newbold and R. K. Tume, *J. Membr. Biol.*, 1979, **48**, 205; J. S. Taylor and D. Hattan, *J. Biol. Chem.*, 1979, **254**, 4402; N. Ronzani, A. Migala, and W. Hasselbach, *Eur. J. Biochem.*, 1979, **101**, 593; B. Rossi, F. de A. Leone, C. Gache, and M. Lazdunski, *J. Biol. Chem.*, 1979, **254**, 2302.

[253] L. R. Jones, H. R. Besch, J. W. Fleming, M. M. McConnaughey, and A. M. Watanabe, *J. Biol Chem.*, 1979, **254**, 530; E. Van Alstyne, D. K. Bartschat, N. V. Wellsmith, S. L. Poe, W. P. Schilling, and G. E. Lindenmeyer, *Biochim. Biophys. Acta*, 1979, **553**, 388.

[254] J. Mas-Oliva, A. J. Williams, and W. G. Naylor, *Biochem. Biophys. Res. Commun.*, 1979, **87**, 441.

[255] J. P. Chovan, E. C. Kulakowski, B. W. Benson, and S. W. Schaffer, *Biochim. Biophys. Acta*, 1979, **551**, 129.

[256] A. Ziegelhoffer, M. B. Anand-Srivastava, R. L. Khandelwal, and N. S. Dhalla, *Biochem. Biophys. Res. Commun.*, 1979, **89**, 1073.

[257] B. J. R. Pitts, *J. Biol. Chem.*, 1979, **254**, 6232; J. P. Reeves and J. L. Sutko, *Proc. Natl. Acad. Sci. USA*, 1979, **76**, 590.

[258] A. Ting, J. W. Lee, and G. A. Vidaver, *Biochim. Biophys. Acta*, 1979, **555**, 239; W. K. Yeung, G. Weisman, and G. A. Vidaver, *ibid.*, p. 249; A. M. Brown, *ibid.*, 1979, **554**, 195; P. V. Gulak, G. M. Boriskina, and Yu. V. Postnov, *Experientia*, 1979, **35**, 1470.

[259] V. Niggli, J. T. Penniston, and E. Carafoli, *J. Biol. Chem.*, 1979, **254**, 9955.

[260] H. Haaker and E. Racker, *J. Biol. Chem.*, 1979, **254**, 6598.

Ca^{2+}-Mg^{2+}-ATPase have been used[261-263] to stimulate the enzyme from other sources. It appears probable that this protein is similar to calmodulin, and it is noteworthy that calmodulin stimulates the Ca^{2+}-Mg^{2+}-ATPase of erythrocyte ghosts, forming a complex which survives solubilization and reconstitution.[264]

The activity of the Ca^{2+}-Mg^{2+}-ATPase of red blood cell membrane is also enhanced by a modulator protein.[265] ATP activates the enzyme at two distinct sites, the one of lower affinity being a regulatory site.[266] The stoicheiometry of the pump (*i.e.* [Ca^{2+}]/[ATP]) is still in dispute.[267] ATP has been synthesized by running the pump backwards,[268] and complications arising from the use of EGTA have been assessed.[269] A rather similar ATP-driven calcium pump is found in squid axon.[270] This activity is inhibited by vanadate, which has no effect on the Na^+–Ca^{2+} exchange, so confirming that these two efflux processes involve different mechanisms.

Synaptic vesicles from the electric organ of *Torpedo californica*[271] and from rat brain[272] are claimed to have a Ca^{2+}-Mg^{2+}-ATPase, although uptake of Ca^{2+} in synaptic vesicles from other sources was attributed to contamination by microsomes.[273] A Ca^{2+}-ATPase has been demonstrated in neutrophil plasma membrane.[274]

The plasma membrane of *Saccharomyces cerevisiae* has a Mg^{2+}-ATPase which is distinct from the Class I ATPases described in the following section and which shows the formation of vanadate-sensitive phosphorylated intermediates, suggesting that it is similar in mechanism to the Na^+-K^+-ATPase and Ca^{2+}-Mg^{2+}-ATPase.[275]

An Mg^{2+}-ATPase (function unknown) has been purified[276] from plasma membranes of sheep kidney medulla, and it contains two peptides, of mol. wt 150 000 and 77 000. The larger protein is phosphorylated. The kinetics of activation by Mg^{2+} and Mn^{2+} are reported, with results of e.p.r. studies, in the latter case. Substantial similarities between this enzyme and Na^+-K^+-ATPase and Ca^{2+}-Mg^{2+}-ATPase have been noted. A similar conclusion has been drawn for a Mg^{2+}-ATPase from the rod outer segment disc membrane.[277] It is probable then that both these enzymes are Class II ATPases.

261 K. S. Au, *Int. J. Biochem.*, 1979, **10**, 637.
262 K. S. Au, *Int. J. Biochem.*, 1979, **10**, 687; R. Kobayashi, M. Tawata, and H. Hidaka, *Biochem. Biophys. Res. Commun.*, 1979, **88**, 1037.
263 T. J. Lynch and W. Y. Cheung, *Arch. Biochem. Biophys.*, 1979, **194**, 165.
264 V. Niggli, P. Ronner, E. Carafoli, and J. T. Penniston, *Arch. Biochem. Biophys.*, 1979, **198**, 124.
265 K. D. Philipson and F. Baumgartner, *Biochim. Biophys. Acta*, 1979, **567**, 523.
266 S. Mualem and S. J. D. Karlish, *Nature (London)*, 1979, **277**, 238.
267 B. Sarkadi, I. Szasz, and G. Gardos, *J. Membr. Biol.*, 1979, **46**, 184.
268 A. Wüthrich, H. J. Schatzmann, and P. Romero, *Experientia*, 1979, **35**, 1589.
269 B. Sarkadi, A. Schubert, and G. Gardos, *Experientia*, 1979, **35**, 1044.
270 R. DiPolo and L. Beaugé, *Nature (London)*, 1979, **278**, 271; R. Dipolo, H. R. Rojas, and L. Beaugé, *ibid.*, 1979, **281**, 228.
271 J. E. Rothlein and S. M. Parsons, *Biochem. Biophys. Res. Commun.*, 1979, **88**, 1069.
272 D. Papazian, H. Rahamimoff, and S. M. Goldin, *Proc. Natl. Acad. Sci. USA*, 1979, **76**, 3708.
273 T. Tsudzuki, *J. Biochem. (Tokyo)*, 1979, **86**, 777.
274 C. Schneider, C. Mottola, and D. Romeo, *Biochem. J.*, 1979, **182**, 655.
275 G. R. Willsky, *J. Biol. Chem.*, 1979, **254**, 3326.
276 M. L. Gantzer and C. M. Grisham, *Arch. Biochem. Biophys.*, 1979, **198**, 263, 268.
277 R. Uhl, T. Borys, and E. W. Abrahamson, *FEBS Lett.*, 1979, **107**, 317.

H⁺-K⁺-ATPase.—This enzyme provides the newest example of the membrane-bound ion-translocating ATPases. The gastric H^+-K^+-ATPase from the parietal cell is responsible for the exchange of K^+ for H^+, resulting in increased acidity in the stomach. It is Mg^{2+}-dependent, generates a phosphoprotein, and has been successfully incorporated into vesicles.[278] The successful techniques used in the investigation of Na^+-K^+-ATPase and Ca^{2+}-Mg^{2+}-ATPase are now being applied to it; *e.g.* hydrolysis with trypsin[279] and the study of the partial reactions.[280] Sodium ions probably play a regulatory role.[281]

Proton-translocating Mg^{2+}-ATPases from Bacteria, Mitochondria, *etc.*—These membrane-bound ATPases are associated with the pumping of protons and the coupling of this process to the hydrolysis of ATP. The ATPase complex from the cell membrane of *E. coli* involves an extrinsic protein (that can be purified as an F_1-ATPase with five different subunits) and an intrinsic membrane protein F_0, which contains the proton-translocating function. Accounts of small-angle X-ray scattering[282] and small-angle neutron scattering[283] by the F_1-ATPase are available. The F_0 protein has three subunits, of mol. wt 24 000, 19 000, and 84 000,[284] although the nature of this species changes with growth conditions in aerobic culture.[285] Different states of the F_1-ATPase have been described.[286] The F_1 and F_0 portions of the complex interact through the δ subunit of the F_1-ATPase, the ATPase being released by the proteolytic cleavage of a portion of this subunit.[287]

The coupling factor–ATPase complex from *Rhodospirillum rubrum* has been partially purified,[288] and it contains eight different polypeptides in total. The subunits of F_1-ATPase from thermophilic bacteria PS3 have been analysed and reconstituted.[289,290] The F_1-F_0 complex has been incorporated into macrolipo-somes and ATP has been synthesized by the application of an electrical field.[291] Proton conduction is inhibited by chemical modification of the membrane[292] and by dicyclohexylcarbodi-imide, which is a known inhibitor for translocation of protons.[293] The surface properties of the ATPase from *Streptococcus faecalis*[294]

[278] B.-C. Lee, H. Breitbart, M. Berman, and J. G. Forte, *Biochim. Biophys. Acta*, 1979, **553**, 107.
[279] G. Saccomani, D. W. Dailey, and G. Sachs, *J. Biol. Chem.*, 1979, **254**, 2821.
[280] B. Wallmark and S. Märdh, *J. Biol. Chem.*, 1979, **254**, 11 899.
[281] M. Ljungström, B. Wallmark, and S. Märdh, *Acta Chem. Scand.*, *Ser. B*, 1979, **33**, 618.
[282] H. H. Paradies and U. D. Schmidt, *J. Biol. Chem.*, 1979, **254**, 5257.
[283] M. Satre and G. Zaccai, *FEBS Lett.*, 1979, **102**, 244.
[284] D. L. Foster and R. H. Fillingame, *J. Biol. Chem.*, 1979, **254**, 8230.
[285] H. Kanazawa and M. Futai, *FEBS Lett.*, 1979, **105**, 275.
[286] P. P. Laget, *Arch. Biochem. Biophys.*, 1979, **192**, 474.
[287] P. D. Bragg and C. Hou, *FEBS Lett.*, 1979, **103**, 12.
[288] R. Oren and Z. Gromet-Elhanan, *Biochim. Biophys. Acta*, 1979, **548**, 106; C. Bengis-Garber and Z. Gromet-Elhanan, *Biochemistry*, 1979, **18**, 3577.
[289] M. Yoshida, N. Sone, H. Hirata, Y. Kagawa, and N. Ui, *J. Biol. Chem.*, 1979, **254**, 9525.
[290] Y. Kagawa, N. Sone, H. Hirata, and M. Yoshida, *Trends Biochem. Sci.*, 1979, 31.
[291] M. Rögner, K. Ohno, T. Hanamoto, N. Sone, and Y. Kagawa, *Biochem. Biophys. Res. Commun.*, 1979, **91**, 162.
[292] N. Sone, K. Ikeba, and Y. Kagawa, *FEBS Lett.*, 1979, **97**, 61.
[293] N. Sone, M. Yoshida, H. Hirata, and Y. Kagawa, *J. Biochem. (Tokyo)*, 1979, **85**, 503.
[294] A. V. Babakov, O.P. Terekhov, and A. S. Yanenko, *Biochim. Biophys. Acta*, 1979, **547**, 438.

and the binding properties of the enzyme from *Clostridium pasteurianum*[295] have been investigated.

The ATPase complex from bovine heart mitochondria contains 9 or 10 subunits, five of which comprise the F_1 protein, two of which are required for the attachment of F_1 to the membrane, and two of which form the proton channel in the membrane. Functions have been assigned to each of these subunits.[296] The Mg^{2+} ion binds tightly[297a] to the F_1-ATPase in a 1:1 ratio, while other metals, with the exception of Cd^{2+},[297b] do not bind tightly. The nucleotide appears to bind at a separate site.[297a] The ATP- and ADP-binding sites in mitochondrial F_1 have been characterized.[298] The inactivation of F_1 that is produced by incubation with Mg^{2+} is reversed by addition of MgATP.[299] The effect of the ATPase inhibitor protein (a small heat-stable protein, of mol. wt 10 000) on phosphorylation in submitochondrial particles has been discussed.[300] The role of the lipid has also been explored.[301] Proton-translocating ATPases have also been discovered in chromaffin granules[302] and in rat liver lysosomes,[303] although the latter case is in dispute.[304] These have similar properties to the mitochondrial Mg^{2+}-ATPases.

Several model systems have been reported for proton translocation in biomembranes, including one that is based on keto–enol shifts in hydrogen-bonded peptide groups.[305]

Proton-linked Transport Systems.—The H^+ gradient that is set up by proton-translocating ATPases is probably involved in the uptake of amines in chromaffin cells,[306] of 5-hydroxytryptamine in synaptic vesicles from rat brain,[307] and of lactose in membrane vesicles[308] and whole cells[309] of *E. coli*. While the linking of transfer of H^+ and of electrons in mitochondria and photosynthetic membranes

295 D. J. Clarke and J. G. Morris, *Eur. J. Biochem.*, 1979, **98**, 613.
296 M. Alfonzo and E. Racker, *Can. J. Biochem.*, 1979, **57**, 1351.
297 (a) A. E. Senior, *J. Biol. Chem.*, 1979, **254**, 11 319; (b) H. Rauchová and Z. Drahota, *Int. J. Biochem.*, 1979, **10**, 735.
298 E. C. Slater, A. Kemp, I. van der Kraan, J. L. M. Muller, O. A. Roveri, G. J. Verschoor, R. J. Wagenvoord, and J. P. M. Wielders, *FEBS Lett.*, 1979, **103**, 7.
299 D. D. Hackney, *Biochem. Biophys. Res. Commun.*, 1979, **91**, 233.
300 D. A. Harris, V. Von Tscharner, and G. K. Radda, *Biochim. Biophys. Acta*, 1979, **548**, 72.
301 C. C. Cunningham and G. Sinthusek, *Biochim. Biophys. Acta*, 1979, **550**, 150; G. Parenti-Castelli, A. M. Sechi, L. Landi, L. Cabrini, S. Mascarello, and G. Lenaz, *Biochim. Biophys. Acta*, 1979, **547**, 161.
302 R. M. Buckland, G. K. Radda, and L. M. Wakefield, *FEBS Lett.*, 1979, **103**, 323.
303 P. Dell'Antone, *Biochem. Biophys. Res. Commun.*, 1979, **86**, 180; D. L. Schneider, ibid., 1979, **87**, 559; J. L. Mego, *FEBS Lett.*, 1979, **107**, 113.
304 M. Hollemans, D.-J. Reijngoud, and J. M. Tager, *Biochim. Biophys. Acta*, 1979, **551**, 55.
305 J. Dilger and S. McLaughlin, *J. Membr. Biol.*, 1979, **46**, 359; P. Läuger, *Biochim. Biophys. Acta*, 1979, **552**, 143; J. Gutknecht and A. Walter, *J. Membr. Biol.*, 1979, **43**, 59; C. Kayalar, ibid., 1979, **45**, 37.
306 R. G. Johnson, D. Pfister, S. E. Carty, and A. Scarpa, *J. Biol. Chem.*, 1979, **254**, 10 963; D. Njus and G. K. Radda, *Biochem. J.*, 1979, **180**, 579.
307 R. Maron, B. I. Kanner, and S. Schuldiner, *FEBS Lett.*, 1979, **98**, 237.
308 E. Padan, L. Patel, and H. R. Kaback, *Proc. Natl. Acad. Sci. USA*, 1979, **76**, 6221; G. J. Kaczrowski and H. R. Kaback, *Biochemistry*, 1979, **18**, 3691.
309 D. Zilberstein, S. Schuldiner, and E. Padan, *Biochemistry*, 1979, **18**, 669; I. R. Booth, W. J. Mitchell, and W. A. Hamilton, *Biochem. J.*, 1979, **182**, 617.

will not be discussed, some references are given for mitochondria,[310] chloroplasts,[311] purple membranes,[312] and *Paracoccus denitrificans*[313] and to a useful general review.[314] A promising method for measuring membrane potentials in *E. coli* and *S. lactis* (*viz.* the uptake of Tl^+) has been shown to be invalid, in that Tl^+ is accumulated actively in these species.[315]

Transport of Ca^{2+} in Mitochondria. Unequivocal evidence[316] has been presented to show that, in respiring mitochondria, uptake (and efflux) of calcium occurs by a $Ca^{2+}/2H^+$ antiport. No evidence could be found for a $Ca_2^{4+} \cdot HPO_4^{2-}$ symport. Future studies on such systems should be facilitated by the development of a calcium-ion-specific electrode for kinetic studies.[317]

Heart mitochondria differ from most other mitochondria in that release of Ca^{2+} is also induced by Na^+, due to the presence of a specific Na^+–Ca^{2+} antiporter. This distinction in Ca^{2+}-efflux processes is revealed in their sensitivities to the lanthanides.[318] Addition of Na^+ causes the release of Ca^{2+} from the mitochondria of brown adipose tissue.[319]

The factors that control the uptake and efflux of calcium continue to be much studied. Uptake of Ca^{2+} may be affected by cellular lipid metabolites.[320] Rat liver mitochondria that are depleted of the Ca^{2+}-binding glycoprotein show 10—20% of the normal uptake activity. This is restored by the addition of the glycoprotein in the presence of Mg^{2+}.[321]

Efflux is dependent upon a permeability that is controlled by the binding of ADP or ATP to the inner membrane, and it depends upon the maintenance of certain thiol groups in a reduced form. Efflux from heart mitochondria (in the absence of Na^+) is inhibited by bongkrekic acid, which promotes the tighter binding of nucleotides to the membrane.[322] The local anaesthetics nupercaine and tetracaine inhibit the ruthenium-red-insensitive efflux of Ca^{2+} but have no

[310] E. Sigel and E. Carafoli, *J. Biol. Chem.*, 1979, **254**, 10 572; A. Alexandre and A. L. Lehninger, *ibid.*, p. 11 555; A. Villalobo and A. L. Lehninger, *ibid.*, p. 4352; T. Pozzan, V. Miconi, F. Di Virgilio, M. Bragadin, and G. F. Azzoni, *Proc. Natl. Acad. Sci. USA*, 1979, **76**, 2123, *J. Biol. Chem.*, 1979, **254**, 10 200; M. Wikstrom and K. Krab, *Biochim. Biophys. Acta*, 1979, **549**, 177; 1979, **548**, 1.

[311] J. Barber and G. F. W. Searle, *FEBS Lett.*, 1979, **103**, 241; L. F. Olsen and R. P. Cox, *ibid.*, p. 253; J. T. Duniec and S. W. Thorne, *ibid.*, 1979, **105**, 1.

[312] K. Hashimoto and M. Nishimura, *J. Biochem.* (*Tokyo*), 1979, **85**, 57; J. K. Lanyi, *Microbiol. Rev.*, 1978, **42**, 682; C. L. Bashford, M. Baltschiffsky, and R. C. Prince, *FEBS Lett.*, 1979, **97**, 55; N. A. Dencher and M. P. Heyn, *ibid.*, 1979, **108**, 307; K. Petty, J. B. Jackson, and P. L. Dutton, *Biochim. Biophys. Acta*, 1979, **546**, 17.

[313] E. M. Meijer, J. W. Van der Zwaan, and A. H. Stouthamer, *FEMS Microbiol. Lett.*, 1979, **5**, 369.

[314] D. B. Kell, *Biochim. Biophys. Acta*, 1979, **549**, 55.

[315] E. R. Kashket, *J. Biol. Chem.*, 1979, **254**, 8129; P. D. Damper, W. Epstein, B. P. Rosen, and E. N. Sorensen, *Biochemistry*, 1979, **18**, 4165.

[316] G. Fisken, B. Reynafarje, and A. L. Lehninger, *J. Biol. Chem.*, 1979, **254**, 6288; G. Fisken and A. L. Lehninger, *ibid.*, p. 6236; A. J. Williams and C. H. Fry, *FEBS Lett.*, 1979, **97**, 288.

[317] R. K. Yamazaki, D. L. Mickey, and M. Story, *Anal. Biochem.*, 1979, **93**, 430.

[318] E. Carafoli, *FEBS Lett.*, 1979, **104**, 1; M. Crompton, I. Heid, C. Baschera, and E. Carafoli,, *ibid.*, p. 352.

[319] M. H. M. Al-Shaikhaly, J. Nedergaard, and B. Cannon, *Proc. Natl. Acad. Sci. USA*, 1979, **76**, 2350.

[320] P. E. Wolkowicz and J. M. Wood, *FEBS Lett.*, 1979, **101**, 63.

[321] G. Sandri, G. Sottocasa, E. Panfili, and G. Luit, *Biochim. Biophys. Acta*, 1979, **558**, 214.

[322] E. J. Harris, M. H. M. Al-Shaikhaly, and H. Baum, *Biochem. J.*, 1979, **182**, 455; E. J. Harris, *ibid.*, 1979, **178**, 673.

effect on its uptake.[323] Tetracaine has also been used to explore the role of phospholipase (which is inhibited by tetracaine),[324] while the action of N-ethylmaleimide plus Ca^{2+} on mitochondria is also affected by intra-mitochondrial phospholipase.[325]

Mitochondria that have been incubated in the presence of low levels of $[Ca^{2+}]$ accumulate calcium rapidly, but they remain in the aggregated (*i.e.* shrunken) condition and are coupled strongly. They then undergo a sudden increase in permeability of the inner membrane, allowing the osmotic support species to enter the matrix, this giving the normal, uncoupled configuration. This is due to binding of Ca^{2+}. This process has been studied in some detail.[326] The rate of the Ca^{2+}-induced transition was affected (inhibited or catalysed) by a number of reagents; *e.g.*, inhibited by any agent which caused reduction of endogenous NAD. Under some conditions this Ca^{2+}-induced membrane transition results in efflux of Ca^{2+}, due to inhibition of uptake of Ca^{2+} by uncoupling, and the stimulation of efflux of Ca^{2+} as a result of the change in permeability.[326]

An attempt has been made to compare the rate of the H^+ pump with that of the Ca^{2+}-transport system in rat liver mitochondria. This shows that transport of Ca^{2+} is limited by the former step. The system was studied further by replacing H^+ extrusion with K^+ diffusion *via* valinomycin. This meant that the rate of transport of Ca^{2+} could be made rate-limiting. The low activation energy and the high turnover number that are associated with this process suggest that transport takes place *via* a pore rather than a carrier.[327]

The effect of $[Ca^{2+}]$ on the transport of adenine nucleotides[328] and pyruvate,[329] the transport of phosphate,[330] and the transport of Ca^{2+} in isolated rat liver cells[331] and inside-out vesicles of rat liver mitochondria[332] have been investigated.

Transport of K^+ in Mitochondria. The possibility of a K^+/H^+ antiport in mitochondria has been explored.[333] The inhibitor dicyclohexylcarbodi-imide decreases the rate of flux of K^+ into rat liver mitochondria, possibly through its action on proton translocation. Loss of K^+ is also induced by uptake of the tetraethylammonium cation.[334]

Miscellaneous Cation/Proton Antiport Systems. Calcium/proton and sodium/proton antiport systems have been studied for *E. coli*. Two independent reports

[323] A. P. Dawson, M. J. Selwyn, and D. V. Fulton, *Nature (London)*, 1979, **277**, 485.

[324] D. Siliprandi, M. Rugolo, F. Zoccarato, A. Toninello, and N. Siliprandi, *Biochem. Biophys. Res. Commun.*, 1979, **88**, 188.

[325] D. R. Pfeiffer, P. C. Schmid, M. C. Beatrice, and H. H. O. Schmid, *J. Biol. Chem.*, 1979, **254**, 11 485.

[326] D. R. Hunter and R. A. Hawarth, *Arch. Biochem. Biophys.*, 1979, **195**, 453, 460, 468.

[327] M. Bragadin, T. Pozzan, and G. F. Azzoni, *FEBS Lett.*, 1979, **104**, 347; *Biochemistry* 1979, **18**, 5972.

[328] A. Gómez-Puyou, M. Tuena de Gómez-Puyou, M. Klapp, and E. Carafoli, *Arch. Biochem. Biophys.*, 1979, **194**, 399.

[329] G. Stepien, R. Debise, and R. Durand, *Biochimie*, 1979, **61**, 861.

[330] A. Lavat, M. Guérin, and B. Guérin, *Arch. Biochem. Biophys.*, 1979, **194**, 405; J. P. Wehrle and P. L. Pedersen, *J. Biol. Chem.*, 1979, **254**, 7269; P. V. Blair, *Biochem. Biophys. Res. Commun.*, 1979, **88**, 537.

[331] H. Krell, H. Baur, and E. Pfaff, *Eur. J. Biochem.*, 1979, **101**, 349.

[332] H. R. Lötscher, K. Schwerzmann, and E. Carafoli, *FEBS Lett.*, 1979, **99**, 194.

[333] L. M. Gauthier and J. J. Diwan, *Biochem. Biophys. Res. Commun.*, 1979, **87**, 1072.

[334] K. D. Garlid, *Biochem. Biophys. Res. Commun.*, 1979, **87**, 842.

confirm the Ca^{2+}/H^+ process to be electrogenic ($H^+:Ca^{2+} > 2:1$) while the Na^+/H^+ antiport is an electroneutral process.[335] Active transport of Ca^{2+} across the plasma membrane of *Neurospora crassa* involves a Ca^{2+}/H^+ antiporter which functions to pump Ca^{2+} out of the intact cell.[336] Other aspects of cation transport in micro-organisms will not be considered in this Report.

5 Cations of Groups IA and IIA as Activators or Inhibitors of Enzymes

An analytical method has been developed for K^+ in the range 10^{-7} to 10^{-12} mol dm^{-3}, based on its activation of pyruvate kinase.[337] The titration of enzyme-bound ATP with Mg^{2+}, using the chemical shifts of the β phosphorus group as a parameter, showed that more than two Mg^{2+} ions are required for saturation, suggesting that there is a metal-binding site on the enzyme in addition to that on ATP (at the catalytic site). The ^{31}P resonance of enzyme-bound phosphoenol-pyruvate is a useful probe of structural and dynamic features of the active site.[338] The Δ-isomer of β,γ-bidentate CrATP is the only form of this complex which activates the phosphoryl-transfer reaction of pyruvate kinase, while it is the most active isomer in the pyruvate-kinase-catalysed enolization of pyruvate. This latter reaction was also promoted by all the bidentate and terdentate isomers of CrATP. The paramagnetic effect of the bidentate CrATP isomers on the relaxation rate of water protons is smaller than those of the terdentate isomers, a fact which has been interpreted in terms of an intramolecular hydrogen-bonding interaction in the case of the bidentate isomers.[339]

Mitochondrial creatine kinase has been purified; its properties are rather different from those of the creatine kinases isolated from mammalian sources.[340] The effects of pH and free $[Mg^{2+}]$ on the equilibrium constant for a number of reactions involving the transfer and the hydrolysis of phosphate have been calculated (for reactions of creatine kinase, myokinase, glucose-6-phosphatase, fructose-1,6-diphosphatase, and phosphoglycerate kinase).[341] A cyclic-nucleotide-independent protein kinase is activated by Ca^{2+} and membrane phospholipid.[342] The binding of metal–nucleotide substrates and analogues to the catalytic cAMP-dependent protein kinase induces the appearance of an additional, tight, inhibitory binding site for bivalent cations on the enzyme. This has been explored, using β,γ-bidentate $[Co^{III}(NH_3)_4(ATP)]$, which competes for the MgATP site. From the paramagnetic effects of Mn^{2+} that is bound at the inhibitory site on the longi-tudinal relaxation times of the protons and the phosphorus nuclei of $[Co(NH_3)_4$-

[335] T. Tsuchiya and K. Takeda, *J. Biochem. (Tokyo)*, 1979, **85**, 943; 1979, **86**, 225; R. N. Brey and B. P. Rosen, *J. Biol. Chem.*, 1979, **254**, 1957; J, C. Beck and B. P. Rosen, *Arch. Biochem. Biophys.*, 1979, **194**, 208.

[336] P. Stroobant and G. A. Scarborough, *Proc. Natl. Acad. Sci. USA*, 1979, **76**, 3102.

[337] W. H. Outlaw and O. H. Lowry, *Anal. Biochem.*, 1979, **92**, 370.

[338] B. D. Nageswara Rao, F. J. Kayne, and M. Cohn, *J. Biol. Chem.*, 1979, **154**, 2689.

[339] D. Dunaway-Mariano, J. L. Benovic, W. W. Cleland, R. K. Gupta, and A. S. Mildvan, *Biochemistry*, 1979, **18**, 4347.

[340] N. Hall, P. Addis, and M. Deluca, *Biochemistry*, 1979, **18**, 1745.

[341] J. W. R. Lawson and R. L. Veech, *J. Biol. Chem.*, 1979, **254**, 6528; N. W. Cornell, M. Leadbetter and R. L. Veech, *ibid.*, p. 6522.

[342] Y. Takai, A. Kishimoto, Y. Iwasa, Y. Kowahara, T. Mori, and Y. Nishizuka, *J. Biol. Chem.*, 1979, **254**, 3692.

(ATP)], nine distances from Mn^{2+} to $[Co(NH_3)_4(ATP)]$ were determined, which indicates either that there is bidentate α,γ- or terdentate α,β,γ-co-ordination of the triphosphate chain of both the Δ- and the Λ-stereoisomers of $[Co(NH_3)_4$-(ATP)] by the enzyme-bound Mn^{2+}.[343]

The interaction of CrADP with yeast hexokinase has been studied by ^{1}H and [31]P n.m.r. spectroscopy. Distance measurements suggest the absence of a direct co-ordination of the phosphoryl group of glucose 6-phosphate by the nucleotide-bound metal on hexokinase, but indicate that there is van der Waals contact between a phosphoryl oxygen of glucose 6-phosphate and the hydration sphere of the nucleotide-bound metal. These distances are consistent with a model that assumes molecular contact between the phosphorus of glucose 6-phosphate and a β-phosphoryl oxygen of ADP, suggesting an associative phosphoryl-transfer process.[344]

Phosphofructokinase is a four-subunit enzyme in the active form, and it undergoes a reversible dissociation into an inactive form of two subunits. Several studies have been reported on the binding of nucleotides under conditions where the oligomeric structure of the enzyme is known. Binding of ATP to a site that is responsible for inhibition has been measured by the quenching of fluorescence of the protein. The affinity for ATP at this site is determined by the protonation of two ionizable groups per subunit.[345,346] Direct binding experiments, in conjunction with these fluorescence studies, show there to be one active site and one inhibitory site per subunit, while a third nucleotide site is specific for cAMP, AMP, and ADP.[345] The use of temperature-jump[347] and fluorescence stopped-flow techniques[348] suggests that cAMP converts the enzyme into the active conformation.

The binding of phosphofructokinase to erythrocyte membranes probably involves an integral membrane protein, as the removal of other proteins does not diminish the binding capacity of the membrane. Binding is also inhibited by proteins which are known to bind to Band 3 protein. These results suggest that phosphofructokinase may not be found exclusively in the cell cytoplasm.[349] It has been purified from skeletal muscle as a phosphoprotein, and accordingly an attempt has been made to correlate this phenomenon with the functional state of muscle in order to assess the physiological function of the phosphoprotein. It has been found that the number of phosphorylated sites increases during contraction.[350] Phosphofructokinase from the thermophile *Thermus X-1* is activated by univalent cations in the sequence $Tl^+ > NH_4^+ > K^+ > Rb^+ > Cs^+$, while Na^+ and Li^+ have no effect.[351]

The structure of horse muscle phosphoglycerate kinase shows that the metal–ADP or –ATP substrates are bound to one of two widely separated domains of

343 J. Granot, H. Kondo, R. N. Armstrong, A. S. Mildvan, and E. T. Kaiser, *Biochemistry*, 1979, **18**, 2339; R. N. Armstrong, *ibid.*, p. 1230.
344 R. L. Pederson and R. K. Gupta, *Biophys. J.*, 1979, **27**, 1.
345 D. W. Pettigrew and C. Frieden, *J. Biol. Chem.*, 1979, **254**, 1887.
346 N. M. Wolfman and G. G. Hammes, *J. Biol. Chem.*, 1979, **254**, 12 289.
347 N. M. Wolfman, A. C. Storer, and G. G. Hammes, *Biochemistry*, 1979, **18**, 2451.
348 M. Laurent, F. J. Seydoux, and P. Dessen, *J. Biol. Chem.*, 1979, **254**, 7515.
349 T. Higashi, C. S. Richards, and K. Uyeda, *J. Biol. Chem.*, 1979, **254**, 9542.
350 H. W. Hofer and B. Sørensen-Ziganke, *Biochem. Biophys. Res. Commun.*, 1979, **90** 204.
351 E. Stellwagen and S. T. Thompson, *Biochim. Biophys. Acta*, 1979, **569**, 6.

the enzyme, which provides an environment that seems to be unsuitable for the binding of phosphoglycerate. In fact the most plausible site for the phosphoglycerate substrate is on the other domain, about 10 Å from the ATP-binding site. This has led to the hypothesis that a 'hinge-bending' mechanism holds for the enzyme, thus bringing the two substrates together in a water-free environment.[352] A similar conclusion was reached on the basis of small-angle X-ray scattering on solutions of yeast phosphoglycerate kinase, and supported by a comparison with the mechanism of hexokinase. Indeed it was suggested that this type of conformational change may prove to be a rather general phenomenon in kinase mechanisms.[353]

Phosphoribosylpyrophosphate synthetase catalyses a reaction in which nucleophilic substitution occurs at the β-phosphorus atom of ATP. This reaction has been examined, using α,β,γ-[Cr(ATP)].[354] Distance measurements, based on the n.m.r. relaxation rates of anomeric protons and of the phosphorus atom in ribose 5-phosphate, between Cr^{3+} and ribose 5-phosphate, suggested a model in which both α- and β-anomers of ribose 5-phosphate bind to the enzyme near to the bound Cr(ATP), and that H-1 of the β-anomer is ~ 1.0 Å closer to the bound Cr(ATP) than is H-1 of the α-anomer. As a result, the 1-OH group of the α-anomer is ~ 1.6 Å closer to the bound Cr(ATP) than the 1-OH group of the β-anomer. Hence the lower limit for the distance from the 1-OH of α-ribose 5-phosphate to the β-phosphorus of Cr(ATP) approximates to van der Waals contact, which is consistent with a direct S_N2 nucleophilic displacement of AMP from ATP by ribose 5-phosphate. The data also established the close proximity of a bound inorganic phosphate moiety 7.1 ± 0.7 Å from the Cr(ATP) site. This may represent the phosphate group that is required for the enzyme to show its activity. Other Mg^{2+}-activated enzymes, associated with transfer of a phosphate-containing group, that have received attention are phosphoribosyl transferase,[355] bovine thymus poly(ADP-ribose) polymerase,[356] and the phenylalanyl-tRNA synthetase of *E. coli* K.[357]

The activity of fructose 1,6-biphosphatase depends upon the presence of univalent and bivalent cations, but it appears that there might also be a physiological role for Zn^{2+}.[358] Eight sites for binding bivalent cations are available on the enzyme (two per subunit). One set of four sites is associated with Zn^{2+}, while the other set is detected only in the presence of substrate and appears to be identical to the sites for activating-Mg^{2+}, and at which Zn^{2+} binds at higher concentration. Other reports on this enzyme have described fluorescence studies on the effect of pH and [Mg^{2+}] on the conformation of the enzyme from spinach

[352] R. D. Banks, C. C. F. Blake, P. R. Evans, R. Haser, D. W. Rice, G. W. Hardy, M. Merrett, and A. W. Phillips, *Nature (London)*, 1979, **279**, 773.
[353] C. A. Pickover, D. B. McKay, D. M. Engleman, and T. A. Steitz, *J. Biol. Chem.*, 1979, **254**, 11 323; W. S. Bennet and T. A. Steitz, *Proc. Natl. Acad. Sci. USA*, 1978, **75**, 4848; C. M. Anderson, F. H. Zucker, and T. A. Steitz, *Science*, 1979, **204**, 375.
[354] T. M. Li, R. L. Switzer, and A. S. Mildvan, *Arch. Biochem. Biophys.*, 1979, **193**, 1.
[355] J. Victor, A. Leo-Mensah, and D. L. Sloan, *Biochemistry*, 1979, **18**, 3597.
[356] Y. Tanaka, T. Hashida, H. Yoshihara, and K. Yoshihara, *J. Biol. Chem.*, 1979, **254**, 12 433.
[357] J. Pimmer and E. Holler, *Biochemistry*, 1979, **18**, 3714.
[358] S. Pentremoli, B. Sparatore, F. Salamino, E. Mellori, and B. L. Horecker, *Arch. Biochem. Biophys.*, 1979, **194**, 481.

chloroplasts,[359] [^{18}O]phosphate–H_2O exchange,[360] the use of a modified AMP grouping,[361] and extensive kinetic studies.[362]

A Ca^{2+}-dependent cyclic nucleotide phosphodiesterase has been purified.[363] Two forms have been isolated from the soluble fraction of the rabbit heart, one of which is sensitive to Ca^{2+} plus the Ca^{2+}-binding regulatory protein.[364] The enzyme from rabbit lung differs from enzymes from other sources in that calmodulin is very tightly bound to it.[365] The phosphodiesterase[366] from the brain in inhibited by certain heat-stable factors from the brain;[367] these may bind calmodulin.

The roles of bivalent cations, metal–ATP complexes, and prostaglandin E_1 in the control of adenylate cyclase[368] and guanylate cyclase[369] have been studied. The ions Mg^{2+} and Mn^{2+} appear to exert substantially different effects on two enzymes that have been isolated from *E. coli, viz. β*-galactosidase[370] and NAD^+-specific malic enzyme.[371]

Examples of specific effects of Ca^{2+} on enzymes that have been discussed are the modulation of the *rec BC* enzyme of *E. coli* K-12,[372] the control of cardiolipin-associated erythrocyte acetylcholine esterase,[373] and the activation of nucleoside triphosphate pyrophosphohydrolase,[374] phospholipase A_2,[375a] and a neutral proteinase from skeletal muscle.[375b]

Alkali-metal ions activate or inhibit intestinal brush-border sucrase, depending upon the pH;[376] they also activate AMP nucleosidase from *Azotobacter vinelandii*,[377] kirromycin-activated EF-Tu GTPase,[378] and NAD^+-dependent glyceraldehyde 3-phosphate dehydrogenase.[379]

359 T. Takebe, H. Ishikawa, M. Miyakawa, K. Takenaka, and S. Nikai, *J. Biochem. (Tokyo)*, 1979, **85**, 203.
360 T. R. Sharp and S. J. Benkovic, *Biochemistry*, 1979, **18**, 2910.
361 R. B. Maccioni, E. Hubert, and J. C. Slebe, *FEBS Lett.*, 1979, **102**, 29.
362 J. P. Casazza, S. R. Stone, and H. J. Fromm, *J. Biol. Chem.*, 1979, **254**, 4661.
363 R. L. Kincaid and M. Vaughan. *Proc. Natl. Acad. Sci. USA*, 1979, **76**, 4903.
364 V. A. Tkachuk, V. G. Lazarevich, M. Yu. Nenshikov, and S. E. Severin, *Biochemistry (Engl. Transl.)*, 1979, **43**, 1276.
365 R. K. Sharma and E. Wirch, *Biochem. Biophys. Res. Commun.*, 1979, **91**, 338.
366 H. M. Thérien and W. E. Mushynski, *Biochim. Biophys. Acta*, 1979, **585**, 201.
367 T. Kanamori, C. R. Creveling, and J. W. Daly, *Biochim. Biophys. Acta*, 1979, **582**, 434.
368 R. A. Johnson, W. Sauer, and K. H. Jakobs, *J. Biol. Chem.*, 1979, **254**, 1094; J. Premont, G. Guillon, and J. Bockaert, *Biochem. Biophys. Res. Commun.*, 1979, **90**, 513; D. Stengel, M.-L. Lacombe, M.-C. Billon and J. Hanoune, *FEBS Lett.*, 1979, **107**, 105; V. Stolc, *Biochim. Biophys. Acta*, 1979, **569**, 267; J.-W. Wei, N. Narayanan, and P. V. Sulakhe, *Int. J. Biochem.*, 1979, **10**, 109.
369 T. Takenawa and B. Sacktor, *Biochim. Biophys. Acta*, 1979, **566**, 371.
370 R. E. Huber, C. Parfett, H. Woulfe-Flanagan, and D. J. Thompson, *Biochemistry*, 1979, **18**, 4090.
371 J. A. Milne and R. A. Cook, *Biochemistry*, 1979, **18**, 3604
372 J. Rosamond, K. M. Telander, and S. Linn, *J. Biol. Chem.*, 1979, **254**, 8646.
373 G. Beauregard and B. D. Roufogalis, *Biochim. Biophys. Acta*, 1979, **557**, 102.
374 C. Torp-Pedersen, H. Flodgaard, and T. Saermark, *Biochim. Biophys. Acta*, 1979, **571**, 94.
375 (a) E. Frei and P. Zahler, *Biochim. Biophys. Acta*, 1979, **550**, 450; (b) K. Suzuki, S. Ishiura, T. Katamoto, H. Sugita, and K. Imahori, *FEBS Lett.*, 1979, **104**, 355; J-L. Azanza, J. Raymond, J-M. Robin, P. Cottin, and A. Ducastaing, *Biochem. J.*, 1979, **183**, 339.
376 F. Alvarado and A. Mahmood, *J. Biol. Chem.*, 1979, **254**, 9534.
377 M. Yoshino, K. Murakami, and K. Tsushima, *Biochim. Biophys. Acta*, 1979, **570**, 118.
378 G. Sander, M. Okonek, J.-B. Crechet, R. Ivell, V. Bocchini, and A. Parmeggiani, *FEBS Lett.*, 1979, **98**, 111.
379 O. P. Malhotra, D. K. Srivastava, and Srinivasan, *Arch. Biochem. Biophys.*, 1979, **197**, 302.

6 Control of Intracellular Processes by Calcium

Calmodulin (Calcium-binding Regulatory Protein).—Calmodulin is known to be involved in the regulation of a range of cellular enzyme systems and most types of cell motility. This section reviews generally the properties of calmodulin, with the exception of its role in the absorption of calcium in the intestine and in secretion of calcium. Calmodulin is a monomeric species, of molecular weight about 17 000. The binding of Ca^{2+} confers resistance to denaturation. There are four sites for Ca^{2+}, and the binding of Ca^{2+} to any of these causes a conformational change. It appears that the extent of occupancy of the sites varies, depending upon the process being activated, and this therefore may be of biological significance.

Calmodulin from the marine invertebrate *Renilla reniformis* is very similar to those isolated from mammalian systems, despite their diverse origins.[380] Other examples of the characterization of these Ca^{2+}-binding regulatory proteins involve those isolated from the unfertilized egg of the sea urchin *Arbacia punctulata*,[381] from *Tetrahymena pyriformis*,[382] from chicken fibroblasts,[383] from *Blastocladiella emersonii*,[384] and from brain, muscle, and other tissues.[385] Procedures for purifying calmodulin (affinity chromatography) have been improved,[381,386] while a calmodulin–Sepharose affinity column has been used to fractionate partly purified adenylate cyclase into two separate forms in the presence of Ca^{2+}, *i.e.* into sensitive and insensitive forms.[387]

The conformational changes that are induced on binding Ca^{2+}, as followed by 250 MHz ^1H n.m.r.,[388] are similar to those found for skeletal troponin-C, in accord with the large sequence homology between the two proteins and their cross-reacting in their respective biological systems.[389] Modified calmodulins that are nitrated[390,391] at tyrosyl residues 99 or 138 and alkylated[391] at five methionine residues have been prepared. The nitrated calmodulin undergoes similar conformational changes to calmodulin on binding of Ca^{2+}, but the spectrum of the tyrosyl residue 138 was affected particularly, indicating that the conformational changes selectively affected the microenvironment around Tyr-138. Addition of Mg^{2+} resulted in non-specific effects. The alkylation of calmodulin does affect its conformational behaviour, and the protein only binds Ca^{2+} to a limited extent.

[380] H. P. Jones, J. C. Matthews, and M. J. Cornier, *Biochemistry*, 1979, **18**, 55; T. C. Vanaman and F. Sharief, *Fed. Proc.*, 1979, **38**, 788.

[381] J. F. Head, S. Mader, and B. Kaminir, *J. Cell Biol.*, 1979, **80**, 211.

[382] G. A. Jamieson, T. C. Vanaman, and J. J. Blum, *Proc. Natl. Acad. Sci. USA*, 1979, **76**, 6471.

[383] L. J. Van Eldik and D. M. Wattersen, *J. Biol. Chem.*, 1979, **254**, 10 250.

[384] S. L. Gomes, L. Mennucci, and J. C. da C. Maia, *FEBS Lett.*, 1979, **99**, 39.

[385] R. J. A. Grand and S. V. Perry, *Biochem. J.*, 1979, **183**, 285; R. J. A. Grand, S. V. Perry, and R. A. Weeks, *ibid.*, 1979, **177**, 521; T. Yamauchi and H. Fujisawa, *Biochem. Biophys. Res. Commun.*, 1979, **90**, 1172.

[386] H. Charbonneau and M. J. Cornier, *Biochem. Biophys. Res. Commun.*, 1979, **90**, 1039.

[387] K. R. Westcott, D. C. La Porte, and D. R. Storm. *Proc. Natl. Acad. Sci. USA*, 1979, **76**, 204.

[388] K. Seamon, *Biochem. Biophys. Res. Commun.*, 1979, **86**, 1256.

[389] M. J. Yerna, D. J.Hartshorne, and R. D. Goldman, *Biochemistry*, 1979, **18**, 673; P. Cohen, C. Picton, and C. B. Klee, *FEBS Lett.*, 1979, **104**, 25.

[390] P. G. Richman and C. B. Klee, *J. Biol. Chem.*, 1979, **254**, 5372.

[391] M. Walsh, F. R. Stevens, K. Oikawa, and C. M. Kay, *Can. J. Biochem.*, 1979, **57**, 267.

Calmodulin is a poor antigen (probably because of its small size and its occurrence in a wide range of cells and tissues from various species). However, the DNB derivative of calmodulin [(DNB)$_3$-calmodulin] is highly antigenic, thus allowing the production of antibodies in rabbits after injection with (DNB)$_3$-calmodulin and the development of a highly specific radioimmunoassay for calmodulin. The sensitivity for (DNB)$_3$-calmodulin and calmodulin is approximately 0.2 and 2 pmol, respectively.[392] An alternative radioimmunoassay is also available.[393]

The myosin light-chain kinase in non-muscle cells is calmodulin-dependent.[394] In association with Ca^{2+}, calmodulin regulates the calcium pump of sarcoplasmic reticulum,[395] phosphorylase kinase from rabbit skeletal muscle,[396] myosin light-chain kinase,[397] the calcium pump of the red blood cell membrane,[398] human platelet phospholipase A_2,[399] guanylate cyclase from *Tetrahymena pyriformis*,[400] adenylate cyclase,[387] glycogen synthase kinase 2 from rabbit skeletal muscle,[401] and phosphodiesterase.[364,365,402] It also appears that calmodulin has a role in the assembly and disassembly of microtubules, processes long known to require Ca^{2+}. Thus the Ca^{2+}-calmodulin complex prevents the assembly of microtubule protein *in vitro* and promotes the complete disassembly of microtubules, as monitored by viscometry.[403] The cytoplasm of macrophages contains numerous actin filaments and undergoes reversible gel-sol transformations, which is controlled by a Ca^{2+}-dependent regulatory protein, gelsolin.[404] This list presents a remarkable collection of calmodulin-mediated processes, and there should be considerable developments in this area in the future. Some Ca^{2+}-calmodulin-enzyme processes are inhibited by proteins that bind calmodulin (see cyclic nucleotide phosphodiesterase[366]). This inhibitory protein, which is found in nervous tissue, also binds four Ca^{2+} per mole, and has been named calcineurin.[405] It is composed of two subunits, *i.e.* calcineurin A (mol. wt 61 000), which binds calmodulin in a Ca^{2+}-dependent reaction, and calcineurin B (15 000),

392 R. W. Wallace and W. Y. Cheung, *J. Biol. Chem.*, 1979, **254**, 6564.
393 J. G. Chafouleas, J. R. Dedman, R. P. Munjaal, and A. R. Means, *J. Biol. Chem.*, 1979, **254**, 10 262.
394 D. R. Hathway and R. S. Adelstein, *Proc. Natl. Acad. Sci. USA*, 1979, **76**, 1653; M. J. Yerna, R. Dabrowska, D. J. Hartshorne, and R. D. Goldman, *ibid.*, p. 184.
395 C. J. Le Peuch, J. Haiech, and J. G. Demaille, *Biochemistry*, 1979, **18**, 5150; H. J. Moeschler, D. A. Malengik, S. Pocinwong, O. Ataba, G. L. Kerrick, and E. H. Fischer, *Biochimie*, 1979, **61**, 615.
396 S. Shenolikar, P. T. W. Cohen, P. Cohen, A. C. Nairn, and S. V. Perry, *Eur. J. Biochem.*, 1979, **100**, 329.
397 A. C. Nairn and S. V. Perry, *Biochem. J.*, 1979, **179**, 89; C. J. Le Peuch, C. Ferraz, M. P. Walsh, J. G. Demaille, and E. H. Fischer, *Biochemistry*, 1979, **18**, 5267.
398 F. L. Larsen and F. F. Vincenzi, *Science*, 1979, **204**, 306.
399 P. Y.-K. Wong and W. Y. Cheung, *Biochem. Biophys. Res. Commun.*, 1979, **90**, 473.
400 S. Nagao, Y. Suzuki, Y. Watanabe, and Y. Nozawa, *Biochem. Biophys. Res. Commun.*, 1979, **90**, 261.
401 D. B. Rylatt, N. Embi, and P. Cohen, *FEBS Lett.*, 1979, **98**, 76.
402 C. B. Klee, T. H. Crouch, and M. H. Krinks, *Biochemistry*, 1979, **18**, 722; R. L. Kincaid and M. Vaughan, *Proc. Natl. Acad. Sci. USA*, 1979, **76**, 4903.
403 E. Nishida, H. Kumagai, I. Ohtsuki, and H. Kai, *J. Biochem. (Tokyo)*, 1979, **85**, 1257; H. Kumagai and E. Nishida, *ibid.*, p. 1267.
404 H. L. Yin and T. P. Stossel, *Nature (London)*, 1979, **281**, 583.
405 C. B. Klee, T. H. Crouch, and M. H. Krinks, *Proc. Natl. Acad. Sci. USA*, 1979, **76**, 6270.

which binds Ca^{2+}. Calcineurin may have a role in the regulation of free Ca^{2+} in the nervous system.[405]

Calcium-binding Proteins involved in Muscle Contraction.—*The Parvalbumins*. A procedure is now available for the rapid isolation and purification of parvalbumin from chicken leg muscles. Four polypeptides that were isolated with it have a low affinity for Ca^{2+}, and they probably do not have a role in Ca^{2+} regulation.[406] Five parvalbumins have been isolated from the lungfish,[407] and details of a new large-scale purification procedure for muscular parvalbumins have been published.[408] Treatment of pike parvalbumin III with CNBr gives fragment 38–108, which contains two Ca^{2+}-binding domains and binds Ca^{2+} well. Loss of Ca^{2+} is associated with a change in the c.d. spectrum.[409] Parvalbumin had been thought to function as a Ca^{2+}-dependent activator of phosphodiesterase, but it is now appreciated that this is due to the presence of a small amount of carp muscle calcium-dependent regulator protein. Purified parvalbumin only causes a non-specific stimulation of the phosphodiesterase.[410]

The binding of cations to parvalbumins has been much studied. Parvalbumins from different genetic lineages have different cation-binding properties,[411] but always have Ca^{2+}–Mg^{2+} high-affinity sites. This has led to the postulate of a physiological role for these proteins. In resting muscle, the parvalbumins (Pa) will be present as $PaMg_2$ (or at least largely so). At the onset of contraction, Ca^{2+} will be liberated from the S.R., but the low rate of dissociation of Mg^{2+} from these high-affinity Ca^{2+}–Mg^{2+} sites means that Ca^{2+} will bind preferentially at the Ca^{2+}-specific sites on troponin C, and so trigger contraction. Later, these high-affinity sites on parvalbumin will be able to compete for Ca^{2+} (as the Mg^{2+} will be dissociated) and so remove the Ca^{2+} from the troponin C, thus inducing or assisting relaxation.[411] An interesting implication of this hypothesis is that the much-studied conformational change[412] that occurs on the binding of Ca^{2+} to ion-free parvalbumin is of no physiological significance. The only conformational changes of importance would be those associated with the substitution of Ca^{2+} for Mg^{2+}. There is some uncertainty about this matter. One report suggests that no difference in conformation could be detected, that the Mg^{2+}–Ca^{2+} exchange involved little structural change, and that in any case the kinetic data suggest that parvalbumin often stays in the Mg form during the contraction-relaxation cycle.[413] However, other workers[414] claim to have observed such conformational differences between the Mg and Ca forms, using ^1H n.m.r., and have measured the exchange between these conformational states in solution. This has allowed a detailed analysis of the calcium flux *in vivo* during contraction. Dissociation of Mg^{2+} from Pa occurs with an 'off' rate-constant of *ca* 10 s^{-1}; thus

[406] C. W. Heizmann and E. R. Strehler, *J. Biol. Chem.*, 1979, **254**, 4296.
[407] C. Gerday, B. Joris, N. Gerardin-Otthiers, S. Collin, and G. Hamoir, *Biochimie*, 1979, **61**, 589.
[408] J. Haiech, J. Derancourt, J.-F. Pechère, and J. G. Demaille, *Biochimie*, 1979, **61**, 583.
[409] E. E. Maximov and Yu. V. Mitin, *Biochimie.*, 1979, **61**, 751.
[410] N. C. Le Donne and C. J. Coffee, *J. Biol. Chem.*, 1979, **254**, 4317.
[411] J. Haiech, J. Derancourt, J.-F. Pechère, and J. G. Demaille, *Biochemistry*, 1979, **18**, 2752.
[412] A. Cavé, M. Pages, Ph. Morin, and C. M. Dobson, *Biochimie*, 1979, **61**, 607.
[413] J. A. Cox, D. R. Winge, and E. A. Stein, *Biochemie*, 1979, **61**, 601.
[414] W. J. Birdsall, B. A. Levine, R. J. P. Williams, J. G. Demaille, J. Haiech, and J.-F. Pechère, *Biochimie*, 1979, **61**, 741.

the uptake of Ca^{2+} at this site is controlled, and allows saturation of the troponin C sites as described earlier. This also serves as a model for the role of parvalbumin in nervous tissue when synaptic transmission is triggered by the influx of Ca^{2+}.

The binding of Mg^{2+} to Pa has been studied by ^{25}Mg and ^{114}Cd n.m.r.,[415] and the binding of lanthanides and Mn^{2+} by ^1H n.m.r.[416] There has been some controversy over the binding of Na^+, as studied by ^{24}Na n.m.r.[417]

Troponin C. The binding of Ca^{2+} to troponin C (TN-C) induces a conformational change, which is transmitted through the outer subunits of troponin (troponin I and troponin T) to tropomyosin, with a resulting change in its conformation from a blocking position in the grooves of the actin filament to one that allows the interaction of myosin and actin. The exchange rates of Ca^{2+} at TN-C are important, as binding of calcium must occur before contraction can take place. The Ca^{2+}-induced changes in the fluorescence of dansylaziridine-labelled TN-C were used[418] to monitor (by stopped-flow techniques) the structural changes resulting from the binding of Ca^{2+} at both Ca^{2+}-specific sites and Ca^{2+}–Mg^{2+} sites. These changes all took place within the mixing time of the instrument. The Ca^{2+} ion was removed from the Ca^{2+}-specific sites with a $t_{1/2}$ of 2–3 ms, and from the Ca^{2+}–Mg^{2+} sites with a $t_{1/2}$ of about 700 ms; the latter is a reaction that is much too slow to be associated with regulation. Clearly, however, all the reactions associated with the Ca^{2+}-specific sites are fast enough to be involved in the regulation of muscle contraction. The conclusions about the Ca^{2+}–Mg^{2+} site were independently confirmed[419] by the study of the kinetics of the conformational change that is induced by binding and removal of Mg^{2+}. Cadmium-113 n.m.r. has been used to study TN-C.[420]

The interactions between the subunits of the troponin complex and their properties have been studied as follows: TN-I–TN-T interactions,[421] a comparison of troponin from various sources,[422] a review of tropomyosins,[423] a peptide analogue of the actomyosin ATPase-inhibitory region of TN-I,[424] and Ca^{2+}-induced conformational changes in the actin–tropomyosin–troponin system.[425]

Regulation of Smooth Muscle Activity. The phosphorylation of the myosin light-chain is necessary for control of actin–myosin interaction in smooth muscle. This is catalysed by a Ca^{2+}-dependent kinase, a subunit of which is probably a Ca^{2+}-dependent regulatory protein. This phosphorylation reaction has been

415 A. Cavé, J. Parello, T. Drakenberg, E. Thulin, and B. Lindman, *FEBS Lett.*, 1979, **100**, 148.
416 A. Cavé, M.-F. Daures, J. Parello, A. Saint-Yves, and R. Sempere, *Biochimie*, 1979, **61**, 755; L. Lee, B. D. Sykes, and E. D. Birnbaum, *FEBS Lett.*, 1979, **98**, 169.
417 J. Parello, P. Reimarsson, E. Thulin, and B. Lindman, *FEBS Lett.*, 1979, **100**, 152; C. Gerday, J. Grandjean, and P. Laszlo, *ibid.*, 1979, **105**, 384.
418 J. D. Johnson, S. C. Charlton, and J. D. Potter, *J. Biol. Chem.*, 1979, **254**, 3497.
419 T. Iio, K. Mihashi, and H. Kondo, *J. Biochem. (Tokyo)*, 1979, **85**, 97.
420 S. Forsén, E. Thulin, and H. Lilja, *FEBS Lett.*, 1979, **104**, 123.
421 M. T. Hincke, W. D. McCubbin, and C. M. Kay, *Can. J. Biochem.*, 1979, **57**, 768; J. Horwitz, B. Bullard, and D. Mercola. *J. Biol. Chem.*, 1979, **254**, 350.
422 D. M. Byers, W. D. McCubbin, and C. M. Kay, *FEBS Lett.*, 1979, **104**, 106; K. Kohama, *J. Biochem.*, (*Tokyo*), 1979, **86**, 811.
423 L. B. Smillie, *Trends Biochem. Sci.*, 1979, 151.
424 J. A. Talbot and R. S. Hodges, *J. Biol. Chem.*, 1979, **254**, 3720.
425 T. Ohyashiki and T. Sekine, *Biochim. Biophys. Acta*, 1979, **576**, 51; T. Mikawa, *Nature* (*London*), 1979, **278**, 473; *J. Biochem. (Tokyo)*, 1979, **85**, 879.

demonstrated under physiological conditions,[426] while the effect of a range of physiologically active reagents on the phosphorylation reaction has been tested.[427] Myosin light-chain kinases from several sources have been studied,[428] while the effect of inhibitors that are known to bind to calmodulin has demonstrated the role of the Ca^{2+}-dependent regulatory protein.[429] A preparation of chicken gizzard heavy meromyosin is now available that retains the two light-chain components of the parent myosin and its characteristic reactions.[430] Other aspects of muscle activity will not be discussed.

Secretion of Enzymes and Hormones.—The important role of Ca^{2+} and calmodulin is demonstrated by the fact that calmodulin has now been implicated in the secretion of insulin,[431] which is inhibited by trifluoperazine, a known inhibitor of the reactivity of calmodulin. Furthermore, Ca^{2+} is implicated in the secretion of renin,[432] of protein from the pancreas,[433] and of amylase,[434] in the synthesis of prostaglandins,[435] and in adrenocorticotropin-induced steroidogenesis by adrenal cells.[436] The release of granule-associated enzymes from neutrophils (polymorphonuclear neutrophilic leukocytes) may require Ca^{2+}, although this suggestion is controversial.[437]

Synexin, a protein of the adrenal medulla, causes Ca^{2+}-dependent aggregation of isolated chromaffin granules by fusing the granule membranes. It has now been found that isolated synexin undergoes a Ca^{2+}-activated self-association, made up of first side-by-side and then end-to-end alignment of rods, of dimensions 150 by 50 Å, which are oligomers of the synexin molecule. It has been suggested that not only is synexin a receptor for Ca^{2+} in the granule-aggregation reaction, but it is also an important receptor for Ca^{2+} in exocytosis. These rod-shaped particles may be involved in binding to membranes and drawing them into the pentalaminar fusion complex which is the initial morphological feature that is characteristic of exocytosis.[438] The transport of catecholamines into the granules is probably linked to the proton gradient, the amine becoming protonated inside

[426] J. T. Barron, M. Bárány, and K. Bárány, *J. Biol. Chem.*, 1979, **254**, 4954.
[427] K. Bárány, M. Bárány, J. M. Gillis, and M. J. Kushmerick, *J. Biol. Chem.*, 1979, **254**, 3617; P. E. Hoar, W. G. L. Kerrick, and P. S. Cassidy, *Science*, 1979, **204**, 501; S. J. Koppi and M. Bárány, *J. Biol. Chem.*, 1979, **254**, 12 007.
[428] M. P. Walsh, B. Vallet, F. Autrie, and J. G. Demaille, *J. Biol. Chem.*, 1979, **254**, 12 136; E. A. Lebowitz and R. Cooke, *J. Biochem. (Tokyo)*, 1979, **85**, 1489; A. A. Depaoli-Roach, J. B. Gibbs, and P. J. Roach, *FEBS Lett.*, 1979, **105**, 321.
[429] H. Hidaka, M. Naka, and T. Yamaki, *Biochem. Biophys. Res. Commun.*, 1979, **90**, 694.
[430] H. Onishi and S. Watanabe, *J. Biochem. (Tokyo)*, 1979, **85**, 457.
[431] M. C. Sugden, M. R. Christie, and S. J. H. Ashcroft, *FEBS Lett.*, 1979, **105**, 95; H. J. Hahn, E. Gylfe, and B. Hellman, *ibid.*, 1979, **103**, 348.
[432] E. Harada, G. E. Lester, and R. P. Rubin, *Biochim. Biophys. Acta*, 1979, **583**, 20.
[433] G. Scheele and A. Haymovits, *J. Biol. Chem.*, 1979, **254**, 10 346; V. V. A. M. Schreurs, H. G. P. Swarts, J. J. H. H. M. de Pont, and S. L. Bonting, *Biochim. Biophys. Acta*, 1979, **583**, 208.
[434] E. L. Watson, J. A. Williams, and I. A. Siegel, *Am. J. Physiol.*, 1979, **236**, C233; J. D Gardener, C. L. Costenbader, and E. R. Uhleman, *ibid.*, p. E745.
[435] A. Erman and A. Raz, *Biochem. J.*, 1979, **182**, 821.
[436] J.-P. Perchellet and R. K. Sharma, *Science*, 1979, **204**, 1259.
[437] R. J. Smith and S. S. Iden, *Biochem. Biophys. Res. Commun.*, 1979, **91**, 263; R. J. Petroski, P. H. Naccache, E. I. Becker, and R. I. Sha'afi, *Am. J. Physiol.*, 1979, **237**, 43.
[438] C. E. Creutz, C. J. Pazoles, and H. B. Pollard, *J. Biol. Chem.*, 1979, **254**, 553.

the granule and the decrease in [H^+] being met by a proton-translocating ATPase system.[439]

The secretion of histamine from mast cells, which is well known to be stimulated by the Ca^{2+}-ionophore A-23187, is also stimulated by the new ionophore ionomycin, which transports Ca^{2+} into the cell.[440] The effect of ATP on Ca^{2+}-dependent secretion has also been studied.[441]

Synaptic vesicles have a Ca^{2+}-dependent protein kinase which is stimulated by calmodulin[442] and which probably participates in the Ca^{2+}-stimulated release of neurotransmitters. The introduction of Ca^{2+} into the cholinergic presynaptic cytosol is accompanied by phosphorylation of specific proteins.[443]

Some Miscellaneous Examples of Calcium-controlled Processes.—The role of Ca^{2+} in the following processes has been discussed: the control of flagellar activity of *Euglena gracilis*,[444] sporulation of *Bacillus megaterium*,[445] development of oosphere initials in *Saprolegnia diclina*,[446] together with its distribution in *Physarum polycephalum*.[447]

7 Calcification and Mobilization

The physiological role of vitamin D in stimulating the intestinal absorption of Ca^{2+} probably involves the enhancement of the passive entry of Ca^{2+} into the cell across the brush-border membrane. The Ca^{2+} is then pumped out of the cell, across the basolateral cell membrane, by an active Na^+-dependent process. The study of the effects of 1-hydroxy-vitamin D_3 on brush-border membrane vesicles suggests that permeability to Ca^{2+} is controlled by a change in the lipid structure of the membrane.[448] Several analogues of 1,25-dihydroxy-vitamin D_3 have been prepared, some of which are highly active; metabolites have also been studied.[449] Certain of these compounds are more effective in influencing the mobilization of calcium in bone than intestinal transport of calcium.[450]

Two helpful mini-reviews are available to resolve the complexity of vitamin D

[439] R. G. Johnson and A. Scarpa, *J. Biol. Chem.*, 1979, **254**, 3750; R. W. Holz, *ibid.*, p. 6703; R. Maron, H. Fiskes, B. I. Kanner, and S. Schuldiner, *Biochemistry*, 1979, **18**, 4781; H. B. Pollard, H. Shindo, C. E. Creutz, C. J. Pazoles, and J. S. Cohen, *J. Biol. Chem.*, 1979, **254**, 1170.

[440] J. S. Bennett, S. Cockcroft, and B. D. Gomperts, *Nature (London)*, 1979, **282**, 851.

[441] S. Cockcroft and B. D. Gomperts, *J. Physiol.*, 1979, **296**, 229.

[442] R. J. De Lorenzo, S. D. Freedman, W. B. Yoke, and S. C. Maurer, *Proc. Natl. Acad. Sci. USA*, 1979, **76**, 1838.

[443] D. M. Michaelson and S. Avissar, *J. Biol. Chem.*, 1979, **254**, 12 542.

[444] M. J. Doughty and B. Diehn, *Biochim. Biophys. Acta*, 1979, **588**, 148.

[445] C. Hogarth and D. J. Ellar, *Biochem. J.*, 1979, **178**, 627.

[446] J. Fletcher, *J. Gen. Microbiol.*, 1979, **113**, 315.

[447] R. P. Holmes and P. R. Stewart, *J. Gen. Microbiol.*, 1979, **113**, 275.

[448] H. Rassmussen, O. Fontaine, E. E. Max, and D. B. P. Goodman, *J. Biol. Chem.*, 1979, **254**, 2993.

[449] J. L. Napoli, W. S. Mellon, M. A. Fivizzani, H. K. Schnoes, and H. F. Deluca, *J. Biol. Chem.*, 1979, **254**, 2017; Y. Tanaka, H. F. Deluca, Y. Kobayashi, T. Taguchi, N. Ikekawa, and M. Morisaki, *ibid.*, p. 7163; A. W. Norman, R. L. Johnson, R. Corradino, and W. H. Okamura, *ibid.*, pp. 11 445, 11 450; S. Edelstein, M. Sheves, Y. Mazur, A. Bar, and S. Hurwitz, *FEBS Lett.*, 1979, **97**, 241; R. P. Esveet, H. K. Schnoes, and H. F. Deluca, *Biochemistry*, 1979, **18**, 3971; P. D. Siebart, N. Ohnuma, and A. W. Norman, *Biochem. Biophys. Res. Commun.*, 1979, **91**, 827.

[450] J. L. Napoli, M. A. Fivizzani, H. K. Schnoes, and H. F. Deluca, *Biochemistry*, 1979, **18**, 1643; R. Brommage and W. I. Neuman, *Am. J. Physiol.*, 1979, **237**, E113.

metabolism.[451] Vitamin D is hydroxylated in the liver and further hydroxylated to 1,25-$(OH)_2$-D_3 in the kidneys. In the intestine it is associated with a cytoplasmic protein receptor, the resulting complex moving to the nucleus and resulting in the synthesis of the calcium-binding protein (CBP). The receptor from chick intestine has been purified,[452] and the receptors from chick kidney and pancreas[453] and mammalian sources[454] have also been studied. The amino-acid sequence of the porcine intestinal calcium-binding protein has been determined.[455] A CBP has also been isolated from mammalian duodenum, and has two Ca^{2+}-binding sites (mol. wt 9000–12 000).[456] It has been suggested that the CBP is not located on the basolateral membrane but possibly has an intracellular locale.[457]

The importance of proteins with γ-carboxyglutamic acid residues (Gla protein) is becoming increasingly recognized. Its biosynthesis involves a vitamin-K-dependent[458] carboxylation of certain glutamic acid residues. Several Gla proteins from bone have been studied. These show a low affinity for Ca^{2+}, a high affinity for hydroxyapatite, and an ability to inhibit the crystallization of hydroxyapatite from supersaturated solutions of calcium and phosphate. The γ carboxy-group can be removed by thermal decomposition.[459]

γ-Carboxyglutamic acid proteins have been identified in atheromatous aortae,[460] epiphyseal growth-plate cartilage,[461] and calcified turkey tendon.[462] Inhibition of formation of hydroxyapatite is also brought about by proteoglycans,[463] which are synthesized in cartilage. The Ca^{2+} ion plays an important regulatory role in this process, but does not affect aggregation.[464]

The phosphohydrolytic activity of calcifying tissues continues to attract attention.[465] The calcinogenic plant *Trisetum flavescens* causes severe calcification of soft tissue upon ingestion, but the identity of the vitamin D_3 species that is involved in this is uncertain.[466] Oestrogen stimulated the 25-hydroxylation of cholecaliferol in quail, a result that may be of physiological significance in the support

[451] A. W. Norman and H. L. Henry, *Trends Biochem. Sci.*, 1979, 14; A. Hay, *Nature (London)*, 1979, **278**, 509.

[452] J. W. Pike and M. R. Haussler, *Proc. Natl. Acad. Sci. USA*, 1979, **76**, 5485.

[453] S. Christakos and A. W. Norman, *Biochem. Biophys. Res. Commun.*, 1979, **89**, 56.

[454] D. Feldman, T. A. McCain, M. A. Hirst, T. L. Chen, and K. W. Colston, *J. Biol. Chem.*, 1979, **254**, 10 378; T. L. Chen, M. A. Hirst, and D. Feldman, *ibid.*, 1979, **254**, 7491.

[455] T. Hofmann, M. Kawakami, A. J. W. Hitchman, J. E. Harrison, and K. J. Dorrington, *Can. J. Biochem.*, 1979, **57**, 737.

[456] A. Miller, T.-H. Ueng, and F. Bronner, *FEBS Lett.*, 1979, **103**, 319.

[457] J. J. Feher and R. H. Wasserman. *Biochim. Biophys. Acta*, 1979, **585**, 599.

[458] P. A. Friedman, P. V. Hauschka, M. A. Skia, and J. K. Wallace, *Biochim. Biophys. Acta*, 1979, **583**, 261.

[459] J. W. Poser and P. A. Price, *J. Biol. Chem.*, 1979, **254**, 431.

[460] Z. Deyl, K. Macek, O. Vancikova, and M. Adam, *Biochim. Biophys. Acta*, 1979, **581**, 307.

[461] M. J. Glimcher, D. Kossiva, and A. Roufosse, *Calcif. Tissue Int.*, 1979, **27**, 187.

[462] M. J. Glimcher, D. Brickley-Parsons, and D. Kossiva, *Calcif. Tissue Int.*, 1979, **27**, 281.

[463] N. C. Blumenthal, A. S. Posner, L. D. Silverman, and L. C. Rosenberg, *Calcif. Tissue Int.*, 1979, **27**, 75.

[464] M. J. Palmoski and K. D. Brandt, *Biochem. J.*, 1979, **182**, 399.

[465] S. E. Kahn and C. Arsenis, *Biochim. Biophys. Acta*, 1979, **569**, 52; M. L. Thomas and W. K. Ramp, *Calcif. Tissue Int.*, 1979, **27**, 137; G. Grantröm, M. Jontill, and A. Linde, *ibid.*, p. 211; T. R. Anderson and S. U. Torerud, *ibid.*, p. 219.

[466] W. Rambeck, W. Oesterhelt, M. Vecchi, and H. Zucker, *Biochim. Biophys. Res. Commun.*, 1979, **87**, 743.

of calcification of egg shell.[467] The proceedings of a conference on matrix calcification have been published.[468]

8 Clotting Mechanisms of Blood

Most attention has been given to the vitamin-K-dependent blood-clotting factors for which Ca^{2+} serves as an essential cofactor. These γ-carboxyglutamic-acid-containing factors are prothrombin and Factors VII, IX, and X. Prothrombin reacts with Factor X to give thrombin in a reaction that depends upon lipid. The carboxylation of glutamyl residues in prothrombin by vitamin K is hindered by some copper complexes[469] which scavenge for superoxide. The interaction of lanthanides with the metal-binding sites of the γ-carboxyglutamyl-rich fragment 12–44 from bovine prothrombin has been studied.[470] This gives one high-affinity site and four to six sites that have lower affinity for Gd^{3+}, as shown by n.m.r. spectroscopy. It has been suggested that the low-affinity sites involve single or paired adjacent γ-carboxyglutamyl residues, while the high-affinity site involves two residues from different parts of the protein.

The binding of phospholipid to prothrombin fragment 1 (*N*-terminal 156 residues) is a function of $[Ca^{2+}]$ in the range 0.6–2.2 mmol dm^{-3} but then is independent of $[Ca^{2+}]$ at higher concentrations. The Mg^{2+} ion is some twenty times less effective in inducing this interaction. A model has been proposed in which as many as twenty Ca^{2+}-bridges link the twenty carboxylate groups of the ten γ-carboxyglutamyl residues of fragment 1 to the phosphate groups of the phospholipids.[471]

The binding of Mg^{2+} and Ca^{2+} to fragment 1 has been followed by several techniques, including ^{25}Mg n.m.r. and ^{43}Ca n.m.r.[472] Both metals cause conformational changes on binding, but only Ca^{2+} shows positive co-operativity. The metal-ion-dependent conformational change has been studied further *via* fluorescence quenching, and an equilibrium constant has been calculated. It has been suggested[473] that, in the absence of metal, there exists an equilibrium between two forms of bovine fragment 1, one of which can react rapidly with Ca^{2+} and subsequently with phospholipid. The other form cannot react with Ca^{2+} in a manner that gives a phospholipid-binding form of the protein. Interconversion of these two forms has been suggested to involve the isomerization of a proline residue.

The kinetics of activation of prothrombin by coagulation Factor X have been studied. It has been established that Ca^{2+} can bind to Factor X in the absence

[467] R. A. Nicholson M. Akhtar and T. G. Taylor, *Biochem. J.*, 1979, **182**, 745.
[468] Proc. 2nd. Conf. on Matrix Calcf., ed. H. C. Anderson and D. S. Howell, Masson, New York, 1978.
[469] M. P. Esnouf, M. R. Green, H. A. O. Hill, and S. J. Walter, *FEBS Lett.*, 1979, **107**, 140.
[470] B. C. Furie, M. Blumenstein, and B. Furie, *J. Biol. Chem.*, 1979, **254**, 12 521.
[471] F. A. Dombrose, S. N. Gitel, K. Zawalich, and C. M. Jackson, *J. Biol. Chem.*, 1979, **254**, 5027.
[472] P. Robertson, K. A. Koehler, and R. G. Hiskey, *Biochem. Biophys. Res. Commun.*, 1979, **86**, 265; H. C. Marsh, P. Robertson, M. E. Scott, K. A. Koehler, and R. G. Hiskey, *J. Biol. Chem.*, 1979, **154**, 10 268; M. E. Scott, K. A. Koehler, and R. G. Hiskey, *Biochem. J.*, 1979, **177**, 879.
[473] H. C. Marsh, M. E. Scott, R. G. Hiskey, and K. A. Koehler, *Biochem. J.*, 1979, **183**, 513.

of the other reactants.[474] The binding of Ca^{2+} to the activation products of bovine Factor IX has been reported:[475] at pH 7.4, Factor IXa possessed at least two strong Ca^{2+}-binding sites and an additional eleven weaker sites; Factor IXaα contained a similar number of sites, while Factor IXaβ contained two strong sites and seven weaker sites. The Ca^{2+}-binding properties of these proteins are therefore similar to those of their precursor, Factor XI.

Calcium has an important effect on the structure of fibrinogen, which has three tightly bound Ca^{2+} ions per molecule.[476] The mobilities of the constituent chains from fragment D of fibrinogen suggest that Ca^{2+} forms an intra-chain bridge towards the C-terminus of each γ-chain.[477] The influence of Ca^{2+} on the reaction between thrombin and fibrinogen has been attributed to the formation of a Ca^{2+}-dependent dimer, with resulting enhancement of the clotting process.[478]

[474] D. P. Kosow and C. L. Orthner, *J. Biol. Chem.*, 1979, **254**, 9448.
[475] G. W. Amphlett, R. Byrne, and F. J. Castellino, *J. Biol. Chem.*, 1979, **254**, 6333.
[476] W. Nieuwenhuizen, A. Vermond, W. J. Nooijen, and F. Haverkate, *FEBS Lett.*, 1979, **98**, 257.
[477] J. N. S. Lawrie and G. Kemp, *Biochim. Biophys. Acta*, 1979, **577**, 415.
[478] G. Marguerie, Y. Benabid, and M. Suscillon, *Biochim. Biophys. Acta*, 1979, **579**, 134.

3
Transport and Storage of Transition Metals

R. R. CRICHTON & J.-C. MARESCHAL

Introduction.—In the present Report we have included not only the transition metals but also the elements of Groups IIA and IIB which, with the exception of Cu^{II}, do not conform to the definition of transition metals but which have properties sufficiently similar to be considered with them. Since the first Report in this series,[1] there has been considerable progress in a number of areas that were reviewed therein. We have, of course, devoted a large part of our text to iron and, to a lesser extent, to copper. However, we have tried to pay particular attention to other transition metals which play an important role in the biosphere. At this point in time the fabric of our society is increasingly troubled by the real or imagined toxic effects of all sorts of chemicals, and pollution has become of as much popular interest as genetic engineering. It seems appropriate therefore not only to point to the progress that has been made in our understanding of the metabolism of iron and copper, but also to draw attention to the other transition metals which are slowly but surely manifesting their (albeit often undesired) presence in living organisms and their influence, often accompanied by pathological consequences, on the metabolism of the essential transition metals.

A number of important reviews and conference proceedings have appeared since the last Report and we have chosen to include these at the beginning of the appropriate sections. Our report has three major sections: the structure and function of proteins of iron storage and transport; the exchange of iron between storage and transport forms, as well as microbial iron metabolism; and finally, the transport and storage of copper, zinc, and other transition metals and their interactions with the metabolism of iron and of copper.

1 Structure and Function of Proteins of Iron Storage and Transport

The proteins of iron storage and transport, ferritin and transferrin, continue to be the focus of considerable activity. In this section we discuss the structure and function of these two proteins and review recent studies on iron binding and release as well as their biosynthesis and metabolism. A number of review articles have appeared since the last Report.[2-7]

[1] P. M. Harrison and A. Treffry, in 'Inorganic Biochemistry', ed. H. A. O. Hill (Specialist Periodical Reports), The Chemical Society, London, 1979, Vol. 1, p.120.
[2] H. N. Munro and M. C. Linder, *Physiol. Rev.*, 1978, **58**, 317.
[3] P. Aisen and I. Listowsky, *Annu. Rev. Biochem.*, 1980, **49**, 357.
[4] M. Worwood, *CRC Crit. Rev. Clin. Lab. Sci.*, 1979, 171.
[5] D. M. Marcus, N. Zinberg, and I. Listowsky, *Immunol. Ser., ISS Immunodiagn. Cancer*, 1979, **9**, 473.

Ferritin.—*Structure.* Progress has been made in the determination of the primary structure of apoferritin (the protein component of ferritin), of the three-dimensional structure of horse spleen apoferritin, and of the iron-containing core of ferritin, together with some clarification of the nature and origins of isoferritins (the term commonly employed in the literature to describe the microheterogeneity observed in ferritin preparations by isoelectric focussing).

The iron core. The ferritin molecule consists of a protein shell surrounding an iron core which contains a variable amount of iron (from 0 to 4500 iron atoms per molecule), mostly as ferric oxyhydroxide, but with some phosphate, this latter component being loosely bound at surface sites on the iron micelle. In the isolation of ferritin, a heat-denaturation step is routinely included (typically 10 minutes at 70—80 °C). Ferritin prepared from the same horse spleen with or without heating showed similar reactivity on reduction or chelation of the core iron.[8] Ferritin has a broad absorption band in the u.v. which tails into the visible region of the spectrum. It has been found that the visible and u.v. extinction coefficients of the iron core of ferritin increase as the iron content per ferritin molecule increases, and that the absorption spectrum undergoes a concomitant red shift.[9] The ^{57}Fe Mössbauer resonance of the core has been used to derive the micelle particle size distributions[10] and to identify large amounts of ferritin-like iron in the erythrocytes of patients with thalassaemia, sickle-cell anaemia, and haemoglobin Hammersmith.[11] The pK_a of the phosphate groups on the surface of the iron core has been estimated to be 7.0 ± 0.2, using a ^1H n.m.r. probe method.[12] Extended X-ray absorption fine-structure (EXAFS) analyses of ferritin and iron–dextran (imferon) show that the near-neighbour environment around the average iron atom in both complexes is identical within experimental error for the first three shells, and the authors concluded that iron–dextran may be a useful model compound for studies on the ferritin core.[13] Using the EXAFS technique, horse spleen ferritin was compared to an iron–glycine complex,[14] and it was concluded that, at room temperature, the iron atoms of the ferritin core are surrounded by 6.4 ± 0.6 oxygens at a distance of 1.95 ± 0.02 Å, most probably in a distorted octahedral arrangement. Each iron atom also has 7 ± 1 iron neighbours at a distance of 3.29 ± 0.05 Å. On the basis of these results (Figure 1a), it has been proposed that the ferritin core is a layered arrangement, with iron at the interstices between two close-packed layers of oxygens with approximately six-fold rotational symmetry. These compact O–Fe–O layers are only weakly bound to adjacent layers. To account for the presence of phosphate in the iron core, the authors suggest (Figure 1b) that phosphate ions terminate

[6] J. W. Halliday and L. W. Powell, *Prog. Hematol.*, 1979, **11**, 229.

[7] G. A. Clegg, J. E. Fitton, P. M. Harrison, and A. Treffry, *Prog. Biophys. Mol. Biol.*, 1980, **36**, 56.

[8] M. L. Bertrand and D. C. Harris, *Experientia*, 1979, **35**, 300.

[9] M. E. May and W. W. Fish, *Arch. Biochem. Biophys.*, 1978, **190**, 720.

[10] J. M. Williams, D. P. Danson, and Chr. Janot, *Phys. Med. Biol.*, 1978, **23**, 835.

[11] E. R. Bauminger, S. G. Cohen, S. Ofer, and E. A. Rachmilewitz, *Proc. Natl. Acad. Sci. USA*, 1979, **76**, 939.

[12] N. Imai, H. Terada, Y. Arata, and S. Fujiwara, *Bull. Chem. Soc. Jpn.*, 1978, **51**, 2538.

[13] E. C. Theil, D. E. Sayers, and M. A. Brown, *J. Biol. Chem.*, 1979, **254**, 8132.

[14] S. M. Heald, E. A. Stern, B. Bunker, E. M. Holt, and S. L. Holt, *J. Am. Chem. Soc.*, 1979, **101**, 67.

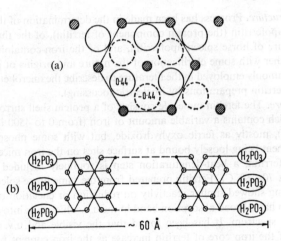

Figure 1 (a) *Idealized model structure for the ferritin core, constructed using data from EXAFS measurements. First-shell Fe—O distances are 1.95 Å and second-shell Fe–Fe distances are 3.29 Å.* ● Fe; ○ OH *or* O. (b) *One way of using the phosphorus atoms to terminate the two-dimensional sheets of Figure 1a into a strip.*
(Reproduced by permission from *J. Am. Chem. Soc.*, 1979, **101**, 67)

the sheet into a strip. The length of the strip depends on the amount of iron in the micelle (and presumably, although this is not indicated by the authors, the phosphate interacts with the protein shell) and the width of the strip 'naturally' accounts for the size of the micelle of about 70 Å. By high-resolution electron microscopy of negatively stained horse spleen ferritin, the effects of freezing and thawing, boiling and cooling, and drying and re-hydration on the disruption of the protein shell have been studied, and it has been found that small portions of the disrupted shells remain closely associated with their core crystallites.[15] In parallel with previous results,[16,17] it is concluded that there is a strong attachment of the iron cores (perhaps through the phosphate) to the protein shell.

Structural studies on apoferritins. Since the last Report, there has been considerable progress on the structure of the apoferritin molecule. The *X*-ray analysis of the horse spleen apoferritin molecule is at a resolution of 2.8 Å and the fitting of the amino-acid sequence (see below) to the electron-density map is in progress. The present status of the structure determination can be summarized as follows. The structure originally proposed by Harrison and co-workers[1,18] remains essentially the same; some features of the structure have become more clearly defined in the newer maps. The original electron-density map at 2.8 Å resolution[18] showed that each subunit consists of a bundle of four long α-helices (A–D),

15 W. H. Massover, *J. Mol. Biol.*, 1978, **123**, 721.
16 W. H. Massover and J. M. Cowley, in 'Proteins of Iron Storage and Transport in Biochemistry and Medicine' ed. R. R. Crichton, North Holland, Amsterdam, 1975, p. 237.
17 H. B. Stuhrmann, J. Haas, K. Ibel, M. H. J. Koch, and R. R. Crichton, *J. Mol. Biol.*, 1976, **100**, 399.
18 S. H. Banyard, D. K. Stammers, and P. M. Harrison, *Nature (London)*, 1978, **271**, 282.

lying parallel or antiparallel to one another, together with a shorter helix (E), situated roughly at right angles to the helices A–D. The connections between helices corresponded in the original map to regions where the electron density was weakest, and hence where it was difficult to trace the course of the polypeptide chain unambiguously. Analysis of newer maps (at the same resolution, but calculated from more extensive data sets) establishes the following points.[19] A short section of electron density that was not accounted for in the first model has been found which indicates a non-helical region of about ten residues preceding the A helix (compared with four in the first interpretation). The helices A, B, C, and D are well established and contain 27, 25, 28, and 20 amino-acid residues respectively. The helical turn (DE in the first interpretation) is considered as a helix of seven residues and the E helix contains about ten residues. The total number of helical residues (117 of a total of 174 residues) accounts for about two-thirds of the main chain density, in good agreement with c.d. and o.r.d. data.[20,21] The helix B seems to have an irregularity near the middle and the N-terminal regions of helices C and D seem to be best fitted by one or two turns of 3_{10} helix. There is good density for the sharp turn connecting C and D and for the long loop L. A new short region of inter-subunit antiparallel pleated sheet has been identified which joins the loops L for neighbouring diad-related subunits around the two-fold axis. Four hydrogen bonds are formed, involving a total of five residues in each subunit. Alternative ways of joining the defined regions which best fit the electron density have been proposed (Figure 2). In each of these, an extended N-terminal sequence of about ten amino-acid residues precedes the A helix. The C-terminal end of A may then either be connected to the N-terminus of B (as described in ref. 18) or of C. The peptide segments are joined subsequently in the order (a) A→B→L→C→D→P→E or

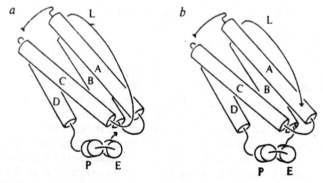

Figure 2 *Alternative conformations of the apoferritin subunit: (a) A–B–L–C–D–P–E and (b) A–C–D–P–E–B–L.*
[Reproduced by permission from *Nature (London)*, 1980, **288**, 298]

[19] G. A. Clegg, R. F. D. Stansfield, P. E. Bourne, and P. M. Harrison, *Nature (London)*, 1980, **288**, 298.
[20] I. Listowsky, J. J. Betheil, and S. Englard, *Biochemistry*, 1967, **6**, 1341.
[21] G. C. Wood and R. R. Crichton, *Biochim. Biophys. Acta*, 1971, **229**, 243.

(*b*) A→C→D→P→E→B→L. The overall chain length (in agreement with amino-acid sequence data) is a few residues longer than that described in ref. 18. The packing of the four long helices in the apoferritin subunit resembles that in a number of other proteins.[19] The positions, of the binding sites for heavy metals have been described.[22]

While the interpretation of the X-ray structure of horse spleen apoferritin has been advancing, corresponding progress has been made in the determination of the primary structure of the protein. The determination of the primary structure of apoferritin has been rendered difficult by a number of technical problems, essentially related to the intractability of the protein to fractionation procedures even after digestion, presumably due to the strong associative forces between subunits and also between subunit fragments. Without entering into details of the methods employed, which are outlined in ref. 23, the complete amino-acid sequence of horse spleen apoferritin has been established (Figure 3). In total, the protein contains 174 amino-acids, which corresponds to a molecular weight of the subunit of 19 824, and to a molecular weight for the apoferritin monomer of 476 000. When the amino-acid composition of the protein (Table 1) calculated from the sequence is compared with published values, corrected for the new sub-unit molecular weight (which is 6% greater than that determined by physical methods[24,25]) excellent agreement is observed. The secondary structure has been predicted[23] and would seem to favour the order given in Figure 2a rather than that in Figure 2b. It is interesting to note that a continuous sequence of five charged amino-acid residues occurs at a point which corresponds roughly to the middle of helix B in model 2a (residues 56–60 in the sequence) and may explain the irregularity observed in the X-ray structure of this helix (see above). The site of the proteolytic cleavage which gives rise to the fragments B and C of molecular weight 7–8000 and 11–12 000 cannot be identified with certainty, but, if we assume the order of helical regions proposed in Figure 2a, we conclude that it would have to be after residue 72 at the beginning of the loop L (which is exposed on the outer surface of the molecule). An interesting possibility would be a cleavage at the Phe-Gln peptide bond (residues 78–79), since this would give fragments of about the correct size and also generate a C-terminal fragment B with an N-terminal Gln. The tendency of Gln to cyclize would then explain why we find no free N-terminal residue in B. Another argument which favours Figure 2a rather than 2b is that, although in earlier studies it had been reported that the C-terminus was accessible to carboxypeptidase action (and thus on the outside of the protein shell), more recent work[19] suggests that the C-terminus of the protein is inaccessible to carboxypeptidases and thus, consistent with Figure 2a, is in the interior of the protein. The fitting of the sequence to the electron-density map is in progress in Sheffield, and we may expect to have a structure for horse spleen apoferritin in a not too distant future.

[22] G. A. Clegg, R. F. D. Stansfield, P. E. Bourne, and P. M. Harrison, *Biochem. Soc. Trans.*, 1980, **8**, 654.
[23] M. Heusterspreute and R. R. Crichton, *FEBS Lett.*, 1981, **129**, 322.
[24] C. F. A. Bryce and R. R. Crichton, *J. Biol. Chem.*, 1971, **246**, 4198.
[25] I. Björk and W. W. Fish, *Biochemistry*, 1971, **10**, 2844.

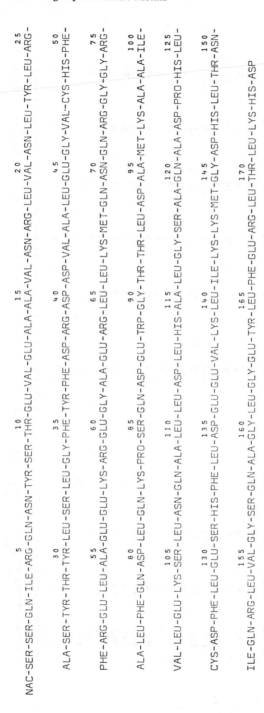

NAC-SER-SER-GLN-ILE-ARG-GLN-ASN-TYR-SER-THR-GLU-VAL-GLU-ALA-ALA-VAL-ASN-ARG-LEU-VAL-ASN-LEU-TYR-LEU-ARG-

ALA-SER-TYR-THR-TYR-LEU-SER-LEU-GLY-PHE-TYR-PHE-ASP-ARG-ASP-ASP-VAL-ALA-LEU-GLU-GLY-VAL-CYS-HIS-PHE-

PHE-ARG-GLU-LEU-ALA-GLU-GLU-LYS-ARG-GLU-GLY-ALA-GLU-ARG-LEU-LEU-LYS-MET-GLN-ASN-GLN-ARG-GLY-GLY-ARG-

ALA-LEU-PHE-GLN-ASP-LEU-GLN-LYS-PRO-SER-GLN-ASP-GLU-TRP-GLY-THR-THR-LEU-ASP-ALA-MET-LYS-ALA-ALA-ILE-

VAL-LEU-GLU-LYS-SER-LEU-ASN-GLN-ALA-LEU-LEU-ASP-LEU-HIS-ALA-LEU-GLY-SER-ALA-GLN-ALA-ASP-PRO-HIS-LEU-

CYS-ASP-PHE-LEU-GLU-SER-HIS-PHE-LEU-ASP-GLU-GLU-VAL-LYS-LEU-ILE-LYS-LYS-MET-GLY-ASP-HIS-LEU-THR-ASN-

ILE-GLN-ARG-LEU-VAL-GLY-SER-GLN-ALA-GLY-LEU-GLY-GLU-TYR-LEU-PHE-GLU-ARG-LEU-THR-LEU-LYS-HIS-ASP

Figure 3 The primary structure of horse spleen apoferritins
(Reproduced by permission from *FEBS Lett.*, 1981, **129**, 322)

Table 1 *Amino-acid compositions of human and horse ferritins*[a]

Residue	Horse					Human				
	Ho S	L	Ho L	Ho H	H	Hu S	Hu L	L	Hu H	H
Asx	18	18	19	19–20	20	21	21	18	20	22
Thr	6	6	6	8	7	7	7	8	8	7
Ser	9	9	9	12	9	8	10	10	8	12
Glx	26	25	26–27	21	24	24	26	23	26	24
Pro	2	n.d.	3	7	n.d.	4	3	n.d.	6	n.d.
Gly	11	10	11	11	12	12	11	17	12	11
Ala	15	16	15	13–14	16	15	15	14	16	13
Val	8	8	7	8	8	7	7	8	8	7
Met	3	3	3	5	3	3	3	2	2	3
Ile	4	3	3–4	5	6	4	3	6	5	6
Leu	28	28	24	21–22	21	25	25	21	25	25
Tyr	6	7	5–6	5	6	6	5	6	5	8
Phe	8	8	8	8	7	8	8	7	6	7
Trp	1	n.d.	2	n.d.	n.d.	2	2	n.d.	n.d.	n.d.
His	6	5	7	9	6	7	6	6	5	7
Lys	9	10	10	14	11	11	11	11	9	11
Arg	11	10	9	7	9	10	9	10	8	9
Cys	2	n.d.	3	n.d.	n.d.	2	2	n.d.	n.d.	n.d.

(*a*) The results are presented as numbers of residues of each amino-acid per 174 residues in the apoferritin subunit. The composition of horse spleen ferritin is based on the sequence shown in ref. 23, those of other ferritins are from ref. 26, and those of the H and L subunits are from ref. 46. Ho S refers to horse spleen, Ho L is horse liver, Ho H is horse heart, and the corresponding abbreviations for the human ferritins refer to the same sources

Structural studies on other human and horse ferritins are in progress.[26,27] As an inspection of the amino-acid compositions of Table 1 shows, liver and spleen ferritins from a given species would be expected to be extremely similar (if not identical), whereas heart ferritin shows quite significant differences. This is borne out by sequence determination. No differences have been found between the sequences of horse liver apoferritin and horse spleen apoferritin. The homology between human and horse spleen apoferritin is most striking, differences being observed in only 10% of the residues determined to date, and most of these are conservative. Recent observations by electron microscopy have established that apoferritin subunits can be clearly resolved[28] and indicate that apoferritin molecules can dissociate their subunits upon dilution,[29] whereas earlier results, using high dilution, did not detect any free subunits.[30,31] It may turn out that there is a dynamic equilibrium between subunits and the apoferritin monomer, which enormously favours the intact apoferritin shell. A correlation has been established between iron content of human spleen ferritin and the amounts of dimeric, trimeric, and heavier ferritin oligomers.[32] The binding of hydrophobic compounds to apoferritin subunits induces the polymerization of the subunits; in contrast, the compounds used (8-anilinonaphthalene-1-sulphonic acid and 1-benzylindazole-3-oxyacetate) did not bind to the native protein.[33] Procedures for removing Cd from crystallized horse spleen ferritin have been described[34] and a quantitative method for determination of tissue ferritin by electroimmunoassay, sensitive to 0.1 μg of ferritin protein, has been developed.[35] Radial immunodiffusion, electroimmunoassay, and an enzyme-linked immuno-assay have been compared for liver ferritin estimations.[36]

Microheterogeneity of ferritins. The apparent microheterogeneity of ferritin as revealed by isoelectric focusing and the contradictory hypotheses either that (i) ferritins are distinct homopolymers, and specific to any given tissue of a given species,[37] or that (ii) ferritins are heteropolymers formed from two subunits of different molecular weight and different charge (called L and H on account of their distribution in liver and heart respectively), and that each tissue has a characteristic distribution of these two subunits,[38] has again generated a large volume of literature.

It has become apparent that for human and horse ferritins a number of correlations can be established between the iron content of the ferritin, the pI, their content

[26] R. R. Crichton, J. M. Mathijs, C. G. M. Magnusson, M. Heusterspreute, C. Wustefeld, and C. F. A. Bryce, *Protides Biol. Fluids, Proc. Colloq.*, 1979, **27**, 71.

[27] M. Heusterspreute, C. Wustefeld, J. M. Mathijs, J. C. Mareschal, G. Charlier, and R. R. Crichton, *Protides Biol. Fluids, Proc. Colloq.*, 1980, **28**, 91.

[28] W. M. Massover, *Proc. 9th Internat. Congr. Electron Micr.*, 1978, **2**, 182.

[29] W. M. Massover, *Biochem. Biophys. Res. Commun.*, 1980, **96**, 1427.

[30] R. Jaenicke and P. Bartmann, *Biochem. Biophys. Res. Commun.*, 1972, **49**, 884.

[31] R. R. Crichton, R. Eason, A. Barclay, and C. F. A. Bryce, *Biochem. J.*, 1973, **131**, 855.

[32] A. Samarel and M. M. Bern, *Lab. Invest.*, 1978, **39**, 10.

[33] S. Stefanini, A. Finazzi-Agro, E. Chiancone, and E. Antonini, *FEBS Lett.*, 1979, **100**, 296.

[34] J. Hegenauer, P. Saltman, and L. Hatlen, *Biochem. J.*, 1979, **177**, 693.

[35] N. Carmel and A. M. Konijn, *Anal. Biochem.*, 1978, **85**, 499.

[36] F. M. J. Zuyderhoudt, F. L. A. Willekens, and W. O. Schreuder, *Anal. Biochem.*, 1979, **98**, 204.

[37] R. R. Crichton, J. A. Miller, R. L. C. Cumming, and C. F. A. Bryce, *Biochem. J.*, 1973, **131**, 51.

[38] J. W. Drysdale, *Ciba Symp.*, 1977, **51**, 41.

of H and L subunits, the immunological reactivity (for human ferritins only), and the capacity to accumulate iron.[39-42] However, whereas for human liver and spleen isoferritins the highest iron content is found in the most acidic ferritins,[39,42] for horse liver and spleen ferritins the most basic proteins have the highest iron content.[40] For human liver and spleen ferritins, an increased content of the H subunit was also found in acidic ferritins.[41,42] The reactivity of human liver ferritin with antibody against spleen ferritin increased with increasing p*I*, but the inverse situation was found with antibodies against heart ferritin.[39] Caution has been advised in the interpretation of isoelectric focusing profiles and of SDS gel electrophoresis patterns. It has been suggested, after examination of the isoelectric focusing patterns of several ferritins and albumins, that the discrete bands need not be attributed to discrete species of heteropolymer, but may result from discrete steps in the ampholine pH gradients.[40,43,44] A novel interpretation of the SDS gel electrophoresis patterns of ferritins and apoferritins has been proposed which implies that the H band, of mol. wt 22 000, could be derived from dimerization of the B fragment that is produced by proteolysis of the protein in the course of isolation.[45]

Further evidence in support of the heterosubunit model[38] has been presented. Thus for rat, horse, and human ferritins, subunit analyses indicate that ferritins from all three species are composed of only two types of subunit, namely H and L, with molecular weights of 21 000 and 19 000 respectively.[46] The relative amounts of the two subunits vary throughout the isoferritin spectrum. Amino-acid analyses and tryptic peptide maps suggest extensive sequence homologies between the H and L subunits. The amino-acid compositions of the H and L subunits for horse and human ferritins are given for comparison in Table 1. The amino-acid composition of the L subunit of horse (corrected for a molecular weight of 19 700) is in excellent agreement with that found by sequence analysis. Some of the discrepancies between the horse H subunit and the horse heart apoferritin amino-acid composition could be explained by the presence of H and L subunits in roughly equal amounts in the heart preparation (for example Thr, Glu, and Arg). The low values for Ser and Met may be due to destruction during hydrolysis. The situation is less clear for the human ferritins, where, despite striking similarities between the L subunit and the liver and spleen ferritin compositions, there are a number of large differences (the Gly, Ile, Asp, and Leu values). Again, some of the differences between the H subunit and the heart ferritin are compatible with the presence of both L and H subunits in the preparation. Hopefully, as knowledge of the primary structure of horse and human ferritins progresses, and as methods for obtaining the H subunit in sufficient amounts to permit sequence analysis develop, the controversy regarding the microheterogeneity of ferritin will find an appropriate molecular explanation.

[39] M. Wagstaff, M. Worwood, and A. Jacobs, *Biochem. J.*, 1978, **173**, 969.
[40] S. M. Russell and P. M. Harrison, *Biochem. J.*, 1978, **175**, 91.
[41] D. J. Lavoie, K. Ishikawa, and I. Listowsky, *Biochemistry*, 1978, **17**, 5448.
[42] A. Bomford, M. Berger, Y. Lis, and R. Williams, *Biochem. Biophys. Res. Commun.*, 1978, **83**, 334.
[43] S. M. Russell, P. M. Harrison, and S. Shinjo, *Br. J. Haematol*,, 1978, **38**, 296.
[44] S. Shinjo and P. M. Harrison, *FEBS Lett.*, 1979, **105**, 353.
[45] C. F. A. Bryce, C. G. M. Magnusson, and R. R. Crichton, *FEBS Lett.*, 1978, **96**, 257.
[46] P. Arosio, T. G. Adelman, and J. W. Drysdale, *J. Biol. Chem.*, 1978, **253**, 4451.

The separation of H and L subunits in the work described above[46] was carried out by preparative SDS-gel electrophoresis. More recently, a gel chromatography procedure in acid medium[47] and a preparative electrochromatography method[48] have been described for the isolation of these two subunits. The formation of intra-species hybrid apoferritins from horse heart and horse spleen ferritin subunits and of inter-species hybrids from horse spleen–human liver subunits and from rat liver–horse spleen subunits has been described.[49]

The isolation and characterization of human placental ferritin has confirmed that in its isoelectric focusing profile, its subunit composition, its amino-acid composition, and its immunological behaviour it is intermediate between human liver or spleen ferritin and human heart ferritin.[50-52] The significance and biochemical relevance of ferritin in malignant deseases have been reviewed[53-55] and a comparison of normal human liver and hepatocellular carcinoma ferritins in terms of their iron contents, gel migration rates, and isoelectric focusing profiles has been established.[56] Isoferritins from human tissues have been separated after ^{59}Fe-tagging of the protein; the preparative procedure involves affinity chromatography and hydrophobic chromatography.[57]

Serum ferritin continues to attract much attention in view of its clinical significance.[4] Some recent studies on the binding of serum ferritin to concanavalin A[58] and an analysis of the diminution in the microheterogeneity that is observed after treatment of serum ferritin with neuraminidase[59] confirm that serum ferritin is glycosylated. A radioimmunoassay for acidic ferritin purified from HeLa cells showed that the HeLa/spleen-type ferritin ratio was always low in sera from patients with malignant disease, and it was concluded that there was little application for antibodies to HeLa or heart ferritin in the diagnosis or monitoring of cancer.[60] The characterization of red-cell and serum ferritins in haematological disorders[61] and the analysis of ferritin from acute monocytic leukaemia cells[62] have been reported.

Purification and microheterogeneities have also been reported for rat and mouse ferritins[63-65] and, in parallel with earlier studies on ferritin in yolk-sac-

[47] P. Arosio, *FEBS Lett.*, 1979, **104**, 51.
[48] S. Otsuka and I. Listowsky, *Anal. Biochem.*, 1980, **102**, 419.
[49] S. Otsuka, I. Listowsky, Y. Niitsu, and I. Urushizaki, *J. Biol. Chem.*, 1980, **255**, 6234.
[50] D. J. Lavoie, D. M. Marcus, S. Otsuka, and I. Listowsky, *Biochim. Biophys. Acta*, 1979, **579**, 359.
[51] P. J. Brown, P. M. Johnson, A. O. Ogbimi, and J. A. Tappin, *Biochem. J.*, 1979, **182**, 763.
[52] J.-C. Mareschal, B. Dublet, C. Wustefeld, G. Charlier, and R. R. Crichton, *Clin. Chim. Acta*, 1981, **111**, 99.
[53] J. W. Drysdale, 'Carcino Embyonic Proteins', ed. F. G. Lehmann, Elsevier/North-Holland, Amsterdam, 1979, Vol. 1, p. 249.
[54] J. W. Drysdale and E. Alpert, *Scand. J. Immunol.*, 1978, **8**, 65.
[55] B. M. Jones, A. Jacobs, and M. Worwood, *Protides Biol. Fluids, Proc. Colloq.*, 1979, **27**, 247.
[56] S. Bullock, A. Bomford, and R. Williams, *Biochem. J.*, 1980, **185**, 639.
[57] M. Page and L. Theriault, *Chromatogr. Symp. Ser.*, 1979, **1**, 273.
[58] M. Worwood, S. J. Cragg, M. Wagstaff, and A. Jacobs, *Clin. Sci.*, 1979, **56**, 83.
[59] S. J. Cragg, M. Wagstaff, and M. Worwood, *Clin. Sci.*, 1980, **58**, 259.
[60] B. M. Jones, M. Worwood, and A. Jacobs, *Clin. Chim. Acta*, 1980, **106**, 203.
[61] H. Yamada, *Acta Haematol. Jpn.*, 1978, **41**, 1334.
[62] Y. Yoda and T. Abe, *Cancer*, 1980, **46**, 289.
[63] Y. Eguchi, *J. Nippon Med. Sch.*, 1980, **47**, 302.
[64] W. H. Massover, *Biochim. Biophys. Acta*, 1978, **532**, 202.
[65] W. H. Massover, *Biochim. Biophys. Acta*, 1979, **579**, 169.

derived red blood cells of embryonic mice,[66] ferritin has been identified as the main yolk component in the oocytes of the pulmonate snail *Planorbis corneus* L.[67] Perhaps most surprisingly, ferritin appears finally to have been established as being present in bacteria, albeit associated with a haem-containing cytochrome. Thus the *b*-type cytochrome from the nitrogen-fixing bacterium *Azotobacter vinelandii*, originally described as containing large amounts of non-haem iron, has been characterized as a bacterioferritin.[68] The molecule has an iron content of 13–20% by weight, a subunit molecular weight of 17 000, a molecular diameter comparable to that of animal ferritins, and an electron-dense core. It contains one protoporphyrin IX per two subunits. The authors speculate that the bacterioferritin acts as an electron store or as a specific iron store for nitrogenase. Confirmation of these results has recently been provided by a Chinese group.[69] An iron-storage compound of molecular weight 300 000, containing electron-dense particles, in *E. coli* was characterized by Mössbauer spectroscopy; the Mössbauer spectra of the isolated iron-storage protein are different from that of ferritin.[70] Similar Mössbauer spectra in cells grown in ^{57}Fe-enriched media confirm the presence of an iron-storage protein in *Proteus mirabilis* and in *Mycoplasma capricolum*.[71]

Function, Metabolism, and Biosynthesis of Ferritin. Since the last Report, four reviews have appeared on deposition and mobilization of iron in ferritin.[72–75] In this section we discuss recent studies on the incorporation of iron and of phosphate into ferritin, on the release of iron from ferritin, and on the metabolism of ferritin, including its biosynthesis. We emphasize at the outset that the function of ferritin is to store iron in a form which is stable, soluble, and non-toxic under normal physiological conditions (that is, predominantly as a hydrolysed ferric oxy-hydroxide), from which iron can be made available as and when it is required within the cell (as we will see in a later section, we might also add 'where it is required'), and which can accommodate increases in availability of iron, both by an inherent capacity for iron storage and by iron-mediated synthesis of the apoprotein component. An understanding of the mechanisms of deposition and release of iron *in vitro* should help in our interpretation of studies of the mobilization of storage iron in intact cells, described in a later section of this review.

Deposition of iron and of phosphate in ferritin. The formation of ferritin from apoferritin proceeds best with Fe^{II} in the presence of a suitable oxidizing agent.

[66] E. C. Theil, *Br. J. Haematol.*, 1976, **33**, 437.
[67] W. Bottke and I. Sinha, *Wilhelm Roux's Arch. Dev. Biol.*, 1979, **186**, 71.
[68] E. I. Steifel and G. D. Watt, *Nature (London)*, 1979, **279**, 81.
[69] L. Jiudi, W. Jiwen, Z. Zepu, T. Ya, and D. Bei, *Sci. Sin.*, 1980, **23**, 897.
[70] E. R. Bauminger, S. G. Cohen, D. P. E. Dickson, A. Levy, S. Ofer, and J. Yariv, *Biochim. Biophys. Acta*, 1980, **623**, 237.
[71] E. R. Bauminger, S. G. Cohen, F. Labenski de Kanter, A. Levy, S. Ofer, and S. Rottem, *J. Phys. (Paris)*, 1980, **41**, C1-491.
[72] R. R. Crichton, in 'Transport by Proteins', ed. G. Blauer and H. Sund, de Gruyter, Berlin and New York, 1978, p. 243.
[73] P. M. Harrison, S. H. Banyard, G. A. Clegg, D. K. Stammers, and A. Treffry, in 'Transport by Proteins', ed. G. Blauer and H. Sund, de Gruyter, Berlin and New York, 1978, p. 259.
[74] R. R. Crichton, *Ciba Symp.*, 1979, **65**, 57.
[75] R. R. Crichton, F. Roman, F. Roland, E. P. Pâques, A. Pâques, and E. Vandamme, *J. Mol. Catal.*, 1980, **7**, 267.

With molecular dioxygen as oxidant, it was found that four Fe^{2+} are oxidized per oxygen molecule[76,77] and that catalase does not affect the time course nor the stoicheiometry of the reaction. Variable stoicheiometry of oxidation of Fe^{II} in the presence or absence of apoferritin has been observed, dependent on buffer, pH, and Fe concentration, and the results have been interpreted as favouring the 'crystal growth' model of ferritin formation.[78] The oxidant specificity has been examined, and it was concluded that while oxidation is more rapid with O_2 at the initial stages of oxidation of Fe^{II}, oxygen or KIO_3 is equally effective when some iron is already present in the ferritin molecule.[79] The kinetics of iron deposition in apoferritin have been studied in a number of buffers, both iron-complexing and non-complexing.[80,81] The conclusions of these studies are that, at high pH, deposition of ferritin iron is a second-order reaction, corresponding to the fixation of two iron atoms per catalytic site, as proposed in the mechanism of ref. 82, whereas at lower pH the rate-limiting step seems to be the deprotonation of a group (or groups) of $pK \approx 6$. Imidazole, especially at pH values of 7 and above, is a potent inhibitor of deposition of iron. A kinetic model was developed which accounts well for ferritin formation at pH values below 6.5 and above 7.0 in non-complexing buffers, but which does not account for the kinetics observed at intermediate pH values. Inhibition of iron deposition by a number of other metal ions, such as Co^{2+}, Zn^{2+}, Tb^{2+}, VO^{2+}, and Ni^{2+}, is consistent with a competitive mode of inhibition; the stoicheiometry of metal binding has been established.[83] The presence of an iron core has been found to reduce the inhibition of uptake of Fe^{II} by Zn^{II}, to protect the sites of iron uptake from chemical modification, and to diminish the effect of modification on the rate of uptake of iron.[84]

The uptake of Fe^{III} chelates by ferritin has been studied, and it was concluded that some Fe^{III} can be taken up *in vitro*,[85] notably from ferric citrate and ferric nitrilotriacetate. Similar results have been found with the ferric–pyrophosphate complex,[86] although the amounts of iron that accumulated in ferritin were somewhat less. The incorporation and release of inorganic phosphate from ferritin has been examined,[87] and it was concluded that much of the P_i is adsorbed on the surface of the iron core and that P_i is not present at the intracellular site of iron incorporation, but is added after incorporation of iron. Correlations have been established between the surface charge, the iron content, and the rate of iron uptake of ferritin.[39,40,88]

[76] G. Melino, S. Stefanini, E. Chiancone, E. Antonini, and A. Finazzi-Agro, *FEBS Lett.*, 1978, **86**, 136.
[77] M. Wauters, A. M. Michelson, and R. R. Crichton, *FEBS Lett.*, 1978, **91**, 276.
[78] A. Treffry, J. M. Sowerby, and P. M. Harrison, *FEBS Lett.*, 1978, **95**, 221.
[79] A. Treffry, J. M. Sowerby, and P. M. Harrison, *FEBS Lett.*, 1979, **100**, 33.
[80] E. P. Pâques, A. Pâques, and R. R. Crichton, *J. Mol. Catal.*, 1979, **5**, 363.
[81] E. P. Pâques, A. Pâques, and R. R. Crichton, *Eur. J. Biochem.*, 1980, **107**, 447.
[82] R. R. Crichton and F. Roman, *J. Mol. Catal.*, 1978, **4**, 75.
[83] E. Vandamme and R. R. Crichton, *Arch. Int. Physiol. Biochim.*, 1979, **88**, B108.
[84] A. Treffry and P. M. Harrison, *Biochem. Soc. Trans.*, 1981, **8**, 655.
[85] A. Treffry and P. M. Harrison, *Biochem. J.*, 1979, **181**, 709.
[86] K. Konopka, J.-C. Mareschal, and R. R. Crichton, *Biochem. Biophys. Res. Commun.*, 1980, **96**, 1408.
[87] A. Treffry and P. M. Harrison, *Biochem. J.*, 1978, **171**, 313.
[88] A. Treffry and P. M. Harrison, *Biochem. Soc. Trans.*, 1980, **8**, 656.

Mobilization of iron from ferritin. The kinetics of reduction of the iron cores of ferritin, of the isolated cores after removal of the protein shell, and of a low-molecular-weight hydroxy-iron polymer have been studied.[89] It was concluded that the rate of reduction is determined both by the presence of the protein shell and the surface area of the crystallite, and that a fragmentation of the core crystallite occurs during the first stages of reduction of ferritin and isolated cores.

An extensive study of the release of ferritin iron by dihydroflavins, using the NAD(P)H:flavin oxidoreductase from *Beneckea harveyi* to maintain a constant concentration of dihydroflavin, has been reported.[90] The advantages of this system (performed under strictly anaerobic conditions) when compared with other studies using chemically reduced flavins[91] or dihydroflavin that is produced by non-enzymic reduction of FMN by NAD(P)H[74] are to eliminate kinetic artefacts due to oxidation of dihydroflavin by oxygen, to maintain a constant concentration of dihydroflavin throughout the assay, and to permit observation of the kinetics of complete release of iron. Reproducible sigmoidal kinetics were observed with an initial slow phase of 2 min; this was attributed to hindered access of the reductant to the ferric core by the protein shell, which is particularly apparent during initial release of iron. An extended mathematical model has been developed in an attempt to account for this hypothesis. The stoicheiometry of iron release is, as expected, two equivalents of Fe^{III} per equivalent of dihydroflavin. On the basis of experiments with Sepharose-linked dihydroflavins, it was concluded that the flavin must penetrate the protein shell prior to reduction of iron. No evidence was found for any specific flavin-binding site, although the dihydroflavins display saturation kinetics with very high K_m values (60 μmol l^{-1} for $FMNH_2$). The rates of release of iron by a number of analogues show that electron transfer is rate-determining for iron release by dihydroflavin whereas diffusion of the reduced flavin through the protein shell is slow in the release of iron by $FMNH_2$.

The release of ferritin iron by the iron chelators desferrioxamine B, rhodotorulic acid, 2,3-dihydroxybenzoate, 2,2′-bipyridyl, and pyridine-2-aldehyde-2-pyridylhydrazone (paphy) has been studied.[92,93] Ferritin that has been prepared by thermal denaturation releases its iron less effectively than ferritin prepared by a modified procedure that avoids this step. Desferrioxamine B and rhodotorulic acid are the most effective in releasing iron from both preparations. When FMN is added, in the presence of light, iron release by desferrioxamine B, rhodotorulic acid, and 2,3-dihydroxybenzoate was effectively blocked, whereas both bipyridyl and paphy showed a marked stimulation. A substantial increase in iron release was also observed for bipyridyl and paphy with ascorbate; a less important increase was observed for rhodotorulic acid. It is suggested that dihydroflavin can react with the peroxo-bridge of the catalytic site to generate a flavin hydroperoxide and reduced iron. If the chelator used is too bulky to have access to the Fe^{II}, it will not be able to remove the iron. In the case of bipyridyl and paphy the chelators can not only complex the Fe^{II} but may be oxidized by the flavin hydroperoxide to re-initiate a subsequent catalytic cycle.

[89] S. Stefanini, E. Chiancone, and E. Antonini, *Biochim. Biophys. Acta*, 1978, **542**, 170.
[90] T. Jones, R. Spencer, and C. Walsh, *Biochemistry*, 1978, **17**, 4011.
[91] S. Sirivech, E. Frieden, and S. Osaki, *Biochem. J.*, 1974, **143**, 311.
[92] R. R. Crichton, F. Roman, and F. Roland, *FEBS Lett.*, 1980, **110**, 271.
[93] R. R. Crichton, F. Roman, and F. Roland, *J. Inorg. Biochem.*, 1980, **13**, 305.

Some properties of a partially purified ferritin reductase from rat liver have been reported.[94]

Metabolic effects of ferritin, and its biosynthesis. The presence of an α-2H isoferritin in the sera of tumour-bearing patients, together with data which establish a hyperferritinaemia in cancer patients that is associated with variations of this protein level during cancer and pregnancy, suggested an immuno-suppressive activity of this ferritin. This activity has been confirmed for the α-2H isoferritin in mice,[95] and a suppressive effect of human spleen ferritin on lymphocyte function *in vitro* has been reported.[96] A study of ferritin- and apoferritin-induced complex glomerulonephritis in mice has appeared.[97] The growth of *Neisseria meningitidis* is inhibited by reduced ferritin as well as by other iron chelates.[98]

The effect of maturation on ferritin synthesis by erythroid precursor cells has been studied. In the course of cellular maturation, a progressive reduction in ferritin synthesis is observed (20 times as much ferritin is synthesized in undifferentiated cells than in reticulocytes). The accumulation of newly formed apoferritin is followed by an increase in cellular iron uptake and increased incorporation of iron in ferritin. With continuing cellular maturation, ferritin iron and transferrin-derived iron are utilized for haemoglobin synthesis.[99,100] The circulating red blood cells formed in bullfrog larvae and chick and mouse embryos also contain large amounts of ferritin. The synthesis of ferritin is coordinated with differentiation, and the special role of red-cell ferritin in developing animals has been extensively analysed.[101-103] It was suggested that the presence of a pool of untranslated ferritin mRNA in liver cell sap[104] plays a key role in the regulation of ferritin synthesis.[105] A study of the distribution of specific mRNAs for ferritin and albumin between microsomes and cell sap has shown that, unlike ferritin, there is little free albumin mRNA in cell sap, and it remains unchanged by induction of albumin synthesis in nephrosis.[106] This tends to confirm that the pool of cytoplasmic ferritin mRNA, which is transferred to the microsomes when iron is administered, functions as a reserve. In terms of the site of action of iron in the stimulation of ferritin biosynthesis, recent results suggest[107] that, although the effect is post-transcriptional, it is unlikely that ribosomes bind iron. Ferritin synthesis in human T lymphocytes has been studied, and it was concluded that such cells may contribute to the synthesis and secretion

[94] Z. Zaman and R. L. Verwilghen, *Biochem. Soc. Trans.*, 1979, **7**, 201.
[95] D. Buffe, C. Rimbaut, D. Erard, and G. P. Tilz, *Scand. J. Immunol.*, 1978, **8**, 633.
[96] Y. Matzner, C. Hershko, A. Polliack, A. M. Konijn, and G. Izak, *Br. J. Haematol.*, 1979, **42**, 345.
[97] G. L. Hagstrom, P. M. Bloom, M. N. Yum, K. J. Lavelle, and F. C. Luft, *Nephron*, 1979, **24**, 127.
[98] G. A. Calver, C. P. Kenny, and D. J. Kushner, *Infect. Immun.*, 1979, **25**, 880.
[99] A. M. Konijn, C. Hershko, and G. Izak, *Isr. J. Med. Sci.*, 1978, **14**, 1181.
[100] A. M. Konijn, C. Hershko, and G. Izak, *Am. J. Hematol.*, 1979, **6**, 373.
[101] E. C. Theil, *J. Biol. Chem.*, 1978, **253**, 2902.
[102] J. E. Brown and E. C. Theil, *J. Biol. Chem.*, 1978, **253**, 2673.
[103] E. C. Theil and G. M. Tosky, *Dev. Biol.*, 1979, **69**, 666.
[104] J. Zähringer, B. S. Baliga, and H. N. Munro, *Biochem. Biophys. Res. Commun.*, 1976, **68**, 1088.
[105] J. Zähringer, B. S. Baliga, and H. N. Munro, *Proc. Natl. Acad. Sci. USA*, 1976, **73**, 857
[106] J. Zähringer, B. S. Baliga, and H. N. Munro, *FEBS Lett.*, 1979, **108**, 317.
[107] R. Fagard, H. Van Tan, and R. Saddi, *Biochimie*, 1978, **60**, 517.

of ferritin molecules with a high content of H subunits.[108] Administration of iron induces preferential synthesis of L-rich isoferritins in rat liver, and a faster degradation of H-rich isoferritins also contributes to the increase in L-rich forms. Turnover rates of isoferritins are markedly different, and vary with subunit composition.[109] This conclusion has been challenged,[110] and it has been suggested that isoferritins of low pI become more basic by post-translational modification.

Transferrin.—*Structure*. Work on the structure of transferrins continues to advance, with particular interest being directed to comparative studies. Spectroscopic investigations of both the metal- and anion-binding sites, as well as a detailed analysis of the thermodynamics of binding of iron to the two sites, have been reported, and the method of Makey and Seal[111] for separation of the two mono-ferritransferrins as well as the diferri- and apo-forms has been applied to analyse the distribution of iron between the two iron-binding sites *in vivo*. Transferrin structure and function and the delivery of iron to cells have been reviewed.[3]

Primary and tertiary structure of transferrins. Sequence studies of human serum transferrin[112] have shown that the *N*- and *C*-terminal halves show a high degree of homology (40%), indicating that the protein is probably derived from an ancestral precursor: since each of the two homologous domains is associated with a single metal-binding site, and (to date) no transferrin has been isolated which contains only one metal-binding site, it seems reasonable to assume that the presence of two iron sites confers an evolutionary advantage on the transferrin molecule. A single iron-binding fragment, of $M_r = 35\,600$, has been prepared from human transferrin by proteolysis with thermolysin.[113] The fragment is derived from the *N*-terminal domain of the molecule (it contains the CNBr fragments corresponding to those present in this region, and is devoid of carbohydrate). The site of cleavage between the *N*- and *C*-terminal domains was identified as a Pro-Val bond (residues 345 and 346 in the sequence). From studies of the stability of the ferric complex of the fragment, it was suggested that it may represent the acid-stable iron-binding site of human transferrin. Initial studies indicate that the fragment does not bind to transferrin receptors and thus does not contain the receptor-binding site of transferrin. The sequence and the cystein bridges of the *N*-terminal CNBr peptide of human transferrin have been determined.[114] Extensive sequence homologies have been found between human lactoferrin and serum transferrin, and the existence of internal homology in

[108] M. H. Dörner, A. Silverstone, K. Nishiya, A. de Sostoa, G. Munn, and M. de Sousa, *Science*, 1980, **209**, 1019.
[109] Y. Kohgo, M. Yokota, and J. W. Drysdale, *J. Biol. Chem.*, 1980, **255**, 5195.
[110] A. Treffry and P. M. Harrison, *Biochim. Biophys. Acta*, 1980, **610**, 421.
[111] D. G. Makey and U. S. Seal, *Biochim. Biophys. Acta*, 1976, **453**, 250.
[112] R. T. A. McGillivray, E. Mendez, and K. Brew, in 'Proteins of Iron Metabolism', ed. E. B. Brown, P. Aisen, J. Fielding, and R. R. Crichton, Grune and Stratton, New York, 1977, p. 133.
[113] J. Lineback-Zins and K. Brew, *J. Biol. Chem.*, 1980, **255**, 708.
[114] J. E. L. Brune, S. R. Martin, B. S. Boyd, R. M. Palmour, and H. E. Sutton, *Tex. Rep. Biol. Med.*, 1978, **36**, 47.

human lactoferrin was established.[115] The amino-acid sequences around the two polysaccharide-attachment sites of human lactoferrin have been determined.[116] The location of the two prosthetic sugar groups was quite different in serum transferrin and lactoferrin, and no sequence homology could be established between the glycopeptides of the two human transferrins. A close relationship in the amino-acid sequence of one of the glycopeptides from human lactoferrin and a glycopeptide from hen ovotransferrin was found.

The X-ray structure determination of diferric rabbit serum transferrin has advanced to the point where an electron-density map has been calculated at 6 Å resolution.[117] The molecule consists of two lobes of roughly equal size, inclined at about 30° to one another, which may correspond to the two domains, each containing an iron-binding site, that are referred to above. Preliminary X-ray studies of crystals of human transferrin have been published.[118] Density measurements indicate four transferrin molecules per unit cell. For rabbit and human transferrins, the diffraction data extend to a resolution of 3.0—2.6 Å.

The carbohydrate component. The structure of the carbohydrate components of human serum transferrin, each of which is attached through a 4-N-(2-acetamido-2-deoxy-β-D-glucopyranosyl)-L-asparagine link, was established by chemical and enzymatic methods, and has been confirmed by high-resolution n.m.r. spectroscopy and mass spectrometry.[119-121] The two carbohydrate chains, which have identical structures, are located at residues 415 and 608, in the C-terminal domain of the protein,[112] A theoretical study suggests that they may take up one of two possible conformations, either in a Y shape or a T shape.[122] The structure of the carbohydrate chain of rabbit serum transferrin is identical to that proposed for human transferrin,[123,124] but whereas the Lille group report only one glycan per polypeptide chain,[123] the Kansas group report two glycans per molecule.[124,125] The two glycopeptides from chicken ovotransferrin differ in amino-acid composition but contain the same carbohydrate moiety, which is quite different from the human and rabbit glycans in that the glycan chain ends in N-acetylglucosamine residues and is devoid of galactose and sialic acid.[126] It was concluded that there is only one oligosaccharide per molecule. The structure of the

[115] M.-H. Metz-Boutigue, J. Jollès, J. Mazurier, G. Spik, J. Montreuil, and P. Jollès, *Biochimie*, 1978, **60**, 557.

[116] M.-H. Metz-Boutigue, J. Jollès, P. Jollès, J. Mazurier, G. Spik, and J. Montreuil, *Biochim. Biophys. Acta*, 1980, **622**, 308.

[117] B. Gorinsky, C. Horsburgh, P. F. Lindley, D. S. Moss, M. Parkar, and J. L. Watson, *Nature (London)*, 1979, **281**, 157.

[118] L. J. DeLucas, F. L. Suddath, R. A. Gams, and C. E. Bugg, *J. Mol. Biol.*, 1978, **123**, 285.

[119] G. Spik, B. Bayard, B. Fournet, G. Strecker, S. Bouquelet, and J. Montreuil, *FEBS Lett.*, 1975, **50**, 296.

[120] L. Dorland, J. Haverkamp, B. L. Schut, J. F. G. Vliegenthart, G. Spik, G. Strecker, B. Fournet, and J. Montreuil, *FEBS Lett.*, 1977, **77**, 15.

[121] K.-A. Karlsson, I. Pascher, B. E. Samuelsson, J. Finne, T. Krusius, and H. Rauvala, *FEBS Lett.*, 1978, **94**, 413.

[122] J. Montreuil, B. Fournet, G. Spik, and G. Strecker, *C.R. Hebd. Seances Acad. Sci., Ser. D*, 1978, **287**, 837.

[123] D. Leger, V. Tordera, G. Spik, L. Dorland, J. Haverkamp, and J. F. G. Vliegenthart, *FEBS Lett.*, 1978, **93**, 255.

[124] D. K. Strickland, J. W. Hamilton, and B. G. Hudson, *Biochemistry*, 1979, **18**, 2549.

[125] D. K. Strickland and B. G. Hudson, *Biochemistry*, 1978, **17**, 3411.

[126] L. Dorland, J. Haverkamp, J. F. G. Vliegenthart, G. Spik, B. Fournet, and J. Montreuil, *Eur. J. Biochem.*, 1979, **100**, 569.

glycan group of human lactoferrin has also been determined;[127] it lacks one of the terminal sialic acid residues and has instead a mannose residue attached to the penultimate N-acetylglycosamine residue on this antenna of the still conserved 'biantennary' structure. The structure of bovine serum transferrin glycans has also been studied,[128,129] and is characterized by the absence of sialic acid and a high mannose content. It has been proposed that human transferrin can have a bi- or a tri-antennary structure and that greater structural heterogeneity exists in the glycans of human transferrin than was previously thought.[130]

Binding of iron to transferrin, and spectroscopic studies. The thermodynamic (or stoicheiometric) equilibrium constants for binding of Fe^{III} to transferrin have been determined; at pH 7.4 and atmospheric $p(CO_2)$ the apparent stability constants K_1' and K_2' are 4.7×10^{20} and 2.4×10^{19} l mol^{-1}, respectively.[131] Using the urea gel electrophoresis method,[111] the relative concentrations of the two species of monoferric transferrin were measured, and thus the four stoicheiometric site constants for binding of iron to transferrin were estimated. Essentially sequential binding to the sites in the order A site (the C-terminal domain) and then B site (N-terminal domain) is observed at pH 6.7, and even at pH 7.4; the binding to the A site is six times stronger than to the B site. A weakly negative site–site interaction for iron binding is observed which is more pronounced at pH 6.7 than at physiological pH. As a result of discussions at the 4th International Conference on Proteins of Iron Metabolism, it has been proposed that the four forms of human transferrin be identified as (i) Tf for the iron-free apo-transferrin, (ii) Fe_NTf for the monoferric transferrin with Fe^{III} bound in the N-terminal domain (previously described as the B site), (iii) $TfFe_C$ for the mono-ferric transferrin with Fe^{III} bound in the C-terminal domain (previously the A site), and (iv) Fe_2Tf (Fe_NTfFe_C) for the diferric transferrin with Fe^{III} on both sites.[132]

By combining electrophoresis in urea–polyacrylamide gels with immuno-electrophoresis using specific anti(human transferrin) antiserum, the concentration of each of these four forms in human serum has been estimated.[133] It was found that the more weakly binding and acid-labile N-terminal site (Fe_NTf) is predominantly occupied in normal human serum (ratios of Fe_NTf to $TfFe_C$ ranging from 2.6 to 30 were observed) and it was concluded that, at physiological pH, the distribution of iron between the two binding sites of transferrin in serum is neither random nor an equilibrium distribution (under these conditions the C-terminal site binds iron six times more strongly than the N-terminal site). These results have been confirmed[134] and it has been noted that, on dialysis of serum against buffer at pH 7.4, or on storage of serum at $-15\,°C$, there is a marked preference for the C-terminal site; the original preference for the N-terminal

127 G. Spik and J. Mazurier, in 'Proteins of Iron Metabolism' ed. E. B. Brown, P. Aisen, J. Fielding and R. R. Crichton, Grune and Stratton, New York, 1977, p. 143.
128 M. W. C. Hatton, E. Regoeczi, and H. Kaur, *Can. J. Biochem.*, 1978, **56**, 339.
129 J. Montreuil, *Adv. Carbohydr. Chem. Biochem.*, 1980, **37**, 157.
130 M. W. C. Hatton, L. März, L. R. Berry, M. T. Debanne, and E. Regoeczi, *Biochem. J.*, 1979, **181**, 633.
131 P. Aisen, A. Leibman, and J. Zweier, *J. Biol. Chem.*, 1978, **253**, 1930.
132 E. Frieden and P. Aisen, *Trends Biochem. Sci.*, 1980, **5**, January, p. XI.
133 A. Leibman and P. Aisen, *Blood*, 1979, **53**, 1058.
134 J. Williams and K. Moreton, *Biochem. J.*, 1980, **185**, 483.

site can be restored by addition of a diffusable fraction. Differences in the distribution of iron between the two sites when human apotransferrin is labelled with iron, both as a function of the iron chelate used and of the pH, have been reported[131,135,136] and the two monomeric human transferrins have been isolated in preparative yields by isoelectric focussing.[137] The characterization of mono-ferric fragments from bovine transferrin[138] and differences in iron distribution between *N*- and *C*-terminal sites in ovotransferrin as a function of pH[139] have been reported.

Over the pH range 6.0—7.1, copper binds only to the *C*-terminal (A) site of human transferrin, whereas as the pH exceeds 7.2 the *N*-terminal site also becomes occupied;[140] the two sites can be distinguished by e.p.r., and it was concluded that the *C*-terminal site has greater affinity for Fe^{3+} than the *N*-terminal site at pH 7.4. The application of pulsed e.p.r. spectroscopy to Cu^{II}–transferrin complexes as a function of pH, with carbonate as anion, indicates that an imidazole ligand is co-ordinated to the Cu^{II}, consistent with the nitrogen superhyperfine structure in the e.p.r. spectrum.[141] It is further demonstrated that in the modulation pattern of the spin-echo envelope of the Cu^{II}–transferrin–[^{13}C]oxalate, ^{13}C superhyperfine structure is observed, confirming that the metal ion is co-ordinated directly to its associated anion. A similar conclusion was derived from e.p.r. studies with a spin-labelled oxalate anion.[142] Ternary complexes of VO^{2+}–transferrin–xylenol orange and VO^{2+}–transferrin–semi-xylenol orange have been prepared; the e.p.r. spectra also indicate that the anion is directly bound to the metal.[143] E.p.r. studies of vanadyl ovotransferrin together with u.v. difference spectroscopy suggest that tyrosyl oxygens (2 or 3 per metal ion) and possibly one nitrogen are the equatorial donor atoms to the VO^{2+} group.[144] From studies of proton release and u.v. difference spectroscopy of human transferrin[145] it was concluded that two tyrosine ligands are attached to Cu^{2+} and Zn^{2+} and three tyrosines to Fe^{3+}, Cu^{3+}, Al^{3+}, and VO^{2+}. Mixed metal (VO^{2+} and Fe^{3+}) complexes of ovotransferrin have been studied by e.p.r.,[146] and a Pt complex of human transferrin has been prepared in which the Pt is probably in the (+2) oxidation state, suggesting that transferrin may be involved in transport of metals of the third transition series.[147]

Determination of transferrin, and its microheterogeneity. A fluorimetric method

[135] R. W. Evans and J. Williams, *Biochem. J.*, 1978, **173**, 543.
[136] H. G. van Eijk, W. L. van Noort, M. J. Kroos, and C. van der Heul, *J. Clin. Chem. Clin. Biochem.*, 1978, **16**, 557.
[137] H. G. van Eijk, W. L. van Noort, M. J. Kroos, and C. van der Heul, *J. Clin. Chem. Clin. Biochem.*, 1980, **18**, 563.
[138] J. H. Brock, F. R. Arzabe, N. E. Richardson, and E. V. Deverson, *Biochem. J.*, 1978, **171**, 73.
[139] J. Williams, R. W. Evans, and K. Moreton, *Biochem. J.*, 1978, **173**, 535.
[140] J. L. Zweier, *J. Biol. Chem.*, 1978, **253**, 7616.
[141] J. L. Zweier, P. Aisen, J. Peisach, and W. B. Mims, *J. Biol. Chem.*, 1979, **254**, 3512.
[142] R. C. Najarian, D. C. Harris, and P. Aisen, *J. Biol. Chem.*, 1978, **253**, 38.
[143] D. C. Harris and M. H. Gelb, *Biochim. Biophys. Acta*, 1980, **623**, 1.
[144] J. D. Casey and N. D. Chasteen, *J. Inorg. Biochem.*, 1980, **13**, 111.
[145] M. H. Gelb and D. C. Harris, *Arch. Biochem. Biophys.*, 1980, **200**, 93.
[146] J. D. Casey and N. D. Chasteen, *J. Inorg. Biochem.*, 1980, **13**, 127.
[147] R. Stjernholm, F. W. Warner, J. W. Robinson, E. Ezekiel, and N. Katayama, *Bioinorg. Chem.*, 1978, **9**, 277.

for determination of serum transferrin has been described[148] as well as a method for the preparation of highly ^{59}Fe-labelled mouse transferrin.[149] The micro-heterogeneity of human transferrin continues to generate a large literature (for example, refs. 150 and 151).

Function of Transferrin. The binding of iron to transferrin and its release from the protein *in vitro*, the functional equivalence (or non-equivalence), at the cellular level, of iron that is bound to the *N*- and *C*-terminal binding sites of transferrin, together with a consideration of some aspects of transferrin metabolism and biosynthesis are reviewed in this section.

Binding and release of iron. The distribution of binding of iron between the two sites of transferrin has become accessible to analysis by the Makey–Seal method,[111] which has been elegantly applied,[131] as described above, to estimate the intrinsic site constants for binding of iron. It also enables a distinction to be drawn between thermodynamic and kinetic accessibility factors in the binding of iron to the two sites when different forms of iron are used.[152] With human trans-ferrin, both this approach[131,135] and the use of enzymic digestion to prepare stable *N*- and *C*-terminal fragments[135] confirm that the *N*-terminal site (to which iron binds less strongly) is the preferred binding site when ferric citrate, ferric chloride, ferric oxalate, ferrous ascorbate, or ferrous ammonium sulphate is the source of iron, whereas the *C*-terminal site is preferred when ferric nitrilotri-acetate is used. The distribution of iron between the two binding sites of ovo-transferrin has also been analysed[139] as a function of the pH and the nature of the iron donor. The role of caeruloplasmin in the formation of ferric transferrin from ferrous ammonium sulphate has been examined.[153] The sequential saturation and desaturation of human lactotransferrin with iron has also been reported.[154]

While the mechanism of uptake of iron by transferrin seems well established (most probably, uptake from a low-molecular-weight Fe^{III}, or perhaps Fe^{II}, chelate) the release of iron from transferrin is much less clearly understood. At least three mechanisms have been considered, and a fourth would appear also to merit consideration, albeit in pathological and non-physiological conditions. Transferrin can release one or both of its iron atoms in reponse to a diminution of pH, and this mechanism could have considerable physiological significance if the transferrin molecules were interiorized in the cell and could come into contact with the acidic pH which prevails in lysosomes. However, iron can also be released from transferrin by phosphate compounds, of which pyrophosphate and organic polyphosphates, such as ATP and GTP, are the most effective.[155,156] Reduction of transferrin iron has also been considered as a possible mechanism, and a kinetic analysis of the reduction of iron and of its complexation by batho-

148 D. Thompson and L. B. Roberts, *Biochem. Med.*, 1978, **20**, 128.
149 J. Hola, J. Vacha, and K. Bohacek, *Acta Haematol.*, 1979, **61**, 55.
150 P. Kühnl and W. Spielmann, *Hum. Genet.*, 1978, **43**, 91.
151 N. L. Anderson and N. G. Anderson, *Biochem. Biophys. Res. Commun.*, 1979, **88**, 258.
152 P. Aisen and A. Leibmann, in 'Transport by Proteins' ed. G. Blauer and H. Sund, de Gruyter, Berlin and New York, 1978, p. 277.
153 J.-C. Mareschal, R. Rama, and R. R. Crichton, *FEBS Lett.*, 1980, **110**, 268.
154 J. Mazurier and G. Spik, *Biochim. Biophys. Acta*, 1980, **629**, 399.
155 E. H. Morgan, *Biochim. Biophys. Acta*, 1979, **580**, 312.
156 F. J. Carver and E. Frieden, *Biochemistry*, 1978, **17**, 167.

phenanthroline sulphonate has led to the proposal of a reaction sequence in which the iron is reduced in the rate-limiting step, followed by rapid release of the iron and complexation by the Fe^{II} chelator; the interaction between chelating agents and reductants has also been analysed.[157] Reductive mobilization of iron from transferrin was also observed with 2-formylpyridine thiosemicarbazone.[158] The release of iron from human and rat transferrin by desferrioxamine, mediated by diphosphoglycerate, can be resolved into two components, corresponding to the release of iron from the two binding sites at markedly different rates.[159] These studies, in parallel with earlier results,[160] imply that another possible mechanism of release of iron from transferrin is displacement of bicarbonate by another anion, followed by exchange of the iron between transferrin and a chelator such as desferrioxamine, which has a greater affinity for iron but which is unable, for kinetic reasons, to release transferrin iron in the absence of an enhancement of lability of the iron on the binding sites. The effects of pyrophosphate and organic polyphosphates may be attributed to their labilization and complexation of the transferrin iron by displacement of bicarbonate. Exchange of iron between the binding sites of rat transferrin, mediated by acid pH and citrate, has been observed but it was concluded that, *in vivo*, there is little scrambling of transferrin iron.[161] However, a fourth mechanism of release of iron may be invoked in the presence of bacterial and synthetic iron chelators. It has been reported that meningococci (notably *Neisseria meningitidis*) are able to mobilize transferrin iron for their growth requirements.[162] Further, the kinetics of removal of iron from transferrin *in vitro* by the synthetic chelators *N,N′,N″*-tris-(5-sulpho-2,3-dihydroxybenzoyl)-1,5,10-triazadecane, 1,3,5-tris-[(2,3-dihydroxybenzoyl)amino-methyl]benzene, and the natural siderophore enterobactin have been studied.[163] The authors conclude that these chelators (all catecholates) are both kinetically and thermodynamically capable of removing iron from transferrin at physiologically reasonable concentrations. This difference in the kinetic rather than the thermodynamic factors distinguishes the catecholate ligands from the hydroxamate-based chelators, such as desferrioxamine, which are kinetically hindered in the mobilization of transferrin iron. The transfer of iron from transferrin to 2,2′-bipyridyl by rabbit reticulocyte cytosol has been observed and the factor that is responsible for transfer of iron has been identified as haemoglobin.[164] The kinetics of removal of iron from human transferrin by EDTA in the presence of salts and detergents have been studied.[165]

Studies of the exchange of iron between transferrin and ferritin *in vitro* have been carried out in the presence of chelators and reductants, and it was concluded that the exchange (under the conditions used) is always from ferritin to transferrin.[166] The possibility that pyrophosphate may mediate the transfer of iron

[157] N. Kojima and G. W. Bates, *J. Biol. Chem.*, 1979, **254**, 8847.
[158] E. Ankel and D. H. Petering, *Biochem. Pharmacol.*, 1980, **29**, 1833.
[159] E. H. Morgan, H. Huebers, and C. Finch, *Blood*, 1978, **52**, 1219.
[160] S. Pollack, G. Vanderhoff, and F. Lasky, *Biochim. Biophys. Acta*, 1977, **497**, 481.
[161] S. Okada, M. D. Rossmann, and E. B. Brown, *Biochim. Biophys. Acta*, 1978, **543**, 72.
[162] F. S. Archibald and I. W. DeVoe, *FEMS Microbiol. Lett.*, 1979, **6**, 159.
[163] C. J. Carrano and K. N. Raymond, *J. Am. Chem. Soc.*, 1979, **101**, 5401.
[164] A. Egyed, A. May, and A. Jacobs, *Biochim. Biophys. Acta*, 1980, **629**, 391.
[165] D. A. Baldwin, *Biochim. Biophys. Acta*, 1980, **623**, 183.
[166] D. C. Harris, *Biochemistry*, 1978, **17**, 3071.

from transferrin to ferritin, particularly at low pyrophosphate concentrations (0.1 mmol l⁻¹), is suggested from recent studies.[86] However, to what extent such studies have any real physiological significance remains uncertain, since the bulk of cellular ferritin is in the cytosol and, even if transferrin enters the cell, it will be contained in an endocytic vacuole or in a secondary lysosome (see Section 2).

Metabolism, biological effects, and biosynthesis of transferrin. A review of transferrin metabolism, particularly in the liver,[167] has appeared. Hepatocytes have a receptor for glycoproteins which have a fucosyl residue in $\alpha 1 \rightarrow 3$ linkage to N-acetylglucosamine. It is observed that human lactoferrin is rapidly cleared by hepatocytes whereas human transferrin and asialotransferrin are not eliminated from the circulation. When fucose is incorporated into asialotransferrin by an $\alpha 1 \rightarrow 3$ N-acetylglucosamine fucosyl transferase, the fucosyltransferrin is rapidly cleared.[168] It is not clear why the asialotransferrin, with its terminal galactose residues, is not cleared by the liver like other asialo-glycoproteins. The turnover of human lactoferrin in humans and in rabbits has been measured,[169-171] and, although the absolute values and turnover rates differ, it is clear that lactoferrin is a normal constituent of human serum and that its turnover is relatively rapid; it is predominantly cleared from the circulation by the liver. Lactoferrin appears to have the capacity to act as a negative feedback regulator of myelopoiesis, inhibiting the production of colony-stimulatory activity for immunoglobulin-A-like antigen-positive subpopulations of human blood monocytes.[172,173] The iron in human milk (0.3–0.7 μg ml⁻¹) is mostly bound to the lipid fraction and to a low-molecular-weight fraction.[174] Only a small amount is bound to lactoferrin, and the iron saturation of lactoferrin is very low, which may be important for the bacteriostatic properties of the protein; it has been suggested that lactoferrin extracts iron from the environment, and thus prevents bacterial growth.[175] In human serum, the low-molecular-weight iron fraction corresponds to less than 0.05 μg per 100 ml of serum.[176] The lactoferrin content and the intracellular distribution of lactoferrin in peripheral human blood cells have been analysed[177-179] and an immunoperoxidase staining technique has been used to

167 A. S. Tavill and A. G. Morton, in 'Metals and the Liver' ed. L. W. Powell, Marcel Dekker, New York, 1978, p. 93.
168 J.-P. Priels, S. V. Pizzo, L. R. Glasgow, J. C. Paulson, and R. L. Hill, *Proc. Natl. Acad. Sci. USA*, 1978, **75**, 2215.
169 G. Fillet, J. van Snick, and P. Masson, *Proc. XVII Congress, Int. Soc. Hematol.*, 1978, p. 583.
170 R. M. Bennett and T. Kokocinski, *Clin. Sci.*, 1979, **57**, 453.
171 H. Karle, N. E. Hansen, J. Malmquist, A. K. Karle, and I. Larsson, *Scand. J. Haematol.*, 1979, **23**, 303.
172 H. E. Broxmeyer, M. DeSousa, A. Smithyman, P. Ralph, J. Hamilton, J. I. Kurland, and J. Bognacki, *Blood*, 1980, **55**, 324.
173 H. E. Broxmeyer, *J. Clin. Invest.*, 1979, **64**, 1717.
174 G. B. Fransson and B. Lönnerdal, *J. Pediatr.*, 1980, **96**, 380.
175 J. J. Bullen, H. J. Rogers, and L. Leigh, *Br. Med. J.*, 1972, **1**, 69.
176 D. Hahn and A. M. Ganzoni, *Biochim. Biophys. Acta*, 1980, **627**, 250.
177 R. M. Bennett and T. Kokocinski, *Br. J. Haematol.*, 1978, **39**, 509.
178 K. B. Pryzwansky, L. E. Martin, and J. K. Spitznagel, *J. Reticuloendothel. Soc.*, 1978, **24**, 295.
179 Y. T. Konttinen and S. Reitamo, *Br. J. Haematol.*, 1979, **43**, 481.

study the distribution of transferrin, ferritin, and lactoferrin in human tissues.[180] Transferrin has been found to stimulate the growth of SV3T3 cells[181] and can replace serum for the growth of mitogen-stimulated T lymphocytes *in vitro*.[182] Serum can be replaced by transferrin, insulin, albumin, and other factors for the growth of SV40-transformed mouse fibroblast and mouse myeloid leukaemia cells.[183,184] It has been suggested that serum transferrin may play a role in the stimulation of the synthesis of DNA in the liver.[185]

Serum transferrin and ovotransferrin are synthesized in liver and oviduct respectively. They are products of the same gene, but the regulation of their synthesis differs in the two tissues. Thus steroids induce a marked stimulation of ovotransferrin synthesis (6–8-fold) whereas a much less pronounced effect was observed on transferrin synthesis. In both tissues the response was correlated with an increase in synthesis of transferrin mRNA.[186] Ovotransferrin that has been synthesized in a cell-free system contains a nineteen-residue *N*-terminal extension.[187] A similar twenty-residue *N*-terminal extension was found in a study of the synthesis and secretion of rat transferrin.[188] In iron-deficient chicks, transferrin mRNA increased 2–3-fold; when the iron stores were rapidly replenished, both the rate of transferrin synthesis and the level of transferrin mRNA returned to normal in 2–3 days.[189] Differences in the regulatory mechanisms of expression of the transferrin gene in iron deficiency and during treatment with an oestrogen are suggested in recent studies.[190] Analyses of the sequence of the transferrin gene in chicken liver and oviduct DNA, using restriction endonucleases, have been presented.[191] A significant positive correlation between hepatic uptake of iron and the rate of synthesis of transferrin was observed.[192] The synthesis and secretion of transferrin in cultured rat and mouse cells has been described.[188,193,194]

Other Iron-storage Proteins.—The iron complex of phosvitin, the principal phosphoglycoprotein of egg yolk, is a source of dietary iron and an important reserve of iron in the eggs of lower vertebrates. The protein contains about 135 residues of phosphoserine. An equilibrium constant of 10^{18} has been reported

[180] D. Y. Mason and C. R. Taylor, *J. Clin. Pathol.*, 1978, **31**, 316.
[181] D. V. Young, F. W. Cox, III, S. Chipman, and S. C. Hartman, *Exp. Cell Res.*, 1979, **118**, 410.
[182] M.-L. Dillner-Centerlind, S. Hammarström, and P. Perlmann, *Eur. J. Immunol.*, 1979, **9**, 942.
[183] G. A. Rockwell, G. H. Sato, and D. B. McClure, *J. Cell Physiol.*, 1980, **103**, 323.
[184] Y. Honma, T. Kasukabe, J. Okabe, and M. Hozumi, *Exp. Cell Res.*, 1979, **124**, 421.
[185] T. Okazaki, *Acta Haematol. Jpn.*, 1978, **41**, 1267.
[186] D. C. Lee, G. S. McKnight, and R. D. Palmiter, *J. Biol. Chem.*, 1978, **253**, 3494.
[187] S. N. Thibodeau, D. C. Lee, and R. D. Palmiter, *J. Biol. Chem.*, 1978, **253**, 3771.
[188] G. Schreiber, H. Dryburgh, A. Millership, Y. Matsuda, A. Inglis, J. Phillips, K. Edwards, and J. Maggs, *J. Biol. Chem.*, 1979, **254**, 12 013.
[189] G. S. McKnight, D. C. Lee, D. Hemmaplardh, C. A. Finch, and R. D. Palmiter, *J. Biol. Chem.*, 1980, **255**, 144.
[190] G. S. McKnight, D. C. Lee, and R. D. Palmiter, *J. Biol. Chem.*, 1980, **255**, 148.
[191] D. C. Lee, G. S. McKnight, and R. D. Palmiter, *J. Biol. Chem.*, 1980, **255**, 1442.
[192] A. G. Morton and A. S. Tavill, *Br. J. Haematol.*, 1978, **39**, 497.
[193] G. C. T. Yeoh, J. A. Wassenburg, E. Edkins, and I. T. Oliver, *Biochim. Biophys. Acta*, 1979, **565**, 347.
[194] J. Papaconstantinou, R. E. Hill, W. H. Gibson, and E. Y. Rao, *Differentiation*, 1978, **10**, 139.

for formation of the Fe^{3+}–phosvitin complex, in which the most probable binding sites would be composed of clusters of diphosphorylserine residues.[195] A similar conclusion that pairs of phosphate groups bind one iron atom each was reached from ultrafiltration, c.d., and sedimentation studies.[196] The ferrous complex of phosvitin was characterized for the first time, and conformational changes that occurred when iron was bound were studied.

2 The Movement of Iron across Membranes

While the transport and storage forms of iron are well characterized in mammalian cells, and their structure, function, and biology are slowly but surely yielding their secrets to the onslaught of sophisticated techniques applied to their analysis, there remains nevertheless a grey area separating these two forms of iron, namely the passage of iron from the extracellular transport form to the intracellular storage form, and the reverse of this process. In prokaryotic cells the movement of iron into cells and its release are somewhat better characterized, which explains why we have included in this section a consideration of microbial transport and metabolism of iron. We have also chosen to include an analysis of the absorption and metabolism of iron in mammals. A number of review articles have appeared; these are indicated at the beginning of each subsection. In view of the fact that this subject was not specifically treated in the first Report, we have included some references to earlier work.

The Uptake of Transferrin Iron by Cells.—The uptake of transferrin iron by reticulocytes, where it supplies the iron required for haemoglobin biosynthesis, and by other mammalian cells has been reviewed.[1,3] It is clear that transferrin must deliver iron to many different cells and, although the immature red blood cell (the reticulocyte) has been the principal object of study in this domain, on account of the avidity of its uptake of iron, the delivery of transferrin iron to other cell types has also been studied. While it is dangerous to extrapolate from studies on reticulocytes to other cell types, in view of the extraordinary capacity of the reticulocyte to accumulate iron for haem synthesis, it is nonetheless tempting to suppose that, once an appropriate system was developed for the intracellular accumulation of iron, it would be logical that it might be utilized by other cells with lesser needs for iron.

Since the pioneering studies of Jandl and Katz,[197] it seems likely that reticulocytes, and presumably all other mammalian cells, have receptor molecules on their plasma membranes which bind transferrin. Subsequent to binding of the transferrin molecule, iron is taken up by the cells and the transferrin, now impoverished in its iron content, is released into the extracellular medium.

The way in which transferrin iron gets into the cell is not known with certainty, but a number of mechanisms can be envisaged (the arguments for and against these mechanisms are discussed in more detail later), and they are outlined in Figure 4. Given that transferrin receptors exist on the plasma membrane of a number of different cell types, we could consider the possibility that, after binding

[195] J. Hegenauer, P. Saltman, and G. Nace, *Biochemistry*, 1979, **18**, 3865.
[196] G. Taborsky, *J. Biol. Chem.*, 1980, **255**, 2976.
[197] J. H. Jandl and J. H. Katz, *J. Clin. Invest.*, 1963, **42**, 314.

PLASMA MEMBRANE

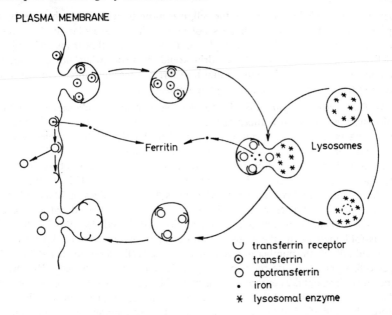

Figure 4 *Possible mechanisms of uptake of iron by cells*

of transferrin to its receptor, the iron is directly released at the plasma membrane by an (as yet) unidentified mechanism involving, for example, pyrophosphate or an organic polyphosphate or an enzyme system (for which there is to date no positive evidence). The iron thus released would penetrate the cell, associated with an appropriate carrier, and would be made available for the iron requirements of the cell; the iron-poor transferrin would be released from the membrane. A second possibility, still assuming binding of transferrin to a membrane receptor, would be the internalization of the transferrin–receptor complex by endocytosis. Once enveloped in an endocytic vacuole, the iron could be released by a mechanism similar to that proposed for the plasma-membrane-mediated process described above, or could occur subsequent to the fusion of the endocytic vacuole with a lysosome, assisted by the low pH, a factor which would favour the release of at least one of the two iron atoms bound to the transferrin molecule. The transferrin, having released some (if not all) of its iron, could then be recycled, together with the plasma membrane receptor, back to the cell surface and be released back into the extracellular medium. The iron could pass from the secondary lysosome into the cytoplasm in the form of an appropriate low-molecular-weight chelate. A third mechanism, independent of the existence of plasma membrane receptors for transferrin, would involve fluid endocytosis of transferrin from the extracellular medium. In this case, fusion of the endocytic vacuole with a lysosome would result in release of the iron, accompanied by proteolytic digestion of the transferrin protein.

Before considering the experimental evidence in favour of or in contradiction with these mechanisms, it is perhaps appropriate to emphasize that, if the transferrin molecule is interiorized in the cell, it is more than probable that it is by a process of endocytosis, which does not permit the extracellular fluid to enter directly into contact with the cytoplasm. Thus, the idea that transferrin might deliver its iron directly to mitochondria would seem somewhat illusory, and, as already emphasized, it is unlikely that transferrin comes into direct contact with ferritin (except in the interior of a phagolysosome).

The Transferrin Receptor. For the best-studied system, namely the rabbit reticulocyte, the maximum number of transferrin receptors has been calculated as being 300 000, with an association constant of 10^7 l mol^{-1}.[198] As the cells mature, the number of receptors decreases while their affinity remains unchanged.[199] For rat reticulocytes in different states of iron deficiency, the numbers of transferrin-binding sites were estimated at between 66 000 and 120 000 for iron-deficient and phenylhydrazine-treated animals.[200] Mean values of 22 000–80 000 iron atoms taken up per minute and per cell were reported. Friend erythroleukaemic cells that had been induced to differentiation with dimethyl sulphoxide showed a six-fold increase in the level of transferrin receptors compared with non-induced cells;[201] the number of receptors in induced cells was estimated at 440 000 per cell, whereas mouse reticulocytes had 86 000 receptors per cell. However, the rate of iron uptake per receptor was two-fold greater in mouse reticulocytes.[201] Association constants of the same order of magnitude as for rabbit reticulocytes were observed $(0.35–0.42 \times 10^7$ l mol$^{-1})$ and the iron uptake was much more rapid that that found for rabbit reticulocytes (values of 37.3×10^5 molecules per cell for induced Friend cells and of 17.1×10^5 molecules per cell for mouse reticulocytes). These results confirm the conclusion of Hu *et al.*[202] that the differentiation of Friend cells is accompanied by induction of transferrin receptors, but are in contradiction with the results of Glass *et al.*,[203] who found that the number of transferrin receptors was not significantly different between differentiated and non-differentiated Friend cells in culture. In contrast, the latter group found greatly increased uptake of iron in induced cells, being more than doubled 24 hours after induction and rising to a maximum value of eight-fold after 72 hours.

The characterization of the transferrin receptor from rabbit reticulocytes has been reported from a number of laboratories.[204–212] Components of molecular

[198] F. M. Van Bockxmeer, G. K. Yates, and E. H. Morgan, *Eur. J. Biochem.*, 1978, **92**, 147.
[199] F. M. Van Bockxmeer and E. H. Morgan, *Biochim. Biophys. Acta*, 1979, **584**, 76.
[200] C. Black, J. Glass, M. T. Nunez, and S. H. Robinson, *J. Lab. Clin. Med.*, 1979, **93**, 645.
[201] G. C. T. Yeoh and E. H. Morgan, *Cell. Differ.*, 1979, **8**, 331.
[202] H-Y. Y. Hu, J. Gardner, P. Aisen, and A. I. Skoultchi, *Science*, 1977, **197**, 559.
[203] J. Glass, M. T. Nunez, S. Fischer, and S. H. Robinson, *Biochim. Biophys. Acta*, 1978, **542**, 154.
[204] H-Y. Y. Hu and P. Aisen, *J. Supramol. Struct.*, 1978, **8**, 349.
[205] D. P. Witt and R. C. Woodworth, *Biochemistry*, 1978, **17**, 3913.
[206] D. A. Sly, D. Grohlich, and A. Bezkorovainy, *Cell Surf. Carbohydr.*, 1978, 255.
[207] J. Glass, M. T. Nunez, and S. H. Robinson, *Biochim. Biophys. Acta*, 1980, **598**, 293.
[208] M. T. Nunez, S. Fischer, J. Glass, and L. Lavidor, *Biochim. Biophys. Acta*, 1977, **490**, 87.
[209] A. Liebman and P. Aisen, *Biochemistry*, 1977, **16**, 1268.
[210] N. D. Light, *Biochim. Biophys. Acta*, 1977, **495**, 46.
[211] N. D. Light, *Biochem. Biophys. Res. Commun.*, 1978, **81**, 261.
[212] B. Ecarot-Charrier, V.L. Grey, A. Wilczynska, and H. M. Schulman, *Can. J. Biochem.*, 1980, **58**, 418.

weight 176 000 and 95 000 (which may be monomeric and dimeric forms of the receptor) are cross-linked with transferrin by dimethyl suberimidate.[204-209] Lactoperoxidase iodination and immunoprecipitation implies a glycoprotein of molecular weight 190 000 as the putative transferrin receptor.[205,212] A more complex subunit composition, for a receptor of molecular weight 180 000—200 000, has been proposed.[206] A subunit molecular weight of 78 000 has recently been reported,[207] and even lower values for the receptor polypeptide[210,211] may represent contamination or degradation products. These data would seem to confirm the original observations of Speyer and Fielding that the transferrin receptor has a molecular weight of around 200 000,[213] and is probably a glycoprotein.

Binding of transferrin and uptake of iron by rat reticulocytes are inhibited by incubation with F_{ab} fragments of an antibody against the membrane receptor for transferrin.[214] In rat reticulocyte ghosts, a membrane receptor for transferrin of $M_r \approx 230 000$ has been identified,[215] while in rat kidney a glycoprotein of $M_r \approx 170 000$ was tentatively identified as a subunit of the transferrin receptor.[216] In human placenta, a receptor glycoprotein of M_r 90 000 on sodium dodecyl sulphate gel electrophoresis has been identified[217,218] with a K_a of 2.2—3.6 × 10^7 l mol^{-1}.[217,219] A similar cell-surface glycoprotein subunit which binds transferrin has been identified on human choriocarcinoma cells, splenic B lymphocytes, nasopharyngeal carcinoma cells, and embryonic lung fibroblasts,[220] Transferrin receptors have also been identified on human B and T lymphoblastoid cell lines,[221] on human reticulocytes,[222] and on rat embryonic fibroblasts in culture.[223] Interestingly, in a study with mouse bone-marrow cells, it was found that cholesterol depletion, which increased membrane-lipid fluidity, resulted in a substantial increase in the number of available transferrin receptors per cell while cholesterol enhancement had the opposite effect.[224] On the basis of these results, it is proposed that the transferrin receptor is a cross-membrane protein with a substantial portion facing the cytoplasm. Substantial increases in the amounts of transferrin receptors have been reported in transformed lymphoid cell lines.[225]

The uptake of iron from different sources by different cell types appears to be quite variable. Thus Chang liver cells take up thirty times as much iron from

[213] B. E. Speyer and J. Fielding, *Biochim. Biophys. Acta*, 1974, **332**, 192.
[214] C. Van Der Heul, M. J. Kroos, and H. G. Van Eijk, *Biochim. Biophys. Acta*, 1978, **511**, 430.
[215] C. Van Der Heul, M. J. Kroos, C. M. H. De Jeu-Jaspars, and H. G. Van Eijk, *Biochim. Biophys. Acta*, 1980, **601**, 572.
[216] J. A. Fernandez-Pol and D. J. Klos, *Biochemistry*, 1980, **19**, 3904.
[217] H. G. Wada, P. E. Hass, and H. H. Sussmann, *J. Biol. Chem.*, 1979, **254**, 12 629.
[218] P. A. Seligman, R. B. Schleicher, and R. H. Allen, *J. Biol. Chem.*, 1979, **254**, 9943.
[219] T. T. Loh, D. A. Higuchi, F. M. Van Bockxmeer, C. H. Smith, and E. B. Brown, *J. Clin. Invest.*, 1980, **65**, 1182.
[220] T. A. Hamilton, H. G. Wada, and H. H. Sussmann, *Proc. Natl. Acad. Sci. USA*, 1979, **76**, 6406.
[221] J. W. Larrick and P. Cresswell, *Biochim. Biophys. Acta*, 1979, **583**, 483.
[222] M. Steiner, *Biochem. Biophys. Res. Commun.*, 1980, **94**, 861.
[223] J.-N. Octave, Y.-J. Schneider, R. R. Crichton, and A. Trouet, *Eur. J. Biochem.*, 1981, **115**, 611.
[224] C. Muller and M. Shinitzky, *Br. J. Haematol.*, 1979, **42**, 355.
[225] J. W. Larrick and P. Cresswell, *J. Supramol. Struct.*, 1979, **11**, 579.

ferric nitrilotriacetate as from ferric citrate or transferrin.[226] Whereas rat embryonic fibroblasts in culture take up iron more effectively from transferrin than from ferric citrate,[223] the situation is inverse in cultured rat hepatocytes.[227] Differences in iron uptake from transferrin and from iron chelates in erythroid cells have also been reported.[228]

The Mechanism of Uptake of Transferrin Iron and its Intracellular Distribution. In the introduction to this section we outlined briefly the possible mechanisms which might be invoked to explain how transferrin delivers its iron to cells. While the mechanism (or mechanisms) which operate remain unclear, some progress has been made; this is briefly reviewed here. In cultured rat embryo fibroblasts, it has been shown that a number of drugs which affect endocytosis and lysosomal function greatly reduce the uptake of iron from transferrin without affecting the accumulation of transferrin protein by the cells.[223,229] Short-term kinetic experiments at 37 °C indicate that iron is rapidly taken up by the cells and the iron-depleted transferrin is released intact to the extracellular medium. On the basis of the results, it has been suggested[223] that uptake of iron involves both receptor-mediated and fluid-phase endocytosis. About half of the iron taken up by the cells is in ferritin and the rest in a low-molecular-weight cytoplasmic component and in mitochondria (presumably in haem). In human reticulocytes the major iron-containing cytosol components have been studied, and from pulse-chase experiments it has been concluded that cytosol ferritin is an obligatory intermediate in intracellular transport of iron.[230] Mobilization of iron from mouse reticulocyte membranes (prepared by pre-incubation with mouse transferrin) by reticulocyte cytosol and by ferritin was observed,[231] and it was concluded that ferritin plays an active role in the uptake and incorporation of iron into haem. A more recent report from this group[232] implies that ferritin, transferrin, and a cytoplasmic component of $M_r \approx 60\,000$ are intermediates in the transport of iron in rabbit reticulocytes from the plasma membrane to the mitochondria. In contrast, whole-animal studies did not show any evidence for transferrin iron or transferrin protein in lysosomal or mitochondrial fractions of rat liver,[233,234] and it was concluded that hepatic uptake of iron did not involve endocytosis. Denatured transferrin was, however, taken up by endocytosis. The iron that was accumulated by the liver cells was incorporated into ferritin of increasing density. Asialotransferrins were taken up by rat liver by endocytosis.[235] Some preliminary studies on hepatic uptake of iron have been reported.[227,236] It was concluded that transferrin can deliver iron to ascites tumour cells, and

[226] G. P. White and A. Jacobs, *Biochim. Biophys. Acta*, 1978, **543**, 217.
[227] J.-C. Sibille, Y.-J. Schneider, J.-N. Octave, A. Trouet, and R. R. Crichton, *Arch. Int. Physiol. Biochim.*, 1981, **89**, B35.
[228] A. Barnekow and G. Winkelmann, *Biochim. Biophys. Acta*, 1978, **543**, 523.
[229] J.-N. Octave, Y.-J. Schneider, P. Hoffmann, A. Trouet, and R. R. Crichton, *FEBS Lett.*, 1980, **108**, 127.
[230] B. E. Speyer and J. Fielding, *Br. J. Haematol.*, 1979, **42**, 255.
[231] M. T. Nunez, J. Glass, and S. H. Robinson, *Biochim. Biophys. Acta*, 1978, **509**, 170.
[232] M. T. Nunez, E. S. Cole, and J. Glass, *Blood*, 1980, **55**, 1051.
[233] J. P. Milsom and R. G. Batey, *Biochem. J.*, 1979, **182**, 117.
[234] R. G. Batey, K. Williams, and J. P. Milsom, *Am. J. Physiol.*, 1980, **238**, G30.
[235] E. Regoeczi, P. Taylor, M. T. Debanne, L. März, and M. W. C. Hatton, *Biochem. J.*, 1979, **184**, 399.
[236] D. Grohlich, C. G. D. Morley, and A. Bezkorovainy, *Int. J. Biochem.*, 1979, **10**, 797.

preliminary results suggested that transferrin iron is released at the cell surface.[237]

The hypothesis of Fletcher and Huehns[238] whereby one iron-binding site of transferrin (the C-terminal site) would deliver its iron mainly to red-cell precursors while the other (the N-terminal site) would deliver its iron mainly to iron-storage pools continues to evoke controversy, although it would seem that, if differences do exist, they are small.[159,239-243] The interaction of heterologous transferrins and monoferric transferrin fragments with reticulocytes has been analysed,[244-246] differences in iron uptake between human and rabbit transferrins have been reported,[247] and the effects of some incubation variables on uptake of iron have been studied.[248] Using immunofluorescence and immunohistological procedures, transferrin receptors have been demonstrated in human placental trophoblasts and in a number of other human cell lines.[249-252] In addition to transferrin, the haem–haemopexin complex can also transport iron to the liver, where it is conserved in ferritin.[253]

Intracellular Metabolism and Release of Iron.—Intracellular iron that is in transport is predominantly in the form of an as yet uncharacterized transit (or chelatable) iron pool, presumably of low molecular weight,[254] and of cytoplasmic ferritin. These two forms are presumed to be in equilibrium with one another and also with a mitochondrial iron pool which is used, in part at least, for haem synthesis. Recently,[255] a mitochondrial non-haem, non-FeS iron pool, of which approximately half is in the Fe^{2+} form, has been detected, and it is suggested that this represents a transit pool for haem synthesis. In copper deficiency, an iron pool accumulates in mitochondria which can be used for haem synthesis in anaerobic conditions or in the presence of electron-transport substrates.[256] We may well pose the question of where the iron for haem synthesis comes from. Although studies on the uptake of transferrin iron by isolated rat liver mito-

[237] R. T. Parmley, F. Ostroy, R. A. Gams, and L. DeLucas, *J. Histochem. Cytochem.*, 1979, **27**, 681.

[238] J. Fletcher and E. R. Huehns, *Nature (London)*, 1968, **218**, 1211.

[239] P. Pootrakul, A. Christensen, B. Josephson, and C. A. Finch, *Blood*, 1977, **49**, 957.

[240] H. Huebers, E. Huebers, E. Csiba, and C. A. Finch, *J. Clin. Invest.*, 1978, **62**, 944.

[241] S. Okada, B. Jarvis, and E. B. Brown, *J. Lab. Clin. Med.*, 1979, **93**, 189.

[242] N. J. Verhoef, M. J. Kottenhagen, H. J. M. Mulder, P. J. N. Noordeloos, and B. Leijnse, *Acta Haematol.*, 1978, **60**, 210.

[243] C. Van Der Heul, M. J. Kroos, W. L. van Noort, and H. G. Van Eijk, *Clin. Sci.*, 1981, **60**, 185.

[244] I. Esparza and J. H. Brock, *Biochim. Biophys. Acta*, 1980, **624**, 479.

[245] I. Esparza and J. H. Brock, *Biochim. Biophys. Acta*, 1980, **622**, 297.

[246] N. J. Verhoef, H. C. M. Kester, P. J. Noordeloos, and B. Leijnse, *Int. J. Biochem.*, 1979, **10**, 595.

[247] J. V. Princiotto and E. J. Zapolski, *Biochim. Biophys. Acta*, 1978, **539**, 81.

[248] J. Martinez-Medellin and L. Benavides, *Biochim. Biophys. Acta*, 1979, **584**, 84.

[249] W. P. Faulk and G. M. P. Galbraith, *Proc. R. Soc. London, Ser. B*, 1979, **204**, 83.

[250] G. M. P. Galbraith, R. M. Galbraith, and W. P. Faulk, *Placenta*, 1980, **1**, 33.

[251] G. M. P. Galbraith, R. M. Galbraith, A. Temple, and W. P. Faulk, *Blood*, 1980, **55**, 240.

[252] G. M. P. Galbraith, R. M. Galbraith, and W. P. Faulk, *Cell. Immunol.*, 1980, **49**, 215.

[253] D. M. Davies, A. Smith, U. Muller-Eberhard, and W. T. Morgan, *Biochem. Biophys. Res. Commun.*, 1979, **91**, 1504.

[254] A. Jacobs, *Ciba Symp.*, 1977, **51**, 91.

[255] A. Tangeras, T. Flatmark, D. Bäckström, and A. Ehrenberg, *Biochim. Biophys. Acta*, 1980, **589**, 162.

[256] D. M. Williams, A. J. Barbuto, C. L. Atkin, and G. R. Lee, *Prog. Clin. Biol. Res.*, 1978, **21**, 539.

chondria have established that transferrin can furnish iron for haem synthesis *in vitro* and that chelators and phosphate compounds can stimulate the accumulation of iron by the mitochondria,[257-259] such a mechanism does not seem likely *in vivo* (see the section on movement of iron across membranes). It has been shown that ferritin can supply iron for the ferrochelatase reaction by isolated rat liver mitochondria in the presence of succinate and FMN (or FAD) at steady-state rates that are essentially the same as those observed with $FeCl_3$.[260] The release of iron from ferritin is dependent on the integrity of the mitochondrial membrane, and was not observed with lysosomes, microsomes, or cytosol. Further studies indicate that, in the presence of an appropriate Fe^{2+} chelator (*e.g.* bathophenanthroline, but, to a much lesser extent, bathophenanthroline sulphate), ferritin iron can be mobilized by rat mitochondria and mitoplasts, dependent on the presence of a respiratory substrate (preferentially succinate) and a low-molecular-weight electron mediator; these observations reinforce the view that ferritin may function as an intermediate in the transport of cytosolic iron to mitochondria.[261]

The mobilization of intracellular iron has been studied in perfused rat liver[262] and in cultured rat fibroblasts and hepatocytes.[263] In the perfusion experiments, the effect of apotransferrin on iron efflux was concentration-dependent, and maximal release was observed at physiological apotransferrin concentrations; it was concluded that iron release represents the sum of two processes, namely the net release of iron by apotransferrin and exchange of iron between iron pools in the plasma and liver. Important differences in release of iron from fibroblasts when compared to hepatocytes were found in the latter study.

The mobilization of iron from iron-overloaded patients represents a major area of activity in view of the problems associated with the management of thalassemia. Without entering into details, in thalassemia major, a disequilibrium in the synthesis of the two polypeptide chains of haemoglobin provokes an anaemia which is treated by transfusion therapy and which, in the long term, results in a fatal iron overload of the subjects. Whereas early clinical studies indicated that continuous treatment with the chelating agent desferrioxamine B (administered daily by intravenous injection) retarded development of the consequences of iron overload,[264] the advent of techniques for assuring continuous infusion of the drug over long periods of time have established that subjects can be maintained in a state of negative iron balance by chelation therapy,[265] and have provoked a major effort to find chelators which can be administered orally and which are less costly than desferrioxamine B. Although it could be argued that this does not represent a normal route of iron transport, we have included a brief review of this subject here.

[257] K. Konopka, *FEBS Lett.*, 1978, **92**, 308.
[258] K. Konopka and E. Turska, *FEBS Lett.*, 1979, **105**, 85.
[259] K. Konopka and I. Romslo, *Eur. J. Biochem.*, 1980, **107**, 433.
[260] R. Ulvik and I. Romslo, *Biochim. Biophys. Acta*, 1978, **541**, 251.
[261] R. Ulvik and I. Romslo, *Biochim. Biophys. Acta*, 1979, **588**, 256.
[262] E. Baker, A. G. Morton, and A. S. Tavill, *Br. J. Haematol.*, 1980, **45**, 607.
[263] R. Rama, J.-N. Octave, Y.-J. Schneider, J.-C. Sibille, J. Limet, A. Trouet, and R. R. Crichton, *Arch. Int. Physiol. Biochim.*, 1981, **89**, B32.
[264] M. Barry, D. M. Flynn, E. A. Letsky, and R. A. Risdon, *Br. Med. J.*, 1974, **2**, 16.
[265] R. D. Propper, S. B. Shurin, and D. G. Nathan, *New Engl. J. Med.*, 1976, **294**, 1421.

The development of new iron-chelating drugs has been reviewed.[266-270] While many interesting clinical studies on desferrioxamine B have appeared, studies on animal models of iron overload suggest that two alternative pathways of chelation of iron by the drug exist, one being intracellular and the other extracellular,[271] and liposome-entrapped desferrioxamine was shown to be more effective (when administered intravenously) than the free drug.[272] Of the potential new chelators that have been identified by screening in rat and mouse models, two promising candidates, *i.e.* 2,3-dihydroxybenzoate[273] (which has the great advantage of being active when administered orally) and rhodotorulic acid (which can be produced in much greater yields than desferrioxamine B), have undergone clinical trials. Both have been eliminated, the former on account of its lack of significant chelation of iron[274] and the latter on account of painful local reactions after intramuscular or subcutaneous administration.[275] A number of other potential candidates have been identified in animal studies, including ethylenediamine-*NN'*-bis-(2-hydroxyphenylacetic acid),[270] isoniazid pyridoxalhydrazone,[276,277] and the thiosemicarbazone, dithiocarbazonate, and 2'-pyridylhydrazone of pyrazinecarboxaldehyde.[278] The use of subcutaneous infusions of desferrioxamine B to attain negative iron balance in regularly transfused patients with sickle-cell anaemia has been reported.[279]

Absorption of Iron.—A number of important studies on the absorption of iron have appeared and the subject has recently been reviewed.[280,281] The use of isolated rat duodenal cells in suspension[282] has shown that iron uptake from $FeCl_3$ is dependent on temperature, pH, and iron and cell concentrations, is unaffected by inhibitors of the respiratory chain, but is blocked by iron-chelating agents. Uptake of iron by iron-deficient cells occurs at the same rate as by normal cells, suggesting that the regulation of iron uptake is not dependent on iron status

[266] Proceedings of the Symposium on Development of Iron Chelators for Clinical Use, ed. W. F. Anderson and M. C. Hiller, U.S. Department of Health Education and Welfare, N.I.H., Bethesda, 1975.
[267] A. Jacobs, *Br. J. Haematol.*, 1979, **43**, 1.
[268] R. W. Grady, J. H. Graziano, H. A. Akers, and A. Cerami, *J. Pharmacol. Exp. Therap.*, 1976, **196**, 478.
[269] R. W. Grady and A. Cerami, *Annu. Rep. Med. Chem.*, 1978, **13**, 219.
[270] C. G. Pitt, G. Gupta, W. E. Estes, H. Rosenkrantz, J. J. Metterville, A. L. Crumbliss, R. A. Palmer, K. W. Nordquest, K. A. Sprinkle Hardy, D. R. Whitcomb, B. R. Byers, J. E. L. Arceneaux, C. G. Gaines, and C. V. Sciortino, *J. Pharmacol. Exp. Therap.*, 1979, **208**, 12.
[271] C. Hershko, *Blood*, 1978, **51**, 415.
[272] S. P. Young, E. Baker, and E. R. Huehns, *Br. J. Haematol.*, 1979, **41**, 357.
[273] J. H. Graziano, R. W. Grady, and A. Cerami, *J. Pharmacol. Exp. Therap.*, 1974, **190**, 570.
[274] C. M. Peterson, J. H. Graziano, R. W. Grady, R. L. Jones, H. V. Vlassara, V. Canala, D. R. Miller, and A. Cerami, *Br. J. Haematol.*, 1976, **33**, 477.
[275] R. W. Grady, C. M. Peterson, R. L. Jones, J. H. Graziano, K. K. Bhargava, V. A. Berdoukas, G. Kokkini, D. Loukopoulos, and A. Cerami, *J. Pharmacol. Exp. Therap.*, 1979, **209**, 342.
[276] T. Hoy, J. Humphrys, A. Jacobs, A. Williams, and P. Ponka, *Br. J. Haematol.*, 1979, **43**, 443.
[277] M. Cikrt, P. Ponka, E. Necas, and J. Neuwirt, *Br. J. Haematol.*, 1980, **45**, 275.
[278] N. E. Spingarn and A. C. Sartorelli, *J. Med. Chem.*, 1979, **22**, 1314.
[279] A. Cohen and E. Schwartz, *Am. J. Hematol.*, 1979, **7**, 69.
[280] J. D. Cook, *Fed. Proc.*, 1977, **36**, 2028.
[281] J. J. M. Marx, *Haematologica*, 1979, **64**, 479.
[282] M. A. Savin and J. D. Cook, *Gastroenterol.*, 1978, **75**, 688.

and that the initial uptake of iron by the mucosa is a passive process. Studies on mucosal biopsies of human duodenum suggest, in contrast, that iron deficiency reversibly induces brush-border iron carriers, and that the entry of iron into the enterocyte rather than cellular retention of iron is the regulatory step in the control of iron absorption.[283] This same system has been used to study the uptake of iron from a range of homologous transferrin-like proteins, and it was tentatively implied that human lactoferrin may have specific receptors on the brush border of human enterocytes.[284] The localization of ferritin, lactoferrin, and transferrin in human intestine by microscopy, using immunohistological methods, has been reported.[285] The duodenal ferritin content has been correlated with serum ferritin levels,[286] and an analysis of the subcellular distribution of iron in guinea-pig enterocytes has suggested a high degree of association with mitochondria.[287] In anaemia due to protein deficiency, the typical increase of iron incorporation in the mucosal transferrin fraction was not observed.[288] A study of the passage of iron out of the intestinal mucosa of the rat seems to underline the importance of establishing the amount of iron remaining in the intestinal lumen after administration of an iron test dose when analysing the kinetics of absorption of iron.[289]

A large number of clinical studies on iron absorption have appeared and a brief resumé is given here. A circadian variation in iron absorption in normal human subjects was observed, which indicates that iron absorption is greater in the evening[290] – we might conclude that red wine with dinner is clearly indicated! However, the effects of alcohol on iron absorption do not appear to be dependent on the iron content of the alcoholic beverage,[291,292] and it has been suggested that, in the absence of a deficiency of folic acid, chronic alcohol ingestion does not modify absorption whereas if there is folic acid deficiency, the absorption of iron is increased by increasing plasma iron turnover and by a direct effect on mucosal cell membranes.[292] The potential of Fe^{III}–EDTA as an iron fortification has been discussed,[293] as has the bioavailability of an iron–polymaltose complex[294] and hemiglobin.[295] Clinical studies on absorption of iron in normal subjects[296–299] as a function of the mode of administration have been

283 T. M. Cox and T. J. Peters, *Br. J. Haematol.*, 1980, **44**, 75.
284 T. M. Cox, J. Mazurier, G. Spik, J. Montreuil, and T. J. Peters, *Biochim. Biophys. Acta*, 1979, **588**, 120.
285 K. Isobe, T. Sakurami, and Y. Isobe, *Acta Haematol. Jpn.*, 1978, **41**, 1328.
286 J. W. Halliday, U. Mack, and L. W. Powell, *Arch. Intern. Med.*, 1978, **138**, 1109.
287 J. M. P. Hopkins and T. J. Peters, *Clin. Sci.*, 1979, **56**, 179.
288 F. A. El-Shobaki and W. Rummel, *Res. Exp. Med.*, 1978, **173**, 119.
289 A. B. R. Thomson and L. S. Valberg, *Can. J. Physiol. Pharmacol.*, 1980, **58**, 129.
290 B. Tarquini, S. Romano, M. De Scalzi, V. de Leonardis, F. Benvenuti, E. Chegai, T. Comparini, R. Moretti, and M. Cagnoni, *Adv. Biosci.*, 1979, **19**, 347.
291 A. Celada, H. Rudolf, and A. Donath, *Am. J. Hematol.*, 1978, **5**, 225.
292 A. Celada, H. Rudolf, and A. Donath, *Blood*, 1979, **54**, 906.
293 C. Martinez-Torres, E. L. Romano, M. Renzi, and M. Layrisse, *Am. J. Clin. Nutr.*, 1979, **32**, 809.
294 P. Jacobs and L. Wormald, *J. Med. (Cincinnati)*, 1979, **10**, 279.
295 E. E. Gabbe, H. C. Heinrich, J. Brüggemann, and A. A. Pfau, *Nutr. Metab.*, 1979, **23**, 17.
296 E. R. Monsen and J. D. Cook, *Am. J. Clin. Nutr.*, 1979, **32**, 804.
297 L. Hallberg, E. Björn-Rasmussen, and L. Rossander, *Br. J. Nutr.*, 1979, **41**, 283.
298 L. Hallberg, E. Björn-Rasmussen, L. Howard, and L. Rossander, *Scand. J. Gastroenterol.*, 1979, **14**, 769.
299 L. Rossander, L. Hallberg, and E. Björn-Rasmussen, *Am. J. Clin. Nutr.*, 1979, **32**, 2484.

reported, as well as on iron uptake by infants.[300-302] Iron absorption is stimulated by chlorinated aromatic hydrocarbons.[303] The roles of phytate, of vitamin A, of oxalate, and of phosphate compounds in iron absorption have been examined [304-309] and the inhibition of iron absorption by tea in thalassemics has been underlined.[310] Evidence that erythropoietin enhances iron transport across the intestinal tract,[311] that pectin reduces iron absorption in haemochromatosis,[312] and on the effects of Mn and Pb on iron uptake[313,314] have appeared. Studies on the sites of absorption and translocation of iron in barley roots[315] and of recycling of metabolized iron by a marine dinoflagellate[316] have been reported.

Mammalian Metabolism of Iron.—Recent advances in iron metabolism and in iron metabolism and deficiency during pregnancy have been reviewed.[317,318] Ferrokinetic studies imply that the main factors determining iron turnover in tissues are the level of iron stores and the erythropoietic activity of bone marrow.[319] A compartmental analysis of iron metabolism and placental transfer of iron in near-term rhesus monkeys has been carried out[320] and the transfer of maternal liver iron to foetal liver in pregnant rats has been studied.[321] No short-term effects of adrenaline on iron metabolism were observed in rats.[322]

In iron-deficient rats, an impaired work performance is observed, and muscle activity was associated with higher concentrations of lactate in the blood than in iron-replete rats. It was concluded that the increased lactate formation was due to depletion in the mitochondrial iron-containing enzyme α-glycerophosphate dehydrogenase, which is an essential enzyme for coupling the regeneration of cytoplasmic NADH *via* the glycerol phosphate shuttle.[323] This is the first clear-cut case of the metabolic consequences of iron deficiency at the level of the activity of iron-containing enzymes. The supplementation of cow-milk, which has a

[300] M. J. Dauncey, C. G. Davis, J. C. L. Shaw, and J. Urman, *Pediatr. Res.*, 1978, **12**, 899.
[301] U. M. Saarinen and M. A. Siimes, *Pediatr. Res.*, 1979, **13**, 143.
[302] K. Furugouri and A. Kawabata, *J. Anim. Sci.*, 1979, **49**, 715.
[303] J. Manis and G. Kim, *Am. J. Physiol.*, 1979, **5**, E763.
[304] D. A. Lipschitz, K. M. Simpson, J. D. Cook, and E. R. Morris, *J. Nutr.*, 1979, **109**, 1154.
[305] R. Ellis and E. R. Morris, *Nutr. Rep. Int.*. 1979, **20**, 739.
[306] M. Liebman and J. Driskell, *Nutr. Rep. Int.*, 1979, **19**, 281.
[307] L. A. Mejia, R. E. Hodges, and R. B. Rucker, *J. Nutr.*, 1979, **109**, 129.
[308] D. R. Van Campen and R. M. Welch, *J. Nutr.*, 1980, **110**, 1618.
[309] A. W. Mahoney and D. G. Hendricks, *J. Food Sci.*, 1978, **43**, 1473.
[310] P. A. deAlarcon, M. E. Donovan, G. B. Forbes, S. A. Landaw, and J. A. Stockman, *New Engl. J. Med.*, 1979, **300**, 5.
[311] A. Gutnisky, E. Speziale, M. F. Gimeno, and A. L. Gimeno, *Experientia*, 1979, **35**, 623.
[312] L. Monnier, C. Colette, C. Ribot, and J. Mirouze, *Ann. Biol. Anim. Biochem. Biophys.*, 1979, **19**, 775.
[313] N. Gruden, *Nutr. Rep. Int.*, 1979, **19**, 69.
[314] P. R. Flanagan, D. L. Hamilton, J. Haist, and L. S. Valberg, *Gastroenterology*, 1979, **77**, 1074.
[315] D. T. Clarkson and J. Sanderson, *Plant Physiol.*, 1978, **61**, 731.
[316] B. E. Frey and L. F. Small, *J. Phycol.*, 1979, **15**, 405.
[317] L. W. Powell, J. W. Halliday, and M. L. Bassett, *Aust. N.Z. J. Med.*, 1979, **9**, 578.
[318] J. G. McFee, *Clin. Obstet. Gynecol.*, 1979, **22**, 799.
[319] I. Cavill, C. Ricketts, and A. Jacobs, *Clin. Sci.*, 1979, **56**, 223.
[320] J. P. van Dijk, *Eur. J. Obstet., Gynecol. Reprod. Biol.*, 1977, **7**, 127.
[321] H. G. Van Eijk, M. J. Kroos, C. Van Der Heul, N. C. Verhoef, C. M. H. de Jeu-Jaspars, and H. C. S. Wallenburg, *Eur. J. Obstet., Gynecol. Reprod. Biol.*, 1980, **10**, 389.
[322] R. Flos and A. Armario, *Rev. Esp. Fisiol.*, 1978, **34**, 167.
[323] C. A. Finch, P. D. Gollnick, M. P. Hlastala, L. R. Miller, E. Dillmann, and B. Mackler, *J. Clin. Invest.*, 1979, **64**, 129.

very low natural content of iron and copper, by these transition metals has been recommended for infants who are at risk of iron deficiency. However, the contamination of milk with metal ions leads to the development of a characteristic oxidized flavour, probably attributable to lipid peroxidation. The use of Fe^{III} chelates of nitrilotriacetate and lactobionate produces less peroxidation of the lipids, presumably by rapid donation of iron to the phosphorylserine residues of casein, which effectively removes the iron from the lipid phase.[324-326] Ochratoxin A, one of a family of toxins produced by *Aspergillus* and *Penicilinium* species, induces iron-deficiency anaemia in chicks.[327] However, in mice, iron-deficient animals were protected from hepatocellular damage and other toxic effects of 2,3,7,8-tetrachlorodibenzo-*p*-dioxin.[328]

An animal model of haemochromatosis which would reflect the endocrine and cardiac disturbances found in Man has not been found to date. However, Awai[329] has now reported that intraperitoneal injection of ferric nitrilotriacetate causes diabetes both in rats and in rabbits. The possibility that iron overload provokes lysosomal disruption was suggested by the finding of increased activities of lysosomal enzymes in liver biopsies of iron-overloaded patients.[330] Further analysis shows a close correlation between lysosomal fragility, as determined by measurements of latent N-acetyl-β-glucosaminidase, and the haemosiderin content of the tissue;[331] the authors suggest that haemosiderin is responsible for the lysosomal disruption, and thus for the tissue damage in iron overload. Studies of hepatic iron clearance from serum in treated haemochromatosis suggested the existence of a cellular abnormality of hepatic metabolism of iron in this disease.[332]

Microbial Metabolism of Iron and Chelators of Iron.—The requirements of all living cells for iron are such that, in prokaryotes as well as in eukaryotes, transport and storage systems have been developed. In the absence of transferrin, microorganisms synthesize a number of chelators which are used to mobilize biologically available iron (predominantly as poorly soluble ferric iron) from their environment. Typically, these organisms excrete into the extracellular medium low-molecular-weight complexants (hydroxamates, thiohydroxamates, or catechol derivatives) which chelate Fe^{III} in an octahedral high-spin complex[333] and which are thereafter transported into the microbial cell, usually by a receptor-mediated process. Thereafter the iron is released from the complexant either by reduction (mediated by a specific enzyme system) or by hydrolysis of the iron–chelate complex. A number of reviews on the siderophores themselves, on their

324   J. Hegenauer, P. Saltman, D. Ludwig, L. Ripley, and P. Bajo, *J. Agric. Food Chem.*, 1979, 27, 860.
325   J. Hegenauer, P. Saltman, and D. Ludwig, *J. Agric. Food Chem.*, 1979, 27, 868.
326   J. Hegenauer, P. Saltman, D. Ludwig, L. Ripley, and A. Ley, *J. Agric. Food Chem.*, 1979, 27, 1294.
327   W. E. Huff, C. F. Chang, M. F. Warren, and P. B. Hamilton, *Appl. Environ. Microbiol.* 1979, 37, 601.
328   G. D. Sweeney, K. G. Jones, F. M. Cole, D. Basford, and F. Krestynski, *Science*, 1979, 204, 332.
329   M. Awai, *Acta Haematol. Jpn.*, 1978, 41 ,1293.
330   C. A. Seymour and T. J. Peters, *Br. J. Haematol.*, 1978, 40, 239.
331   C. Selden, M. Owen, J. M. P. Hopkins, and T. J. Peters, *Br. J. Haematol.*, 1980, 44, 593.
332   R. G. Batey, J. E. Pettit, A. W. Nicholas, S. Sherlock, and A. V. Hoffbrand, *Gastroenterology*, 1978, 75, 856.
333   K. N. Raymond, K. Abu-Dari, and S. R. Sofen, *ACS Symp. Ser.*, 1980, 119, 133.

uptake by microbial cells, on their stereochemistry, and on methods of studying iron transport into bacterial cells have appeared.[333-336]

The best-studied system is *Escherichia coli*, where several different pathways for the uptake of iron have been established. It is interesting to note that siderophores share the same receptors as bacteriophages and colicins on the surface of *E. coli*. Thus, the specific receptor for its catechol siderophore enterobactin, the product of the cbr gene, is also the receptor for colicin B. *E. coli* can also take up iron from hydroxamate siderophores at an outer membrane receptor (ton A), which also serves as a common binding site for albomycin, colicin M, and the bacterio-phages T1, T5, and φ80. *E. coli* also has a system for uptake of iron from ferric citrate.

The outer membrane receptor of ferric enterobactin has been solubilized[337] and evidence of competition between colicin B and ferric enterobactin for the receptor *in vitro* has been established. The receptor is virtually absent from a colicin-B-resistant mutant. A membrane protein of molecular weight 76 000 has been identified as the binding protein that is essential for the transport of ferric enterochelin iron.[338] Modification of the receptor protein, either in whole membranes or in Triton-X-100-solubilized membranes, resulting in the conversion of the subunit molecular weight from 81 000 to 74 000, and being inhibited by benzamidine, was observed.[339] From mutant and other analyses, the receptor was found to be identical with component a, which is a well-known major protein of the outer membrane.[340] The ferric form of a carboxylic analogue of enterobac-tin, *cis*-1,5,9-tris-(2,3-dihydroxybenzamido)cyclododecane, supports the growth of siderophore auxotrophs, apparently without hydrolysis to dihydroxybenzoate and re-synthesis to enterobactin.[337] This would seem to question the role (pre-viously suggested) of an esterase in the release of iron from enterobactin. An iron(II) enterobactin has been identified[341] as the major form at pH 4; formation of the iron(II) complex is presumed to occur by electron-transfer interactions of the iron with catechol rings and possibly adjacent water molecules. From the titration curve for iron enterobactin proton,[342] iron(II) enterobactin would be predicted to have four additional protons compared to iron(III) enterobactin, and thus to have a net charge of zero. If the environment between the cell wall and the cytoplasmic membrane were acidic, a proportion of the iron enterobactin would be in the iron(II) state. This would eliminate the problem of transporting a tribasic anion through the cytoplasmic membrane.

The transport of ferrichrome for *E. coli* strains that are deficient in the synthesis of enterobactin has been studied, using inner membrane vesicles;[343] since com-

[334] J. B. Neilands, *Ciba Symp.*, 1977, **51**, 107.

[335] K. N. Raymond and C. J. Carrano, *Acc. Chem. Res.*, 1979, **12**, 183.

[336] H. Rosenberg, *Methods Enzymol.*, 1979, **56**, 388.

[337] W. C. Hollifield and J. B. Neilands, *Biochemistry*, 1978, **17**, 1922.

[338] A. Boyd and I. B. Holland, *FEBS Lett.*, 1977, **76**, 20.

[339] W. C. Hollifield, E. H. Fiss, and J. B. Neilands, *Biochem. Biophys. Res. Commun.*, 1978, **83**, 739.

[340] E. H. Fiss, W. C. Hollifield, and J. B. Neilands, *Biochem. Biophys. Res. Commun.*, 1979, **91**, 29.

[341] R. C. Hider, J. Silver, J. B. Neilands, I. E. G. Morrison, and L. V. C. Rees, *FEBS Lett.*, 1979, **102**, 325.

[342] S. Salama, J. D. Stong, J. B. Neilands, and T. G. Spiro, *Biochemistry*, 1978, **17**, 3781.

[343] R. S. Negrin and J. B. Neilands, *J. Biol. Chem.*, 1978, **253**, 2339.

parable kinetics were found in vesicles prepared from *E. coli* RW 193 and AN 193 (the latter lacking a functional outer-membrane receptor for ferrichrome), it was concluded that there is a discrete uptake system in the cytoplasmic membrane. From studies with [^{51}Cr]desferri-ferrichrome it was suggested that penetration of the inner membrane is the rate-limiting step in ferrichrome transport, which was also found to be dependent on the presence of an energized membrane. It was proposed that transport of ferrichrome in these vesicles proceeds by a symport mechanism, possibly involving cations in addition to iron. More recent studies suggest that the energetics of ferrichrome transport can be explained by a unique structural feature of the iron-bound form of the siderophore (Figure 5)

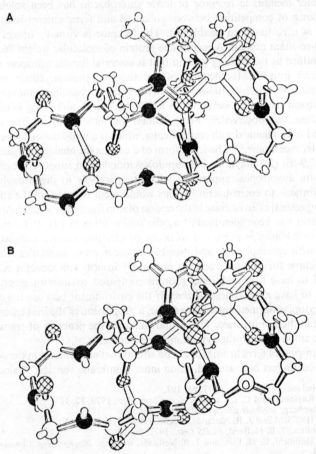

Figure 5 (*A*) *A Nicholson model of ferrichrome, based on the X-ray structure of ferrichrysin* (*B*) *A Nicholson model of the proposed structure for the ferrichrome–alkaline-earth-metal complex. Hydrogen-bonds and metal–oxygen bonds are not shown*
● *Nitrogen;* ○ *carbon;* ◉ *oxygen*
(Reproduced by permission from *Naturwissenschaften*, 1980, **67**, 136)

which enables the complexation of an alkaline-earth cation.[344] The tonA-specified protein of molecular weight 78 000 is the receptor for the ferrichrome complex.[345,346] Studies with *E. coli* mutants with reduced amounts of major outer-membrane proteins show that, while no significant differences in iron uptake of strains with markedly reduced amounts of protein I (mol. wt 36 000) are observed, reduction of the initial velocity of iron uptake (up to 40%) is observed in *E. coli* that is deficient in outer-membrane protein II.[347] Further genetic studies, using mutants that are resistant to lethal agents which bind to the ton A protein, identified two additional genes, one of which is required for the utilization of rhodotorulic acid but not ferrichrome; it has been suggested that the other gene codes for a transport component in the cytoplasmic membrane that is necessary for the uptake of all hydroxamate siderophores following their receptor-mediated passage across the outer membrane.[348]

Enterobactin is not re-utilized, the cyclic triester bonds of ferric enterobactin being cleaved by a specific esterase[349] (called by Raymond and Carrano[335] 'the American approach to iron transport'). In contrast, it is assumed that the desferri-ferrichromes could be used repeatedly to transport iron into *E. coli*; after transport of the ferric complex into the cell and release of iron by reduction, the ligand could be excreted for another cycle (the 'European approach'). From a recent study by Hartmann and Braun,[350] it appears that there is no 'European' system for iron transport, since ferrichrome is modified into a form (probably acetylated on one of the N-hydroxyl groups) which has little affinity for iron. This modification occurs at the same rate as the rate of transport of iron, and, since desferri-ferrichrome is rapidly converted into the modified form, it seems likely that this is the true substrate for the modification reaction, and not the ferrichrome. The authors assume that ferric iron is reduced and released before modification. The redox potential of ferrichrome has been estimated to be about -450 mV at pH 7, which lies within the range of physiological reductants.[351] In *Ustilago sphaerogena* (a smut fungus), an NADH-dependent ferrichrome reductase has been identified which catalyses the dissociation of the iron–ferrichrome complex by reduction of the metal. A spectrophotometric assay, based on trapping of Fe^{2+} by ferrazine, was developed.[352] Similar activities have been reported in *Pseudomonas aeruginosa*[353] and in *Bacillus megaterium*;[354] in the pseudomonad, a second activity, specific for ferric citrate, was reported. The uptake of ferrichrome and the release of iron in *E. coli* do not depend on

[344] R. C. Hider, A. F. Drake, R. Kuroda, and J. B. Neilands, *Naturwissenschaften*, 1980, **67**, 136.

[345] K. Hantke and V. Braun, *FEBS Lett.*, 1975, **49**, 301.

[346] M. Luckey, R. Wayne, and J. B. Neilands, *Biochem. Biophys. Res. Commun.*, 1975, **64**, 687.

[347] J. W. Coulton and V. Braun, *J. Gen. Microbiol.*, 1979, **110**, 211.

[348] R. J. Kadner, K. Heller, J. W. Coulton, and V. Braun, *J. Bacteriol.*, 1980, **143**, 256.

[349] K. T. Greenwood and R. K. J. Luke, *Biochim. Biophys. Acta*, 1978, **525**, 209.

[350] A. Hartmann and V. Braun, *J. Bacteriol.*, 1980, **143**, 246.

[351] S. R. Cooper, J. V. McArdle, and K. N. Raymond, *Proc. Natl. Acad. Sci. USA*, 1978, **75**, 3551.

[352] J. G. Straka and T. Emery, *Biochim. Biophys. Acta*, 1979, **569**, 277.

[353] C. D. Cox, *J. Bacteriol.*, 1980, **141**, 199.

[354] J. E. L. Arceneaux and B. R. Byers, *J. Bacteriol.*, 1980, **141**, 715.

cytochromes.[355] A high-affinity iron-uptake system in *E. coli*, associated with the presence of plasmids which specify the production of colicin V, has been described.[356,357] Iron uptake from ferric pyochelin and ferric citrate by *Pseudomonas aeruginosa*[358] and from ferriexochelin by *Mycobacterium* strains[359,360] has been studied. The mechanism of iron uptake from rhodotorulic acid by the yeast *Rhodotorula pilimanae* (using radioactively labelled Fe and chelator, and kinetically inert chromic-substituted chelator complexes) appears to involve the mediation of iron transport to the cell without actual transport of iron into the cell by the chelator. It is suggested that, at the cell surface, rhodotorulic acid exchanges the ferric iron with a membrane-bound chelating agent, which completes the active transport of iron into the cell.[361] Thus, as indicated in Figure 6, three mechanisms for the uptake of iron in micro-organisms may be envisaged,[335] of which finally only the one just described for rhodotorulic acid in yeast may be truly 'European' or conservative.

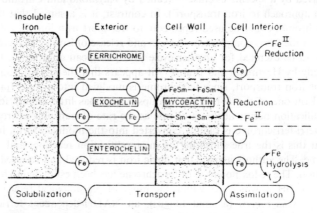

Figure 6 *Three limiting mechanisms for microbial transport of iron* (Reproduced by permission from *Acc. Chem. Res.*, 1979, **12**, 183)

A rapid column-chromatographic method, using a polyamide support, has been described for isolation of catechol-type siderophores.[362] Evidence for an oxazoline ring in the catecholamide spermidine siderophore from *Paracoccus denitrificans* has been found and a revised structure proposed.[363] A similar structural feature was found in agrobactin, a siderophore produced by *Agrobacterium tumefaciens*.[364] Siderophores (about 5—10 mg per kg) were detected in mould-ripened cheese. Oregon Blue cheese apparently contained ferrichrome,

[355] B. Eberspächer and V. Braun, *FEMS Microbiol. Lett.*, 1980, **7**, 61.
[356] S. J. Stuart, K. T. Greenwood, and R. K. J. Luke, *J. Bacteriol.*, 1980, **143**, 35.
[357] P. H. Williams and P. J. Warner, *Infect. Immun.*, 1980, **29**, 411.
[358] C. D. Cox, *J. Bacteriol.*, 1980, **142**, 581.
[359] M. C. Stephenson and C. Ratledge, *J. Gen. Microbiol.*, 1979, **110**, 193.
[360] M. C. Stephenson and C. Ratledge, *J. Gen. Microbiol.*, 1980, **116**, 521.
[361] C. J. Carrano and K. N. Raymond, *J. Bacteriol.*, 1978, **136**, 69.
[362] A. V. Robinson, *Anal. Biochem.*, 1979, **95**, 364.
[363] T. Peterson and J. B. Neilands, *Tetrahedron Lett.*, 1979, 4805.
[364] S. A. Ong, T. Peterson, and J. B. Neilands, *J. Biol. Chem.*, 1979, **254**, 1860.

while *Penicillium roqueforti* excreted copro'gen.[365] The structures of a number of other siderophores have been reported[366-368] and there have been reports on siderophore synthesis and biological activity.[369-372] A low-molecular-weight iron-binding factor from mammalian cells that potentiates microbial growth has been partially purified and characterized.[373] The effectiveness of iron chelates in plant nutrition has been studied with iron-deficient soybeans.[374]

3 Transport and Storage of Other Transition Metals

Introduction.—A number of books and review articles on the biology and biochemistry of trace elements have appeared recently.[375-378] For elements such as cobalt, nickel, manganese, chromium, molybdenum, *etc.*, the uptake, effects on animal and plant growth, excretion, toxicity, and deficiency are well documented. However, the transport and storage forms of these elements remain, to some extent, unknown. Most of the transition elements can bind specifically or non-specifically with transferrin and other serum proteins and can be stored in the form of soluble complexes or in ferritin. In consequence, the trace elements play an important role in iron metabolism when their concentrations deviate from normal levels. This point of view will be discussed in more detail in the last section of this Report. In the following sections we have examined particularly the transport and storage of copper, zinc, cadmium, mercury, vanadium, and some actinides. Some recent data on metallothionein are also summarized.

Copper.—Copper is transported from the gut to the liver in the form of Cu–albumin, Cu–histidine, or mixed complexes. The binding of copper to mucosal transferrin has been reported in the case of iron deficiency in rats.[379] In the liver, copper is transferred to caeruloplasmin, which contains 95% of the circulating copper in the blood of mammals or other species and which constitutes the source of copper for cytochrome oxidase and other copper-containing enzymes.[380] Since the last Report[1] appeared, progress on the spectroscopic characterization of the copper-binding sites, primary structure, and functional properties of

[365] S. A. Ong and J. B. Neilands, *J. Agric. Food Chem.*, 1979, **27**, 990.
[366] T. Emery, *Biochim. Biophys. Acta*, 1980, **629**, 389.
[367] T. Takemoto, K. Nomoto, S. Fushiya, R. Ouchi, G. Kusano, H. Hikino, S. Takagi, Y. Matsuura, and M. Kakudo, *Proc. Jpn. Acad., Ser. B*, 1978, **54**, 469.
[368] S. Fushiya, Y. Sato, and S. Nozoe, *Tetrahedron Lett.*, 1980, **21**, 3071.
[369] J.-M. Meyer and M. A. Abdallah, *J. Gen. Microbiol.*, 1980, **118**, 125.
[370] R. D. Perry and C. L. San Clemente, *J. Bacteriol.*, 1979, **140**, 1129.
[371] J. H. Marcelis, H. J. Den Daas-Slagt, and J. A. A. Hoogkamp-Korstanje, *Antonie van Leeuwenhoek; J. Microbiol. Serol.*, 1978, **44**, 257.
[372] J. E. Armstrong and C. Van Baalen, *J. Gen. Microbiol.*, 1979, **111**, 253.
[373] R. L. Jones, C. M. Peterson, R. W. Grady, and A. Cerami, *J. Exp. Med.*, 1980, **151**, 418
[374] A. Wallace, R. T. Mueller, and G. V. Alexander, *Commun. Soil Sci. Plant Anal.*, 1976, **7**, 1.
[375] 'Trace Element Metabolism in Man and Animals', ed. M. Kirchgessner, Tech. Univ. Munich Freising, Weihenstephen, Germany, 1978.
[376] R. J. P. Williams, *Chem. Br.*, 1979, **15**, 506.
[377] 'Environmental Chemistry of the Elements', ed. H. J. M. Bowen, Academic Press, New York, 1979.
[378] I. G. Macara, *Trends Biochem. Sci.*, 1980, **5**, 92.
[379] F. A. El-Shobaki and W. Rummel, *Res. Exp. Med.*, 1979, **174**, 187.
[380] E. Frieden, in ref. 375, p.8.

caeruloplasmin has been reported. Two recent reviews on caeruloplasmin biochemistry have also appeared.[381,382]

Caeruloplasmin. The copper-binding sites. The copper atoms in caeruloplasmin
have been classified in three distinct types, in terms of their spectroscopic and
paramagnetic properties.[383] The type I, blue, copper is characterized by an intense
electronic absorption band at 600 nm and by a narrow e.p.r. hyperfine structure.
The type II, colourless, copper presents e.p.r. hyperfine structure similar to that
found in tetragonal copper complexes. The type III copper is non-detectable by
e.p.r. and is associated with an absorption band at 300 nm. Caeruloplasmin has
seven copper-binding sites: two type I, one type II, and four type III.

The distorted tetrahedral CuN_2SS^* (N = His, S = Cys, S* = Met) co-ordination
geometry for the type I site has been re-examined and confirmed in native caeruloplasmin. The ligand-field parameters that were obtained for the type I copper in
caeruloplasmin are remarkably similar to those extracted from an analysis of the
d–d spectrum of plastocyanin. Further support for this comes from the very close
similarities between caeruloplasmin and plastocyanin (or azurin) in the charge-
transfer region of the spectra (absorption, c.d., and m.c.d.) of type I copper. There
are also spectroscopic similarities between caeruloplasmin and tree laccase. Type
I copper of caeruloplasmin is known to differ in its reactivity. Given the close electronic structural similarities of the two type I coppers, Dawson *et al.*[384] suggest
that the above-mentioned differences in reactivity reflect variation in the degree to
which the redox centres are buried in the protein and/or the placement of each
type I site relative to the other copper. Type II and III sites have been examined
in ascorbate-treated caeruloplasmin. Dialysis of native caeruloplasmin against
ascorbate results in loss of the type I copper, thereby permitting spectroscopic
examination of the type II and type III copper. Visible absorption and c.d. spectra
data of ascorbate-treated caeruloplasmin are consistent with tetradehral or near-
tetrahedral geometries for the type II and type III copper, with four nitrogen
donor ligands, as has been suggested by e.p.r. spectroscopy.[384]

The primary structure of caeruloplasmin and evolutionary data. Caeruloplasmin
has been shown to be a single polypeptide chain, with a molecular weight of
about 130 000, that is readily cleaved to large fragments by proteolytic enzymes.
[385,386] Three fragments, of M_r 20 000, 53 000, and 67 000, have been isolated
and the complete amino-acid sequence of the smallest fragment has been reported.[387] This fragment, of 159 amino-acid residues, is histidine-rich and appears
to be attached to the *C*-terminal end of the caeruloplasmin chain by a 'labile
interdomain peptide bond'. The secondary structure of the *C*-terminal peptide
has been predicted. In the model proposed, approximately 30% of the residue

381 E. Frieden, in 'Copper in the Environment' ed. J. O. Nziagu, Wiley, New York, 197
    p. 241.
382 E. Frieden, *Ciba Symp.*, 1979, **79**, 93.
383 L. Ryden and I. Björk, *Biochemistry*, 1976, **15**, 3411.
384 J. H. Dawson, D. M. Dooley, R. Clark, P. J. Stephens, and H. B. Gray, *J. Am. Chem.
    Soc.*, 1979, **101**, 5046.
385 L. Ryden, *Eur. J. Biochem.*, 1972, **26**, 380.
386 I. B. Kingston, B. L. Kingston, and F. W. Putnam, *Proc. Natl. Acad. Sci. USA*, 1977, 7
    5377.
387 I. B. Kingston, B. L. Kingston, and F. W. Putnam, *Proc. Natl. Acad. Sci. USA*, 197
    76, 1668.

occur in β-sheets, 25% in α-helices, and 45% in β-turns or other structures. Some of the histidine residues are close to one cysteine, which is assumed to be a copper-binding site.

In Nature, copper is associated with a few intensely blue proteins which can be divided mainly into two classes (for a complete review, see ref. 388). In brief, there are, first, the blue oxidases; laccase (found in lacquer trees and fungi), ascorbate oxidase (from plants of the cucumber family), and caeruloplasmin (ferroxidase) (from blood serum of mammals). The oxidases contain from four to eight atoms of copper and have M_r between 60 000 and 140 000. The second group of blue proteins have lower M_r (between 11 000 and 15 000) and contain one copper atom per mole of protein: they are the plastocyanins (found in photosynthetic cells), the azurins (from pseudomonal bacteria and plants), and stellacyanin (found in lacquer trees). Early studies, conducted on the primary sequences of azurins and eight plastocyanins, demonstrated that these two families of proteins were monophyletic.[389] The recent reports on the three-dimensional structures of poplar plastocyanins[390] and *Pseudomonas* azurin[191] have confirmed the conclusions of Ryden and Lundgren. Now that the complete amino-acid sequence of stellacyanin,[392] partial sequences around the copper-binding site of human Cp,[393] and the amino-acid sequence of subunit H of bovine mitochondrial cytochrome *c* oxidase are available, a comparison of all these proteins, regarding their evolutionary relationship, has been reported.[394,395] It is suggested that all of the proteins examined are monophyletic. There is evidence that caeruloplasmin has undergone an internal duplication.[396] In the history of these proteins, the divergence of azurin and plastocyanin from a common ancestor would be the earliest event, stellacyanin and later caeruloplasmin would evolve from plastocyanin, while the cytochrome *c* oxidase subunit would derive from the azurin ancestor. However, Kingston *et al.* stated that there are no statistically significant relationships between caeruloplasmin fragments and azurin, plastocyanin, and superoxide dismutase.[387]

Ferroxidase activity of caeruloplasmin. The ferroxidase activity of caeruloplasmin and its role in the mobilization of iron and in the formation of transferrin have often been questioned. Recently, it has been shown that, *in vitro*, caeruloplasmin at a physiological concentration has no effect on the transfer of iron from ferritin to transferrin. Moreover, apotransferrin has a low affinity for iron that has been oxidized by caeruloplasmin. On the other hand, if Fe^{II} has been first complexed by apotransferrin, then caeruloplasmin enhances the formation of transferrin.[153] In rat hepatocytes, in culture, Rama *et al.* showed that

[388] B. G. Malmström, L.-E. Andreasson, and B. Reinhamman, in 'The Enzymes', ed. P. Boyer, Academic Press, New York, 1975, 3rd edn., Vol. XII, p. 507.

[389] L. Ryden and J.-O. Lundgren, *Nature (London)*, 1976, **261**, 344.

[390] P. M. Colman, H.-C. Freeman, J. M. Guss, M. Murata, V. A. Norris, J. A. M. Ramshaw, and M. P. Venratappa, *Nature (London)*, 1978, **272**, 319.

[391] E. T. Adman, R. E. Stenkamp, L. C. Scikker, and L. H. Jensen, *J. Mol. Biol.*, 1978, **123**, 35.

[392] C. Bergman, E.-K. Gandvik, P. O. Nyman, and L. Strud, *Biochem. Biophys. Res. Commun.*, 1977, **77**, 1052.

[393] G. J. Steppens and G. Guse, *Hoppe-Seyler's Z. Physiol. Chem.*, 1979, **360**, 613.

[394] L. Ryden and J.-O. Lundgren, *Biochimie*, 1979, **61**, 781.

[395] L. Ryden and J.-O. Lundgren, *Protides Biol. Fluids*, 1980, **28**, 87.

[396] F. E. Dwulet and F. W. Putnam, *Protides Biol. Fluids*, 1980, **28**, 83.

caeruloplasmin cannot mobilize iron from intracellular ferritin.[263] The role of caeruloplasmin in the metabolism of iron and in the formation of transferrin is far from being elucidated. It seems that the choice of the experimental system is of great importance. For example, the role of caeruloplasmin in the mobilization of iron in pigs seems to be unique, and cannot be extrapolated to other mammals.

Caeruloplasmin as an antioxidant. Based on its ability to inhibit the autoxidation of lipids that is induced by ascorbic acid or by inorganic iron,[397] Dormandy has suggested that caeruloplasmin may be the major antioxidant in serum.[398] Goldstein also suggested another antioxidant property for caeruloplasmin, *i.e.* its capacity to catalyse the dismutation of superoxide anion radicals (O_2^-) to H_2O_2 and O_2.[399] This function is normally carried out by superoxide dismutase, a copper-containing intracellular enzyme. On a weight basis, the O_2^--scavenging activity of caeruloplasmin is substantially less than that of purified human erythrocyte superoxide dismutase (3000-fold, in the experimental conditions reported). Nevertheless, superoxide dismutase being almost exclusively an intracellular enzyme, caeruloplasmin may be an important circulating scavenger of O_2^- (in particular, under conditions where the levels of caeruloplasmin are elevated, as during the later stages of pregnancy or during acute infections).

Zinc.—It has been shown previously that, in venous plasma, about one-third of the zinc is tightly bound to α_2-macroglobulin and two-thirds is loosely bound to albumin, with traces being bound to the amino-acids histidine and cysteine.[400] However, Evans showed that, in rats, transferrin is the protein that is mainly responsible for the uptake of zinc from the intestinal membrane and its transport through the portal vein to the liver.[401] More recently, from studies on the relative affinity of transferrin and albumin for zinc *in vitro*, it was concluded that albumin binds zinc more firmly than does transferrin.[402] In perfused rat intestine, albumin was identified as the protein that is involved in the removal of zinc from intestinal mucosal cells and its transport, in portal blood, to the liver.[403]

In 1976 it was suggested that, in the rat, the zinc-binding ligand that is responsible for absorption of zinc is prostaglandin E_2 (PGE_2).[404] A prostaglandin-like ligand for zinc has also been found in human milk.[405] While prostaglandins E_2 and F_2 probably act as regulators of the transport of zinc by intestinal mucosa, they do not contain chemical groups that are capable of forming stable metal complexes. It was also suggested that citric acid is the zinc-binding ligand[406] or simply a breakdown product of a zinc metallothionein.[407] Finally, Evans

[397] D. J. An-Timimi and T. L. Dormandy, *Biochem. J.*, 1977, **168**, 283.

[398] T. L. Dormandy, *Lancet*, 1978, **1**, 647.

[399] I. M. Goldstein, H. B. Kaplan, H. J. Edelson, and G. Weissmann, *J. Biol. Chem.*, 1979, **254**, 4040.

[400] R. I. Henkin, in 'Protein–Metal Interactions', ed. M. Friedman, Plenum, New York, 1974, p. 299.

[401] G. W. Evans, *Proc. Exp. Biol. Med.*, 1976, **51**, 775.

[402] P. A. Charlwood, *Biochim. Biophys. Acta*, 1979, **581**, 260.

[403] K. T. Smith, M. L. Failla, and R. J. Cousins, *Biochem. J.*, 1979, **184**, 627.

[404] M. K. Song and N. F. Adham, *Fed. Proc.*, 1976, **35**, 1667.

[405] G. W. Evans and P. E. Johnson, *Lancet*, 1976, **2**, 1310.

[406] L.-S. Hurley, B. Lonnerdal, and A. G. Stanislowski, *Lancet*, 1979, **1**, 677.

[407] R. J. Cousins, *Am. J. Clin. Nutr.*, 1979, **32**, 339.

and Johnson suggested that picolinic acid is the metabolite that is responsible for the binding of zinc in human milk[408] and rat intestine.[409] As recently proposed by Evans,[410] the absorption of zinc proceeds as follows: in the exocrine cells of the pancreas, the metabolism of tryptophan produces picolinic acid, which is secreted in the lumen of the intestine. In the lumen, picolinic acid co-ordinates with zinc to form a complex that facilitates the passage of zinc through the luminal membrane.

Cadmium.—Many aspects of cadmium biochemistry in mammals have been reviewed recently.[411] While cadmium may enhance the catalytic activities of more than 27 enzymes, absorption of high doses may cause testicular damage, placental haemorrhage, and foetal malformation and death.[412] Oral as well as parenteral administration of a sub-toxic amount of cadmium induces the synthesis of metallothionein in liver, kidney, and intestine. Cadmium has a particularly strong affinity for metallothionein. However, because of its potential nephrotoxic effect, Cd-metallothionein is not considered as a transport form of cadmium. In the bile, cadmium has been found to be associated with glutathione.[413] Recently, Cherian has suggested that chelation of cadmium by low-molecular-weight sulphur-containing amino-acids and peptides may be an important mode of transport of cadmium from organ to organ. Weak binding of cadmium to plasma proteins, such as albumin and macroglobulin, cannot be excluded.[414]

Mercury.—In response to the adminstration of mercury salts or elemental mercury vapour, a low-molecular-weight Hg-binding protein is formed in the kidneys of the rat.[415] In contrast to cadmium, mercury does not induce the synthesis of a binding protein in the liver, where the induction by cadmium is high. The same effect was observed for bismuth and gold. The proteins so induced by mercury, bismuth, and gold in the kidney have been called 'Renal Metal Binding Proteins' (RMBP).[416] The stimulating properties of various metals with regard to metallothionein and related proteins are presented in Table 2.[416]

The mercury-induced RMBP have a molecular weight between 7500 and 9500, and they contain 5.7 to 8.9% of Hg. The RMBP display significant differences as compared with metallothioneins.

Metallothioneins.—The administration of copper, zinc, cadmium, or mercury to animals induces the synthesis of a particular type of protein called metallothionein, which plays an important role in the metabolism of these elements. The biochemistry of metallothioneins has been widely discussed at the First International Meeting on Metallothionein and other Low Molecular Weight Metal-Binding Proteins, held in Zürich in July 1978. The papers presented at the meeting and

[408] G. W. Evans and P. E. Johnson, *Fed. Proc.*, 1979, **38**, 703.
[409] G. W. Evans, P. E. Johnson, and E. C. Johnson, *J. Nutr.*, 1979, **109**, 21
[410] G. W. Evans, *Nutr. Rev.*, 1980, **38**, 137.
[411] 'Environmental Health Perspectives', 1979, Vol. 28.
[412] B. L. Vallee, *Experientia, Suppl.*, 1979, **34**, 19.
[413] M. G. Cherian and J. J. Vostal, *J. Toxicol. Environ. Health*, 1977, **2**, 946.
[414] M. G. Cherrian, *Experientia, Suppl.*, 1979, **34**, 339.
[415] M. Nordberg and Y. Kojima, *Experientia, Suppl.*, 1979, **34**, 91.
[416] J. K. Piotrowski, J. A. Szymanska, E. M. Mogilnicka, and A. J. Zelazowski, *Experientia, Suppl.*, 1979, **34**, 363.

Table 2 *Stimulating properties of various metals with regard to the levels of* '*metal-lothionein-like protein*' *in the liver and kidneys of rats:* (+), *positive effect;* (−), *no effect*

Metal	Liver	Kidney
Cd	++++	+++
Zn	+++	++
Cu	+	+
Co, Ag	+	−
Hg	−	+++
Bi	−	++
Au	+	+++
Others[a]	−	−

(a) The other metals are Rb, Cs, Be, Mg, Sr, Ba, Tl, Sn, Pb, V, As, Sb, Cr, Sc, and U

some up-to-date reports of the participants have been published.[417] Nevertheless, we give here a summary of recent developments on the question.

Historically, the term 'metallothionein' has been used to designate the cadmium-, zinc-, and copper-containing sulphur-rich protein from equine renal cortex.[418] Thus, a so-called metallothionein must satisfy several criteria:[412]

(1) The metal content is very high, usually comprising Cd, Zn, or Cu.
(2) The cysteine content constitutes about 30—35% of the total amino-acid composition.
(3) Aromatic amino-acids, histidine, and disulphide are usually absent. The mammalian proteins contain only a single methionine residue.
(4) The ratio of metal ions bound to —SH groups is one-third. Generally, there are 7 gram atoms of M^{2+} per 20 to 21 —SH groups of metallothionein.
(5) The metallothionein exhibits no protein absorption band near 280 nm.
(6) The M_r of these proteins range from 6000 to 7000.

Occurrence of Metallothioneins. The metallothioneins have been found in sixteen vertebrate species and in marine invertebrates.[419] Prinz and Weser purified a copper-containing metallothionein from *Saccharomyces cerevisiae*.[420] Another copper-binding protein was isolated from *Neurospora crassa*.[421] The first unequivocal demonstration of a metallothionein in a vascular plant has been reported recently.[422] The amount of metallothionein in different species and tissues is variable. The concentration has been reported to increase up to 40-fold by the induction of its biosynthesis by certain metals such as cadmium or zinc. In newborn rat liver (1 to 4 days old) the concentration of Zn- and Cu-metallothionein is 20 times that in 70-day-old adult rats.[423]

417 'Metallothionein' (*Experientia, Supplementum*, 1979, Vol. 34), ed. J. H. R. Kägi and M. Nordberg.
418 J. H. R. Kägi and B. L. Vallee, *J. Biol. Chem.*, 1960, **235**, 3460.
419 R. W. Olafson, R. G. Suin, and A. Kearns, *Experientia, Suppl.*, 1979, **34**, 197.
420 R. Prin and U. Weser, *Hoppe-Seyler's Z. Physiol. Chem.*, 1975, **356**, 767.
421 K. Lerch, *Experientia, Suppl.*, 1979, **34**, 173.
422 W. E. Rauser and N. R. Curvetto, *Nature* (*London*), 1980, **287**, 563.
423 K. L. Wong and C. D. Klaassen, *J. Biol. Chem.*, 1979, **254**, 12 399.

Primary Structure and Evolutional Aspects of Metallothioneins. Comparative studies of amino-acid sequences of twelve mammalian metallothioneins have been reported.[424] They are very similar, and they all contain 61 residues, with *N*-acetylmethionine and alanine at the *N*- and *C*-termini, respectively. The cysteinyl residues are distributed fairly uniformly along the polypeptide chain (Figure 7), with a predominance of -Cys-X-Cys- sequences.

Figure 7 *The mode of distribution of cysteinyl residues in metallothioneins*
(Reproduced by permission from *Experientia, Suppl.*, 1979, **34**, 57)

Metallothioneins from *Neurospora crassa* contain only 25 amino-acid residues, but the primary structure is quite similar to the *N*-terminal part of the mammalian proteins.[421] These data indicate that the gene that codes for the *Neurospora crassa* metallothionein is evolutionarily related to the gene of the vertebrate metallothioneins. Equine metallothionein exists in two major variants (metallothionein-1A and metallothionein-2A), which show remarkable similarities. Some allelic polymorphic variants also occur in Man, horse, and rabbit.

The Metal-binding Site in Metallothioneins. The -Cys-X-Cys- sequences indicate the primary metal-binding sites. Moreover, most metal ions can interact with a third cysteinyl residue located elsewhere in the molecule. The resulting trithiolate complexes (trimercaptide) $[(metal)^{2+}(Cys^-)_3]^-$ are responsible for the overall negative charge of all metallothioneins at neutral pH. On lowering the pH, the metal ions are released, yielding the apometallothionein, which is stable at low pH but which polymerizes by forming disulphide bridges at neutral pH (reviewed in ref. 424).

Optical Properties and Spatial Structure of Metallothioneins. Because of the lack of aromatic amino-acids and of cysteine, there is no protein absorption above 270 nm. Metallothioneins exhibit a broad absorption peak, with the maximum at 190 nm. Absorptions due to the metal–thiolate complexes are manifested as shoulders at 250 nm (Cd), 220 nm (Zn), and 270 nm (Cu).[425,426] According to theoretical predictions from the amino-acid sequence of the peptide chain, the α-helical confirmation is forbidden, and β-structure is almost impossible. Circular dichroism and n.m.r. studies on both the metal-containing and metal-free protein confirmed the above prediction.[426,427] However, the fact that metallothioneins

[424] M. Nordberg and Y. Kojima, *Experientia, Suppl.*, 1979, **34**, 57.
[425] U. Weser and H. Rupp, *Experientia, Suppl.*, 1979, **34**, 221.
[426] R. H. O. Bühler and J. H. R. Kägi, *Experientia, Suppl.*, 1979, **34**, 211.
[427] A. Galdes, H. A. O. Hill, J. H. R. Kägi, M. Vasak, I. Bremner, and B. W. Young, *Experientia, Suppl.*, 1979, **34**, 241.

are stable to tryptic digestion and the slow exchange of many peptide hydrogens of metallothionein with the solvent indicate that the protein exists as a tight and well-defined tertiary structure.

Metabolism of Metallothioneins. Biosynthesis of metallothionein shows a maximum rate between 4 and 10 hours after the administration of Cd or Zn. The induction of metallothioneins by the steroid hormone dexamethasone has been demonstrated in HeLa cells[428] and in adrenalectomized rats.[429] Glucocorticoid hormones lead to an intracellular accumulation of zinc.[430] Nutritional status, environmental stress, route of administration, strain, and sex can influence the hepatic accumulation of metallothionein. Alkylating agents such as bromobenzene, diethyl maleate, iodomethane, and sodium iodoacetate can also induce the synthesis of metallothionein.[431]

Actinomycin D, cordycepin, and cycloheximide, when administered to animals prior to an injection of either zinc or cadmium, blocked the induction of synthesis of hepatic metallothionein. When these inhibitors were administered *after* induction by each metal, the process was not inhibited.[424] These results indicate that zinc and cadmium are able to induce the synthesis of metallothionein by causing changes in the intracellular concentration of metallothionein mRNA. The mechanism by which the metals may be bound to the translation product of metallothionein mRNA is still an open question.

The biological half-life of metallothionein is very variable. Measurements of the degradation of zinc-metallothionein gave values of between 4 and 30 hours, depending upon the mode of induction.[424] The half-life of the hepatic Cd-induced protein has been reported to be 2.8 to 6.8 days.

The Role of Metallothionein in the Metabolism and Toxicity of Metals. From the earliest publication,[418] it has been speculated that metallothionein could be involved either in catalysis, storage, immune phenomena, or detoxification of metals. Metallothioneins probably participate in other cellular functions. For example, the recent observation that GTP binds to metallothionein could reflect a role in nucleic acid metabolism.[412] Metallothioneins and related proteins play an important role in copper metabolism. These proteins function in the regulation of absorption of copper and its handling by the liver.[432] A soluble copper-metallothionein has been purified and characterized from human foetal liver.[433] Copper-metallothionein also accumulates in large amounts in the kidneys of animals under both physiological and pathological conditions.[434]

Nickel, Manganese, and Cobalt.—The transport of nickel, manganese, and cobalt in animals is not well documented. These elements can bind to plasma transferrin and to conalbumin.[435] Decsy and Sunderman found that a large dose of $^{63}Ni^{2+}$ was needed for rapid labelling of rabbit serum α_1-macroglobulin *in vitro* and

428 M. Karin and H. R. Herschman, *Science*, 1979, **204**, 176.
429 R. R. Etzel, S. G. Shapiro, and R. J. Cousins, *Biochem. Biophys. Res. Commun.*, 1979, **89**, 1120.
430 M. L. Failla and R. J. Cousins, *Biochim. Biophys. Acta*, 1978, **518**, 435.
431 F. N. Kotsanis and C. D. Klaason, *Toxicol. Appl. Pharmacol.*, 1979, **51**, 19.
432 G. W. Evans, *Experientia, Suppl.*, 1979, **34**, 321.
433 L. Ryden and H. F. Deutsch, *J. Biol. Chem.*, 1978, **253**, 519.
434 I. Bremmen, *Experientia, Suppl.*, 1979, **34**, 273.
435 A. T. Tan and R. C. Woodworth, *Biochemistry*, 1969, **8**, 3711.

in vivo.[436] It has been suggested that nickel may form complexes with citrate.[437] In plants, the uptake and translocation of nickel apparently occur on a negatively charged citratonickelate complex, with $[Ni(H_2O)_6]^{2+}$ as the major cationic constituent.[438] More recently, it has been shown that the leaves of various nickel-accumulating plants show a strong correlation between the levels of nickel and of citric acid.[439]

Vanadium.—While vanadium is required for growth by the rat and chick, and is also suspected to be an essential nutrient for the human body, much attention has been given to this metal since Rancitelli *et al.*[440] reported that its concentration in fuel oil is high. In consequence of the production of electrical power by power plants that use coal and oil as the primary energy source, vanadium is mobilized into the atmosphere in significant amounts. Due to the low sensitivity of the classical analytical methods for vanadium, little is known on vanadium in animals and Man. Recently, it has been shown that vanadate (the $+5$ oxidation state of vanadium) acts as a specific regulator of (Na,K)-ATPase activity.[441,442] The fate of vanadate taken up by red cells has been studied, using e.p.r. spectroscopy.[444] The e.p.r. signal indicated that most of the cytoplasmic vanadate is reduced to the $+4$ oxidation state and is associated with haemoglobin. However, neither sodium vanadate nor vanadyl sulphate binds to purified haemoglobin *in vitro*. The transport and storage of vanadium in mammals have been intensively investigated by the group of Girardi, at the Joint Research Centre of the Commission of the European Communities (at Ispra). Using the neutron-activation analysis technique, they have been able to show that the main form of transport of vanadium-48 in normal human subjects is transferrin. No significant amount of vanadium was detected in the red blood cells nor in iron-containing haemoproteins such as haemoglobin, heart myoglobin, and liver cytochromes *b* and *c*.[444] The principal organs that retain vanadium-48 are the kidneys, liver, testicles, and spleen, in which the nuclei are the intracellular organelles that retain the major fraction of vanadium.[445] Vanadium can be taken up by ferritin from plasma transferrin. The mechanism by which vanadium is exchanged between transferrin and ferritin is not clear. No transfer occurs when the two proteins are simply incubated in a buffer at pH 7.4, but it does if some plasma and liver cytosol are present in the incubation mixture.[446]

When bovine milk is incubated with $^{48}VO^{2+}$ and $^{59}Fe^{3+}$, ^{48}V can be incorporated into lactoferrin in the ratio Fe:V = 4:1. The authors suggest that the

[436] M. I. Decsy and F. W. Sunderman, *Bioinorg. Chem.*, 1974, **3**, 95.

[437] F. H. Nielsen, *J. Nutr.*, 1980, **110**, 965.

[438] J. Lee, R. D. Reeves, R. R. Brooks, and T. Jaffre, *Phytochemistry*, 1977, **16**, 1503.

[439] T. Jaffre, W. Kersten, R. R. Brooks, and R. D. Reeves, *Proc. R. Soc. London, Ser. B*, 1979, **205**, 385.

[440] L. A. Rancitelli, K. H. Abel, and W. C. Weiner, *Rep. BNWL-1850, Pt. 3*, Battelle-Northwest, Richland, Wa., 1973, p. 17.

[441] L. Josephson and L. C. Cantley, Jr., *Biochemistry*, 1977, **16**, 4572.

[442] L. C. Cantley, Jr., L. Josephson, R. Warner, M. Yanagisawa, C. Lechene, and G. Guidotti, *J. Biol. Chem.*, 1977, **252**, 7421.

[443] L. C. Cantley, Jr., and P. Aisen, *J. Biol. Chem.*, 1979, **254**, 1781.

[444] E. Sabbioni and E. Marfante, *Proc. XIth Int. Conf. of Biochemistry*, 1979, Toronto Canada, pp. 13–R122.

[445] E. Sabbioni and E. Marafante, *Bioinorg. Chem.*, 1978, **9**, 389.

[446] E. Sabbioni, J. Rade, and F. Bertolero, *J. Inorg. Biochem.*, 1980, **12**, 307.

iron-containing protein of milk has a possible role in vanadium metabolism of the neonate.[447] The only organisms that are known to accumulate high concentrations of vanadium are the fungus *Amanita muscaria* and certain species of tunicate or sea-squirt. The latter can store vanadate at concentrations as high as 1M within specialized blood cells.[378] It has been demonstrated that these blood cells do not have a respiratory function and that the association of vanadium with a protein (haemovanadin) is an artefact. Vanadium occurs unbound and dissolved in a solution of sulphuric acid as V^{3+}.[448] A speculative mechanism for the accumulation of vanadium in the tunicate *Ascidia nigra* has been proposed.[449]

Actinides.—The uptake by the liver of polyvalent ions such as those of protactinium, gallium, plutonium, palladium, americium, curium, and uranium has been studied. The importance of phospholipids as potential ionophores that might control the intracellular uptake of these cations has been shown.[450,451] In the case of plutonium, Bulman and Griffin showed that phosphatidic acid and phosphatidylserine can bring about the removal of Pu^{4+} from transferrin and that these lipids may be involved in the transport of Pu^{2+} from transferrin to the interior of the cell,[452] where it could form a complex with low-molecular-weight compounds or could be taken up by ferritin.

Interactions of Metal Ions with the Metabolism of Iron.—Among the transition elements that influence the metabolism of iron, the effects of copper and cadmium are the best known.

Copper interferes with iron metabolism in various manners. Depending on the site of interaction, the effect of copper can be inhibitory or enhancing. Deficiency as well as excess of copper have deleterious effects on iron metabolism. Deficiency leads to anaemia, while an excess causes haemolytic anaemia and methaemoglobinaemia. As reviewed recently by Chou and Rennert,[454] copper interacts with iron at four sites during normal metabolism, namely (i) gastrointestinal absorption, (ii) mobilization from cells, (iii) utilization for haem synthesis, and (iv) re-utilization in reticuloendothelial cells.

Excess copper inhibits the absorption of iron, particularly in the case of iron deficiency. Whether or not the effect is due to the binding of copper to mucosal transferrin needs to be confirmed. Copper deficiency in experimental animals also results in an anaemia which presents many characteristics of bone-marrow failure, with ineffective erythropoiesis.[454]

That the serum copper protein caeruloplasmin might have a role in iron metabolism by affecting the rate of formation of transferrin was first suggested in 1966 by Osaki *et al.*[455] According to their hypothesis, iron is presented to the cell surface in the form of ferrous iron and is oxidized to the ferric form by plasma

[447] E. Sabbioni and J. Rade, *Toxicol. Lett.*, 1980, **5**, 381.
[448] I. G. Macara, G. C. McLeod, and K. Kustin, *Comp. Biochem. Physiol.*, *A*, 1979, **62**, 821.
[449] I. G. Macara, G. C. McLeod, and K. Kustin, *Biochem. J.*, 1979, **181**, 457.
[450] R. A. Bulman, *Int. J. Nucl. Med. Biol.*, 1980, **7**, 295.
[451] R. A. Bulman and R. J. Griffin, *J. Inorg. Biochem.*, 1980, **12**, 89.
[452] R. A. Bulman and R. J. Griffin, *J. Inorg. Nucl. Chem.*, 1979, **41**, 1639.
[453] W.-Y. Chan and O. M. Rennert, *Ann. Clin. Lab. Sci.*, 1980, **10**, 338.
[454] G. R. Lee, D. M. Williams, and G. E. Cartwright, in 'Trace Elements in Human Health and Disease', ed. A. S. Prasad, Academic Press, New York, 1976, Vol. 1, p. 373.
[455] S. Osaki, D. A. Johnson, and E. Frieden, *J. Biol Chem.*, 1966, **241**, 2746.

caeruloplasmin, thus enabling its binding to transferrin. Their hypothesis has not been confirmed by recent results.[153]

The effect of copper excess on iron status of sheep leads to an increase of plasma iron concentration and transferrin saturation and an increase of spleen iron and spleen ferritin content, while the iron content in liver and in bone marrow remains constant.[456]

Cadmium has a marked effect on the metabolism of copper, zinc, and iron. The effects depend on dose and the route of administration of the element. Cadmium antagonizes, and is antagonized by, the uptake of the other metals. There is some evidence that cadmium and iron share the same uptake step into the mucosa, since cadmium, added in the drinking water of mice that were fed a low-iron diet, inhibited the absorption of an intragastric dose of ^{59}Fe by up to 75%. Cadmium not only competes directly with iron for intestinal absorption but, like zinc, may inhibit the uptake and release of iron through binding to ferritin.[457]

Acute administration of cadmium provokes an increase of plasma copper and of synthesis of caeruloplasmin by inhibiting the biliary excretion of copper.[458]

[456] E. C. Theil and K. T. Calvert, *Biochem. J.*, 1978, **170**, 137.
[457] M. D. Stonard and M. Webb, *Chem.-Biol. Interact.*, 1976, **15**, 349.
[458] S. L. Ashby, L. J. King, and D. V. W. Parke, *Environ. Res.*, 1980, **21**, 177.

4
Oxygen-Transport Proteins

M. BRUNORI, B. GIARDINA & H. A. KUIPER

Introduction.—Following the scheme of the chapter in Volume 1,[1] we have focussed our attention on the molecular aspects of the structure and function of oxygen carriers, both natural and synthetic. This review is an attempt to present critically those problems relating structure to function which are still controversial. We have frequently referred to our chapter in Volume 1 to avoid repetition, and because in some respects the two articles are complementary.

A very large number of papers dealing with haemoglobins and haemocyanins have been published in the past few years; more specifically, from 1977 to (May) 1980, over 2000 papers have appeared, and approximately 90% of these have been devoted to haemoglobin. It may be useful to point out that, on the occasion of Max Perutz's 65th birthday, a Symposium on the 'Interaction between Iron and Proteins in Oxygen and Electron Transport' was held at Airlie House (Virginia, USA). The proceedings of this meeting, still at present in the press,[2] contain a great deal of recent information to which we shall frequently refer.

It is clear that to review all pertinent results completely would be a formidable task, and therefore in this Report we have not dealt with some aspects of the work on respiratory proteins; in particular: (i) the pathological aspects of haemoglobinopathies and the screening of new variants; (ii) the properties of haemoglobins in relation to the function and physiology of the red blood cell; (iii) the comparative systematic work dealing with proteins from many different species, unless immediately relevant to the matter under discussion; (iv) the physiological role of haemocyanins in the respiration of molluscs and arthropods. Even within the restricted boundaries which we have chosen to explore and present, we may very well have missed important papers which appeared since the last review article, and we offer our apologies to friends and colleagues if their papers are not mentioned.

The level of understanding and the overall knowledge of haemocyanins are by no means as detailed as they are of haemoglobins, although it begins to appear that some general ideas that were born within the territory of haemoglobins are being transferred successfully to other, more complex, respiratory proteins. Thus we may attempt to give a unified overall impression of the more recent advances

[1] M. Brunori, B. Giardina, and J. V. Bannister, in 'Inorganic Biochemistry' (Specialist Periodical Reports), ed. H. A. O. Hill, The Chemical Society, London, 1979, Vol. 1, p. 159.
[2] Symposium on 'Interaction between iron and proteins in oxygen and electron transport', ed. J. P. Collman, W. A. Eaton, Q. H. Gibson, Chien Ho, J. S. Leigh, E. Margoliash, J. K. Moffat, M. F. Perutz, and W. R. Scheidt, 1980.

in the area of the structure–function relationships in respiratory proteins along the following lines.

(i) The role of a dynamic approach to the molecular interpretation of oxygen transport appears greater than ever, with extension of the time domain into the range of picoseconds and with the application of fluctuation analysis to structural crystallography of macromolecules.

(ii) The very important contribution provided by the work on models, nowadays tailored to requirements by sophisticated organic chemistry and aimed at testing specific features of a structural mechanism, is coming to be appreciated.

(iii) Knowledge of the localization and of the mechanism of dissipation of the free energy of co-operativity is improving, thanks to refinements in structure, the application of sophisticated spectroscopic techniques, and solution studies that exploit the role of the interfaces between subunits in haemoglobin.

(iv) The structural information on the active site in haemocyanins, together with a unifying interpretation of the basic dynamic and equilibrium properties of these systems, may stimulate the proposal of a viable stereochemical model for co-operative binding of oxygen by these giant copper proteins.

This summary, although necessarily approximate, reflects our own impression about the areas in which exciting research is presently expanding.

1 Haemoglobins

Structural Analysis.—*Model Systems.* Research on synthetic iron-containing oxygen carriers as models for the natural oxygen-carrying proteins has undergone significant theoretical and experimental developments.[3–5] A number of physical-organic chemists have been concerned with the design and the synthesis of complexes whose ligands have superstructures which may emulate effects that occur in the natural carriers. The results of this type of approach carry interesting implications for the understanding of the nature of the environment of the haem in haemoproteins and of the possible modes for the control of the binding of ligands to a protein by the protein itself.

The crystal structures of two different 'picket fence' oxygenated complexes have been determined.[6–8] The first contains the unhindered ligand 1-methylimidazole as the fifth ligand of the haem, and it might be regarded as a model of the 'relaxed' (R) structure of liganded haemoglobin; the second contains the sterically hindered ligand 2-methylimidazole in the same position, and it may be regarded as a model of the 'tensed' (T) structure of deoxyhaemoglobin, because repulsion between the methyl group in position 2 and the nitrogen atoms of the porphyrin molecule appears to mimic the restraints imposed by the globins (see p. 145). Analysis of the crystal structure shows that the iron–porphyrin bonds in

[3] R. D. Jones, D. A. Summerville, and F. Basolo, *Chem. Rev.*, 1979, **19**, 139.

[4] J. P. Collman, *Acc. Chem. Res.*, 1977, **10**, 265.

[5] W. R. Scheidt, *Acc. Chem. Res.*, 1977, **10**, 339

[6] J. P. Collman, R. R. Gagne, C. A. Reed, W. T. Robinson, and G. A. Rodley, *Proc. Natl. Acad. Sci. USA*, 1974, **71**, 1326.

[7] G. B. Jameson, F. S. Molinaro, J. A. Ibers, J. P. Collman, J. I. Brauman, E. Rose, and K. S. Suslick, *J. Am. Chem. Soc.*, 1978, **100**, 6769.

[8] G. B. Jameson, G. A. Rodley, W. T. Robinson, R. R. Gagne, C. A. Reed, and J. P. Collman, *Inorg. Chem.*, 1978, **17**, 850.

models that were designed to mimic deoxyhaemoglobin are longer (by 0.09 Å) than in those designed to mimic liganded haemoglobin. In the latter case, the iron and the porphyrin molecule lie in the same plane (within 0.1 Å), while in the former cases the iron is at the apex of a pyramid that has the four porphyrin nitrogens as its base (see Table 1[4-13]). The exact height of the pyramid and the doming of the four nitrogen atoms relative to the rest of the porphyrin molecule

Table 1 *Stereochemistry of model compounds (from refs. 4—13)*

Compound	Co-ordination number	Spin	Fe–N_p/Å	Out-of-plane displacement/Å
[(2-Me-Im)(TPP)Fe^{II}]	5	2	2.086	0.42
[(2-Me-Im)(TpivPP)Fe^{II}]	5	2	2.072	0.40
[(2-Me-Im)(TpivPP)Fe^{II}(O$_2$)]	6	0	1.996	0.09
[(1-Me-Im)(TpivPP)Fe^{II}(O$_2$)]	6	0	1.980	-0.03
[(TPP)Fe^{II}(NO)]	5	$\frac{1}{2}$	2.001	-0.211
[(1-Me-Im)(TPP)Fe^{II}(NO)]	6	$\frac{1}{2}$	2.008	-0.07
[bis(H$_2$O)(TPP)Fe^{III}]	6	$\frac{5}{2}$	2.045	0

TPP is tetraphenylporphyrin; TpivPP is meso-tetra($\alpha,\alpha,\alpha,\alpha$-*o*-pivalamidophenyl)porphin; Me-Im is methylimidazole

depend on the angle that the plane of the imidazole ring makes with the N–Fe–N bonds (see ref. 13). The different geometries arise largely from two effects: (i) the van der Waals repulsion of the axial ligands by the porphyrin nitrogens and (ii) repulsion between orbitals of iron and the π-orbitals of the porphyrin molecule. In this connection, it has been pointed out that repulsion between the fifth ligand and the porphyrin nitrogens is more important than the spin state in determining the displacement of the iron atom from the plane of the porphyrin molecule. However, in contrast with this hypothesis, the displacement of the iron atom observed in [(TPP)Fe^{III}(N$_3^-$)] is more than twice (*i.e.* -0.45 Å) that which is observed in [(TPP)Fe^{II}(NO)] (*i.e.* -0.211 Å).[12,13]

Another important point arises from two six-co-ordinated high-spin structures, *i.e.* [(bis-tetrahydrofuran)(TPP)Fe^{II}] and [(bis-water)(TPP)Fe^{III}]. In both compounds, a planar geometry, due to the balancing effect of the two equal axial ligands which hold the iron atom in the plane of the porphyrin molecule, was observed.[11] It should be mentioned that the strain that is imposed by this planarity makes the Fe–N_p distances shorter than in five-co-ordinated high-spin porphyrins (see Table 1). Analysis of these structures has therefore demonstrated that the porphyrin core may be expanded, because the position of the iron atom relative to the plane of the porphyrin molecule and the length of the Fe—N_p bonds are influenced by the constraints that are exercised on the distal side (see also ref. 14). Since the resonance Raman spectra of ferric porphyrins are essentially

[9] R. W. Scheidt and M. E. Frisse, *J. Am. Chem. Soc.*, 1975, **97**, 17.

[10] R. W. Scheidt and P. L. Piciulo, *J. Am. Chem. Soc.*, 1976, **98**, 1913.

[11] M. E. Kastner, R. W. Scheidt, T. Masiko, and C. A. Reed, *J. Am. Chem. Soc.*, 1978, **100**, 6354.

[12] J. L. Hoard, in 'Porphyrins and Metalloporphyrins', ed. K. M. Smith, Elsevier, Amsterdam, 1975, p. 317.

[13] M. F. Perutz, *Annu. Rev. Biochem.*, 1979, **48**, 327.

[14] W. R. Scheidt, I. A. Cohen, and M. E. Kastner, *Biochemistry*, 1979, **18**, 3546.

equivalent to those obtained for aquo- and fluoro-methaemoglobins and myoglobin, core expansion may be postulated to occur also in the case of haemoproteins.[15, 16] However, it remains to be seen whether the displacement of the iron atom is a consequence of the non-equivalent axial ligands[17] or a reflection of the control of the stereochemistry of the haem by the protein.[13]

Monomeric Haemoproteins. The structural information available on several haemoproteins is increasing in resolution and reliability as compared to previous results, and allows the presentation of a comparative view of the stereochemistry of the active site of these proteins. The high-resolution structural data on spermwhale myoglobin (Mb) and on the oxy-, deoxy-, and met-forms[18-21] were reported and commented on in Volume 1.[1] Also, the structural data on cobalt-substituted Mb[22] were outlined previously.

Some of the geometrical parameters which are more critically important in defining the structure of the active site in deoxy and liganded simple haemoproteins are reported in Table 2. Similar information on the α and β subunits in

Table 2 *Stereochemistry of the haem in different derivatives of erythrocruorin (Ery), myoglobin (Mb), and haemoglobin (Hb) (from ref. 13, modified)*

Protein	Displacement/Å of Fe from haem plane	Fe–N_p/Å[a]	Fe–N_ε/Å[b]	Distance/Å from N_ε to haem plane
deoxy-Ery	0.17	2.02	2.2	2.3
EryO$_2$	0.30	2.04	2.1	2.3
EryCO	0.01	2.01	2.1	2.2
deoxy-Mb	0.55	2.06	2.1	2.6
MbO$_2$	0.26	2.05	2.1	2.3
MbCO	0	—	—	2.3
α-deoxy in Hb	0.60	2.1	2.0	2.6
α-CO in Hb	0.04	—	—	—
β-deoxy in Hb	0.63	2.1	2.2	2.8
β-CO in Hb	0.2	—	—	—

(a) N_p = nitrogen atoms of pyrrole; (b) N_ε = nitrogen atom of His(F8)

tetrameric HbA is also given. At this level of resolution, the values reported do not indicate a simple situation, whereby all the liganded derivatives have features common among themselves and different from the deoxy-form.

The interest in the three-dimensional analysis of the myoglobin from *Aplysia limacina* arises from some peculiar features of this molecule which have been previously summarized.[23] Therefore, the crystal analysis of this protein was

[15] T. G. Spiro and J. M. Burke, *J. Am. Chem. Soc.*, 1976, **98**, 5482.
[16] T. G. Spiro, J. D. Strong, and P. Stein, *J. Am. Chem. Soc.*, 1979, **101**, 2648.
[17] C. A. Reed, in 'Metal Ions in Biological Systems' ed. H. Sigel, Marcel Dekker, New York, 1978, Vol. 7.
[18] T. Takano, *J. Mol. Biol.*, 1977, **110**, 537.
[19] T. Takano, *J. Mol. Biol.*, 1977, **110**, 569.
[20] S. E. V. Phillips, *Nature (London)*, 1978, **273**, 247.
[21] S. E. V. Phillips, *J. Mol. Biol.*, 1980, **142**, 531.
[22] G. A. Petsko, D. Rose, D. Tsernoglou, M. Ikeda-Saito, and T. Yonetani, in 'Frontiers of Biological Energetics', ed P. L. Dutton, J. S. Leigh, and A. Scarpa, Academic Press, New York, 1978, Vol. 2, p. 1011.
[23] M. Brunori, G. M. Giacometti, E. Antonini, and J. Wyman, *J. Mol. Biol.*, 1972, **63**, 139.

started, and data are now available at 5 and at 3.8 Å resolution.[24,25] The overall structure of *Aplysia* myoglobin is similar to that of other myoglobins, although some differences are already evident. In the surroundings of the haem pocket, the metal site is more accessible to the solvent, especially along the Helix E, where the distal histidine is lacking.[26] This enhanced accessibility is probably the result of the peculiar distribution of polar and apolar residues, combined with the different geometry of the distal position.[25] Recent n.m.r. data on *Aplysia* Mb (G. M. Giacometti *et al.*, unpublished) have also provided very good evidence for the absence of a water molecule that is co-ordinated to the ferric ion in the acid met-derivative. This result is of considerable interest and in complete agreement with temperature-jump kinetic data on the reaction of ferric *Aplysia* Mb with azide ion.[27]

Structural studies '*ex novo*' on the various derivatives of erythrocruorin have been carried out at high resolution by the Münich group.[28-30] In this case as well, peculiar features concerning the Helix E and the haem pocket have been reported. In the liganded state, Helix E, which runs along the distal side of the haem, is affected largely by direct contact with the iron-bound ligand. The distal amino-acid residue, *i.e.* Ile(E11), creates a steric hindrance which forces linear ligands (such as cyanide and carbon monoxide) to bind at an angle to the axis of the haem, thus straining the haem, which moves deeper into the globin. Furthermore, in the oxygenated form, a water molecule that is hydrogen-bonded to dioxygen is probably the main cause of: (i) the enhanced bent configuration, with an angle of $\sim 150°$, which is considerably larger than that observed in model compounds (*i.e.* 120°), and (ii) the slight opening of the haem pocket when compared with the deoxy-form. Moreover, the iron atom is displaced slightly more from the plane of the haem in oxy- (~ 0.3 Å) than in deoxy-erythrocruorin (~ 0.2 Å). This unusual finding may be related to the bulkiness of the dioxygen–water complex that is bound to the metal and to the short iron–ligand bond (1.8 Å for O_2 and 2.4 Å for CO).

The results indicate that the changes in the spin state that are associated with the binding of a ligand do not produce pronounced movements of the metal atom with respect to the plane of the porphyrin molecule, and hence the ligation-induced changes in the tertiary structure have to be ascribed mainly to perturbations in the steric interactions between ligand and globin.

The component III of erythrocruorin of *Chironomus thummi thummi*, which is known to have a large Bohr effect for binding of O_2 but not for CO,[31] (see also ref. 1) has been studied by high-resolution n.m.r.,[32] in conjunction with the

[24] L. Ungaretti, M. Bolognesi, E. Cannillo, R. Oberti, and G. Rossi, *Acta Crystallogr.*, Sect. B, 1978, **34**, 3658.
[25] M. Bolognesi, E. Cannillo, P. Ascenzi, M. Brunori, and G. M. Giacometti, in Proceedings of the International Congress of Crystallography, Warzawa, Poland, 1978, paper 4.3.27.
[26] L. Tentori, G. Vivaldi, S. Carta, E. Antonini, and M. Brunori, *Nature (London)*, 1968, **219**, 487.
[27] G. M. Giacometti, P. Ascenzi, M. Bolognesi, and M. Brunori, *J. Mol. Biol.*, 1981, **146**, 363.
[28] R. Huber, O. Epp, W. Steigemann, and H. Formanek, *Eur. J. Biochem.*, 1971, **19**, 42.
[29] E. Weber, W. Steigeman, T. A. Jones, and R. Huber, *J. Mol. Biol.*, 1978, **120**, 327.
[30] W. Steigemann and E. Weber, *J. Mol. Biol.*, 1979, **127**, 309.
[31] G. Steffens, G. Buse, and A. Wollmer, *Eur. J. Biochem.*, 1977, **72**, 201.
[32] W. Ribbing and H. Rüterjans, *Eur. J. Biochem.*, 1980, **108**, 89.

crystallographic information obtained by Steigemann and Weber.[30] It has been established (by n.m.r.) that the Bohr group is His(G2), which, upon binding of O_2, changes its interaction with the carboxyl group of the C-terminal Met(H22).[32]

Neutron-diffraction studies have proven[33] to be capable of defining the position of the hydrogen and deuterium atoms, which are undetected by the X-ray diffraction approach. Schoenborn and collaborators[34-36] have presented the neutron-diffraction data on metMb and MbCO from sperm whale, giving particular attention to bound water. Of the 40 water molecules which were located in MbCO, nine were involved in ^{2}H-bonds with main-chain atoms. From a general viewpoint it is of interest that the ^{2}H-bond angles that are involved in helical regions showed considerable deviations from linearity (with angles as small as 145°).[36a] Very recent neutron-diffraction data on oxymyoglobin[36b] have indicated that the bound dioxygen is hydrogen-bonded to the nitrogen atom of the distal histidine residue.

Human Haemoglobin. The structure of *deoxy-cobalt-haemoglobin*, obtained by reconstitution of globin with Co-porphyrin, has recently been compared with that of native deoxy-HbA by X-ray diffraction.[37] The results indicate that, in deoxy-Hb, the cobalt ion is only 0.3 Å out of the plane of the porphyrin, this distance being 0.6 Å for the iron. However, in the α subunits the haem plane moves itself in the same direction, so that the distance between the metal and the plane of the haem does not change, while in the β subunits the porphyrin maintains its original position, and therefore the cobalt is truly shifted towards the plane. Moreover, concerted movements of some helices (B, E, and G of both subunits) appear in several regions of the molecule. Hence, if the substitution of Co^{2+} for Fe^{2+} is responsible for these perturbations, these structural data indicate that changes at the haem level can induce structural rearrangements over a long distance, as required by theories of co-operativity in haemoglobin.

The refined structure of *human carbonmonoxyhaemoglobin*, determined at 2.7 Å resolution,[38] is generally similar to that of horse methaemoglobin, whose structure is known at even higher resolution.[39] The new structure confirms that carbon monoxide lies off the normal to the plane of the haem in both the α and the β subunits, as previously suggested[40] on the basis of steric hindrance between the oxygen atom of the ligand and the side-chain of His(E7) and Val(E11). Even the new structure provided no reason why the side-chains of these two residues cannot move away from the ligand. Moreover, the Fe—C—O group, which was assumed to have a linear geometry, makes an angle of about 13° with the normal to the plane of the haem and points towards the inside of the haem pocket.[38]

The iron atom lies in the plane of the haem in the α subunits, while in the β subunits it protrudes from the plane of the haem by 0.22 Å. Furthermore, the

[33] B. P. Schoenborn, *Nature (London)*, 1969, **224**, 143.

[34] B. P. Schoenborn, *Cold Spring Harbor Symp. Quant. Biol.*, 1971, **36**, 569.

[35] J. C. Norvell, A. C. Nutes, and B. P. Schoenborn, *Science*, 1975, **190**, 568.

[36] (a) J. C. Hanson and B. P. Schoenborn, *J. Mol. Biol.*, 1981, **153**, 117; (b) S.E.V. Phillips and B. P. Schoenborn, *Nature (London)*, 1981, **292**, 81.

[37] G. Fermi, in ref. 2.

[38] J. M. Baldwin, *J. Mol. Biol.*, 1980, **136**, 103.

[39] R. C. Ladner, E. J. Heidner, and M. F. Perutz, *J. Mol. Biol.*, 1977, **114**, 385.

[40] E. J. Heidner, R. C. Ladner, and M. F. Perutz, *J. Mol. Biol.*, 1976, **104**, 707.

new structure for HbCO confirms that the side-chain of $\beta93$(F9)Cys points away from the surface of the molecule into a crevice in between helices F, G, and H; in deoxyhaemoglobin, this crevice is occupied by the side-chain of $\beta145$(HC2)Tyr.[41] It should be recalled that, in crystalline metHb, the sulphydryl group of $\beta93$-(F9)Cys occupies alternative conformations,[40] the one pointing into a pocket enclosed by helices F, G, and H and the other towards the solvent. The first of these conformations is the only one occupied in HbCO, while in deoxy-Hb the crevice is occupied by $\beta145$Tyr, whose OH group is hydrogen-bonded to the carboxylate of $\beta98$(FG5)Val. This structural information has been correlated[38] to the known difference in reactivity of $\beta93$Cys towards various reagents in liganded and unliganded haemoglobins (see ref. 42).

The newly available data have also allowed a detailed comparison of the structures of the deoxy and liganded forms from the same species, allowing some of the basic structural considerations that are reported below (see p. 145) to be recognized.

Arnone *et al.*[43] have resolved the structure of the *CO derivative of HbH (β_4)*. The results show that the iron–iron distances in β_4CO are more nearly similar to those of liganded than to those of deoxy-HbA. The high degree of similarity between β_4CO and liganded HbA is also confirmed by the comparison of the interfaces between subunits, which have been examined with considerable attention. It seems clear, however, that some regions of the β_4CO molecule have a conformation that is different from that reported either for deoxy- or for liganded haemoglobin, emphasizing once more the role of dissimilar chains in establishing the correct interrelationships between the four subunits.

The X-ray structure of *carbonmonoxyhaemoglobin Zürich* [$\alpha_2\beta_2$63(E7)His→Arg] has been reported by Tucker *et al.*,[44] together with information on the same derivative of *haemoglobin Sydney* [$\alpha_2\beta_2$67(E11)Val→Ala]. The interesting result which appears from examination of the map of HbZhCO is that the bulky side-chain of arginine, which replaces the distal histidine residue in β subunits, moves out of the haem pocket and makes an electrostatic interaction with one of the propionates of the porphyrin. This leaves the haem pocket completely accessible to external ligands, and thus helps in rationalizing the very peculiar reactivity of this abnormal haemoglobin towards gaseous ligands (*e.g.* CO and O_2)[45–48] as well as large molecules (such as the sulphanilamides, which are

[41] G. Fermi, *J. Mol. Biol.*, 1975, **97**, 237.

[42] E. Antonini and M. Brunori, 'Hemoglobin and Myoglobin in their reactions with ligands', North Holland, Amsterdam, 1971.

[43] A. Arnone, P. D. Briley, and P. H. Rogers, in ref. 2.

[44] P. W. Tucker, S. E. V. Phillips, M. F. Perutz, R. Houtchens, and W. S. Caughey, *Proc. Natl. Acad. Sci. USA*, 1978, **75**, 1076.

[45] K. H. Winterhalter, N. M. Anderson, G. Amiconi, E. Antonini, and M. Brunori, *Eur. J. Biochem.*, 1969, **11**, 435.

[46] G. M. Giacometti, E. E. Di Iorio, E. Antonini, M. Brunori, and K. H. Winterhalter, *Eur. J. Biochem.*, 1977, **75**, 267.

[47] G. M. Giacometti, M. Brunori, E. Antonini, E. E. Di Iorio, and K. H. Winterhalter, *J. Biol. Chem.*, 1980, **255**, 6160.

[48] M. Ikeda-Saito, M. Brunori, K. H. Winterhalter, and T. Yonetani, *Biochim. Biophys. Acta*, 1979, **580**, 91.

involved in speeding up the oxidation of the metal of the β subunits and thus in the development of the pathological symptoms in the carriers).[49]

Haem–globin interactions and haem disorder. The crucial role of the interactions between the periphery of the iron porphyrin and the protein moiety in the regulation of the functional properties of haemoglobin (well established since the early work on reconstituted haemoproteins[42, 50]) demands that the orientation of the porphyrin within the haem pocket be known and uniquely determined. This view, which has always been assumed to hold true, has been challenged by La Mar and co-workers[51–54] and by Rüterjans and co-workers.[55, 56] A very careful study, using high-resolution ^{1}H n.m.r. and selective deuteriation of the methyl(s) of the porphyrin, has shown[53] that the haem is in two isomeric states, which differ by a rotation about the α–γ meso axis of the porphyrin by 180°. The existence of these two conformers in a solution of deuteroporphyrin-reconstituted sperm-whale Mb^+CN^- has unequivocally been established.

This type of 'disorder' is reflected in the ^{1}H n.m.r. spectrum of *Aplysia* Mb,[57] one cytochrome,[58] and one of the components of *Chironomus* erythrocruorin,[55, 57] and it may rationalize the multiple resonances that arise from methyl groups that are observed in several cases. The dynamic aspects of this reorientation of haem have proven to be of great interest.[54]

There is no direct evidence for inversion of haem in native proto-Hb; however, recent crystallographic data on reconstituted haemoglobins[59] have clearly shown the presence of two orientations of the haem moiety even in the solid. This isomerism of the haem will have to be carefully considered in the interpretation of structural and dynamic studies of reconstituted as well as native haemoproteins.

The Primary Structure of Various Haemoproteins. The work on sequence analysis of different haemoproteins continues to represent a useful basis for a comparative approach to structure and function.

Descriptions of a number of new sequences have been published in the past few years, including Mb from the Pacific sea whale (*Balaenoptera borealis*),[60]

[49] P. G. Frick, W. H. Hitzig, and K. Betke, *Blood*, 1962, **20**, 261.

[50] A. Rossi-Fanelli, E. Antonini, and A. Caputo, *Adv. Protein Chem.*, 1964, **19**, 73.

[51] G. N. La Mar, M. Overkamp, H. Sick, and K. Gersonde, *Biochemistry*, 1978, **17**, 352.

[52] D. B. Viscio and G. N. La Mar, *J. Am. Chem. Soc.*, 1978, **100**, 8092.

[53] G. N. La Mar, B. L. Budd, D. B. Viscio, K. M. Smith, and K. C. Langry, *Proc. Natl. Acad. Sci. USA*, 1978, **75**, 5755.

[54] G. N. La Mar, in ref. 2.

[55] W. Ribbing, D. Krümpelmann, and H. Rüterjans, *FEBS Lett.*, 1978, **92**, 105.

[56] W. Ribbing and H. Rüterjans, *Eur. J. Biochem.*, 1980, **108**, 79.

[57] K. Wütrich, R. Keller, M. Brunori, G. M. Giacometti, R. Huber, and H. Formanek, *FEBS Lett.*, 1972, **21**, 63.

[58] R. Keller, O. Grondinsky, and K. Wütrich, *Biochim. Biophys. Acta*, 1978, **427**, 497.

[59] K. Moffat, and collaborators, 1981, to be published.

[60] B. N. Jones, T. M. Rothgeb, R. D. England, and F. R. N. Gurd, *Biochim. Biophys. Acta*, 1979, **577**, 464.

several of the components from *Chironomus thummi thummi*,[61-66] two main components of the river lamprey (*Lampetra fluviatilis*),[67] Component III of the Hb of hagfish (*Myxine glutinosa* L.),[68] the β_A and β_B chains of the carp (*Cyprinus carpio* L.),[69] the α and β chains of the main component of the bullfrog (*Rana catesbeiana*),[70,71] the α and β chains of the greylag goose (*Anser anser*),[72] the main haemoglobins of the guinea pig (*Cavia aperea*), of the dromedary (*Camelus dromedarius*), and of the domestic pig,[73,74] the α and β chains of the brown lemur (*Lemur fulvus fulvus*),[75] and a new embryonic haemoglobin from pig θ chains.[76]

Some of these have provided the basis for evolutionary considerations about protein structures.

Ligands of the Iron Atom of Haemoglobins.—*The Binding of Oxygen and of Carbon Monoxide.* The *X*-ray structures of 'picket-fence' iron–porphyrin model compounds[7,8] indicate that oxygen is bound in a bent geometry, as originally suggested by Pauling,[77] with a Fe—O—O angle of $\sim 135°$, whereas carbon monoxide is bound linearly, with a Fe—C—O angle of 180°.

Phillips' data on crystalline oxymyoglobin,[20,21] on which comments have already been made,[1] support the Pauling geometry, assigning a Fe—O—O angle of $\sim 120°$. Further support for the bent configuration comes from infrared spectroscopic studies,[78] from theoretical calculations,[79] and from single-crystal spectroscopy[80] (see also ref. 1).

An extensive theoretical analysis of the binding of O_2 and of CO to iron–porphyrin complexes has been presented by Case *et al.*[81] Ground- and excited-state structures have been considered; the FeO_2 unit is well represented by an equal mixture of $[Fe^{2+}(S=0)O_2(S=0)]$ and $[Fe^{2+}(S=1)O_2(S=1)]$, while the FeCO complex corresponds to the $[Fe^{2+}(S=0)CO(S=0)]$ species.

[61] T. Kleinschmidt, H. Van der Mark-Neuwirth, and G. Braunitzer, *Z. Physiol. Chem.*, 1980, **361**, 401.
[62] T. Kleinschmidt and G. Braunitzer, *Liebigs Ann. Chem.*, 1978, 1060.
[63] H. Aschauer, Z. H. Zaidi, and G. Braunitzer, *Hoppe-Seyler's Z. Physiol. Chem.*, 1979, **360**, 1513.
[64] W. Steer and G. Braunitzer, *Z. Naturforsch., Teil. C*, 1979, **34**, 882.
[65] D. Sladíc-Simíc, T. Kleinschmidt, and G. Braunitzer, *Hoppe-Seyler's Z. Physiol. Chem.*, 1979, **360**, 115.
[66] R. Lalthantluanga and G. Braunitzer, *Hoppe-Seyler's Z. Physiol. Chem.*, 1979, **360**, 99.
[67] M. Zelenik, V. Rudloff, and G. Braunitzer, *Hoppe-Seyler's Z. Physiol. Chem.*, 1979, **360**, 1879.
[68] G. Liljeqvist, G. Braunitzer, and S. Paléus, *Hoppe-Seyler's Z. Physiol. Chem.*, 1979, **360**, 125.
[69] B. Grujic-Injac, G. Braunitzer, and A. Stangl, *Hoppe-Seyler's Z. Physiol. Chem.*, 1979, **360**, 609.
[70] T. Maruyama, K. W. K. Watt, and A. Riggs, *J. Biol. Chem.*, 1980, **255**, 3285.
[71] K. W. K. Watt, T. Maruyama, and A. Riggs, *J. Biol. Chem.*, 1980, **255**, 3294.
[72] G. Braunitzer and W. Oberthür, *Hoppe-Seyler's Z. Physiol. Chem.*, 1979, **360**, 679.
[73] G. Braunitzer, B. Schrank, A. Stangl, and M. Wiesner, *Hoppe-Seyler's Z. Physiol. Chem.*, 1979, **360**, 1941.
[74] G. Braunitzer, B. Schrank, A. Stangl, and U. Scheithaner, *Hoppe-Seyler's Z. Physiol. Chem.*, 1978, **359**, 137.
[75] T. Maita, M. Setoguchi, G. Matsuda, and M. Goodman, *J. Biochem.*, 1979, **85**, 755.
[76] W. Steer and G. Braunitzer, *Hoppe-Seyler's Z. Physiol. Chem.*, 1980, **361**, 1165.
[77] L. Pauling and C. E. Coryell, *Proc. Natl. Acad. Sci. USA*, 1936, **22**, 210.
[78] C. H. Barlow, J. C. Maxwell, W. J. Wallace, and W. S. Caughey, *Biochem. Biophys. Res. Commun.*, 1973, **55**, 91.
[79] W. A. Goddard and B. D. Olafson, *Proc. Natl. Acad. Sci. USA*, 1975, **72**, 2335.
[80] W. A. Eaton, J. Hofrichter, and L. K. Hanson, Proceedings of the Taniguchi International Symposium on Biophysics, Lake Biva, Japan, November 1974, p. 51.
[81] D. A. Case, B. H. Huynh, and M. Karplus, *J. Am. Chem. Soc.*, 1979, **101**, 4433,

More theoretical work, stimulated by the increasingly detailed knowledge of the structural parameters, has been carried out by Otsuka and Seno.[82]

Carbon monoxide that is bound to model compounds has a linear configuration,[12] while it appears that in haemoproteins the CO is tilted off-axis, in a slightly bent configuration.[28, 35, 38, 40, 83, 84] This distortion has been attributed to the bulky amino-acid residues on the distal side, and its functional significance is also discussed below (see p. 154).

Recently, the configuration of the haem–carbonyl group in crystalline sperm-whale MbCO has been investigated by infrared spectroscopy.[85] This study shows that the single-crystal i.r. spectrum of MbCO is not compatible with that obtained in solution. Thus the proportions of the three, spectrally distinct, haem–CO conformers of myoglobin that are observed by i.r. spectroscopy[86, 87] can be made to vary by incorporating in the MbCO crystals small amounts of either metMb or deoxy-Mb. This suggests that the incorporation of these other derivatives into the MbCO crystal may impose a structural perturbation on neighbouring MbCO molecules which affects the stretching mode of bound CO.

The problem of magnetic properties of haemoglobin and its derivatives was extensively reviewed in Volume 1.[1] The results of Cerdonio and collaborators on the paramagnetic properties of HbO_2[88, 89] have been followed by additional work that indicates that the carbon-monoxide derivative may, under some conditions, also be paramagnetic, the observed magnetic moment being dependent on chloride ion.[90, 91]

Although understanding the significance of the paramagnetic state of HbO_2 may demand a more complete study and an independent check of the so-called 'diamagnetic correction' (*i.e.* the correction that has to be introduced to take into account the diamagnetism of the globin), it seems at present that no new experimental and/or theoretical arguments have been provided to refute the observations of Cerdonio *et al.*,[88] after Pauling's initial comments.[92]

Lang and co-workers[93, 94] have re-investigated the electronic structure of the deoxyhaem moiety, using high-magnetic-field Mössbauer spectroscopy. New data have been collected, over a wide temperature range, for deoxymyoglobin, deoxy-

[82] J. Otsuka and Y. Seno in ref. 2.

[83] E. A. Padlan and W. E. Love, *J. Biol. Chem.*, 1975, **249**, 4067.

[84] R. Huber, O. Epp, and H. Formanek, *J. Mol. Biol.*, 1970, **52**, 349.

[85] M. W. Makinen, R. A. Houtchens, and W. S. Caughey, *Proc. Natl. Acad. Sci. USA*, 1979, 76, 6042.

[86] S. McCoy and W. S. Caughey, in 'Probes of Structure and Function of Macromolecules and Membranes', ed. B. Chance, T. Yonetani, and A. S. Mildvan, Academic Press, New York, 1971, Vol. 2, p. 289.

[87] J. O. Alben, in 'The Porphyrins', ed. D. Dolphin, Academic Press, New York, 1978, Vol. 3, p. 323.

[88] M. Cerdonio, A. Congiu-Castellano, F. Mogno, B. Pispisa, G. L. Romani, and S. Vitale, *Proc. Natl. Acad. Sci. USA*, 1977, **74**, 398.

[89] M. Cerdonio, A. Congiu-Castellano, L. Calabrese, S. Morante, B. Pispisa, and S. Vitale, *Proc. Natl. Acad. Sci. USA*, 1978, **75**, 4916.

[90] M. Cerdonio, S. Morante, S. Vitale, A. De Young, and R. W. Noble, *Proc. Natl. Acad. Sci. USA*, 1980, **77**, 1462.

[91] M. Cerdonio, S. Morante, S. Vitale, G. M. Giacometti, and M. Brunori, in ref. 2.

[92] L. Pauling, *Proc. Natl. Acad. Sci. USA*, 1977, **74**, 2612.

[93] T. A. Kent, K. Spartalian, G. Lang, T. Yonetani, C. A. Reed, and J. P. Collman, *Biochim. Biophys. Acta*, 1979, **580**, 245.

[94] T. A. Kent, K. Spartalian, and G. Lang, *J. Chem. Phys.*, 1979, **71**, 4899.

haemoglobin, and the isolated deoxygenated α and β chains, as well as a number of synthetic analogues [*e.g.* (2-methylimidazole)(meso-tetraphenylporphyrin)-iron(II)]. The differences between the isolated subunits and the tetramer are indeed very small, although the new data indicate a definite effect of assembly on the α subunits.

The temperature dependence of the Mössbauer spectra of oxyhaemoglobin[95] (which suggested that there are different orientations of the bound ligand, separated by small barriers[96]) has been differently interpreted by Bacci *et al.*[97]

Local Effects on Ligand Affinity of Haemoglobins. The great variability in affinity for oxygen that is displayed by natural oxygen carriers[42] has been thought to arise from several factors, involving stereochemical and electronic effects related to the interaction of the metalloporphyrin with the proximal and distal sites, as well as the haem pocket at large. The design and synthesis of model compounds that are capable of reversibly binding O_2 and of closely emulating natural oxygen carriers has provided[3] a test of the significance of these various effects.

Figure 1 incorporates the features of one of these successful model compounds,[98] and it serves as a scheme to outline the different factors which may

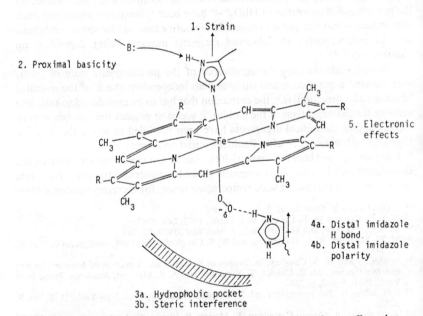

Figure 1 *Scheme of the ligand-binding site and of the principal regulatory effects that are supposed to be operational in haemoproteins, according to studies with model compounds*

(Reproduced by permission from *Proc. Natl. Acad. Sci. USA*, 1975, **72**, 1166)

95 G. Lang and W. Marshall, *Proc. Phys. Soc.*, 1966, **87**, 3.
96 G. Lang, *Q. Rev. Biophys.*, 1970, **3**, 1.
97 M. Bacci, M. Cerdonio, and S. Vitale, *Biophys. Chem.*, 1979, **10**, 113.
98 C. K. Chang and T. G. Traylor, *Proc. Natl. Acad. Sci. USA*, 1975, **72**, 1166.

affect the affinity for ligands and which may be tested by suitable modelling. Experiments on such model compounds may help in attempts to discriminate between the relative contributions of the various effects that are involved in the ligand-binding process. In this perspective, important synthetic compounds are the so-called 'picket-fence',[4,6,99] 'capped',[100,101] 'basket handle',[102] and 'tail-base' porphyrins.[103-106]

'Picket-fence' porphyrins represent a beautiful example of a model system that will carry oxygen at room temperature, and they have been made able to reproduce the O_2 affinities of unstrained haemoproteins.[107] Thus, solutions of an iron-(II) complex of a porphyrin that has three pivalamido pickets lying on one side of the porphyrin ring and an appended imidazole moiety that is capable of co-ordinating to the metal centre show $P_{1/2}$ values (*ca* 0.60 mmHg) that are comparable with those of Mb (*ca* 0.70 mmHg) or the isolated α (0.46 mmHg) and β (0.40 mmHg) chains of haemoglobin.[42] This shows that the oxygen affinities of relaxed, simple haemoproteins may be reproduced exclusively with the ferrous porphyrin–imidazole system. Thus, in contrast to the case of carbon monoxide,[108] special interactions between the protein and the bound ligand are not needed to reproduce the oxygen affinities of some of the non-co-operative haemoproteins. Furthermore, the affinity for O_2 of an iron(II) complex of a 'picket fence' porphyrin with hindered imidazole as the axial base[107] was found to mimic that of the T state of haemoglobin. This low affinity for a ligand has been ascribed to the severe steric restriction involving the 2-methyl group of the imidazole and the porphyrin ring, which moves towards the plane of the porphyrin molecule when the iron atom becomes six-co-ordinate (see above).

A kinetic study, in different solvents, of the reactions of O_2 and of CO with two 'tail-base' haem compounds, *i.e.* pyrrohaem *N*-[3-(1-imidazolyl)propyl]-amide and pyrrohaem *N*-[3-(3-pyridyl)propyl] ester,[103-106] allowed the following conclusions to be drawn: (i) 'tail-base' compounds are a reasonable model for the binding of O_2 to simple haemoproteins or to the R state of haemoglobin; (ii) the protein moiety, which is required to account for the high stability of the haem towards autoxidation, is not necessary for reversible binding of O_2; (iii) variations in affinities for O_2 among different haemoproteins result more from polarity of the environment than from stereochemical factors. In relation to the latter point, it has been postulated[98] that a dipole–dipole interaction between the Fe^+—O—O^- dipole and the distal imidazole may modulate the affinity for oxygen through structural changes which affect the relative positions

[99] J. P. Collman, J. I. Brauman, K. M. Doxsee, T. R. Halbert, S. E. Hayes, and K. S. Suslick, *J. Am. Chem. Soc.*, 1978, **100**, 2761.
[100] J. Almog, J. E. Baldwin, R. L. Dyer, and M. Peters, *J. Am. Chem. Soc.*, 1975, **97**, 226.
[101] J. Almog, J. E. Baldwin, and J. Huff, *J. Am. Chem. Soc.*, 1975, **97**, 227.
[102] M. Momentau, B. Loock, J. Mispelter, and E. Bisagni, *Nouv. J. Chim.*, 1979, **3**, 77.
[103] C. K. Chang and T. G. Traylor, *J. Am. Chem. Soc.*, 1973, **95**, 5810.
[104] C. K. Chang and T. G. Traylor, *Biochem. Biophys. Res. Commun.*, 1975, **62**, 729.
[105] J. Geibel, C. K. Chang, and T. G. Traylor, *J. Am. Chem. Soc.*, 1975, **97**, 5924.
[106] J. Cannon, J. Geibel, M. Whipple, and T. G. Traylor, *J. Am. Chem. Soc.*, 1976, **98**, 3395.
[107] J. P. Collman, J. I. Brauman, K. M. Doxsee, T. R. Halbert, and K. S. Suslick, *Proc. Natl. Acad. Sci. USA*, 1978, **75**, 564.
[108] J. P. Collman, J. I. Brauman, and K. M. Doxsee, *Proc. Natl. Acad. Sci. USA*, 1979, **76**, 6035.

of these two dipoles. Introduction of proximal base strain in the 'tail-base' complexes decreases the affinity for O_2 and hence may contribute to the differences between the R and T states of haemoglobin.[109]

In the case of the binding of CO at low pH, the formation of the six-co-ordinate complex proceeds *via* the binding of carbon monoxide to a porphyrin that is four-co-ordinate, followed by co-ordination of the appended imidazole to the liganded metal ion.[109] This mechanism may itself be involved in the regulation of ligand affinity in haemoproteins, as suggested by double-mixing kinetic experiments performed on myoglobins and haemoglobin chains at low pH[110] (see also Vol. 1).

In conclusion, the regulation of the affinities for oxygen and carbon monoxide seems to involve different structural mechanisms. In the case of carbon monoxide, a substantial role is played by hindrance effects at the distal side, in view of the tendency of CO to bind with a linear geometry. This is illustrated by kinetic experiments performed on the so called 'pagoda-porphyrin', where the second-order rate constant for combination of CO ($l' \simeq 10^3$ l mol^{-1} s^{-1}) is drastically reduced as compared to chelated haems without steric effects ($l' \simeq 10^7$ l mol^{-1} s^{-1}).[42, 111] However, a variable polarity effect within the distal pocket cannot be excluded[112, 113] even for CO.

In the case of oxygen, the role of the haem pocket in preventing oxidation is overwhelming, while modulation of the affinity for O_2 may be attributed to different degrees of strain involving the proximal imidazole and its movement towards the haem plane.[13]

The great importance of the stereochemical environment of the haem in determining the ligand affinity is also shown by results obtained on mutant haemoglobins in which the distal and proximal residues are substituted. The classical case of HbM (for Met) has been amply described by several authors (see ref. 42); recently, other abnormal haemoglobins have been studied in detail, information on their *X*-ray structures being included. Kincaid *et al.*[114] have reported the Fourier-transform i.r. spectrum of the CO stretching band of $Hb_{Kansas}CO$ [$\alpha_2\beta_2102(G4)Asn \rightarrow Thr$],[115] which is known to be stabilized in the T quaternary state upon addition of inositol hexaphosphate (IHP).[116, 117] The liganded T state of $Hb_{Kansas}CO$ displays a stretching frequency that is shifted towards lower values by 0.7 cm^{-1}, an effect opposite to that reported by Ascoli

109 J. Geibel, J. Cannon, D. Campbell, and T. G. Traylor, *J. Am. Chem. Soc.*, 1978, **100**, 3575.
110 G. M. Giacometti, T. G. Traylor, P. Ascenzi, M. Brunori, and E. Antonini, *J. Biol. Chem.*, 1977, **252**, 7447.
111 T. G. Traylor, D. Campbell, and S. Tsuchiya, *J. Am. Chem. Soc.*, 1979, **101**, 4748.
112 S. J. Cole, G. L. Curthoys, and E. A. Magnusson, *J. Am. Chem. Soc.*, 1971, **93**, 2153.
113 W. S. Brinigar, C. K. Chang, J. Geibel, and T. G. Traylor, *J. Am. Chem. Soc.*, 1974, **96**, 5597.
114 J. R. Kincaid, T. G. Spiro, J. S. Valentine, D. D. Saperstein, and A. J. Rein, *Inorg. Chim. Acta*, 1979, **33**, L181.
115 J. Bonaventura and A. Riggs, *J. Biol. Chem.*, 1968, **243**, 980.
116 S. Ogawa, S. Mayer, and R. G. Shulman, *Biochem. Biophys. Res. Commun.*, 1972, **49**, 1485.
117 R. G. Shulman, J. J. Hopfield, and S. Ogawa, *Q. Rev. Biophys.*, 1975, **8**, 325.

et al.[118] for the same derivative of trout HbIV ($\Delta \nu = -0.4$ cm^{-1}, going from R to T). The observed shift is in a direction opposite to that expected for weakening the proximal imidazole bond, but is consistent with an enhanced interaction with the distal imidazole in the T quaternary state. However, the differences between the two systems do not exclude more complex interpretations.

In the case of Hbzürich [$\alpha_2\beta_263$(E7)His→Arg], the i.r. stretch of bound CO corresponds to that of unconstrained porphyrin,[119] the accessibility of the haem pocket is greater,[44] and the kinetics of combination with carbon monoxide are faster.[46, 47] It appears then that mutant haemoglobins having distal residues which do not sterically force carbon monoxide 'off the axis' have higher 'on' rates and higher affinity as compared with HbA. Thus the effect of distal constraints which lower the affinity of normal haemoglobin for CO seems well founded, and has also been assigned a physiological significance as a sort of molecular protection towards poisoning by CO. The differences that exist in the modes of binding of O_2 and of CO in haemoproteins are also consistent with recent results on HbZh.[47]

Spectroscopic Studies of Chain Differences in Haemoglobins. Among the low-spin compounds of HbA, the NO derivative is unique in so far as its quaternary structure may be switched to the low-affinity form, even in the presence of the ligand, by the addition of IHP or by a decrease in pH.[120-126] This transition is accompanied by marked changes in electronic, i.r., and e.p.r. spectra. The e.p.r. spectrum of HbNO displays a superhyperfine structure in the region corresponding to the magnetic field being parallel to the direction that is normal to the plane of the haem moiety, and this represents a useful spectroscopic tool to investigate alterations of the ligand–metal complex that are eventually associated with quaternary structural changes. By comparison with the results obtained on model compounds and on several haemoproteins,[127, 128] it has been proposed that, upon addition of IHP to HbNO, the haem iron undergoes a transition from hexaco-ordination to pentaco-ordination, with rupture or extreme weakening of the covalent bond between the iron and N_ε of the proximal histidine. The low-pH (or IHP-saturated) species displays a three-line e.p.r. pattern, corresponding to only one nitrogen nucleus interacting with the unpaired electron of NO; on the other hand the deprotonated form gives rise to a nine-line spectrum, indicating interaction of the unpaired electron with two inequivalent nitrogen nuclei, *i.e.* that of

[118] F. Ascoli, E. Gratton, F. Riva, P. Fasella, and M. Brunori, *Biochim. Biophys. Acta*, 1978, **533**, 534.
[119] W. S. Caughey, J. O. Alben, S. McCoy, S. H. Boyer, S. Charache, and P. Hathaway, *Biochemistry*, 1969, **8**, 59.
[120] E. Trittelvitz, H. Sick, K. Gersonde, and H. Rüterjans, *Eur. J. Biochem.*, 1973, **35**, 122.
[121] R. Cassoly, *C.R. Hebd. Seances Acad. Sci., Ser. D*, 1974, **278**, 1417.
[122] J. M. Salhany, *FEBS Lett.*, 1974, **49**, 84.
[123] M. F. Perutz, J. V. Kilmartin, K. Nagai, A. Szabo, and S. R. Simon, *Biochemistry*, 1976, **15**, 378.
[124] A. Szabo and M. F. Perutz, *Biochemistry*, 1976, **15**, 4427.
[125] J. C. Maxwell and W. S. Caughey, *Biochemistry*, 1976, **15**, 388.
[126] M. Chevion, A. Stern, J. Peisach, W. E. Blumberg, and S. R. Simon, *Biochemistry*, 1978, **17**, 1745.
[127] H. Kon, *Biochim. Biophys. Acta*, 1975, **379**, 103.
[128] T. Yonetani, H. Yamamoto, J. E. Ernan, J. S. Leigh, jr., and G. H. Reed, *J. Biol. Chem.*, 1972, **247**, 2447.

the NO itself and that of the proximal imidazole. Different experiments have indicated that the perturbation involves only the α subunits of HbA (see ref. 124), very distant from the binding site for organic phosphate. The problem of the rupture of the bond upon addition of IHP is, however, debated, and alternative hypotheses have been proposed. Thus studies on model compounds[129] and energetic considerations[126] have been interpreted as indicating that the nitrosyliron(II) complex is always hexaco-ordinated, the perturbation in the hyperfine structure being related to the protonation of the nitrogen atom N-1 of the proximal imidazole rather than to that of N_ε. The same authors[126] outline the difference between the α and β subunits in the tetramer, since the latter chains, even in the low-affinity form, exhibit a nine-line pattern, due to the weaker crystal field.

The mechanism is obviously complex and may possibly involve the protonation of either of the two nitrogen atoms (N-1 and N_ε), depending on the protein and the experimental conditions. Recent spectroscopic and kinetic results on *Aplysia* MbNO, which is stable down to very low pH (~ 3), are, however, fully consistent with the rupture of the iron–imidazole bond, with the concomitant protonation of the N_ε of the proximal histidine residue.[130]

Inequivalence of the α and β subunits has been demonstrated also by resonance Raman spectroscopy[131] on valency-hybrid haemoglobins. Thus, $\alpha_2^{\text{deoxy}}\beta_2^{+\text{CN}}$ and $\alpha_2^{+\text{CN}}\beta_2^{\text{deoxy}}$ are in the R structure when stripped, but they can be converted into the T structure by addition of IHP.[117, 132, 133] Upon conversion from R into T, the vibrational wavenumber of the $Fe^{II}-N_\varepsilon(\text{His-F8})$ bond changes from 223 to 207 cm^{-1} in the α^{deoxy} subunits and from 224 to 220 cm^{-1} in the β^{deoxy} subunits. It is concluded that this bond is stretched three times more in the α subunits (0.024 Å) than in the β subunits (0.0085 Å) by the allosteric transition, suggesting that the strain energy that is developed in this bond is larger in the α subunits.

Asher and Schuster[134, 135] have investigated the properties of a number of ferric derivatives of Mb and Hb by resonance Raman excitation. In particular, they have examined differences in the geometry of haem between human metHbF, the isolated α and β chains of HbA, and metMbF. Different out-of-plane distances for iron that were observed for the α and β chains were preserved in the tetramer, the iron atom being displaced more from the plane of the haem moiety in the α than in the β subunits (by 0.02 Å). Moreover, the pH dependence of the spectral properties indicates that there is coupling with the protonation of the distal imidazole.

Dynamics and Structure of Haemoglobins.—*Assessment of Structural Fluctuations from X-Ray Crystallography.* The notion that proteins are fluctuating structures[136]

[129] Y. Henry, J. Peisach, and W. E. Blumberg, *Biophys. J.*, 1975, **15**, 286.
[130] P. Ascenzi, G. M. Giacometti, E. Antonini, G. Rotilio, and M. Brunori, *J. Biol. Chem.*, 1981, **256**, 53.
[131] K. Nagai and T. Kitagawa, *Proc. Natl. Acad. Sci. USA*, 1980, **77**, 2033.
[132] S. Ogawa and R. G. Shulman, *J. Mol. Biol.*, 1972, **70**, 315.
[133] K. Nagai, *J. Mol. Biol.*, 1977, **111**, 41.
[134] S. A. Asher and T. M. Schuster, *Biochemistry*, 1979, **24**, 5377.
[135] S. A. Asher and T. M. Schuster, in ref. 2.
[136] G. Careri, in 'Quantum Statistical Mechanics in the Natural Sciences' ed. B. Kursunoglu, S. L. Mintz, and S. M. Widmayer, Plenum, New York, 1974, p. 15.

is nowadays on a firm experimental basis,[137-140] and the significance of protein dynamics in relation to protein function is being actively investigated.

Different groups have recently collected single-crystal *X*-ray diffraction data, obtained at different temperatures (*e.g.* from 220 to 300 K). The data on Mb obtained by Frauenfelder, Petsko, and Tsernoglou[141] have been analysed in terms of the mean-square displacement $\langle x^2 \rangle$ of all the atoms (in the backbone or the side-chains) except hydrogen, as a function of temperature. The results have been interpreted in terms of regions of the molecule characterized by different mobilities.

The correlation between the flexibility of the molecular domains and the overall thermodynamic stability or the biological properties of the macromolecule represents an area of very active research which extends the domain of crystallography towards dynamics. In relation to low-temperature kinetic investigations (see below), Frauenfelder and co-workers[141] have suggested that the pathway for the diffusion of a ligand into the domain of the active site may occur through the 'semi-liquid' sections of the molecule, as identified through the greater values of the mean-square displacement of individual atoms.

Low-temperature Kinetics of Haemoglobins. Advantage has been taken of the photosensitivity of the liganded complexes of ferrous haemoproteins[1,42,142] to investigate the ligand-recombination process since the early work of Gibson. In the past few years this approach has been systematically applied to the study of dynamics at temperatures below 300 K, employing a number of haemoproteins, such as myoglobin, the isolated chains of human haemoglobin (as well as cytochrome-*c* oxidase and cytochrome *P*-450).[138,143-146]

In the case of MbCO, the duration of recombination (followed by monitoring the changes in absorbance of the porphyrins) varies from microseconds to kiloseconds when the temperature is varied from 300 to 60 K. The basic observation[138] is that there is a non-exponential re-binding of carbon monoxide when photolysed by a pulsed laser in the lower temperature range. In a series of papers, Frauenfelder and collaborators (see ref. 147) have proposed an interesting idea in an attempt to find a correlation between the structural features of the macromolecule and the dynamic processes which determine the binding of the ligand to the metal. The non-exponential re-binding of ligands that is observed at lower temperatures has been interpreted in terms of a number (possibly *very large*) of 'conformational sub-states' of the macromolecule; these are differently populated

[137] J. R. Lekowicz and G. Weber, *Biochemistry*, 1973, **12**, 4171.
[138] R. M. Austin, K. W. Beeson, L. Eisenstein, H. Frauenfelder, and I. C. Gunsalus, *Biochemistry*, 1975, **14**, 5355.
[139] K. Wüthrich and G. Wagner, *Nature (London)*, 1978, **275**, 247.
[140] R. J. P. Williams, *Proc. R. Soc. London, Ser. B*, 1978, **200**, 353.
[141] H. Frauenfelder, G. A. Petsko, and D. Tsernoglou, *Nature (London)*, 1979, **280**, 558.
[142] Q. H. Gibson, *Biochem. J.*, 1959, **71**, 293.
[143] N. Alberding, R. H. Austin, S. S. Chan, L. Eisenstein, H. Frauenfelder, I. C. Gunsalus, and T. M. Nordlund, *J. Chem. Phys.*, 1976, **65**, 4701.
[144] N. Alberding, S. S. Chan, L. Eisenstein, H. Frauenfelder, A. Good, I. C. Gunsalus, T. M. Nordlund, M. F. Perutz, A. M. Reynolds, and L. B. Sorensen, *Biochemistry*, 1978, **17**, 43.
[145] B. Chance, C. Saronio, and J. S. Leigh, *J. Biol. Chem.*, 1975, **250**, 9226.
[146] M. Sharrock and T. Yonetani, *Biochim. Biophys. Acta*, 1976, **434**, 333.
[147] H. Frauenfelder, *Methods Enzymol.*, 1978, **54**, 506, and references cited therein.

at the various temperatures and are characterized by different rates of binding of the ligand.

The differences between conformational sub-states are therefore, by definition, functional, since their existence is defined on the basis of their different rates of ligand binding. With this perspective, the distinction between 'conformational states' (such as the R and T states of haemoglobin, which display the same basic function *i.e.* binding of O_2, but which are characterized by different equilibrium and kinetic parameters) and 'conformational sub-states' seems to be quantitative, and not qualitative.

Analysis of the results led Frauenfelder and co-workers[147] to propose that the ligand has to overcome a number of energy barriers in its trajectory from the solvent to the metal; therefore, after photolysis of the ligand–metal bond, the ligand may be trapped in a number of states before diffusing away from the protein into the bulk. Figure 2 shows the activation-energy profile that corresponds to the binding of CO and O_2 to haem and to Mb.[148] It is presumed from the kinetic data that the structure of the active site, with specific amino-acid residues which protrude into the haem pocket, may be correlated to the dynamic barriers. The effect of viscosity of the solvent has also been investigated recently.[148]

Recent results from i.r. spectroscopy substantiate this hypothesis.[149] Low-temperature kinetics have been monitored by using the stretching frequency of CO, which differs in the free state (2140 cm^{-1}) and in the bound state (1945 cm^{-1} in Mb). After photolysis, a new state of CO, characterized by a new stretching frequency, has been observed. This new signal has been attributed to a complex between the carbonyl and some of the protein residues that coat the interior of the haem pocket. The results also demonstrate molecular tunnelling of CO in the lower temperature range.

Photochemical intermediates, monitored by following the state of the metal, have been observed at low temperatures.

Marcolin *et al.*[150] have collected Mössbauer data of the low-temperature photodissociation product of ^{57}Fe-enriched MbCO. Three conclusions were reached on the basis of data analysis: (i) the photoproduct (Mb*) that is populated at low temperatures (5 K) has been identified as a ferrous high-spin complex that is characterized by Mössbauer parameters different from those of deoxy-Mb; (ii) the time-dependent change of the signal characteristic of Mb* suggests that the photodissociated myoglobin exists in a number of slightly different conformations ('conformational substates') that have different rates of recombination with CO; (iii) in the temperature range below 46 K, the kinetics of recombination are independent of temperature, consistent with tunnelling of the CO molecule through a potential barrier (see above). The direct evidence for a

[148] D. Beece, L. Eisenstein, H. Frauenfelder, D. Good, M. C. Marden, L. Reinisch, A. M. Reynolds, L. B. Sorensen, and K. T. Yue, *Biochemistry*, 1980, **19**, 5147.

[149] J. O. Alben, D. Beece, S. F. Bowne, L. Eisenstein, H. Frauenfelder, D. Good, M. C. Marden, P. P. Moh, L. Reinisch, A. H. Reynolds, and K. T. Yue, *Phys. Rev. Lett.*, 1980, **44**, 1157.

[150] H. E. Marcolin, R. Reschke, and A. Trautwein, *Eur. J. Biochem.*, 1979, **96**, 119.

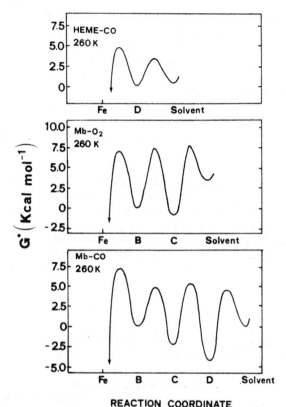

REACTION COORDINATE

Figure 2 *The Gibbs activation energy, G*, at 260 K, as a function of reaction co-ordinate for the pathway of a ligand from the bulk solvent to the haem iron (Fe), at a solvent viscosity of 10 cP. The energy scale is arbitrarily normalized to zero in the well B for MbO₂ and MbCO, and in the well D in the case of haem–CO.*

(Modified from *Biochemistry*, 1980, **19**, 5147)

photochemical intermediate even in Mb is a strong support for the mechanism proposed some years ago by Phillipson *et al.*[151]

Yonetani and co-workers[152–154] have carried out a whole series of experiments, using several cobalt-substituted haemoproteins, which provided relevant information. Cobalt has the advantage of being e.p.r.-detectable both in the deoxy (Co^{II}) and in the oxygenated derivatives. Moreover, it has been shown that Co-Hb maintains the basic functional features of native Fe-Hb (*i.e.* co-operative binding of O₂, Bohr effect, *etc.*).[155] Low-temperature (5 K) photolysis of oxy-Co-

[151] P. E. Phillipson, B. J. Ackerson, and J. Wyman, *Proc. Natl. Acad. Sci. USA*, 1973, **70**, 1550.

[152] T. Yonetani, H. Yamamoto, and T. Iizuka, *J. Biol. Chem.*, 1974, **249**, 2168.

[153] M. Ikeda-Saito, H. Yamamoto, F. G. Kayne, and T. Yonetani, *J. Biol. Chem.*, 1977, **252**, 4482.

[154] M. Ikeda-Saito, M. Brunori, and T. Yonetani, *Biochim. Biophys. Acta*, 1978, **533**, 173.

[155] K. Imai, T. Yonetani, and M. Ikeda-Saito, *J. Mol. Biol.*, 1977, **109**, 83.

Mb yields a species whose e.p.r. signal is different from that of oxy-Co-Mb or that of deoxy-Co-Mb. The spectroscopic features of this new intermediate change with the nature of the amino-acid residue in the distal position. Thus the properties of this photochemically induced intermediate have been correlated with the interactions of the metal–ligand complex with residues on the distal side, which vary from one protein to another.

Pico- and Nano-second Kinetics of Haemoglobin: Geminate Recombination. The availability of very short laser pulses (of duration of ns or ps) has opened a new window for kinetic studies of haemoglobin and other haemoproteins at room temperature. Earlier investigations by Alpert *et al.*[156] on the photolysis of carbon monoxide haemoproteins identified a very fast transient, with a half-time of ∼ 50 ns. The original interpretation, related to a local conformational change, has been revised in the light of more recent work. Duddell *et al.*[157, 158] and Alpert *et al.*[159] have published details of laser photolysis experiments showing that the fast initial transient is related to re-binding of the photolysed ligand, which remains trapped within the haem pocket. The amplitude of this very fast process decreases at higher temperatures (*e.g.* above 20 °C) and the observed rate is independent of the CO concentration, indicating a reaction that occurs within the pocket.

Time-resolved resonance Raman spectroscopy of both Hb and Mb[160] has provided unequivocal evidence for this type of effect. In the case of MbCO at room temperature, essentially 100% of the photodissociated ligand diffuses out in the bulk, and recombination occurs with (slow) bimolecular kinetics (in agreement with flash photolysis and rapid-mixing kinetics) (see ref. 42). With Hb, a significant fraction of the ligand remains trapped, and it re-binds in a geminate-recombination mode, with a half-time of less than 100 ns. This difference has been correlated with a barrier which, in Hb, (partially) hinders the diffusion of the ligand into the bulk, and it seems to account for the difference in quantum yield for photodissociation of CO between liganded Hb ($\phi \simeq 0.4$) and liganded Mb ($\phi = 1$).[42]

These observations demand a redefinition of quantum yield for ligand photo-dissociation as obtained by more conventional pulse methods, and a re-interpretation of some of the results and comparisons reported in the literature.[161] This type of phenomenon is observed for other haemoproteins, and the work has been extended to other ligands as well (*e.g.* O_2 and NO).[162, 163]

The study of molecular dynamics in the picosecond time range has been investigated by absorption[164, 165] and resonance Raman[166] spectroscopies. The

156 B. Alpert, R. Banerjee, and L. Lindqvist, *Proc. Natl. Acad. Sci. USA*, 1974, **71**, 558.
157 D. A. Duddell, R. J. Morris, and J. T. Richards, *Biochim. Biophys. Acta*, 1980, **621**, 1.
158 D. A. Duddell, R. J. Morris, N. J. Muttucumaru, and J. T. Richards, *Photochem. Photobiol.*, 1980, **31**, 479.
159 B. Alpert, S. El Mohsni, L. Lindqvist, and T. Tfibel, *Chem. Phys. Lett.*, 1979, **64**, 11.
160 J. M. Friedman and K. B. Lyons, *Nature (London)*, 1980, **284**, 570; see also ref. 2.
161 M. Brunori and G. M. Giacometti, *Methods Enzymol.*, 1981, **76**, 582.
162 D. A. Chernoff, R. M. Hochstrasser, and A. W. Steele, in ref. 2.
163 R. J. Morris and Q. H. Gibson, *J. Biol. Chem.*, 1980, **255**, 8050.
164 D. Huppert, K. D. Straub, and P. M. Rentzepis, *Proc. Natl. Acad. Sci. USA*, 1977, **74**, 4139
165 A. M. Reynolds, P. M. Rentzepis, and K. D. Straub, in ref. 2.
166 J. Terner, J. A. Strong, and T. G. Spiro, in ref. 2.

observations in the different wavelength ranges reveal that a short-lived transient is formed within 8 ps in MbCO, but it is not observed in MbO_2.[165] Moreover, in HbCO,[166] the intermediate shows properties that are different from those of the high-spin ($S = 2$) relaxed deoxy-species in so far as the iron is still 'in-plane'.

In the nanosecond time range, fluorescence-decay studies have been employed to probe fast structural changes in apo-Hb[167, 168] and in Hb as well as in the isolated subunits.[169-171]

Haem–Haem Interactions in Haemoglobins.—*The Structural Basis of Co-operativity*. In the formulation of co-operative binding according to the classical allosteric model[172] it is required that the intrinsic ligand affinity of one subunit in the T quaternary state be lower than that of the same site in the R quaternary state, and that in the absence of the ligand T_0 is preferentially populated ($[T_0] \gg [R_0]$). Deviations from a simple behaviour that is based, for example, on differences in the affinities of the T state have been pointed out; thus the shift in the lower asymptote of the ligand-binding curve or the presence of intermediate conformational states have been taken to indicate that a simple two-state allosteric model is inadequate.[173-175]

It is beyond our present scope to review critically all the arguments for or against the applicability of a simple two-state model to haemoglobin; on the other hand, we wish to present recent data relevant to the problems of the physical interpretation of the differences in affinity between the extreme states. The contribution of Perutz and collaborators[13, 176-178] to the interpretation of the role of the globin–haem interactions in regulating the affinity of haemoglobin for O_2 is outstanding and well known.

According to a simple formulation, the following is a summary of the relevant questions: (i) why the affinity of the T state for ligands is lower than that of the R state; (ii) where in the molecule is the difference in free-energy of binding between the T and R state stored, and how is it dissipated upon binding a ligand; (iii) why the T_0 state is more stable than the R_0 state in HbA.

Information on the first point has been obtained from structural crystallographic studies, from spectroscopic investigation, and from theoretical calculations. A comprehensive comparison of the atomic co-ordinates of (*a*) human deoxy-, (*b*) human carboxy-, and (*c*) horse met-haemoglobin has been carried out by Baldwin and Chothia.[179] It has been confirmed, first of all, that the two liganded species (*b* and *c*) are very similar, although clear-cut differences among various liganded states of ferrous and ferric haemoglobins are clearly observed.

[167] L. Lindqvist, A. Lopez-Campillo, and B. Alpert, *Photochem. Photobiol.*, 1978, **28**, 417.
[168] E. Bucci, C. Fronticelli, K. Flanigan, J. Perlman, and R. F. Steiner, *Biopolymers*, 1979, **18**, 1261.
[169] B. Alpert and R. Lopez-Delgado, *Nature (London)*, 1976, **263**, 5576.
[170] B. Alpert, D. M. Jameson, and G. Weber, *Photochem. Photobiol.*, 1980, **31**, 1.
[171] M. P. Fontaine, D. M. Jameson, and B. Alpert, *FEBS Lett.*, 1980, **116**, 310.
[172] J. Monod, J. Wyman, and J. P. Changeux, *J. Mol. Biol.*, 1965, **12**, 88.
[173] J. M. Baldwin, *Prog. Biophys. Mol. Biol.*, 1975, **29**, 225.
[174] A. P. Minton, *Science*, 1974, **184**, 577.
[175] G. Viggiano, and H. Chien, *Proc. Natl. Acad. Sci. USA*, 1979, **76**, 3673.
[176] M. E. Perutz, *Nature (London)*, 1970, **228**, 726.
[177] M. F. Perutz, *Nature (London)*, 1972, **237**, 495.
[178] M. F. Perutz, *Proc. R. Soc. London, Ser. B*, 1980, **208**, 135.
[179] J. M. Baldwin and C. Chothia, *J. Mol. Biol.*, 1979, **129**, 175.

The structural changes seen in going from deoxy-[180] to liganded haemoglobin include tertiary and quaternary perturbations, with a shift in the packing of the $\alpha_1\beta_1$ *versus* $\alpha_2\beta_2$ dimers.

The environment of the ligand-binding site in deoxy- and carbonmonoxy-Hb is shown in Figure 3. The distance between C_ε of His(F8) and the nitrogen atom of the pyrrole ring 1 (N_p) is already very close to van der Waals contact in deoxy-Hb, in view of the asymmetric position of the imidazole ring relative to the haem.

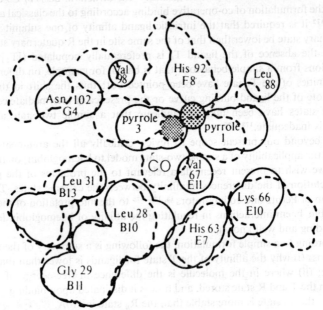

Figure 3 *Structural changes at the level of the haem pocket of the β subunits in going from deoxy- (continuous lines) to carbonmonoxy-haemoglobin (dashed lines). The Figure is of a section through a space-filling model, cut perpendicular to the plane of the haem and passing through the iron atom and the nitrogen atoms of pyrroles 1 and 3. When a ligand is bound, the shift of the haem (by 1.5 Å) and its rotation (by 9°) make room for the bound CO by removing β67(E11)Val, which is not shown in the liganded state.*
(Reproduced by permission from *J. Mol. Biol.*, 1979, **129**, 175)

Upon binding of a ligand (*e.g.* CO), the transition to a low-spin state and the tendency of the metal to move towards the ligand impose a shift on His(F8) which is opposed by the steric repulsion between C_ε of the imidazole and the nitrogen atom of pyrrole 1. This steric effect produces a reorganization involving (i) the haem as a whole, which tilts and shifts in the direction of pyrroles 1–3 (by 0.5 Å in α and by 1.5 Å in β); (ii) the helix F, which is translated across the plane of the haem by 1 Å, and (iii) Val(E11), which in the β chains moves away from the ligand (and thus is not apparent in Figure 3 in the CO derivative). Theoretical

180 G. Fermi, *J. Mol. Biol.*, 1975, **97**, 237.

alculations by Warshel[181] and by Gelin and Karplus[182] had already indicated hat a subunit in a T quaternary state develops a 'strain' when binding of a ligand orces the iron towards the plane of the haem moiety, thus implying that no strain' is present in a deoxy-haem in a T quaternary state (see also ref. 1). There-ore crystallographic evidence and theoretical calculations indicate that the lower igand affinity of a subunit in the T state is determined by steric repulsions be-ween His(F8) and N_p of pyrrole, which oppose the movement of the metal and hus the binding of the ligand.

The finding that, upon quaternary switch, the whole iron–porphyrin complex hifts and tilts indicates that the interactions between the haem periphery and the lobin exert an important role in the control of co-operativity, in addition to the nore localized steric effects involving the fifth ligand of the iron.

Thus a multiplicity of structural perturbations in the close proximity of the igand-binding site is involved in the process of ligand binding, and the emphasis f the structural analysis is not exclusively on the 'in-plane' shift of the metal tom. It appears that the new analysis is consistent with the spectroscopic observ-tion of Eisenberger and co-workers,[183, 184] who reported EXAFS measurements ndicating that the Fe–N_p bond distance remains the same (*i.e.* 2.02 Å, within less han 0.01 Å) when deoxy-Hb goes from T_0 (*i.e.* HbA) to R_0 (deoxy-$Hb_{Kempsey}$).

The difference in the free-energy of oxygen binding between T and R amounts o approx. 3 kcal per site in HbA at neutral pH.[185] On the basis of structural ata, Perutz[177] proposed that ΔF_I is localized at the haem, the Fe–N_ε[His(F8)] ond playing a crucial role in the transmission of stereochemical effects to the rotein as a whole. Although the involvement of the Fe–N_ε bond in the structural erturbations coupled to ligand binding is on a very firm basis, the localization of ΔF_I in this bond has been challenged by several authors, with the formulation of n alternative proposal whereby the free-energy of interaction is spread through-ut the protein.[186, 187] Since the last Report,[1] more spectroscopic data have been btained dealing with this crucial point.

Pertinent information has been obtained by resonance Raman scatter-ng.[131, 188-193] The Fe–O_2 stretching frequency of Hb in the R_4 state (HbAO$_2$) and n the T_4 state (Hb$_{Kansas}O_2$ and HbM$_{Milwaukee}O_2$ in the presence of IHP) is the ame (567 cm^{-1}), indicating[190] that the quaternary structural change in the ganded state is not associated with a strain of the iron–ligand bond. Other

[181] A. Warshel, *Proc. Natl. Acad. Sci. USA*, 1977, **74**, 1789.
[182] B. R. Gelin and M. Karplus, *Proc. Natl. Acad. Sci. USA*, 1977, **74**, 801.
[183] P. Eisenberger, R. G. Shulman, G. S. Brown and S. Ogawa, *Proc. Natl. Acad. Sci. USA*, 1976, **73**, 491.
[184] P. Eisenberger, R. G. Shulman, B. M. Kincaid, G. S. Brown, and S. Ogawa, *Nature (London)*, 1978, **274**, 30.
[185] J. Wyman, *Adv. Protein Chem.*, 1964, **19**, 223.
[186] J. J. Hopfield, *J. Mol. Biol.*, 1973, **77**, 207.
[187] G. Weber, *Biochemistry*, 1972, **11**, 864.
[188] T. G. Spiro, *Biochim. Biophys. Acta*, 1975, **416**, 169.
[189] T. Kitagawa, Y. Ozaki, and Y. Kyogoku, *Adv. Biophys.*, 1978, **11**, 153.
[190] K. Nagai, T. Kitagawa, and M. Morimoto, *J. Mol. Biol.*, 1980, **136**, 271.
[191] M. Coppey, C. de Loze, and B. Alpert, *C.R. Hebd. Seances Acad. Sci., Ser. D*, 1979, **289**, 173.
[192] J. A. Shelnutt, D. L. Rousseau, J. M. Friedman, and S. R. Simon, *Proc. Natl. Acad. Sci. USA*, 1979, **76**, 4409.
[193] D. L. Rousseau, J. A. Shelnutt, J. M. Friedman, and S. R. Simon, in ref. 2.

evidence in agreement with this conclusion was previously reported (see ref. 117). In deoxy-Hb the line at 216 cm^{-1} is associated primarily with the Fe–N$_\varepsilon$[His(F8)] stretching mode, as shown by isotopic (^{54}Fe) frequency shift.[190] When the quaternary structure of unliganded Hb is flipped from T_0 to R_0, this frequency shifts indicating that the Fe–N$_\varepsilon$ bond is stretched in the T_0 state, due to the tension exerted by the globin. This result substantiates the suggestion that the metal–histidine bond is involved in the spread of the conformational changes from the metal–ligand complex to the globin;[13, 176] however, the very small shift that is observed in going from T_0 to R_0 suggests that the strain energy that is developed in this bond is smaller than the free-energy of interaction ($\Delta\nu$ yielding a maximum energy difference of 0.3 kcal per site).[131]

Raman difference spectroscopy with higher resolving power (the detection limit is differences of 0.1 cm^{-1}) has, however, shown[192, 193] that the T_0 to R transition in deoxy-Hb is associated with several small (0.3–2.2 cm^{-1}) difference at various lines from 1357 to 1605 cm^{-1}. The lines in this frequency region of the spectrum (1300 to 1650 cm^{-1}) have been assigned to vibrations of the porphyrin core and correlated with the nature of the axial ligands.

Although more complex interpretations arising from chain differences cannot be excluded, the shift in Raman lines indicates electronic interactions between the haem and the globin. Since the relative amount of charge depletion in the antibonding π^*-orbitals of the porphyrin and the free-energy of co-operativity appear to show a correlation, Rousseau *et al.*[192, 193] concluded that their measurement provide direct evidence for the storage of part of the free-energy of interaction at the haem.

New crystallographic evidence concerning the fundamental role of interface between subunits in determining the relative stability of the two quaternary states has been analysed by Baldwin and Chothia.[179] Perutz has recently reviewed the arguments which support the stereochemical interpretation of the mechanism of haem–haem interactions,[13, 178] and has emphasized the role of the salt bridge in stabilizing the low-affinity quaternary structure in Hb. Some of the salt bridges which are present in deoxy-Hb[176, 177] are contributed by ionizable groups that are involved in the Bohr effect (see below). A comparison of the amino-acid residues involved in the salt bridges with the structural data available on trout HbI, which has no organic phosphate effect, no Bohr effect, and yet co-operative,[194] is useful [see Table 4 (p. 156)]. Such a comparison indicates that the salt bridges cannot by themselves be responsible for the difference in stability between T and R states, since several amino-acid substitutions in trout HbI involve amino-acid residues that contribute to the salt bridges.[195]

The energetics of subunit interactions have been investigated in a series of beautiful papers by Ackers and collaborators.[196–198] These authors have determined, by different techniques, the relationships between ligand binding and stability of monomers, dimers, and tetramers in the liganded and unliganded

194 M. Brunori, *Curr. Top. Cell. Regul.*, 1975, **9**, 1.
195 M. Brunori, B. Giardina, A. Colosimo, M. Coletta, G. Falcioni, and S. J. Gill, in ref.
196 R. Valdes and G. K. Ackers, *J. Biol. Chem.*, 1977, **252**, 74, 88.
197 S. H. C. Ip and G. K. Ackers, *J. Biol. Chem.*, 1977, **252**, 82.
198 R. Valdes and G. K. Ackers, *Proc. Natl. Acad. Sci. USA*, 1978, **75**, 311.

tates. The bulk of these results were discussed previously.[1] However, the more recent investigation indicates that the affinity of the triply liganded tetramer $\alpha_2\beta_2(O_2)_3$ for oxygen is greater than that of the isolated chains,[199] with important implications for modelling the haemoglobin function. Thus, if the properties of the isolated chains[42] cannot be equated with those of the R state, the contention that the α and β subunits in a tetramer are subject to constraints whose disruption allows the ligand affinity of each subunit to approach that of the isolated chains is not strictly valid. Similar findings in the case of haemoglobin H (β_4) indicate that the monomeric species have lower affinity for oxygen than the tetramers.[198] Ackers[199] has therefore proposed that, in haemoglobin, besides the quaternary constraints, there is another effect, called 'quaternary enhancement', which amounts to ~ 0.8 kcal mol^{-1} and which manifests itself at the $\alpha_1\beta_1$ contact. The perturbation of this interface that is associated with the binding of a ligand or/ and heterotropic effectors is directly monitored by following the vibrational absorption of the SH group of $\alpha 104(G11)$Cys.[200]

Yonetani and collaborators (see refs. 153 and 201, and references cited therein) have, over the past few years, systematically investigated the functional and spectroscopic properties of haemoglobin that had been reconstituted with a cobalt–porphyrin complex, and of hybrid tetramers obtained by recombination of a cobalt-containing subunit (either α or β) with an iron-containing partner.

The more recent work has been devoted to the investigation of the differences in functional properties of the individual chains. It should be recalled that, while the iron-containing subunits combine reversibly with both O_2 and CO, the cobalt-containing subunits bind exclusively with O_2, but with an affinity that is very much lower than that of the iron-containing protein.[200] Therefore, in conjunction with the specific spectral features of the cobalt–porphyrin complex, these binding properties allow an attack on the formidable problem of determining independently the individual microscopic constants of a complete Adair scheme, which differentiates among the α and β chains; *i.e.*, that shown in Figure 4. On the assumption that the interactions between the subunits depend on the ligation state, and not on the type of metal (an assumption that is largely but not absolutely justified), the twelve microscopic equilibrium constants were estimated, using six independent equilibrium curves.

Moreover, the problem of the α–β equilibrium differences has been discussed extensively,[200] and also in relation to other results dealing specifically and critically with this problem.[202–207] The authors point out that, in the natural protein (carrying, of course, the iron–porphyrin complex), the β subunits have an affinity for O_2 that is higher than that of the α subunits, the difference becoming smaller as the pH is decreased from 7.9 to 6.5. This conclusion, which is in accord

[199] F. C. Mills and G. K. Ackers, *Proc. Natl. Acad. Sci. USA*, 1979, **76**, 273.
[200] J. O. Alben and G. H. Bare, *J. Biol. Chem.*, 1980, **255**, 3892.
[201] K. Imai, M. Ikeda-Saito, H. Yamamoto, and T. Yonetani, *J. Mol. Biol.*, 1980, **138**, 635.
[202] M. E. Johnson and C. Ho, *Biochemistry*, 1974, **13**, 3653.
[203] H. Yamamoto and T. Yonetani, *J. Biol. Chem.*, 1975, **235**, 7964.
[204] T. H. Huang and A. G. Redfield, *J. Biol. Chem.*, 1976, **251**, 7114.
[205] N. Makino and Y. Sugita, *J. Biol. Chem.*, 1978, **253**, 1174.
[206] T. Asakura and P. W. Lau, *Proc. Natl. Acad. Sci. USA*, 1978, **75**, 5462.
[207] P. W. Lau and T. Asakura, *J. Biol. Chem.*, 1979, **254**, 2595.

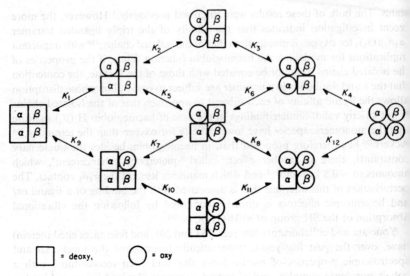

Figure 4 *An extended Adair oxygenation scheme for a tetrameric haemoglobin in which the α and β subunits are treated as non-equivalent, while the $\alpha_1\beta_1$ and $\alpha_1\beta_2$ interfaces are regarded as equivalent. The K's refer to the microscopic equilibrium constants.*
(Reproduced by permission from *J. Mol. Biol.*, 1980, **138**, 635)

with independent kinetic data (see below), is of interest in relation to the structural information obtained from X-ray crystallography.[176, 177] It may be recalled that the oxidation process also shows clear α–β differences in rates and equilibria (see refs. 42, 208, and 209).

The thermodynamic parameters for the binding of O_2 to haemoglobin have been re-analysed by Imai[210] as a function of pH and anion concentration. The dependence of the heat of binding of O_2 on the degree of saturation, already reported previously,[211, 212] is correlated quantitatively with the release of H^+ and of anions (*e.g.* Cl^-, DPG, or inorganic phosphates) which occurs non-linearly with saturation. After proper correction, the intrinsic enthalpy changes associated with the binding of O_2 were found to be uniform. This conclusion appears to be in agreement wih the statement that co-operative effects in ligand binding are entropic in nature,[42, 185] as already outlined previously (see also ref. 1).

The Kinetics of Binding of Ligands to Haemoglobin. Kinetic investigations have been directed towards the phenomenological description of the kinetic basis of co-operativity and of the consistency of the results with a two-state kinetic model. A major contribution to a comprehensive description of kinetics of haemoglobin

208 A. Mansouri and K. H. Winterhalter, *Biochemistry*, 1973, **12**, 4946.
209 A. Tomoda and Y. Yoneyama, *Biochim. Biophys. Acta*, 1979, **581**, 128.
210 K. Imai, *J. Mol. Biol.*, 1979, **133**, 233.
211 K. Imai and T. Yonetani, *J. Biol. Chem.*, 1975, **250**, 7093.
212 H. T. Gaud, B. G. Barisas, and S. J. Gill, *Biochem. Biophys. Res. Commun.*, 1974, **59**, 1389.

was presented by Hopfield, Schulman, and Ogawa,[213] who provided the framework of a kinetic allosteric model. Some of the more recent results were reported in Volume 1 of this series (see also the review by Parkhurst[214]).

A great deal of the earlier investigations[42, 214, 215] contributed to the clarification and quantitation of some basic features of the kinetic description of the binding of ligands by haemoglobin: (i) in agreement with earlier data (see ref. 42), it was proven that the rate of the quaternary ($R_i \rightleftharpoons T_i$) transition is generally very fast compared to ligand binding, and the relaxation time is dependent on the fractional saturation with the ligand;[216, 217] (ii) the contribution of the rate constants for combination ('on') and dissociation ('off') to co-operativity depends on the type of ligand, although the overall free-energy of interaction is often similar for different ligands; (iii) the presence of two different types of chain introduces additional elementary steps, which in some cases appear to have an overwhelming importance in determining the overall kinetics; (iv) the kinetic role of allosteric effectors has been described either in terms of selective stabilization of one quaternary structure (*e.g.* Hb$_{Kansas}$ is stabilized in the T state by organic phosphates)[218, 219] or in terms of tertiary perturbation of structure (*e.g.* chain heterogeneity).

Kinetic investigations published in the past few years have led to a more complete characterization of the dynamic aspects of binding of ligands both by mammalian and non-mammalian haemoglobins.

Haemoglobins from fish have been extensively used, in view of their unusual physico-chemical properties,[1] to test the general applicability of a simple two-state kinetic model. Thus the work of Gibson and co-workers on a Root-effect haemoglobin (Hb from the menhaden, *Brevoortia tyrannus*) has shown that a Monod–Wyman–Changeux (M–W–C) model, with chain heterogeneity, satisfactorily describes their kinetic data as a function of pH[220] and temperature,[221] as well as photostationary conditions.[222] Other Root-effect haemoglobins, which are stabilized in a T quaternary state by lowering the pH (see ref. 194), have also been characterized in their kinetic behaviour towards O_2 and CO.[223]

On the other hand, trout HbI, which is the haemoglobin component from trout blood that is devoid of heterotropic effects[194] (see also below), has proven to be an ideal material for a stringent test of consistency with the simple two-state model. Using this haemoglobin, Brunori *et al.*[224] have shown that the kinetics of binding of CO can be quantitatively described over a large temperature

[213] J. J. Hopfield, R. G. Shulman, and S. Ogawa, *J. Mol. Biol.*, 1971, **61**, 425.
[214] L. J. Parkhurst, *Annu. Rev. Phys. Chem.*, 1979, **30**, 503.
[215] Q. H. Gibson, *Prog. React. Kinet.*, 1964, **2**, 319.
[216] C. A. Sawicki and Q. H. Gibson, *J. Biol. Chem.*, 1977, **252**, 5738, 5783.
[217] G. Ilgenfritz and T. M. Schuster, *J. Biol. Chem.*, 1974, **249**, 2959.
[218] S. Ogawa, A. Mayer, and R. G. Shulman, *Biochem. Biophys. Res. Commun.*, 1972, **49**, 1485.
[219] C. L. Castillo, S. Ogawa, and J. M. Salhany, *Arch. Biochem. Biophys.*, 1978, **185**, 504.
[220] W. A. Saffran and Q. H. Gibson, *J. Biol. Chem.*, 1978, **253**, 3171.
[221] W. A. Saffran and Q. H. Gibson, *J. Biol. Chem.*, 1979, **254**, 1666.
[222] S. J. Torkelson and Q. H. Gibson, *J. Biol. Chem.*, 1978, **253**, 7331.
[223] 'Study of fish bloods and hemoglobins, the α-Helix expedition' *Comp. Biochem. Physiol.*, 1979, **1**, 139, 145, 155, 169, 173, 213, and 227.
[224] M. Brunori, B. Giardina, A. Colosimo, G. Falcioni, and S. J. Gill, *J. Biol. Chem.*, 1980, **255**, 3841.

interval (from below 20 °C to 80 °C); at the higher temperatures ($\geqslant 75$ °C), the R_0 state becomes preferentially populated, and the change in the relative stability of the two quaternary states accounts quantitatively for the kinetic results, providing, in addition, thermodynamic information on the quaternary conformational transition.[225]

It appears, therefore, that the capability of a simple M–W–C allosteric model to describe the kinetics of ligand binding is surprisingly good, although under some conditions it seems to be unsatisfactory. Thus dynamic studies with valency hybrids (*i.e.* $\alpha_2{}^{+CN}\beta_2$ and $\alpha_2\beta_2{}^{+CN}$), with HbM$_{Saskatoon}$ ($\alpha_2\beta_2{}^{63His\rightarrow Tyr}$), and with HbM$_{Milwaukee}$ ($\alpha_2\beta_2{}^{67Val\rightarrow Glu}$), carried out by Makino *et al.*,[226,227] have indicated that at least three states are necessary in a model that is to fit the whole pH dependence of the kinetics of binding of CO to these artificial intermediates. Moreover, the kinetics of binding of the Co–Fe hybrid haemoglobins with O_2 and CO,[228] extensively investigated by stopped-flow and temperature-jump methods, yield relevant results (summarized in Table 3). These kinetic data, taken together

Table 3 *Kinetic constants for the reactions of iron–iron or iron–cobalt hybrid haemoglobins with carbon monoxide and with oxygen (in 0.1M-phosphate, at pH 7, and at 20 °C)*

Reaction	$k'/\text{l mol}^{-1}\,\text{s}^{-1}$	k/s^{-1}
$\alpha(\text{Fe})\beta(\text{Co}) + \text{CO} \rightarrow \alpha(\text{FeCO})\beta(\text{Co})$	6.4×10^4	—
$\alpha(\text{Co})\beta(\text{Fe}) + \text{CO} \rightarrow \alpha(\text{Co})\beta(\text{FeCO})$	1×10^5	—
$\alpha(\text{Fe}^+\text{CN})\beta(\text{Fe}) + \text{CO} \rightarrow \alpha(\text{Fe}^+\text{CN})\beta(\text{FeCO})$	3×10^{5a}	—
$\alpha(\text{Fe}^+\text{CN})\beta(\text{Fe}^+\text{CN}) + \text{CO} \rightarrow \alpha(\text{FeCO})\beta(\text{Fe}^+\text{CN})$	3×10^{5a}	—
$\alpha(\text{FeCO})\beta(\text{Co}) + O_2 \rightarrow \alpha(\text{FeCO})\beta(\text{CoO}_2)$	4.1×10^7	2.1×10^3
$\alpha(\text{Co})\beta(\text{FeCO}) + O_2 \rightarrow \alpha(\text{CoO}_2)\beta(\text{FeCO})$	4.4×10^7	1.4×10^3
$\alpha(\text{Co})\beta(\text{Co}) + O_2 \rightarrow \alpha(\text{CoO}_2)\beta(\text{CoO}_2)^b$	4.8×10^7	8.5×10^3

(*a*) From M. Brunori, G. Amiconi, E. Antonini, J. Wyman, and K. H. Winterhalter, *J. Mol. Biol.*, 1970, **49**, 641; values refer to the slow kinetic component. (*b*) At 15 °C (from H. Yamamoto, F. J. Kayne, and T. Yonetani, *J. Biol. Chem.*, 1974, **249**, 691; calculated for the fast relaxation process)

with the equilibrium analysis,[201] suggest a different role of the α and β subunits in the homotropic and heterotropic interactions in haemoglobin, indicating that chain heterogeneity cannot be altogether neglected even with HbA. The theme of kinetic and equilibrium heterogeneity as between the two types of subunit in haemoglobin has been considered in a number of studies of reactions of ferrous and ferric haemoproteins (see refs. 46, 47, 220, 223, and 229—232). These investigations emphasize the significance of tertiary effects on the kinetics of

225 J. Wyman, S. J. Gill, L. Noll, B. Giardina, A. Colosimo, and M. Brunori, *J. Mol. Biol.*, 1977, **109**, 195.
226 N. Makino, Y. Sugita, and T. Nakamura, *J. Biol. Chem.*, 1979, **254**, 2353.
227 N. Makino, Y. Sugita, and T. Nakamura, *J. Biol. Chem.*, 1979, **254**, 10862.
228 M. Ikeda-Saito and T. Yonetani, *J. Mol. Biol.*, 1980, **138**, 845.
229 A. Raap, J. W. van Leeuwen, T. van Eck-schouten, H. S. Rollema, and S. H. de Bruin, *Eur. J. Biochem.*, 1977, **81**, 619.
230 Y. A. Ilan, Y. Ilan, M. Chevion, and G. Czapski, *Eur. J. Biochem.*, 1980, **103**, 161.
231 P. W. Lau and T. Asakura, *J. Biol. Chem.*, 1980, **255**, 1617.
232 V. S. Sharma, T. S. Vedvick, D. Magde, R. Lith, D. Friedman, M. R. Schmidt, and H. M. Ranney, *J. Biol. Chem.*, 1980, **255**, 5879.

ligand binding. Thus, in the case of rabbit Hb,[232] a kinetic analysis indicates that the β subunits in the deoxy tetramer are characterized by a 'normal' rate constant, and thus determine the overall binding, while the α chains have greatly reduced rate constants for combination of CO and of O_2, possibly due to specific substitutions [Leu→Val at (B10), Leu→Phe at (CD6), and Ser→Thr at (CD7)]

A complete characterization of the mechanism of reaction of Hb$_{\text{zürich}}$ [β63(E7)-His→Arg] with CO and O_2 has been reported by Giacometti *et al.*[46, 47] The striking features emerging from their kinetic analysis are: (i) the abnormal β subunits in a T quaternary structure have a high rate constant for combination of CO [$^{T}k'(\beta) = 1.3 \times 10^7$ 1 mol^{-1} s^{-1}] but an almost normal rate in the R state [$^{R}k'(\beta) = 3 \times 10^7$ 1 mol^{-1} s^{-1}]; (ii) thus the binding of CO is strictly sequential, with the β chains combining rapidly in the T state, then a switch over of quaternary structure, and then saturation of the α chains with the rate constant that is characteristic of the R state [$^{R}k'(\alpha) = 4.7 \times 10^6$ 1 mol^{-1} s^{-1}]; (iii) the mechanism of the reaction with O_2 is, however, not sequential, since binding in the T state is essentially statistical among the abnormal β and the normal α chains. The sequential binding of CO, previously proposed,[46] has been substantiated by performing experiments with a hybrid molecule containing the cobalt–porphyrin complex on the normal α subunits [$\alpha(\text{Co})_2\beta^{\text{Zh}}(\text{Fe})_2$]. The kinetic data for both ligands, as well as the revised equilibrium results,[48] have been correlated with the crystallographic information obtained on Co-HbZh (see above[44]).

Very interesting approaches have been pursued in attempts to identify definite correlations between dynamics of ligand binding and structural features of the macromolecule.

Taking a general approach, Szabo[233] discussed the kinetics of association and dissociation of ligands in haemoglobin and the isolated chains in terms of the properties of the transition state. Starting from basic considerations, the author points out that the absolute values of the rate constants for the reaction with various ligands should be correlated with the structural properties of the transition state, as compared to those of the reactants on the one hand and of the products on the other. A dynamic description of binding (see p. 141) suggests that, in the process of diffusion from the solvent to the metal, the ligand follows a pathway that is characterized by several dynamic barriers (see Figure 2). The favourable energy gain that is coupled with the formation of a bond between ligand and metal may be partly spent in the process of fitting a ligand into the pocket.

A lucid analysis of the role of the steric constraints imposed by (i) the nature of the ligand (whether linear or bent) and (ii) the steric hindrance of the haem pocket has been presented by Moffat *et al.*[234] The approach is based on a correlation between independent crystallographic observations on deoxy-Hb and various liganded derivatives of ferrous and ferric haemoglobins (*e.g.* HbNO, HbCO, Hb^{+}CN^{-}, Hb^{+}N$_3^{-}$, Hb^{+}F^{-}, and Hb^{+}H$_2$O; all with the quaternary R structure) (see also ref. 235). The difference Fourier maps show some specific features in every one of the derivatives examined. Thus the binding of a ligand

[233] A. Szabo, *Proc. Natl. Acad. Sci. USA*, 1978, **75**, 2108.
[234] K. Moffat, J. F. Deatherage, and D. W. Seybert, *Science*, 1979, **206**, 1035.
[235] J. F. Deatherage, S. K. Obendorf, and K. Moffat, *J. Mol. Biol.*, 1979, **134**, 419.

within the R quaternary state induces tertiary conformational changes which are qualitatively similar to those observed in the R–T transition,[179] but much smaller in magnitude. The important point is that steric hindrance effects that are related to the process of accommodating the ligand within the pocket are clearly more severe for ligands whose preferred geometry in the bound state is 'linear' (such as CO) than for those which have a 'bent' geometry (such as O_2 or NO). These preferred geometries are consistent with extensive data on model compounds (see above, p. 136).

Therefore the distortion and rearrangement that are imposed by the binding of 'linear' ligands are reflected in a decrease in the rate constant of association, steric hindrance being manifested in the combination process. Steric effects on the distal side are considerably more important in the T state, and according to Moffat *et al.*[234] they make a significant contribution to the decreased affinity of the T state for ligands (see also above). In the terminology used by Szabo,[233] this implies that the extensive perturbation that is induced by a linear (or a bulky) ligand makes the transition-state structure more 'product-like', and thus slows down the association process. On the other hand, the structural perturbation involved in fitting the ligand is more limited for bent ligands, and the rate of association is considerably closer to diffusion-controlled (*i.e.* the transition state is more 'reactant-like').

This approach is also fully consistent with the kinetics of binding of ligands in Hbzürich.[46,47] Substitution of the distal histidine residue with arginine, which swings away from the haem pocket[44] and reduces steric constraints, leads to a more open and accessible pathway for a linear ligand (reducing the number and/or the height of the barriers?). Very little change in the rate constant for association of O_2 is observed,[47] in agreement with the ideas outlined above.

The influence of steric effects on rate constants for dissociation is stated to be minimal,[234] and in this case the overwhelming contributions seem to be provided by the electronic properties of the iron–ligand bond.[236,237]

The analysis given above provides a structural basis for the well-known fact that the kinetic contribution to co-operativity resides preferentially in the 'on' constants for some ligands (*e.g.* CO) and in the 'off' constants for others (*e.g.* O_2). Some of the conclusions of Moffat *et al.*[234] are, however, in contrast with the analysis given by Reisberg and Olson[238–240] of the equilibrium and the kinetics of binding of thirteen isocyanides to human haemoglobin [methyl, ethyl, n-propyl, n-butyl, n-pentyl, n-hexyl, isopropyl, isobutyl, (+)- and (−)-s-butyl, t-butyl, cyclohexyl, and benzyl]. Following an approach which had been partially exploited previously, with identical results,[241–244] the authors have re-examined

236 D. W. Seybert, K. Moffat, Q. H. Gibson, and C. K. Chang, *J. Biol. Chem.*, 1977, **252**, 4225.
237 M. Sono, P. D. Smith, J. A. McCray, and T. Asakura, *J. Biol. Chem.*, 1976, **251**, 1418.
238 P. I. Reisberg and J. S. Olson, *J. Biol. Chem.*, 1980, **255**, 4144.
239 P. I. Reisberg and J. S. Olson, *J. Biol. Chem.*, 1980, **255**, 4151.
240 P. I. Reisberg and J. S. Olson, *J. Biol. Chem.*, 1980, **255**, 4159.
241 B. Talbot, M. Brunori, E. Antonini, and J. Wyman, *J. Mol. Biol.*, 1971, **58**, 261.
242 M. Brunori, B. Talbot, A. Colosimo, E. Antonini, and J. Wyman, *J. Mol. Biol.*, 1972, **65**, 423.
243 J. S. Olson and Q. H. Gibson, *J. Biol. Chem.*, 1972, **247**, 1713.
244 E. Antonini and M. Brunori, *J. Biol. Chem.*, 1970, **245**, 5412.

and analysed the kinetics of binding of these isocyanides to the isolated α and β chains and to HbA. First of all, the role of functional differences in α *versus* β subunits in the analysis of the overall kinetics and in the approach to equilibrium has been carefully considered. The α subunits in the tetramer are less accessible to ligands, in agreement with the results obtained several years before on the isolated chains.[241] This shows how difficult it may be to predict dynamic behaviour from static structural data, since the crystallographic analysis has shown that access of the ligand to the β chains is 'occluded' by the side-chain of Val-(E11), while the α chains appear to be more accessible.[176, 177] Secondly, the α chains display a greater role in the expression of co-operativity with respect to the β chains, indicating that more 'strain' is induced upon binding of a ligand to the former subunits as compared to the latter. This conclusion is fully consistent with independent data arising (i) from the analysis of the spectral perturbation that is induced upon mixing liganded with unliganded chains and (ii) from ^{1}H n.m.r. data.

The important role attributed to the interactions between the haem and the globin at the periphery or at the fifth position is supported by kinetic results on model compounds (see p. 136).

Thus, although the difference in the free-energy of binding of a ligand to the two allosteric states has been attributed, for all ligands, to protein-dependent effects, the molecular mechanism by which co-operativity is achieved may be significantly different for different ligands. It is surprising and rewarding, however, that the bulk of the available kinetic data is quantitatively consistent with an allosteric model, modified (where necessary) to account for the kinetic heterogeneity between the chains (as in the case of very bulky isocyanides, or for some of the smaller ligands in the presence of organic phosphates).

The kinetics of the reactions of metHb have been investigated by temperature-jump[245, 246] and pulse-radiolysis[247, 248] methods.

The reactions of aquo-metHb, and its reaction with formate, yield information concerning the dynamics of spin equilibria in these haemoproteins. A clear effect of organic phosphates is observed also in the binding of formate, with a perturbation involving preferentially one type of subunit (presumably the β chains).[245]

Heterotropic Effects in Haemoglobins.—*The Bohr Effect.* The heterotropic linkage between molecular oxygen and hydrogen ions, known as the 'Bohr effect', finds its physical basis in the ligand-linked conformational transition which affects the ionization equilibria of amino-acid side-chains.[42, 185, 249] As a result, a substantial difference is observed in the number of protons that are bound to the protein in the ligand-free and ligand-bound forms. In view of the binding of other charged components (anions and cations) of the solution to the protein, the analysis of the Bohr effect is complicated by differential interaction of small and large ions with deoxy- and oxy-haemoglobins.

[245] K. O. Okonjo and G. Ilgenfritz, *Arch. Biochem. Biophys.*, 1978, **189**, 499.
[246] U. Dreyer and G. Ilgenfritz, *Biochem. Biophys. Res. Commun.*, 1979, **87**, 1011.
[247] Y. A. Ilan, A. Samuni, M. Chevion, and G. Czapski, *J. Biol. Chem.*, 1978, **253**, 82.
[248] M. Chevion, Y. A. Ilan, A. Samuni, T. Navok, and G. Czapski, *J. Biol. Chem.*, 1979, **254**, 6370.
[249] D. W. Allen, K. F. Guthe, and J. Wyman, *J. Polym. Sci.*, 1950, **7**, 499.

The identification of specific amino-acid residues of the α and β chains that are involved in the 'acid' and in the 'alkaline' Bohr effects has stimulated massive work, but some aspects of the phenomenon are still unclear. In what follows, we report a summary of the problem as it appears at present.

The 'acid' Bohr effect is the increase of the affinity of HbA for O_2 with pH below $pH \approx 6.$[42] In a recent paper,[250] the Bohr effects of haemoglobins from various species and of several abnormal haemoglobins have been compared, and the results have been interpreted in the light of the structures of oxy- (or met-) and deoxy-Hb.[176, 177] This comparative approach allowed the proposal that, under stripped conditions, about half of the acid Bohr effect is contributed by β143-(H21)His, which lies in between β82(EF6)Lys and β144(HC1)Lys and which has an abnormally low pK in oxyhaemoglobin, owing to interaction with the cationic groups of these two neighbouring lysine residues.

The 'alkaline' Bohr effect, which is the physiologically relevant phenomenon, has been assigned to contributions from several groups (see also ref. 251). The data that are available up to now are summarized in Table 4. Since the relative

Table 4 *Groups of the alkaline Bohr effect*

$HbA_0{}^a$	DPG	Chloride	HbI^b
α1(NA2)Val	no	yes	Ac-Ser
α122(H5)His	no	?	His
β1(NA1)Val	yes	±	Val
β2(NA2)His	yes	yes	Glu
β143(H21)His	yes	±	Ser
β146(HC3)His	no	no	Phe

(a) Conditions were 0.1M-NaCl + 2 mM-DPG; (b) Component I of trout Hb (sequence data from F. Bossa and collaborators, personal communication)

contributions of some of these groups depend on the presence of anions [such as 2,3-diphosphoglycerate (DPG) and/or Cl$^-$], these composite interactions have to be dissected out. In Table 4 we have therefore indicated if a particular ionizable group is anion-linked, and, if it is so linked, the specific anion that is involved (either DPG or Cl$^-$).

In the absence of phosphates and in the presence of 0.1M-Cl$^-$, α1Val and β82Lys both contribute to the alkaline Bohr effect because they bind more chloride in deoxyhaemoglobin than in oxyhaemoglobin. The α1Val residue makes its contribution to the alkaline Bohr effect by binding Cl$^-$, which forms a salt bridge to the guanidinium of α141Arg of the partner α chains; this salt bridge is present in deoxy-Hb and is broken in oxyhaemoglobin. Likewise, β82Lys is involved in the alkaline Bohr effect since its contribution is largely chloride-dependent, in line with several observations (see p. 159). The mechanism by which Cl$^-$ could affect the ionization properties of this residue is by forming a bridge between the two β82Lys that lie in the gap between the two β chains. In this case the protonated amino-groups of both lysine residues could be connected

250 M. F. Perutz, J. V. Kilmartin, K. Nishikura, J. M. Fogg, and P. J. G. Butler, *J. Mol. Biol.*, 1980, **138**, 649.
251 J. V. Kilmartin, *Trends Biochem. Sci.*, 1977, November, p. 247.

by a bridge that involves a hydrated chloride ion, *i.e.* $-\overset{+}{N}H_3-Cl^--H_2O-\overset{+}{N}H_3$ (see ref. 250).

In the presence of organic phosphates (2 mmol l^{-1}), a substantial part of the alkaline Bohr effect is due to the residues present in the dyad axis of the molecule, in the gap between the two β chains[252] (the DPG-binding site). Therefore the contributions of β1Val, β2His, β143His, and β82Lys to the Bohr effect are coupled to the preferential combination of deoxyhaemoglobin with organic phosphates (*e.g.* DPG).

Two more ligand-linked groups, namely β146His and α122His, contribute intrinsically to the Bohr effect and are not dependent on either DPG or Cl^-. In the case of α122His, its implication as one of the unknown alkaline Bohr groups (see ref. 176) was made on the basis of hydrogen–tritium exchange,[253] and supported by the finding that the alkaline Bohr effect in Hb$_{Tacoma}$[254, 255] and in Hb$_{Portland}$[256] is reduced. Finally, it should be reported that the contribution of β146His, which was considered to be one of the main residues involved,[176,177,251,257] has recently been questioned, on the basis of ^1H n.m.r. experiments.[258] The authors claim that, in 0.1M-bis-Tris in D_2O, with 5 to 60mM-Cl^-, and in the absence of phosphate, β146His does not contribute significantly to the alkaline Bohr effect, while it does so in the presence of phosphates.

Finally, one of the haemoglobin components of trout, *i.e.* trout HbI, which has no Bohr effect either in the presence or in the absence of anions,[194, 259] represents a useful control for some of the assignments described above. It was shown some years ago that β146His is substituted by Phe in trout HbI,[260] in complete agreement with the lack of Bohr effect contributed by this residue.[176, 257] In this particular haemoglobin component, the only residues that are not substituted are β1Val and α122His. The complete lack of Bohr effect in this Hb constitutes negative evidence for the role of α122His in the alkaline Bohr effect of human haemoglobin A, in agreement with recent results obtained by the deuterium-exchange method.[261]

The theoretical work carried out by Gurd and collaborators,[262] based on an electrostatic interaction model, is relevant to the problem of the Bohr effect. Employing the atomic co-ordinates of oxy- and deoxy-Hb[180] and potentiometric titration data, the authors conclude from their analysis that the Bohr effect is contributed by a large number of groups, *i.e.* ten per tetramer at high ionic strength and 28 per tetramer at low ionic strength. However, this conclusion

[252] A. Arnone, *Nature (London)*, 1972, **237**, 146.
[253] K. Nishikura, *Biochem. J.*, 1978, **173**, 651.
[254] L. I. Idelson, N. A. Didkowsky, R. Casey, P. A. Lorkin, and H. Lehmann, *Acta Haematol.*, 1974, **52**, 303.
[255] A. Hayashi, T. Suzuki, and G. Stamatoyannopoulos, *Biochim. Biophys. Acta*, 1974, **351**, 453.
[256] S. Tuchinda, K. Nagai, and H. Lehmann, *FEBS Lett.*, 1975, **49**, 390.
[257] J. V. Kilmartin and J. F. Wootton, *Nature (London)*, 1970, **228**, 766.
[258] I. M. Russu, N. T. Ho, and Chien Ho, *Biochemistry*, 1980, **19**, 1043.
[259] M. Brunori, G. Falcioni, G. Fortuna, and B. Giardina, *Arch. Biochem. Biophys.*, 1975, **168**, 512.
[260] D. Barra, F. Bossa, J. Bonaventura, and M. Brunori, *FEBS Lett.*, 1973, **35**, 151.
[261] M. Ohe and A. Kajita, *Biochemistry*, 1980, **19**, 4443.
[262] J. B. Matthew, G. I. H. Hanania, and F. R. N. Gurd, *Biochemistry*, 1979, **18**, 1919, 1928.

is not easily reconciled with crystallographic data and with comparative sequence analysis of haemoglobins from various species (F. Bossa *et al.*, personal communication).

The Interaction of Haemoglobins with Polyanions. The role of organic phosphates, notably 2,3-diphosphoglycerate (DPG), adenosine triphosphate (ATP), and inositol hexaphosphate (IHP), in the regulation of the functional properties of vertebrate haemoglobins has been intensively studied. This effect involves the preferential binding of polyanions to the deoxygenated form of the haemoglobin tetramer, which results in a progressive shift of the oxygen-binding curves towards lower affinities as the concentration of the organic phosphate is increased.[263] The structural work of Arnone,[252, 264] which showed that the binding of DPG occurs in between the two subunits, along the dyad axis of the haemoglobin molecule, has been amply substantiated since it was first proposed. Among the more recent work, it seems pertinent to quote the paper by Perutz and Imai,[265] who have been able to rationalize the different effects exerted by DPG on various mammalian haemoglobins that have been studied over the past few years.[266–270]

One of the questions that are still unsettled concerns the release of anions during the course of oxygenation, since DPG may be released either in a linear mode, in parallel with the fractional saturation with oxygen, or in all-or-none fashion at some point along the ligand-association isotherm ('switch-over'). A careful investigation of the oxygen-dissociation curves of human haemoglobin A, in the presence and absence of Cl^-, DPG, and IHP,[266] has led to the following conclusions: (i) K_4, *i.e.* the last association equilibrium constant of the Adair scheme, is almost insensitive to Cl^- and DPG; (ii) K_1, K_2, and K_3 are non-identically affected by Cl^- and DPG concentration, leading to a non-uniform release at the individual oxygenation steps; (iii) these conclusions are a consequence of the fact that the shape of the oxygen-dissociation curve changes with anion concentration (*i.e.* the binding isotherms seem to converge to a common asymptote at high saturation, but approach different asymptotes at low saturations) (see also ref. 173).

A second point, which is sometimes still questioned, regards the possible existence in oxyhaemoglobin of more than one binding site for organic phosphate per tetramer. It has recently been confirmed by yet another study (Bohr protons released in the presence of DPG)[271] that DPG binds to both oxy- and deoxy-haemoglobin in a 1:1 ratio. The association constants for the binding of DPG to oxy- and deoxy-haemoglobin have been obtained at different pH values; at low pH, the two derivatives show the same affinity for DPG, a fact which strongly supports the proposal that the binding site is the same. The pH dependence of the

263 R. E. Benesch, R. Benesch, R. Renthal, and W. B. Gratzer, *Nature (London), New Biol.*, 1971, **234**, 174.
264 A. Arnone and M. F. Perutz, *Nature (London)*, 1974, **249**, 34.
265 M. F. Perutz and K. Imai, *J. Mol. Biol.*, 1980, **136**, 183.
266 K. Imaizumi, K. Imai, and I. Tyuma, *J. Biochem. (Tokyo)*, 1979, **86**, 1829.
267 C. Bonaventura, B. Sullivan and J. Bonaventura, *J. Biol. Chem.*, 1974, **249**, 3768.
268 M. N. Hamilton and S. J. Edelstein, *J. Biol. Chem.*, 1974, **249**, 1323.
269 D. Petschow, J. Würdinger, R. Baumann, J. Duhm, G. Braunitzer, and C. Bauer, *J. Appl. Physiol.*, 1977, **42**, 139.
270 F. Taketa, *Ann. N.Y. Acad. Sci.*, 1974, **241**, 524.
271 G. G. M. Van Beek and S. H. De Bruin, *Eur. J. Biochem.*, 1979, **100**, 497.

binding of ATP and/or IHP has been investigated by a number of authors (*e.g.* refs. 272—275), employing haemoglobins from different species.

A similar approach was used[276] for the characterization of the ionizable groups which interact with benzenehexacarboxylate (BHC) in HbA and in two mutant forms, *i.e.* Hb$_{Deer\ Lodge}$ ($\alpha_2^A\beta_2^{2His\rightarrow Arg}$) and Hb$_{Providence\ Asp}$ ($\alpha_2^A\beta_2^{82Lys\rightarrow Asp}$), with mutations involving the polyphosphate-binding site. The results show that the liganded form of HbA differs from its deoxy-derivative in the interaction with BHC. In the ligand-bound form, the *N*-terminal amino-group of the β chains and β2His do not appear to experience pK changes upon binding of BHC, and therefore have similar proton-binding behaviour. These findings have been taken to indicate that the site of binding of BHC is not the same in liganded and unliganded haemoglobins. A possible rationalization has been offered.[276]

The thermodynamics of binding of organic phosphates to haemoglobin strongly depend on the quaternary state of the protein,[117] and therefore heats of binding of organic phosphates may be used as a quantitative probe of the ligation state of the protein. This possibility has been explored in a recent calorimetric study[277, 278] of the binding of IHP to deoxy- and carbonmonoxy-HbM$_{Iwate}$; since this abnormal Hb is believed to be frozen in the low-affinity conformation in both ligation states,[173] it is not surprising that the observed heats that are associated with IHP binding are the same for both derivatives. In the case of deoxy-HbA the binding of IHP is associated with a negative enthalpy change (of -7 to -11 kcal per tetramer),[278] which may be accounted for by the exothermic protonation of β143His and α2His, and/or the α-terminal amino-groups involved in the IHP-binding site.[264] Finally, it may be of interest, also from a pharmacological viewpoint, to comment on the elegant work of Beddell *et al.*,[279] who have measured the oxygen-linked effect of a series of compounds synthesized to fit the DPG-binding site (4,4'-diformylbibenzyl-2-oxyacetic acid and its bisulphite adduct). This line of research is aimed at the synthesis of compounds which mimic the physiological effector with possibly more powerful action.

The Influence of Small Ions on Haemoglobins. It is well known[42] that univalent anions affect the affinity of human haemoglobin for O_2, and that below concentrations of the anion of ~ 1mol l^{-1}, a decrease in oxygen affinity is observed upon increase of anion concentration. The direction of the shift implies that chloride ions bind preferentially to deoxyhaemoglobin, as proven directly by n.m.r. techniques.[280] The magnitude of the shift in affinity for O_2 is pH-depend-

[272] G. G. M. Van Beek, E. R. P. Zuiderweg, and S. H. De Bruin, *Eur. J. Biochem.*, 1978, **92**, 309.

[273] A. Hsing Chu and E. Bucci, *J. Biol. Chem.*, 1979, **254**, 371.

[274] H. S. Rollema and C. Bauer, *J. Biol. Chem.*, 1979, **254**, 12038.

[275] G. S. Greaney, M. K. Hobish, and D. A. Powers, *J. Biol. Chem.*, 1980, **255**, 445.

[276] E. Bucci, A. Salahuddin, J. Bonaventura, and C. Bonaventura, *J. Biol. Chem.*, 1978, **253**, 821.

[277] S. J. Gill, H. T. Gaud, and B. G. Barisas, *J. Biol. Chem.*, 1980, **255**, 7855.

[278] L. A. Noll, H. T. Gaud, S. J. Gill, K. Gersonde, and B. G. Barisas, *Biochem. Biophys. Res. Commun.*, 1979, **88**, 1288.

[279] C. R. Beddell, P. J. Goodford, D. K. Stammers, and R. Wootton, *Br. J. Pharmacol.*, 1979, **65**, 535.

[280] J. E. Norne, E. Chiancone, S. Forsen, E. Antonini, and J. Wyman, *FEBS Lett.*, 1978, **94**, 410.

ent. At neutral pH, in 0.1M-KCl, haemoglobin binds a small number of chloride ions, thus suggesting that the effect of chloride is a specific binding phenomenon, and not an aspecific ionic-strength effect.

The chloride-binding sites consist of positively charged residues, whose pK is affected by Cl$^-$, thus possibly contributing to the overall Bohr effect (see Table 4). The two principal groups that are supposed to contribute to the Bohr effect (*i.e.* α1Val and β146His) were thought[176, 177, 179, 251] to form salt bridges with negatively charged carboxyl groups; thus their contribution to the Bohr effect was considered to be independent of chloride concentration.

However, the problem proved to be more complex, and several experimental approaches have been used to identify the number, the location, and the linkage with O_2 and H$^+$ of the chloride-binding sites. The recent and pertinent information is derived from (i) crystallographic studies on normal and modified haemoglobins,[281, 282] (ii) the dependence of the O_2-binding properties and of the Bohr effect on Cl$^-$ concentration (*i.e.* O_2 equilibria and differential titrations),[283–287] and (iii) ^{35}Cl n.m.r.[280, 288]

Unequivocal evidence for the anion-binding sites in deoxy-Hb comes from X-ray-diffraction and solution studies of specifically carbamylated human HbA.[281] This work shows that there are two chloride-binding sites, the first involving α1Val and α141Arg of the opposite chain, and the second involving α1Val and α131Ser of the same chain. The agreement about the first of these sites being oxygen- and proton-linked is very satisfactory,[285, 287, 288] while the evidence is only marginal for the second one.

The presence of O_2-linked anion binding at the level of β82Lys (one of the residues involved in interaction with DPG[252]) is also well established, mainly on the basis of functional and spectroscopic data on abnormal or modified haemoglobins.[283, 287, 289–291] Particularly revealing results have been produced by studying the properties of Hb$_{Providence}$ ($\alpha_2^A\beta_2^{82Lys\rightarrow Asn}$ or $\alpha_2^A\beta_2^{82Lys\rightarrow Asp}$),[283, 289] Hb$_{Suresnes}$ ($\alpha_2^{141Arg\rightarrow His}\beta_2$),[282] and Hb that is either carbamylated at α1Val[292] or digested with carboxypeptidase at α141Arg.[293, 294] Thus, Nigen *et al.*[287] have

[281] S. O'Donnel, R. Mandaro, and T. M. Schuster, *J. Biol. Chem.*, 1979, **254**, 12204.
[282] C. Poyart, E. Bursaux, A. Arnone, J. Bonaventura, and C. Bonaventura, *J. Biol. Chem.*, 1980, **255**, 9465.
[283] J. Bonaventura, C. Bonaventura, B. Sullivan, G. Ferruzzi, P. R. McCurdy, J. Fox, and W. F. Moo-Penn, *J. Biol. Chem.*, 1976, **251**, 7563.
[284] W. F. Moo-Penn, K. C. Bechtel, R. M. Schmidt, M. M. Johnson, D. L. Jue, D. E. Schmidt, W. M. Dunlap, S. J. Opella, J. Bonaventura, and C. Bonaventura, *Biochemistry*, 1977, **16**, 4872.
[285] G. G. M. Van Beek, E. R. P. Zuiderweg, and S. H. De Bruin, *Eur. J. Biochem.*, 1979, **99**, 379.
[286] G. G. M. Van Beek and S. H. De Bruin, *Eur. J. Biochem.*, 1980, **105**, 353.
[287] A. M. Nigen, J. M. Manning, and J. O. Alben, *J. Biol. Chem.*, 1980, **255**, 5525.
[288] E. Chiancone, J. E. Norne, S. Forsén, J. Bonaventura, M. Brunori, E. Antonini, and J. Wyman, *Eur. J. Biochem.*, 1975, **55**, 385.
[289] A. M. Nigen and J. M. Manning, *J. Biol. Chem.*, 1975, **250**, 8248.
[290] A. G. Mauk, M. R. Mauk, and F. Taketa, *Fed. Proc.*, 1976, **15**, 1603.
[291] K. J. Wiechelman, J. Fox, P. R. McCurdy, and C. Ho, *Biochemistry*, 1978, **17**, 79.
[292] J. V. Kilmartin and L. Rossi Bernardi, *Biochem. J.*, 1971, **124**, 31.
[293] J. Bonaventura, C. Bonaventura, M. Brunori, B. Giardina, E. Antonini, F. Bossa, and J. Wyman, *J. Mol. Biol.*, 1974, **82**, 499.
[294] J. V. Kilmartin, J. A. Hewitt, and J. F. Wootton, *J. Mol. Biol.*, 1975, **93**, 203.

recently studied a hybrid haemoglobin, $\alpha_2{}^C\beta_2{}^{Prov}$, constituted of α chains that are specifically carbamylated at the level of $\alpha1$Val and β chains obtained from Hb$_{Providence\ Asp}$.

These data, besides confirming that both residues ($\alpha1$Val and $\beta82$Lys) have a major role in the binding of small inorganic anions, seem to indicate the presence of a third O_2-linked site, which the authors [287] have suggested to be located between α_1131Ser and α_2141Arg. In the case of Hb$_{Providence}$, both mutants (Asn and Asp) were shown to have normal homotropic interactions, while heterotropic effects (including binding of Cl^-) are greatly reduced. [283] The decrease in the effect of anions was obviously attributed to the absence of a positively charged residue at position $\beta82$.

The decreased Bohr effect and the lower affinity for O_2 that are displayed by both mutants have suggested that $\beta82$Lys is involved in the chloride-dependent part of the Bohr effect (see above and Table 4).

The differences in equilibrium and kinetics of binding of ligands as between HbA, Hb$_{Providence\ Asn}$ and Hb$_{Providence\ Asp}$ vanish in 1M-NaCl.

Haemoglobin Suresnes, in which histidine replaces arginine at the C-terminus of the α chains, displays [282] high affinity for O_2, reduced homotropic interactions, and greatly reduced sensitivity to pH and to chloride concentration. The functional properties of this mutant haemoglobin are similar in many respects to those of HbA that has been enzymatically digested with carboxypeptidase B, *i.e.* to haemoglobin in which the C-terminal arginine residue of the α chains has been removed.[42, 293, 294] The X-ray crystallographic analysis of deoxy-Hb$_{Suresnes}$[282] has revealed the loss of the normal inter-subunit salt bridge involving $\alpha127$Lys; thus $\alpha141$His cannot form salt bridges with 126Asp and 127Lys of the other α chain, causing the disruption of the anion-binding site on the α chain. Moreover, both the unliganded and the liganded forms show an increased dissociation into subunits, indicating that the substitution is not only involved in the binding of small anions but also in the stabilization of the tetramer, as revealed by the structural changes that are observed both at quaternary and tertiary levels.

These studies have led to the hypothesis (see ref. 295) that the degree of stabilization of the low-affinity state is sensitive to the density of positive charge in the region where DPG and IHP are bound, *i.e.* in the cluster of positive charges that forms the anion-binding site.

The binding sites in oxyhaemoglobin have not been identified unequivocally as yet, despite numerous studies. A study of the pK values of the α-amino-groups at different ionic concentrations and of the heats of deprotonation of the oxygen-linked groups [286] indicated that the binding sites for chloride ion in oxyhaemoglobin may involve histidine residues. On the basis of studies on abnormal and enzymatically modified haemoglobins, possible candidates for the chloride-binding sites in oxy-Hb are $\alpha122$His, $\alpha103$His, and $\beta97$His. Since these residues interact with non-polar groups through a water molecule,[39] anions would replace water, and this would explain why the binding of Cl^-, Br^-, and I^- ions to oxyhaemoglobin follows the lyotropic series.[286]

[295] C. Bonaventura and J. Bonaventura, in 'Biochemical and Clinical Aspects of Hemoglobin Abnormalities' ed W. S. Caughey, Academic Press, New York, 1977, p. 647.

Coding Sequences and the Function of Haemoglobins. An interesting paper on the evolution of the function of haemoglobins has been presented by Eaton.[296] The hypothesis starts with an analysis of the distribution of variants of the α and β chains, both normal (*i.e.* with no functional modifications) and abnormal. A map relating the coding sequences of the (mouse) α and β globin genes with the functionally significant residues of both chains is presented. According to the author, co-operative binding and heterotropic effects have evolved, starting from the basic property of the reversible binding of O_2 that is related to the central coding sequence (residues coding for the haem crevice are between residues 32 and 99 in the α chains and 31 and 104 in the β chains). The primitive carrier, capable only of reversible binding of O_2, was coded by a gene that contained separate coding sequences. The appearance of co-operative binding of ligands (related to the formation of a dimeric and, afterwards, a tetrameric protein) and of allosteric regulation (Bohr effect, and effects of CO_2 and of DPG) depended on the replacement of the third and the first coding sequences by mutation and recombination processes.

Protein–Protein Association and Phase Changes in Haemoglobins.—*Polysteric Linkage.* It has been shown that, in the case of haemoglobin A, the oxygenated form is largely dissociated into dimers under conditions in which the deoxy-form remains in the tetrameric state.[42] The $\alpha\beta$ dimers are non-co-operative, and have been taken to simulate the R state of tetrameric HbA. The relationships between the state of aggregation of the protein and its functional properties represented an area of active research several years ago; nowadays, most of the basic problems have been solved.

The relationship between dissociation of subunits and binding of O_2 has been considered within the framework of the linkage theory;[185] these effects have been termed polysteric,[297] to underline the similarities with and differences from allosteric phenomena.

A number of papers have dealt with the relationships between binding of ligands and aggregation of subunits, both from the theoretical and experimental viewpoints.[9, 10] An understanding of the regulatory significance of association–dissociation phenomena in haemoglobin demands a complete characterization of the assembly processes and of the functional properties of the subunits within the various and specified states of assembly. This approach should provide information on the role of inter-subunit contacts, namely $\alpha_1\beta_1$ and $\alpha_2\beta_2$ (see ref. 298).

In the case of human haemoglobin A, the relationship between the binding of O_2 and dimer–tetramer dissociation has been carefully explored in a series of papers reporting accurate oxygen-binding isotherms, analysed in conjunction with independently determined values of the dimer–tetramer equilibrium constants for the unliganded and fully oxygenated states.[197-199] A complete description of the fundamental relevant parameters and the thermodynamics of association was given in Volume 1 of this series.[1] Analysis has allowed the resolution of the free-energies for dimer–tetramer association into four definite states, corres-

[296] W. A. Eaton, *Nature (London)*, 1980, **284**, 183.
[297] A. Colosimo, M. Brunori, and J. Wyman, *J. Mol. Biol.*, 1976, **100**, 47.
[298] G. K. Ackers, *Biophys. J.*, 1980, **6**, 331.

ponding to deoxygenated and singly, triply, and fully liganded haemoglobins.[299] The functional significance of this analysis has been outlined above (the 'quaternary enhancement' effect).[199]

The free-energies of inter-subunit contacts and of binding of ligands for dimers and tetramers have been determined[300] for haemoglobin Kansas, whose substitution (*i.e.* $\beta 102\text{Asn}\rightarrow\text{Thr}$) is at the $\alpha_1\beta_2$ inter-subunit contact.[301] It should be recalled that the dimer–tetramer association constants of deoxyhaemoglobins A and Kansas are practically identical, while oxygenated haemoglobin Kansas was shown to be largely dissociated into dimers.[302, 303] From more recent data,[300] it appears that the free-energies of association of fully liganded HbA and Hb$_{\text{Kansas}}$ differ by about 3 kcal per mole of tetramer.

This is in complete agreement with crystallographic data which show major differences between Hb$_{\text{Kansas}}$ and HbA for the CO derivative but not for the deoxy-form.[304] Again, in agreement with the structural data, the 'quaternary' enhancement is not shown in haemoglobin Kansas. The origin of the low affinity of Hb$_{\text{Kansas}}$ for O_2 is not known, but it has been suggested[300] that the amino-acid substitution, being in contact with the edge of the haem, may hinder the ability of the haem to tilt, thereby affecting the affinity for oxygen.

The preparation and characterization of asymmetric haemoglobin hybrids[42] which are formed between two different haemoglobins have been achieved.[305] A sub-zero temperature method has recently been applied[305] to the isolation and characterization of hybrids between liganded haemoglobins A and C, to investigate aspects of allosteric and polysteric effects that are related to the dimer interfaces. It should be mentioned that the experimental approach that was used to isolate mixed HbA–HbC tetramers ($\alpha^A\beta^A\alpha^C\beta^C$) can also be applied to the preparation and isolation of intermediates in the reaction of haemoglobin A at least with ligands which are dramatically slowed down at low temperatures. Along the same lines, the functional characterization of the oxidation intermediates of trout HbI was achieved, thanks to the high stability of the tetramer.[306]

Phase Changes and the Binding of Ligands in Haemoglobins. Wyman introduced the theoretical foundations of the linkage between binding of ligands and changes of state of haemoglobin several years ago.[185] The well-known dependence of solubility of haemoglobin on the state of ligation (*i.e.* the difference in solubility between oxy- and deoxy-Hb[42]) implies that the ligand-binding curve of the protein in solution, when in equilibrium with the solid, will reflect the change of state which occurs as the ligand is removed from the system. These relationships have been called polyphasic linkage.[307] This approach provides a useful framework for the work on HbS (sickle-cell).

[299] F. C. Mills and G. K. Ackers, *J. Biol. Chem.*, 1979, **254**, 2881.
[300] D. H. Atha, M. L. Johnson, and A. F. Riggs, *J. Biol. Chem.*, 1979, **254**, 12390.
[301] J. Bonaventura and A. F. Riggs, *J. Biol. Chem.*, 1968, **243**, 980.
[302] D. H. Atha and A. Riggs, *J. Biol. Chem.*, 1976, **251**, 5537.
[303] Q. H. Gibson, A. Riggs, and T. Imamura, *J. Biol. Chem.*, 1973, **248**, 5976.
[304] L. Anderson, *J. Mol. Biol.*, 1975, **94**, 33.
[305] M. Perrella, M. Samaja, and L. Rossi Bernardi, *J. Biol. Chem.*, 1979, **254**. 8748.
[306] G. Falcioni, A. Colosimo, B. Giardina, and M. Brunori, Proceedings of the Italian Society for Molecular Biology, Como (Italy), 1980, p. 7.
[307] J. Wyman and S. J. Gill, *Proc. Natl. Acad. Sci. USA*, 1980, **77**, 5239.

The gelation of haemoglobin S. Sickle-cell anaemia arises from the substitution of valine for glutamic acid at the sixth position of the β chains of HbS. This replacement, which is on the surface of the molecule, has almost no effect on the binding of O_2 by haemoglobin in solution, but causes a large decrease in the solubility of the deoxy-form, which leads to the aggregation of tetramers into long fibres (*i.e.* gelation). Alignment of these fibres to form larger aggregates is responsible for the distortion of the membranes of the red blood cells, which is the underlying cause of the vasculo-occlusive nature of the disease.

Aggregation of HbS in deoxygenated solutions has been studied with a wide variety of techniques, and has been characterized in detail from both the thermodynamic and the kinetic points of view.[308]

Structural studies indicate the existence of at least two types of fibre, one being composed of six and the other of fourteen filaments of HbS molecules.[309-311] A recent study by electron microscopy and computer-based image reconstruction[312] has indicated that the fibres have an elliptical cross-section and are composed of fourteen packed filaments, with ten outer filaments surrounding four inner ones. Analysis of data at various stages of assembly of the fibres revealed stable intermediates containing ten and six filaments; these intermediates could be characterized in terms of the loss of specific pairs of filaments. From the structures of the six- and ten-filament species, pairs of filaments may be considered as the basic structural units in the stabilization of the complete fibres.[312] However, it should be mentioned that studies[313,314] on a new monoclinic crystalline form of HbS seem to corroborate the idea that the structural unit of the fibre consists of pairs of antiparallel double filaments.

An attractive feature of the pairing model lies in the identification of the stereochemistry of the intermolecular contacts existing within a single fibre. Thus, following this model, the $\beta6$ site would lie between filaments within the pairs, together with the $\beta73$ site also implicated in fibre formation;[315] the $\beta121$ and $\alpha23$ sites would be located between molecules of the same filament, while the residues on the α chain in the vicinity of $\alpha47$ would lie at the contacts within and between filament pairs.[316] Furthermore, the existence of fibres which are arranged in complex patterns, with a feather-like appearance, was recently noted in embedded cells.[317] This finding could be taken as new evidence for the helical nature of the fibres, and could represent an important factor in establishing the sickle-like appearance of the cell.

[308] Proceedings of the Symposium on Molecular and Cellular Aspects of Sickle Cell Disease, DMEW Publication No. 76-1007, NIH, Bethesda, Maryland, USA.

[309] J. T. Finch, M. F. Perutz, J. F. Bertles, and J. Dobler, *Proc. Natl. Acad. Sci. USA*, 1973, **70**, 718.

[310] R. M. Crepeau, G. Dykes, and S. J. Edelstein, *Biochem. Biophys. Res. Commun.*, 1977, **75**, 496.

[311] G. Dykes, R. M. Crepeau, and S. J. Edelstein, *Nature (London)*, 1978, **272**, 506.

[312] G. Dykes, R. M. Crepeau, and S. J. Edelstein, *J. Mol. Biol.*, 1979, **130**, 451.

[313] B. Magdoff-Fairchild and C. C. Chiu, *Proc. Natl. Acad. Sci. USA*, 1979, **76**, 223.

[314] C. C. Chiu and B. Magdoff-Fairchild, *J. Mol. Biol.*, 1980, **136**, 455.

[315] B. C. Wishner, J. C. Hanson, W. M. Ringle, and W. E. Love, in ref. 308, p. 1.

[316] R. E. Benesch, S. Yung, R. Benesch, J. Hack, and R. G. Schneider, *Nature (London)*, 1976, **260**, 219.

[317] S. J. Edelstein and R. M. Crepeau, *J. Mol. Biol.*, 1979, **134**, 851.

The crystallization of deoxyhaemoglobin S. Crystallization of deoxyhaemoglobin S is the result of a progressive alignment and fusion of the fibres. Hence the fibres, which are randomly oriented when initially formed, progressively align, giving rise to more tightly packed fascicles. After several hours, fascicles converge into microcrystals and propagate rapidly throughout the solution, suggesting an autocatalytic process.[318]

Kinetic experiments, using electron microscopy, on the fibre-to-crystal transition have demonstrated[319] that the alignment of fibres, which leads to the formation of nucleation sites for crystal growth, is the rate-limiting step in the crystallization process. It seems unlikely that this process involves a molecular rearrangement within the pairs of filaments; more probably, association of filament pairs should involve the relatively weakly interacting regions between the double strands. Thus the pH-invariant molecular packing within the fibre and crystal structures seems to involve hydrophobic interactions between molecules.[318]

The gelation of deoxyhaemoglobin A. The kinetics of gelation of deoxy-HbS are characterized by a delay time prior to aggregation.[320, 321] The mechanism and the physiopathological significance of the delay time have been amply illustrated in Volume 1.[1] It has recently been shown[322, 323] that, in concentrated phosphate buffer, dilute solutions of deoxy-HbA can also aggregate by a nucleation mechanism similar to that outlined above, and they form a gel after a clearly resolved delay time.[324] The physical nature of this gel, as well as the dependence of the delay time on concentrations of Hb and of phosphate and on temperature, seem to be essentially similar to those observed with deoxy-HbS.[325]

Binding of ligands and phase changes in haemoglobin S. Haemoglobin S provides an ideal system in which to study the regulatory effects arising from a ligand-linked phase transition (solution↔gel). This case of linkage, described as polyphasic,[307] has been investigated with concentrated solutions of HbS;[326, 327] these show a correlation between oxygen saturation and phase separation. In these studies, precise oxygen-binding isotherms, obtained by employing a thin-film apparatus,[328] were coupled with either light-scattering or birefringence measurements. Analysis of the results has allowed phase diagrams of the solution concentration in equilibrium with the polymer phase to be obtained, as a function of the partial pressure of oxygen. Furthermore, oxygen-binding curves for solutions of HbS below the minimum gelling concentration (m.g.c.) were used in conjunction with the phase diagrams to evaluate the oxygen-binding isotherm for the polymer phase. The main results may be summarized as follows: (i) high co-operativity in binding of O_2 is observed for the aggregating solutions, (ii) some

[318] T. E. Wellems and R. Josephs, *J. Mol. Biol.*, 1979, **135**, 651.
[319] S. M. Wilson and M. W. Makinen, *Proc. Natl. Acad. Sci. USA*, 1980, **77**, 944.
[320] J. Hofrichter, P. D. Ross, and W. A. Eaton, *Proc. Natl. Acad. Sci. USA*, 1974, **71**, 4864.
[321] J. Hofrichter, P. D. Ross, and W. A. Eaton, *Proc. Natl. Acad. Sci. USA*, 1976, **73**, 3035.
[322] K. Adachi and T. Asakura, *J. Biol. Chem.*, 1978, **253**, 6641.
[323] K. Adachi, T. Asakura, and M. L. McConnel, *Biochim. Biophys. Acta*, 1979, **580**, 405.
[324] K. Adachi and T. Asakura, *J. Biol. Chem.*, 1979, **254**, 12 273.
[325] K. Adachi and T. Asakura, *J. Biol. Chem.*, 1979, **254**, 7765.
[326] S. J. Gill, R. Skold, L. Fall, T. Shaeffer, R. Spokane, and J. Wyman, *Science*, 1978, **201**, 362.
[327] S. J. Gill, R. C. Benedict, L. Fall, R. Spokane, and J. Wyman, *J. Mol. Biol.*, 1979, **130**, 175.
[328] D. Dolman and S. J. Gill, *Anal. Biochem.*, 1978, **87**, 127.

(reversible) binding of oxygen by the gel is present; (iii) the difference in the over-all oxygen affinity between the solution and the gel that is obtained by this approach is similar to that directly measured by Hofrichter for the binding of carbon monoxide.[329]

It should be recalled that Hofrichter[329] made use of three different techniques to investigate the gelation of partially liganded HbS solutions, *i.e.* (*a*) sedimentation analysis for routine measurements of the solubility of deoxy-Hb in equilibrium with the polymer; (*b*) linear dichroism measurements by a microspectro-photometric method to detect directly the saturation of a single domain of aligned polymers of HbS; and (*c*) birefringence and turbidometric measurements for the kinetics of polymerization. The results obtained clearly showed that the total solubility increases monotonically with increase in the fractional saturation, that the polymer phase contains less than 5% of CO-liganded haems even at supernatant fractional saturations in excess of 70%, and that the polymerization reaction is nearly specific for deoxy-Hb (requiring the unliganded haemoglobin molecule as the main polymerizing component).

The close correspondence in the polymerization process between partially liganded and totally deoxygenated solutions strongly suggests that the conclusions obtained from investigation of deoxy-HbS can also be applied to describe the polymerization phenomenon in the presence of a ligand. Moreover, these results provide additional support for extrapolating the kinetic data '*in vitro*' to physiological conditions, as required by the hypothesis of kinetic control of the sickle-cell syndrome.[330]

In summary, the phase diagrams provide a useful point of reference for a quantitative description of the influence of antisickling agents upon the regions of stable and unstable states. In addition, the relationship between binding of ligands and phase equilibria and the ensuing regulatory effects bring out the underlying parallelism between allosteric, polysteric, and polyphasic control.

2 Haemocyanins

Haemocyanins are the copper-containing respiratory pigments that occur freely dissolved in the haemolymph of molluscs and arthropods.[331] Since the last Report,[1] significant progress has been made in elucidating the structural characteristics of haemocyanins of both arthropods and molluscs and, although not yet available, a three-dimensional structure of arthropod haemocyanin seems within reach. The ligand-binding properties of haemocyanins have been analysed within the framework of a two-state Monod–Wyman–Changeux model, with encouraging results, notwithstanding the complexity of these systems. Moreover, kinetic investigations suggest that the rate of transfer of information within these large systems may be correlated with the dimensions of the functional domains.

The results outlined here are relevant to the understanding of the molecular mechanism of binding of oxygen and the ensuing conformational changes, but provide little information on the oxygen-transport mechanism '*in vivo*'. Correla-

[329] J. Hofrichter, *J. Mol. Biol.*, 1979, **128**, 335.
[330] W. A. Eaton, J. Hofrichter, and P. D. Ross, *Blood*, 1976, **47**, 621.
[331] K. E. Van Holde and E. F. J. van Bruggen, in 'Biological Macromolecules', ed. S. N. Timasheff and G. D. Fasman, Marcel Dekker, New York, 1971, Vol. V, p. 1.

tion of data obtained '*in vitro*' with physiological conditions is at this stage very limited, but deserves full attention in future research.

Structural Information on Haemocyanins.—Since the overall architecture of molluscan and arthropodal haemocyanins is entirely different, they will be dealt with under separate headings. On the other hand, the chemical structure of the active site of the haemocyanins from the two phyla seems rather similar,[332, 333] suggesting that the sometimes large differences observed in the affinities for ligands of haemocyanins of the two phyla may arise from different 'environmental' conditions.

Molluscan Haemocyanins. These appear, by electron microscopy, to be hollow cylindrical molecules with a diameter of approx. 280 Å and a height of 360 Å.[331] The subunits ($M_w \simeq 450\,000$), built up of 8 or 9 covalently bound globular O_2-binding units of $M_w \simeq 50\,000$, are arranged in a helical pattern on the cylinder wall, while both ends of the cylinder are limited by the so-called 'collar'.[334, 335]

Electron microscopy[335, 336] and partial proteolysis and subsequent isolation of functional fragments ($M_w \simeq 50\,000$ or multiples thereof)[337] have definitely proven the domain structure of molluscan haemocyanins.[337–342] The isolated domains are heterogeneous in their spectral and functional properties;[339, 341–345] however, it is not clear whether the observed heterogeneity is a result of the fragmentation procedure or is an intrinsic property, possibly related to differences in the structure of the polypeptide chain. In the case of the α-haemocyanin of *Helix pomatia* and the haemocyanin of *Murex trunculus*, the existence of different polypeptides has been demonstrated.[341, 346] It is interesting that a domain structure has also been demonstrated in one of the large invertebrate haemoglobins,[347] indicating, once more, that similar structure–function relationships may exist between these two classes of oxygen-binding proteins.

[332] T. B. Freedman, J. S. Loehr, and T. M. Loehr, *J. Am. Chem. Soc.*, 1976, **98**, 2809.

[333] A. F. Hepp, R. S. Himmelwright, N. C. Eichman, and E. I. Solomon, *Biochem. Biophys. Res. Commun.*, 1979, **89**, 1050.

[334] J. E. Mellema and A. Klug, *Nature (London)*, 1972, **239**, 146.

[335] J. F. L. van Breemen, G. J. Schuurhuis, and E. F. J. van Bruggen, in 'Structure and Function of Haemocyanin', ed. J. V. Bannister, Springer-Verlag, Berlin, Heidelberg, and New York, 1977, p. 122.

[336] R. J. Siezen and E. F. J. van Bruggen, *J. Mol. Biol.*, 1974, **90**, 77.

[337] R. Lontie, M. De Ley, H. Robberecht, and R. Witters, *Nature (London)*, *New Biol.*, 1973, **242**, 180.

[338] J. V. Bannister, A. Galdes, and W. H. Bannister, *Comp. Biochem. Physiol.*, *B*, 1975, **51**, 1.

[339] M. Brouwer, M. Volters, and E. F. J. van Bruggen, *Biochemistry*, 1976, **15**, 2618.

[340] C. Gielens, G. Preaux, and R. Lontie, in 'Structure and Function of Haemocyanin', ed. J. V. Bannister, Springer-Verlag, Berlin, Heidelberg, and New York, 1977, p. 85.

[341] M. Brouwer, M. Wolters, and E. F. J. van Bruggen, *Arch. Biochem. Biophys.*, 1979, **193**, 487.

[342] C. Gielens, L. J. Verschueren, G. Preaux, and R. Lontie, *Eur. J. Biochem.*, 1980, **103**, 463.

[343] J. Bonaventura, C. Bonaventura, and B. Sullivan, in ref. 3, p. 206.

[344] J. M. van der Laan, R. Torensma, and E. F. J. van Bruggen, in 'Invertebrate Oxygen-binding Proteins', ed. J. Lamy and J. Lamy, Marcel Dekker, New York, 1981, p. 739.

[345] R. Torensma, J. M. van der Laan, E. F. J. van Bruggen, C. Gielens, L. van Paemel, L. J. Verschueren, and R. Lontie, *FEBS Lett.*, 1980, **115**, 213.

[346] M. Brouwer, M. Ryan, J. Bonaventura, and C. Bonaventura, *Biochemistry*, 1978, **17**, 2810.

[347] R. C. Terwilliger, N. B. Terwilliger, C. Bonaventura, and J. Bonaventura, *Biochim. Biophys. Acta*, 1977, **494**, 416.

The development of highly sensitive electrophoretic and immunological techniques, and their application to haemocyanins, has established antigenic relationships between haemocyanin components of the same or of different species. As an example, the haemolymph from *Helix pomatia* contains α-haemocyanin and two forms of β-haemocyanin, *i.e.* a crystallizable fraction (β_c) and a soluble fraction (β_s); β_s and α are immunologically identical but both different from β_c.[348-351] It is common that antigenic similarities are observed not only between components of a given species (α- and β-haemocyanins), but also, and sometimes even more clearly, between the same type of component from different species.

Mild trypsinolysis of gastropod haemocyanins results in the formation of 'tubular' structures, due to end-to-end polymerization of single molecules.[352,353] The collar domains, which may be removed by enzymatic treatment,[352] act as 'antisickling' fragments which may be of physiological significance for the animal. Three-dimensional image reconstruction of a tubular polymer of the β-haemocyanin of *Helix pomatia* clearly showed well-separated and shallow helical grooves between the polypeptide chains of the wall, and identified the domain structure of the subunits. Deoxygenation of this polymeric haemocyanin results in a shortening of the tubules and in an increase of diameter, which proves the O_2-linked conformational change in β-haemocyanin.[354] The system appears to be suitable to study in detail the structural propagation of ligand-linked conformational changes.

Spontaneous aggregation into tubular structures, mostly consisting of coupled half-molecules, has been reported for haemocyanin of *Aplysia limacina*.[355,356] This system may provide new information on the position and contacts of the collar domains within the cylindrical molecule.

Arthropod Haemocyanins. Arthropod haemocyanins are formed of multiples of 16S structures, with squared, rectangular, or hexagonal views in the electron microscope and dimensions of 100—125 Å;[331] each hexamer is built up of six polypeptide chains whose molecular weights are in the range 70–90 000.

Subunit diversity among arthropod haemocyanins has become a common observation since the initial work on the haemocyanin of *Limulus polyphemus*.[357] Improvement of high-resolution chromatographic and electrophoretic methods

[348] C. Gielens, G. Preaux, and R. Lontie *Arch. Int. Physiol. Biochim.*, 1973, **81**, 182.

[349] C. Gielens, G. Preaux, and R. Lontie, *Arch Int. Physiol. Biochim.*, 1979, **87**, 412.

[350] G. Preaux, C. Gielens, L. J. Verschueren, and R. Lontie, in ref. 344, p. 197.

[351] G. Preaux, C. Gielens, and R. Lontie, in 'Metalloproteins: Structure, Molecular Function and Clinical Aspects', ed. U. Weser, Georg Thieme Verlag, Stuttgart and New York, 1979, p. 73.

[352] J. F. L. van Breemen, T. Wichertjes, M. F. J. Muller, R. van Driel, and E. F. J. van Bruggen, *Eur. J. Biochem.*, 1975, **80**, 129.

[353] E. J. Wood, in ref. 335, p. 77.

[354] J. F. L. van Breemen, J. H. Ploegman, and E. F. J. van Bruggen, *Eur. J. Biochem.*, 1979, **100**, 61.

[355] A. Ghiretti-Magaldi, B. Salvato, L. Tallandini, and M. Beltramini, *Comp. Biochem. Physiol.*, A, 1979, **62**, 579.

[356] A. Ghiretti-Magaldi, B. Salvato, G. Tognon, M. Mammi, and G. Zanotti, in ref. 344, p. 393.

[357] B. Sullivan, J. Bonaventura, and C. Bonaventura, *Proc. Natl. Acad. Sci. USA*, 1974, **71**, 2558.

has increased the power of separation and reliability, leaving little doubt that the observed heterogeneity is not artefactual. The application of immunoelectrophoretic techniques (such as crossed immunoelectrophoresis, fused rocket immunoelectrophoresis, and crossed-line immunoelectrophoresis) has contributed in the detection and characterization of subunit diversity among arthropod haemocyanins:[358-364] 5–8 subunits for cheliceratan haemocyanins (Xiphosura, scorpions, and spiders); 2–7 subunits for crustacean haemocyanins; 8 subunits for the haemocyanins of *Androctonus australis* and *Limulus polyphemus*; and 7 for the haemocyanin of *Tachypleus tridentatus*.[362-367]

Some of the isolated fractions of chelicerate haemocyanins (horseshoe crabs, spiders, and scorpions) are made up of two different subunits which form a stable dimer; these have been observed, for example, in the haemocyanin of *Cherax destructor*[368,369] and that of *Jasus edwardsii*.[370] These dimers are essential for association of the subunits beyond hexamer (16S), and may therefore be identified as the bridging structures, seen by electron microscopy, in between the hexameric building blocks.[362,363] Unfortunately, no information is available on the role, if any, of these dimeric subunits in mediating co-operative interactions between the hexameric building blocks. This is interesting because the binding of oxygen by the haemocyanins of *Dugesiella californica* and *Cupiennius salei* is highly co-operative,[371,372] the maximum Hill coefficients of ~9 implying that allosteric interactions are not confined to the hexamer, and extend between hexamers.

Immunochemical techniques have been employed to determine the relative orientation of the various subunits. In the case of the haemocyanin of *Androctonus australis* it was demonstrated that subunits which are present in two copies per 34S native molecule are located internally, while others (present in four copies) occupy external positions.[373]

Antigenic relationships have been observed for subunits of arthropod haemocyanins from the same or from different species (see refs. 362 and 366). Thus close immunochemical relationships exist between the dimeric subunits of the arachnid haemocyanins of *Androctonus australis* and of *Eurypelma californicum* and of the merostomate haemocyanins of *Limulus polyphemus* and of *Tachypleus tridentatus*,

[358] 'Structure and Function of Hemocyanin', ed. J. V. Bannister, Springer-Verlag, Berlin, Heidelberg, and New York, 1977.
[359] H.-J. Schneider, J. Markl, W. Schartau, and B. Linzen, *Hoppe-Seyler's Z. Physiol. Chem.*, 1977, **358**, 1133.
[360] J. Markl, A. Markl, W. Schartau, and B. Linze, *J. Comp. Physiol.*, 1979, **130**, 283.
[361] J. Markl, A. Hofer, G. Bauer, A. Markl, B. Kempter, M. Brenzinger, and B. Linzen, *J. Comp. Physiol.*, 1979, **133**, 167.
[362] J. Lamy, J. Lamy, and J. Weill, *Arch. Biochem. Biophys.*, 1979, **196**, 324.
[363] J. Lamy, J. Lamy, and J. Weill, *Arch. Biochem. Biophys.*, 1979, **193**, 140.
[364] J. Lamy, J. Lamy, M. Leclerc, S Compin, P.-Y. Sizaret, and J. Weill, in ref. 344, p. 181.
[365] M. Hoylaerts, G. Preaux, R. Witters, and R. Lontie *Arch. Int. Physiol. Biochim.*, 1979, **87**, 417.
[366] G. Preaux, M. Hoylaerts, R. Witters, and R. Lontie, in ref. 344, p. 239.
[367] J. Jollès, P. Jollès, J. Lamy, and J. Lamy, in ref. 344, p. 305.
[368] P. D. Jeffrey, D. C. Shaw, and G. B. Treacy, *Biochemistry*, 1976, **15**, 5527.
[369] A. C. Murray and P. D. Jeffrey, *Biochemistry*, 1974, **13**, 3667.
[370] H. A. Robinson and H. D. Ellerton, in ref. 335, p. 55.
[371] B. Linzen, D. Angersbach, R. Loewe, J. Markl, and R. Schmid, in ref. 335, p. 31.
[372] R. Loewe and B. Linzen, *J. Comp. Physiol.*, 1975, **98**, 147.
[373] J. Lamy, J. Lamy, M. Leclerc, P.-Y. Sizaret, and J. Weill, *FEBS Lett.*, 1980, **112**, 45.

suggesting thus that the dimeric subunits may have evolved from a common ancestor.[364] Also in the class of crustacean haemocyanins, antigenic relationships between the subunits have been reported,[374,375] while no correspondence seems to exist between cheliceratan and crustacean species.[364] The observed molecular homologies between the various haemocyanins, as determined from immuno-electrophoretic techniques, are of great value in establishing phylogenetic relationships in arthropod haemocyanins.

No proteolytic fragmentation into active components has been observed for the haemocyanin of the arthropod *Limulus polyphemus*.[343]

Amino-Acid Sequences and X-*Ray Crystallography of Haemocyanins.* The large molecular weight of the polypeptide chains and the chemical heterogeneity are the main barriers to sequence studies. Fragmentation of the chain with proteolytic enzymes (either specific or aspecific limited proteolysis) is a possible way of tackling the problem. For example, treatment of Fraction II of the haemocyanin of *Limulus polyphemus* with V-8 protease from *Staphylococcus aureus* and subsequent cleavage with CNBr yields seven fragments with molecular weights between 7000 and 23 000;[376] limited trypsinolysis of native chains of the haemocyanin of *Eurypelma californicum* yields distinct cleavage patterns for each chain with molecular weights between 10 000 and 45 000.[377]

Sequence homology is observed between certain subunits of haemocyanin from the same species and also between similar subunits from different species. Examples are the four subunits of the haemocyanin of *Androctonus australis*, where Thr is the *N*-terminal residue;[378] subunit IV from the haemocyanin of *Limulus polyphemus* and subunits 2 and 4 from that of *Androctonus australis*;[378] one of the chains of the haemocyanin of *Eurypelma californicum* and the subunits of that of *Androctonus australis*;[377] and two subunits of the haemocyanin of *Panulirus interruptus* (*i.e.* 94K and 90$^\text{I}$K).[379]

The similarities in the *N*-terminal regions of haemocyanin subunits reveal phylogenetic relationships, while homology between subunits from different species may also point to a specific structural role of the various subunits, this having led to the preservation of some sequences during evolution.[378]

The determination of the three-dimensional structure of haemocyanins has for a long time been hampered by the lack of good quality crystals. In case of the haemocyanin of *Panulirus interruptus*, suitable crystals have been obtained and X-ray diffraction patterns up to a resolution of 3.3 Å indicate a point-group symmetry of 32, with the unit cell containing one hexameric molecule.[380] The electron-density maps show a hexagonal or a rectangular shape, depending upon the projection, while the shape of the subunits is kidney-like, suggesting that the polypeptide chains may be folded in a two-domain structure.[380] In the case of the

[374] D. Rochu and J. M. Fine, *Comp. Biochem. Physiol.*, *A*, 1978, **59**, 145.
[375] C. Sevilla and J.-G. Lagarrigue, *Comp. Biochem. Physiol.*, *A*, 1979, **62**, 539.
[376] M. Moore and A. Riggs, in ref. 344, p. 315.
[377] W. Schartau, H.-J. Schneider, W. Strych, F. Eyerle, and B. Linzen, in ref. 344, p. 327.
[378] J. Jollès, P. Jollès, J. Lamy, and J. Lamy, *FEBS Lett.*, 1979, **106**, 289.
[379] J. P. van Eerd and A. Folkerts, in ref. 344, p. 139.
[380] E. J. M. van Schaick, W. G. Schutter, W. P. J. Gaykema, E. F. J. van Bruggen, and W. G. J. Hol, in ref. 344, p. 353.

haemocyanin of *Limulus polyphemus*, a three-dimensional electron-density map of subunit II at 5.5 Å resolution has been obtained.[381] In this case the crystals have a space group *R*32, with only one subunit per asymmetric unit cell.

Spectroscopic Studies of Haemocyanins.—*A Model for the Ligand-binding Site.* The nature of binding of oxygen at the active site of haemocyanins has been the subject of very active research. The use of new, independent, and complementary spectroscopic and biochemical techniques has increased the knowledge about the structure of the active site. In view of the spectral similarities between haemocyanins and tyrosinases,[382, 383] results on haemocyanins may be of general interest in order to elucidate the mechanism of binding and activation of oxygen by binuclear metalloproteins.

Resonance Raman spectroscopy of oxy-haemocyanins (by excitation in the visible region) indicated that dioxygen is bound as a peroxide. Therefore, in agreement with other evidence previously reported,[384] the formal oxidation state of copper in oxyhaemocyanin is II, while that in the deoxy-form has been established as I.[332, 385] The use of labelled oxygen has indicated that, contrary to the situation reported for haemerythrin,[386] the bound oxygen atoms in haemocyanin are equivalent,[387] although asymmetrical binding cannot be completely excluded.[388] Excitation in the 340 nm band gives rise to a variety of resonance Raman bands between 100 and 350 cm^{-1}, some of which result from copper-imidazole stretching vibrations.[332, 389]

The assignment of the 340 nm absorption band in oxy-haemocyanins has been the subject of debate. Thus (i) transfer of charge between imidazole nitrogen(s) and copper(II),[332] (ii) simultaneous pair excitation (SPE),[390] and, more recently, (iii) Cu$\rightarrow$O$_2$ charge transfer for the 340 and 570 nm band, originating from two different π^*-orbitals of O$_2^{2-}$,[389, 391] have been proposed.

The absence of a detectable Cu–O$_2$ stretching mode still remains a problem. The available data suggest an asymmetric arrangement of the imidazole ligands around the copper ions.[389] Co-ordination of the copper atoms by imidazole is fully consistent with a number of previously reported biochemical and photo-

[381] K. A. Magnus and W. E. Love, in ref, 344, p. 363.
[382] N. Eickman, J. A. Larrabee, E. I. Solomon, K. Lerch, and T. G. Spiro, *J. Am. Chem. Soc.*, 1978, **100**, 6529.
[383] H. A. Kuiper, K. Lerch, M. Brunori, and A. Finazzi-Agrò, *FEBS Lett.*, 1980, **111**, 232.
[384] G. Morpurgo and R. J. P. Williams, in 'Physiology and Biochemistry of Haemocyanins', ed. F. Ghiretti, Academic Press, New York, 1968, p. 113.
[385] J. S. Loehr, T. B. Freedman, and T. M. Loehr, *Biochem. Biophys. Res. Commun.*, 1974, **56**, 510.
[386] D. M. Kurtz, Jr., D. E. Shriver, and I. M. Klotz, *J. Am. Chem. Soc.*, 1976, **98**, 5033.
[387] T. J. Thamman, J. S. Loehr, and T. M. Loehr, *J. Am. Chem. Soc.*, 1977, **99**, 4187.
[388] I. M. Klotz, L. L. Duff, D. M. Kurtz, Jr., and D. F. Shriver, in ref. 344, p. 469.
[389] J. A. Larrabee and T. G. Spiro, *J. Am. Chem. Soc.*, 1980, **102**, 4217.
[390] J. A. Larrabee, T. G. Spiro, N. S. Ferris, W. H. Woodruff, W. A. Maltese, and M. S. Kerr, *J. Am. Chem. Soc.*, 1977, **99**, 1979.
[391] N. C. Eickman, R. S. Himmelwright, and E. I. Solomon, *Proc. Natl. Acad. Sci. USA*, 1979, **76**, 2094.

chemical studies,[392-395] as well as e.p.r. studies of NO- or nitrite-treated haemocyanins.[396,397]

Recently, EXAFS studies of the haemocyanin of *Buccinum canaliculatum*[398] have indicated that: (i) the edge structures are consistent with Cu^{II} and Cu^{I} being bound to imidazole moieties in, respectively, oxy- and deoxy-haemocyanin; (ii) the Cu–Cu distance in oxyhaemocyanin is different from that observed for deoxy-haemocyanin (3.67 *versus* 3.39 Å); and (iii) the average distance between the copper atoms and the immediate neighbour is the same in oxy- and deoxy-haemocyanins (~1.95 Å), but the co-ordination number of copper is 4 or 5 for oxy- and 2 or 3 for deoxy-haemocyanin. This very interesting result suggests that two ligands of the copper are lost on deoxygenation, namely the dioxygen molecule and the endogenous protein ligand which bridges the two copper atoms in a site (see Figure 5). EXAFS measurements have also been reported for haemocyanin from the keyhole limpet *Megathura crenulata*.[399] The change in the co-ordination

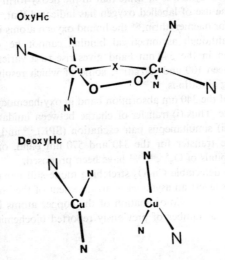

Figure 5 *A model for the binding site of haemocyanin. In oxyhaemocyanin, both copper ions (Cu^{II}) have a square-pyramidal co-ordination group whose basal plane is defined by the bridging dioxygen molecule, the internal oxygen atom (X), and two of the nitrogen atoms of imidazole. In the deoxy-state, both Cu^{I} have a trigonal-planar arrangement, being bound only to the three imidazole moieties.*

Reproduced by permission from *J. Am. Chem. Soc.*, 1980, **102**, 4217)

392 B. Salvato, A. Ghiretti-Magaldi, and F. Ghiretti, *Biochemistry*, 1974, **13**, 4778.
393 Y. Engelborghs and R. Lontie, *Eur. J. Biochem.*, 1973, **39**, 335.
394 E. J. Wood, and W. H. Bannister, *Biochim. Biophys. Acta*, 1968, **154**, 10.
395 G. Jori, S. Cannistraro, F. Ricchelli, B. Salvato, and L. Tallandini, in ref. 344, p. 621.
396 A. J. M. Schoot Uiterkamp, H. van der Deen, H. J. C. Berendsen, and J. E. Boas, *Biochim. Biophys. Acta*, 1974, **372**, 407.
397 H. van der Deen and H. Hoving, *Biochemistry*, 1977, **16**, 3519.
398 J. M. Brown, L. Powers, B. Kincaid, J. A. Larrabee, and T. G. Spiro, *J. Am. Chem. Soc.*, 1980, **102**, 4210.
399 M. S. Co, K. O. Hodgson, T. K. Eccles, and R. Lontie, *J. Am. Chem. Soc.*, 1981, **103**, 984.

number upon deoxygenation is confirmed, while no copper–copper or copper–sulphur distance of less than 2.8 Å has been found.

Results obtained through a combination of low- and room-temperature spectroscopies (e.p.r., i.r., and absorbance) have also led Solomon and co-workers to propose that exogenous (*i.e.* O_2) and endogenous (*i.e.* protein) ligands are involved in bridging the copper atoms in oxyhaemocyanin.[400-404] Comparison of the ligand-binding properties of the half-met-derivative (in which one of the copper atoms was oxidized) and the met-apo-form (one of the copper atoms had been removed) of the haemocyanin of *Busycon canaliculatum* indicated that, like oxygen, other exogenous ligands (*e.g.* CN^-, NO_2^-, N_3^-, SCN^-, F^-, and $CH_3CO_2^-$) may bridge the copper ions at the active site. Furthermore, in cases where exogenous ligands (*e.g.* CN^-, N_3^-, and SCN^-) keep the copper atoms sufficiently far apart, *i.e.* > 5 Å, a second co-ordination position becomes available, presumably through the breakage of the endogenous protein bridge.

A potential candidate for the endogenous bridging ligand (X) would be a phenolate ion, since the absorption spectra of oxy- and met-haemocyanins at low temperatures both display a shoulder around 400 nm, which may reasonably be attributed to a copper–tyrosine energy-transfer transition.[391] The proposed structural model for the active site of haemocyanin is shown in Figure 5. The Cu^{II} ions in oxyhaemocyanin have a square-pyramidal co-ordination group, with a plane through the bridging dioxygen, the internal protein ligand (X), and two of the imidazole rings, the third of these providing the axial ligand. Upon deoxygenation, the endogenous bridging ligand is broken, leaving the two Cu^I ions bound to three imidazoles, presumably in a trigonal-planar arrangement.

The recently reported emission properties of the complexes of haemocyanins with carbon monoxide may represent a useful tool to probe the structure of the active site in haemocyanins.[405] The visible emission of the CO complex of haemocyanins, which has been observed for all molluscan and arthropodal proteins so far examined (about ten species), is associated with a quenching of the intrinsic aromatic fluorescence. Excitation at ~310 nm (where this derivative displays a specific absorption band[406] and at 280 nm (in the aromatic absorption region) may be used to excite the luminescence of carbonmonoxy-haemocyanins.[405]

Although a precise assignment of the transition involved in the luminescence of carbonmonoxyhaemocyanin has not been made, the possibility of a $Cu \rightarrow CO$ charge transfer is likely.[405] Moreover, the very long lifetime for the excited state ($\tau \simeq 100 \ \mu s$) (A. Finazzi-Agrò, L. Zolla, H. K. Kuiper, and M. Brunori, in

[400] R. S. Himmelwright, N. C. Eickman, and E. I. Solomon, *Biochem. Biophys. Res. Commun.*, 1978, **81**, 237.

[401] R. S. Himmelwright, N. C. Eickman, and E. I. Solomon, *Biochem. Biophys. Res. Commun.*, 1978, **81**, 243.

[402] R. S. Himmelwright, N. C. Eickman, and E. I. Solomon, *Biochem. Biophys. Res. Commun.*, 1978, **84**, 300.

[403] R. S. Himmelwright, N. C. Eickman, and E. I. Solomon, *Biochem. Biophys. Res. Commun.*, 1979, **86**, 628.

[404] R. S. Himmelwright, N. C. Eickman, and E. I. Solomon, *J. Am. Chem. Soc.*, 1979, **101**, 1576.

[405] H. A. Kuiper, A. Finazzi-Agrò, E. Antonini, and M. Brunori, *Proc. Natl. Acad. Sci. USA*, 1980, **77**, 2387.

[406] C. Bonaventura, B. Sullivan, J. Bonaventura, and S. Bourne, *Biochemistry*, 1974, **13**, 4784.

the press) suggests that emission may occur from a triplet state, and may not be inconsistent with a charge transfer from the Cu to the bound CO. Selective oxidation or removal of the CO-binding copper atom and its involvement in the observed emission properties are the subjects of active investigation.

X-Ray spectra that have been obtained in the presence of CO indicate that there is a mixture of Cu^I and Cu^{II}; this has been interpreted as being due to oxidation of the copper atom which binds CO.[399] Although residual oxygen, whose affinity is often higher than that of CO,[406] may complicate the spectrum, this interpretation is interesting in view of what is reported above.

It has recently been observed[407] that treating the haemocyanin of *Panulirus interruptus* with mercuric chloride (molar ratio $Hg/Cu = 20$) yields a species which has about one half of the total copper still bound, but which has less than 10% of the residual oxygen-combining capacity (which, of course, is a measure of the number of coupled binuclear metal centres that are still capable of binding O_2). This result, though preliminary, suggests that it may be possible to obtain a species which contains one copper and one mercury atom per site, and which is obviously incapable of binding O_2

A Comparison of the Haemocyanins of Arthropods and of Molluscs. Although the overall architectures of the haemocyanins of arthropods and of molluscs are markedly different, spectroscopic data indicate that the active sites are quite similar. The mode of binding of dioxygen, involving a bridging peroxide, seems to be the same for haemocyanins from members of both phyla, and small differences in (for example) resonance Raman spectra of bound oxygen may simply reflect small perturbations of the local environment.[332] Some spectroscopic differences which have been well known for several years, such as the maximum of absorption of the copper–oxygen band (which is at 346 nm in molluscan and at 337 nm in arthropodal haemocyanins) or some c.d. differences, should, however, not be forgotten, and they have not yet been interpreted in structural terms. The difference in absorption maximum of the near-u.v. band has been assigned to shifts in the d–d transition energies.[333]

Differences in reactivity of the haemocyanins of arthropods and of molluscs towards various ligands are well documented. For example, addition of peroxide to the deoxyhaemocyanin of *Cancer magister* results in the formation of met-haemocyanin,[408, 409] while 'aged' molluscan haemocyanins are regenerated by the action of peroxide.[410] The formation of uncoupled ('sprung') methaemocyanin when N_3^- is added proceeds with different rates for the haemocyanins of the two phyla.

Model Compounds for Haemocyanins. The synthesis of binuclear copper compounds which would mimic the function of haemocyanin is a promising line of research in view of the significant results obtained in the past few years on model

407 H. A. Kuiper, L. Zolla, P. Vecchini, and M. Brunori, *Comp. Biochem. Physiol.*, 1981, 69, 253.

408 G. Felsenfeld and M. P. Printz, *J. Am. Chem. Soc.*, 1959, 81, 6259.

409 N. Makino, H. van der Deen, P. McMahill, D. C. Gould, T. H. Moss, C. Simo, and H. S. Mason, *Biochim. Biophys. Acta*, 1978, 532, 315.

410 R. Lontie and R. Witters, in 'The Biochemistry of Copper', ed. J. Peisach, P. Aisen, and W. E. Blumberg, Academic Press, New York, 1966, p. 455.

compounds for haemoglobin (see above). Data on Cu^I model compounds that have well-defined and reversible binding properties towards O_2 are, however, scarcely available. Wilson and co-workers[411,412] have been successful in preparing a red Cu^I compound that is derived from 2,6-diacetylpyridine and histamine; it binds oxygen in solution with a stoicheiometry of one O_2 molecule per two Cu atoms and has a reversibility of 80%. The green oxygenated form is probably a dioxygen adduct, which is e.p.r.-silent at 100 K. However, the compound does not react with CO and is readily decomposed in the presence of NO. Interestingly, when the imidazole rings are replaced by pyridine groups, the capability to bind O_2 is lost.

Functional Properties of Haemocyanins.—*Equilibria with Dioxygen and Carbon Monoxide*. Haemocyanins may bind oxygen co-operatively, depending on the presence and/or the concentration of various anions and cations (*e.g.* H^+, Ca^{2+}, Na^+, and Cl^-). Hill plots of the O_2-binding curves are characterized by a steep transition from a low to a high oxygen-affinity state of the protein, with maximum Hill coefficients of 7—9 (*e.g.* the β-haemocyanin of *Helix pomatia* and the haemocyanins of *Cupiennius salei* and *Androctonus australis*).[372,413,414] This implies that the ligand-linked transition occurs over a narrow range of ligand concentration, indicating the 'explosive' character of the change in conformation. This phenomenon is very characteristic of allosteric systems with a large number of binding sites, as indicated by theoretical studies.[415—417] Furthermore, the binding curves are characterized by relatively small values for the overall free-energy of interaction per site (ΔF_i), as indicated by the spacing between the two asymptotes of a Hill plot.[185] Values of ΔF_i of the order of 0.9—2.5 kcal mol^{-1} have been reported.[413,418,419] The shape and position of the O_2-binding curves may be affected by the presence of various ionic components, including protons (positive and negative Bohr effects have been reported).[331,413,420,421]

The binding of CO to haemocyanins, studied by direct absorbance or by luminescence, is always non-co-operative ($n \leqslant 1$), while heterotropic interactions are strongly reduced or totally absent.[406,422,423] The profound differences in O_2- and CO-binding behaviour may find a structural basis in the different modes of binding of these ligands at the active site (see below).

[411] M. G. Simmons, and L. J. Wilson, *J. Chem. Soc., Chem. Commun.*, 1978, 634.
[412] L. J. Wilson, C. L. Merrill, M. G. Simmons, and J. M. Trantham, in ref. 344, p. 571.
[413] L. Zolla, H. A. Kuiper, P. Vecchini, E. Antonini, and M. Brunori, *Eur. J. Biochem.*, 1978, **87**, 467.
[414] J. Lamy, J. Lamy, J. Bonaventura, and C. Bonaventura, in ref. 344, p. 785.
[415] J. Wyman, *J. Mol. Biol.*, 1969, **39**, 523.
[416] A. Colosimo, M. Brunori, and J. Wyman, *Biophys. Chem.*, 1974, **2**, 338.
[417] J. P. Changeux, J. Thiéry, Y. Tung, and C. Kittel, *Proc. Natl. Acad. Sci. USA*, 1967, **57**, 335.
[418] M. Brouwer, C. Bonaventura, and J. Bonaventura, *Biochemistry*, 1977, **16**, 3897.
[419] Z. Er-El, N. Shaklai, and E. Daniel, *J. Mol. Biol.*, 1972, **64**, 341.
[420] M. Brouwer, C. Bonaventura, and J. Bonaventura, *Biochemistry*, 1978, **11**, 2148.
[421] H. A. Kuiper, L. Forlani, E. Chiancone, E. Antonini, M. Brunori, and J. Wyman, *Biochemistry*, 1979, **18**, 5849.
[422] H. A. Kuiper, R. Torensma, and E. F. J. van Bruggen, *Eur. J. Biochem.*, 1976, **68**, 425.
[423] M. Brunori, L. Zolla, H. A. Kuiper, and A. Finazzi-Agrò, *J. Mol. Biol.*, 1981, **153**, 1111.

The co-operative oxygen-binding behaviour of haemocyanins has been success-fully analysed in a number of cases within the framework of the Monod–Wyman–Changeux allosteric model.[117,172] Thus a simple formulation of the two-state M–W–C model is sufficient to describe the O_2-binding properties of the α- and β-haemocyanins of *Helix pomatia* and of the haemocyanins of *Limulus polyphemus* and *Murex trunculus*, under specified conditions.[413,418,424-426] Furthermore, this simple analysis indicates that co-operativity in these multi-subunit systems is confined to subgroups of interacting binding sites called 'functional constella-tions'.[413,418,425] Therefore, without more complex assumptions, the allosteric interactions involve a limited number of sites and do not extend significantly over the boundaries of a functional constellation. Allosteric effectors may change the dimensions of a functional constellation, as indicated by the cases of β-haemocyanin of *Helix pomatia* and haemocyanin of *Limulus polyphemus*.[413,418,427] The relationships between the functional constellation and the well-defined structural units of the molecule still demand further elucidation. In the case of the β-haemocyanin of *Helix pomatia*, the size of the allosteric unit has been correlated with the one-tenth species that is obtained by dissociation.[413]

Analysis of the O_2-binding equilibria of the haemocyanin of *Eurypelma californicum* indicates[428] that the functional constellation contains more than nine sites, and thus co-operative interactions are not confined to the $16S$ building blocks of the molecule (L. Zolla, unpublished work).

The action of allosteric effectors on the oxygen-binding equilibria of haemo-cyanins show that these ions, besides shifting the allosteric equilibrium constant L_0, affect specifically the properties of either the T or the R state of the protein. For example, chloride ions change the shape and position of the O_2-binding curve of the haemocyanin of *Limulus polyphemus*, specifically affecting the position of the lower (T-state) asymptote.[418] A qualitatively similar behaviour was observed for the pH dependence of the oxygen-binding equilibrium of the β-haemocyanin of *Helix pomatia*:[413] high pH stabilizes the low-oxygen-affinity (T) state of the protein, but in addition shifts the position of the T-state asymptote itself. In the case of the haemocyanin of *Panulirus interruptus*, calcium ions, besides shifting the allosteric equilibrium, affect both asymptotic values.[421]

Since the concerted M–W–C model[172] does not assume a variation of the binding properties of the extreme states (T and R), a modified allosteric model has been invoked to fit the binding data, in so far that multiple T or R states are allowed.[413,418] The O_2-binding curves of the haemocyanins of *Callianassa californiensis*[429,430] and *Penaeus setiferus*[420] were fitted with the assumption of one R state, one T state, and a symmetrical hybrid state (R_3T_3), the allosteric unit containing six sites. A general treatment which applies specifically to multi-subunit proteins has recently been described by Whitehead.[431]

[424] R. van Driel, *Biochemistry*, 1973, **12**, 2696.
[425] A. Colosimo, M. Brunori, and J. Wyman, in ref. 335, p. 189.
[426] J. V. Bannister, A. Galdes, and W. H. Bannister, in ref. 335, p. 193.
[427] M. Brouwer, C. Bonaventura, and J. Bonaventura, in ref. 344, p. 761.
[428] R. Loewe, *J. Comp. Physiol.*, 1978, **128**, 161.
[429] K. Miller and K. E. Van Holde, *Biochemistry*, 1974, **13**, 1668.
[430] F. Arisaka and K. E. Van Holde, *J. Mol. Biol.*, 1979, **134**, 41.
[431] E. P. Whitehead, *J. Theor. Biol.*, 1980, **87**, 153.

Information about the thermodynamic parameters of binding of oxygen to haemocyanins is still scarcely available. Analysis of the temperature dependence of binding of oxygen to the β-haemocyanin of *Helix pomatia* under conditions of co-operative ligand binding has shown[432] that the heats of oxygenation of the T and R states are similar, indicating that the allosteric transition is mainly entropic in nature. On the other hand, the haemocyanins of *Levantina hierosolima* and *Leirus quinquestriatus* exhibit enthalpy values for the oxygenation of the T and R states that are of different sign $[\Delta H(\text{T}) \approx +3 \text{ kcal mol}^{-1}$ and $\Delta H(\text{R}) \approx -7.5$ kcal mol$^{-1}]$.[419, 433] In this case, the allosteric transition is clearly also enthalpy-driven.

The Binding of Ions and Heterotropic Effects in Haemocyanins. The effects of anions and cations on the O_2-binding properties and the association–dissociation behaviour of haemocyanins are the results of complex interactions between the (multiple) ion-binding sites and the oxygen-binding sites on the protein. In several cases, the effect of the various components on the O_2-binding behaviour is well documented. Positive and negative proton Bohr effects, as well as analogous effects of other ions, have been inferred from oxygen-binding experiments.[413, 420, 421, 430, 433, 434] In general, the number of oxygen-linked binding sites is only a fraction of the total number of ion-binding sites (*e.g.*, in the case of protons, a few oxygen-linked ionizable groups are found).

For the binding of the magnesium ions to the haemocyanin of *Callianassa californiensis*,[430] equilibrium dialysis experiments in the pH range 7—8.2 have indicated that at least seven strongly binding sites are present, while only one of these is O_2-linked. A similar phenomenon has been observed for the calcium-binding sites of the haemocyanin of *Panulirus interruptus*.[421] This may explain why, in several cases, 'direct' measurements of the binding of ions fail to detect the O_2-linked binding sites. Heterogeneity of oxygen-linked ion-binding sites becomes more and more apparent (see references to the binding of Ca^{2+} to the haemocyanins of *Panulirus interruptus*[421] and *Levantina hierosolima*, as well as to the oxygen-linked proton-binding sites of the haemocyanin of *Leirus quinque-striatus*[433]). Competition between Ca^{2+}, Na^+, and H^+ for the same sites has been demonstrated by following the release of protons from (unbuffered) protein upon addition of Ca^{2+} or Na^+ ions in the case of haemocyanins of *Dolabella auricularia*[435] and *Panulirus interruptus*.[421]

The binding of Na^+ ions to the haemocyanin of *Panulirus interruptus*, studied by ^{23}Na n.m.r. spectroscopy, reveals that there are two types of binding sites: the more strongly bound Na^+ ions may compete with Ca^{2+}.[436] Furthermore, these experiments indicate a linkage between the binding of Na^+ and of oxygen, which is in agreement with O_2-binding equilibria. Chlorine-35 n.m.r. measurements indicated that binding of chloride ion is moderately strong for the haemocyanins of *Panulirus interruptus* and *Octopus vulgaris*, but much stronger for that

[432] L. Zolla, H. A. Kuiper, M. Brunori, and E. Antonini, in ref. 344, p. 719.
[433] E. Daniel and A. Klarman, in ref. 344, p. 775.
[434] N. Shaklai, A. Klarman, and E. Daniel, *Biochemistry*, 1975, **14**, 105.
[435] N. Makino, *J. Biochem. (Tokyo)*, 1972, **71**, 987.
[436] J.-E. Norne, H. Gustavsson, S. Forsén, E. Chiancone, H. A. Kuiper, and E. Antonini, *Eur. J. Biochem.*, 1979, **98**, 591.

of *Limulus polyphemus*.[437] The temperature dependence of the n.m.r. signals indicates that chloride ions that are bound to the haemocyanins of *Limulus polyphemus* are slowly exchanging.

The use of lanthanide ions as proper replacement species for Ca^{2+} was shown in the case of the haemocyanin of *Panulirus interruptus*.[438, 439] Binding of terbium (as Tb^{3+}) to the deoxygenated haemocyanin of *Panulirus interruptus* results in an increase in fluorescence yield of the bound ion, owing to the transfer of energy from the tryptophan residues. The addition of other lanthanides or of calcium to haemocyanin that is saturated with terbium ion results in a quenching of the luminescence of the terbium, and indicates that some of the cation-binding sites are grouped in clusters along the polypeptide chain. Unfortunately, the relationship between Tb^{3+} and oxygen-binding sites could not be studied directly by fluorescence, since the addition of oxygen results in an almost complete quenching of the emission from the Tb^{3+} ion.

An extreme example of linkage between proton- and oxygen-binding sites is represented by the pH dependence of the O_2-binding equilibrium of the haemocyanin of *Panulirus interruptus*.[440] Upon lowering the pH, a marked decrease in the saturation of the protein with O_2 is observed, although parallel high-pressure spectrophotometry shows that the O_2-binding capacity does not change over the pH range where the protein becomes progressively unsaturated in air. Thus quaternary as well as tertiary proton Bohr effects determine the shape and position of the O_2-binding isotherm of this haemocyanin. Root effects have already been reported for the haemocyanins of *Callianassa californiensis*,[430] *Octopus vulgaris*,[441] and *Buccinum undatum*.[442]

A semi-quantitative map of the linkage between dioxygen, protons, and calcium and sodium ions has been designed for the haemocyanin of *Panulirus interruptus*.[421] According to Wyman, this topological map illustrates the coupling between various binding sites in terms of pairwise interactions or complete linkage, although more experimentation is needed before it will be possible to decide between the various possibilities.

Kinetics of Binding of Ligands to Haemocyanins. Analysis of the kinetics of the oxygen-binding reaction of haemocyanins within the framework of the M–W–C model[117] has provided estimates of the rate-constants for the allosteric states, and in some cases of the rate(s) of the allosteric transition.[443-447] It is essential to perform the kinetic experiments under conditions that are well defined by equilibrium studies. As discussed above, allosteric effectors may preferentially

437 E. Chiancone, J.-E. Norne, S. Forsén, H. A. Kuiper, and E. Antonini, in ref. 344, p. 647.
438 H. A. Kuiper, A. Finazzi-Agrò, E. Antonini, and M. Brunori, *FEBS Lett.*, 1979, **99**, 317.
439 H. A. Kuiper, L. Zolla, A. Finazzi-Agrò, and M. Brunori, *J. Mol. Biol.*, 1981, **149**, 805.
440 H. A. Kuiper, M. Coletta, L. Zolla, E. Chiancone, and M. Brunori, *Biochim. Biophys. Acta*, 1980, **626**, 412.
441 M. Brunori, *J. Mol. Biol.*, 1971, **55**, 39.
442 O. Brix, G. Lykkeboe, and K. Johansen, *J. Comp. Physiol.*, 1979, **129**, 97.
443 R. van Driel, M. Brunori, and E. Antonini, *J. Mol. Biol.*, 1974, **89**, 103.
444 E. J. Wood, G. R. Cayley, and J. S. Pearson, *J. Mol. Biol.*, 1977, **109**, 1.
445 H. A. Kuiper, E. Antonini, and M. Brunori, *J. Mol. Biol.*, 1977, **116**, 569.
446 R. van Driel, H. A. Kuiper, E. Antonini, and M. Brunori, *J. Mol. Biol.*, 1978, **121**, 431.
447 M. Brunori, H. A. Kuiper, E. Antonini C. Bonaventura and J. Bonaventura, in ref. 344, p. 693.

stabilize one of the allosteric states, which makes investigations of their kinetic properties feasible.

Several years ago, Brunori[441] found that the haemocyanin of *Octopus vulgaris*, under conditions in which the binding of O_2 is non-co-operative and the affinity for O_2 is low (*i.e.* the T state is prevailing both in the presence and absence of ligand), shows a complex reaction mechanism. This observation has subsequently been reproduced for other haemocyanins in the T state (see, for example, ref. 446). It appears, therefore, that the binding of O_2 to a haemocyanin in the R state corresponds to a simple bimolecular process, while the mechanism of reaction of a haemocyanin that is in the T state is complex, involving (at least) two coupled steps (a bimolecular and a unimolecular process).[447]

The idea that has been advanced on the basis of results obtained on the haemocyanin of *Panulirus interruptus*, *i.e.* that co-operativity in the binding of O_2 is kinetically controlled by the rates of dissociation of oxygen,[445] seems now to be of general validity.[447] Data for a fairly large number of haemocyanins under conditions where either the R or the T state prevails indicate that the 'on' rate-constants are very similar for both states, while the 'off' rate-constants differ by a factor of 25—140, depending upon the species (see Table 5).

Table 5 *Rate-constants for the reactions with oxygen of various haemocyanins that are in the R and T states[a]*

Species	$^T k' \times 10^{-6}$/l mol^{-1} s^{-1}	$^T k$/s^{-1}	$^R k' \times 10^{-6}$/l mol^{-1} s^{-1}	$^R k$/s^{-1}
Octopus vulgaris	12	4000	—	—
Panulirus interruptus	46	1500	31	60
Helix pomatia α	1.3	300	3.8	10
Buccinum undatum	—	—	7.8	50
Helix pomatia β	5.0	700	5.0	5
Limulus polyphemus	20	800	7.0	8

(a) Experimental conditions are as detailed in ref. 447

Rate-limiting steps have been observed during the oxygenation of α- and β-haemocyanins of *Helix pomatia* under conditions of co-operative binding of O_2 (*e.g.* see refs. 443, 446, and 448); the observed rates ($k = 20$—150 s^{-1}) have been related to the rate of the allosteric transition. The temporal course of the uptake of Bohr protons that is associated with the binding of O_2 to the β-haemocyanin of *Helix pomatia* has been interpreted along the same lines.[448]

A correlation between the size of the allosteric unit and the rate of the switch of quaternary conformation has been suggested.[446] Whereas the α- and β-haemocyanins of *Helix pomatia* (with, respectively, twelve and fifteen sites that interact within a functional constellation) exhibit rate-limiting isomerization steps during oxygenation, no such slow steps have been detected with the much smaller haemocyanin from the arthropod *Panulirus interruptus* (with six interacting sites).[445] Determination of the rates of allosteric transition for haemocyanins that are characterized by conformational constellations of different size may

[448] H. A. Kuiper, M. Brunori, and E. Antonini, *Biochem. Biophys. Res. Commun.*, 1978, **82**, 1062.

yield information on the 'fundamental' rate-constant for transfer of allosteric information in these systems. Any quantitative description, however, demands that the rate of the $T \rightleftharpoons R$ transition should depend on the degree of ligation of the molecule, as demonstrated for haemoglobin.[449]

Kinetics of the dissociation of CO from its complex with the haemocyanin of *Limulus polyphemus* have been investigated by displacing CO with oxygen, using a stopped-flow method.[406] More recently, an extensive study of the kinetics of reaction with CO has been carried out,[423] using a fluorescence temperature-jump instrument and making use of the luminescence properties of CO–haemocyanin complexes.[405] A preliminary report on the reaction of the α-haemocyanin from *Helix pomatia*[450] showed that there is a single relaxation event, and that it follows bimolecular kinetics. Extension of this work to several other haemocyanins[423] has shown that: (i) the resolved relaxation event corresponds to the perturbation of an equilibrium that is characterized by a negative enthalpy change ($\Delta H = -13.5$ kcal mol^{-1}); (ii) the reciprocal relaxation time is linearly dependent on the concentration of the reagents; (iii) the affinity for CO is dominated by the values of the dissociation velocity constant(s), which range from 10 to 10^4 s^{-1} for the various species; (iv) although the binding is non-co-operative, the overall reaction appears to be complex.

A Stereochemical Model for Haemocyanins.—The amount of quantitative information on the structural and functional properties of haemocyanin is now sufficient for a stereochemical model of the binding of ligands to these multi-subunit proteins to be proposed, in analogy to haemoglobin (see refs. 13, 176, and 177).

The overall structure and molecular architecture of haemocyanins, their primary structure(s), their ligand-binding site, and the stereochemistry of the binuclear copper centre where dioxygen is reversibly bound are each as different as they may possibly be from the corresponding structural features of haemoglobins. In spite of these extreme differences in structure, there are outstanding analogies in the functional properties of these O_2 carriers which stimulate some consideration.

The basic features of the allosteric mechanism,[172] which is based on the existence of two functionally relevant states in chemical equilibrium, are often adequate to describe co-operative effects in the binding of O_2. The overall features which apparently differentiate the ligand-binding properties of these two classes of proteins (*e.g.* values of the overall free-energy of interaction, maximum Hill coefficient, effects of different components of the solution on the slope and position of the O_2-binding curve *etc.*) have been rationalized by a model which proposes that the interaction phenomena extend over a limited number of sites.[416] These 'allosteric units' (or functional constellations) include all of the binding sites for the smaller haemocyanins (*e.g.* the haemocyanins of *Panulirus interruptus* or other arthropods), but only a group of sites for the larger proteins (*e.g.* only sixteen sites out of 160 in β-haemocyanin from *Helix pomatia*).

The kinetics of binding of ligands display noticeable similarities between haemoglobin and haemocyanin. Thus, in either case, the kinetic basis of the

---

449 C. Sawicki and Q. H. Gibson, *J. Biol. Chem.*, 1977, **252**, 7538.
450 H. A. Kuiper, A. Finazzi-Agrò, and M. Brunori, in ref. 344, p. 755.

difference(s) in affinity between the two allosteric states resides in the values of the rate-constants for dissociation of oxygen (see Table 5). Rate-limiting steps that are related to ligand-linked conformational states are not observed in the case of haemoglobin because the relaxation processes for the allosteric equilibrium are always very fast for these tetrameric proteins ($\tau^{-1} \simeq 10^4$ s^{-1}).[449] On the other hand, some of the larger haemocyanins display slow conformational transitions which have been tentatively related to the 'size' of the allosteric unit[446] (see above).

The structure of the active site in haemocyanin offers new and interesting information. Comparison of the binuclear copper centres in oxy- and deoxy-haemocyanins shows dramatic changes. The oxidation state of the two copper atoms in oxyhaemocyanin demands the formation of a peroxo-Cu$^{II}_2$ complex and the stereochemistry of the site involves coupling of the two metal atoms through an endogenous protein ligand (oxygen from a phenolate). In deoxyhaemo-cyanin, the binuclear centre has very different stereochemistry and co-ordina-tion, with breakage of the endogenous bridge.[398] Given the species and the conditions used in these studies, these structural features should refer to a deoxy-protein that is in a T state.

It is therefore not surprising that the binding of O_2 to a haemocyanin that is in the T state demands a two-step reaction mechanism, as shown several years ago by Brunori.[441] The considerable rearrangement that is necessary to fit an oxygen molecule that is symmetrically bound to the two copper atoms may very well be correlated to the slow step in the binding of O_2 to a site in the T state. Therefore, by analogy with haemoglobin, it may be proposed that the lower affinity for oxygen of a site in the T state of haemocyanin is related to the work that must be done so as to optimize the structure of the binuclear copper centre to fit the O_2 molecule. The difference in free-energy of binding to the T and to the R state is generally smaller than in haemoglobin (typically ~ 1.0—2 kcal per site);[416] the larger co-operative effects, exemplified by high values of n, depend on the amplification mechanism that is related to the large number of sites in a functional constellation, which make the allosteric transition essentially 'explosive'.

Finally, it is interesting to point out that the binding of carbon monoxide to haemocyanins is never co-operative, contrary to the case of the binding of CO to haemoglobin. Infrared spectra[451, 452] have shown that, in haemocyanins, CO is bound to only one of the copper atoms, and interactions with the other copper atom in a site are not observed (except possibly in the excited state). In spite of this, kinetic studies[423] indicate that the reaction mechanism may be complex. It therefore seems possible that a reorganization of the binuclear copper centre that occurs upon binding a CO molecule (as suggested by the transfer of charge from Cu$\rightarrow$CO) is not associated with a quaternary switch from the T to the R state, since this ligand does not bind to the second metal atom in a site.

The co-operative effects should therefore be related, according to this hypothe-sis, to the bridging features that are unique to O_2, which acts as a 'clamp' between the two copper atoms. The spectroscopy of a deoxyhaemocyanin in the R state

---

[451] L. Y. Fager and J. O. Alben, *Biochemistry*, 1972, **11**, 4786.
[452] H. van der Deen and H. Hoving, *Biophys. Chem.*, 1979, **9**, 169.

and of an oxyhaemocyanin in the T state will provide interesting findings for or against a stereochemical model in haemocyanins.

Acknowledgements: We express our thanks to Dr J. V. Bannister for helpful discussions and for the literature search, to Dr M. F. Perutz for reading and commenting on the manuscript, and to Mrs. Pina Storari for the arduous work involved in typing and setting out the manuscript.

5
Oxidases and Reductases

A. E. G. CASS

Introduction.—This chapter covers the literature on metal-containing oxidases and reductases published during 1979. The arrangement of the sections follows closely that of Volumes 1 and 2; *i.e.*, by the nature of the metal-ion centre. However, a number of changes have been made, and these reflect the *functional* similarities of the enzymes involved. Thus the papers relating to cytochrome oxidase appear along with those of complexes I—III in a single section on the mitochondrial respiratory chain; this section also includes the papers on the bacterial terminal oxidases. Although this means that this section will cover copper, haem iron, and iron–sulphur centres, it was considered that the close structural and functional integrity of this system makes this a more 'natural' basis for arrangement. In a similar vein, the seleno-enzyme glutathione peroxidase is included with the haemoprotein peroxidases, and the references to the iron- and manganese-containing superoxide dismutases are included with the copper-containing enzyme. Finally, for multi-centred redox enzymes, where there has been no over-riding functional criterion, the classification under one redox metal rather than another has been, to some extent, arbitrary.

1 Copper-containing Enzymes

Superoxide Dismutases (EC 1.15.1.1).—*Copper–Zinc Superoxide Dismutase.* Several reviews on selected aspects of superoxide dismutases have appeared, covering the areas of inflammation therapy and treatment,[1] their occurrence in the nervous system,[2] in cancer,[3] and in oxygen toxicity,[4] and the nature of the various isozymes.[5]

Binding of metal ions to the apoenzyme has been extensively studied; Cass *et al.* have used[6] high-resolution ^{1}H n.m.r. spectroscopy to follow the binding of zinc(II) ions. They concluded that there are two binding sites for this metal ion; the first, presumed to be the zinc site in the holoenzyme, pre-forms the second site on co-ordinating a zinc ion. Very little associated structural change occurs elsewhere in the molecule upon binding a metal ion, and the second binding site for the zinc ions has at least a ten-fold lower affinity for ions of this

[1] A. M. Michelson, *Recherche*, 1979, **10**, 1269.
[2] R. Fried, *J. Neurosci. Res.*, 1979, **4**, 435.
[3] L. W. Oberley and G. R. Buettner, *Cancer Res.*, 1979, **39**, 1141.
[4] J. M. McCord, *Rev. Biochem. Toxicol.*, 1979, **1**, 109.
[5] J. M. McCord, *Isozymes: Curr. Top. Biol. Med. Res.*, 1979, **3**, 1.
[6] A. E. G. Cass, H. A. O. Hill, J. V. Bannister, and W. H. Bannister, *Biochem. J.*, 1979, **177**, 477.

metal. This was assumed to be the site that binds copper ions in the native enzyme. Further evidence in support of the binding of other metals at the copper-binding site comes from optical and electron paramagnetic resonance (e.p.r.) spectra of the cobalt(II)-substituted enzyme.[7] Two cobalt ions bind to each subunit of the apoenzyme; one has spectroscopic and magnetic properties very similar to those previously described for cobalt that is bound at the zinc-binding site. The second cobalt ion has quite different properties, similar to those seen when cobalt binds to the zinc-containing protein. This cobalt ion is presumed to occupy the copper-binding site.

Valentine and her co-workers have described two pH-dependent migrations of the metal ions in superoxide dismutase.[8,9] In the native enzyme, as the pH is lowered below 3.8, zinc is lost whilst copper is retained.[8] This behaviour seems to be a function of the metal sites *per se* rather than of the ion that is present, as copper is lost when it is bound at the zinc-binding site, but not from its native site, under the same conditions. Loss of metal ions from the zinc-binding site is attributed to a pH-dependent conformational change. In contrast, if the enzyme with only a copper ion in each subunit is subjected to an increase in pH, then these ions migrate, to form apo-subunits and subunits that contain a pair of copper ions, with the two metals magnetically coupled.[9] The pK_a for this rearrangement is 8.2. It is reversible, albeit slowly, upon lowering the pH.

Chemical modifications of bovine superoxide dismutase have been carried out with several different reagents by Malinowski and Fridovich.[10–12] Diazonium 1*H*-tetrazole (DHT) partially inactivates the enzyme, with concomitant modification of one histidine, one tyrosine, and ten lysine residues.[10] In the presence of 8M-urea, hybrids formed between DHT-modified and totally inactivated subunits had only half the specific activity of the DHT-modified dimers, supporting the idea that the subunits are catalytically independent. The interaction between the subunits in the enzyme was probed by examining the reactivity of the cysteine and histidine residues towards alkylating agents.[11] No additional modification was observed in the presence of SDS or urea. However, removal of the metal ions by treatment with EDTA, followed by the addition of denaturants, resulted in further alkylation reactions.

A very intriguing modification is that of the residue arginine-141 with $\alpha\beta$-diketones.[12] The reaction of this single residue resulted in a 90% loss of activity, but this was reversible upon removal of the blocking group. This residue is approximately 6 Å from the copper ion, and the modification causes a change in the optical spectrum. The authors have suggested a role for arginine-141 in either superoxide-binding or proton-transfer steps.

Several n.m.r. studies of superoxide dismutase were reported in 1979; Rotilio

[7] L. Calabrese, D. Cocco, and A. Desideri, *FEBS Lett.*, 1979, **106**, 142.

[8] M. W. Pantoliano, P. J. McDonnell, and J. S. Valentine, *J. Am. Chem. Soc.*, 1979, **101**, 6454.

[9] J. S. Valentine, M. W. Pantoliano, P. J. McDonnell, A. R. Burger, and S. J. Lippard, *Proc. Natl. Acad. Sci. USA*, 1979, **76**, 4245.

[10] D. P. Malinowski and I. Fridovich, *Biochemistry*, 1979, **18**, 237.

[11] D. P. Malinowski and I. Fridovich, *Biochemistry*, 1979, **18**, 5055.

[12] D. P. Malinowski and I. Fridovich, *Biochemistry*, 1979, **18**, 5909.

and co-workers have studied the relaxation times of $^{19}F^-$ in the presence of the enzyme.[13, 14] They have described the dependence of both transverse and longitudinal relaxation times on the concentrations of enzyme and fluoride ion and on pH, temperature, and ionic strength.[13] The ion is in slow exchange, on the ^{19}F timescale, and the binding constant (2—3 l mol^{-1}) and activation enthalpy (25 kJ mol^{-1}) are reported. In their second publication[14] the authors show how measurement of relaxation times of $^{19}F^-$ may be used to determine the concentrations of the copper–zinc and manganese superoxide dismutases. Boden et al.[15] have measured the relaxation times of water in solutions of the bovine enzyme as a function of pH, temperature, and frequency. They analysed their results in terms of two binding sites for the water molecules on the copper(II) ion; the first is occupied by a water molecule that is 3.4 Å from the copper, in a roughly axial orientation. Above pH 9 a second site, approximately equatorial, is occupied by a hydroxide ion that is 2.7 Å from the copper.

In a high-resolution ^1H n.m.r. study, Stoesz et al.[16] assigned resonances to the exchangeable imidazole protons of the histidine residues 41, 44, and 69; they reported the pH dependence of the chemical shifts and relaxation times and the solvent-exchange behaviour of these protons. A nuclear Overhauser effect between the protons attached to N-1 and N-3 of histidine-41 and the previously assigned proton at C-2 was observed, as was an effect between the proton attached to N-1 and that to N-3. The remaining assignments to histidines 44 and 69 were based on a comparison of solvent inaccessibility, as judged by the n.m.r. spectra, with that suggested by the crystal structure. In contrast to the relatively ready exchange of protons at N-1 and N-3, exchange rates of protons attached to C-2 of histidine are low, though they are measurable. In a study of the rates of exchange of protons at C-2 with solvent deuterons, it was shown that the rate was much reduced in those histidine residues that were co-ordinated to a metal ion.[17, 18] The exchange behaviour in the apo-, holo-, and zinc-only-forms of yeast and bovine superoxide dismutases showed that, under conditions where the protons attached to C-2 on non-co-ordinated residues had completely exchanged, those on residues that were bound to metal ions had not exchanged. This enabled resonances of copper and zinc ligands to be distinguished, and it was suggested as a general method for the assignment of histidine residues as metal ligands in metalloproteins.[18]

Another comparison between the yeast and bovine enzymes was concerned with their resistance to 8M-urea.[19] Whilst the bovine enzyme was stable, the

[13] P. Viglino, A. Rigo, R. Stevanato, G. A. Ranieri, G. Rotilio, and L. Calabrese, *J. Magn. Reson.*, 1979, **34**, 265.
[14] A. Rigo, P. Viglino, E. Argese, M. Terenzi, and G. Rotilio, *J. Biol. Chem.*, 1979, **254**, 1759.
[15] N. Boden, M. C. Holmes, and P. F. Knowles, *Biochem. J.*, 1979, **177**, 303.
[16] J. D. Stoesz, D. P. Malinowski, and A. G. Redfield, *Biochemistry*, 1979, **18**, 4669.
[17] A. E. G. Cass, H. A. O. Hill, W. H. Bannister, and J. V. Bannister, *Biochem. Soc. Trans.*, 1979, **9**, 716.
[18] A. E. G. Cass, H. A. O. Hill, J. V. Bannister, W. H. Bannister, V. Hasemann, and J. T. Johansen, *Biochem. J.*, 1979, **183**, 127.
[19] D. Barra, F. Bossa, F. Marmocchi, F. Martini, A. Rigo, and G. Rotilio, *Biochem. Biophys. Res. Commun.*, 1979, **86**, 1199.

yeast enzyme exhibited a reversible biphasic loss of activity and a concomitant change in the e.p.r. spectrum of copper towards a more axial signal. Lawrence and Sawyer[20] investigated the electrochemistry of the copper and manganese superoxide dismutases, using a variety of mediators. The manganese enzyme from *Bacillus stearothermophilus* had a mid-point potential (E_m) of $+210$ mV at pH 8 and a value of ΔE_m of -50 mV per pH unit; these are consistent with the uptake of a single proton upon reduction. In contrast, the enzyme from *Escherichia coli* had an E_m of $+320$ mV and showed a complex pH dependence. The electrochemistry of the copper enzyme was also complex, and the results were interpreted in terms of an interaction between the two copper centres.

The reaction of bovine superoxide dismutases with several radicals that had been generated by pulse radiolysis revealed that, although some reduced the copper(II) centre, the reaction was not catalytic.[21] The high specificity of the enzyme for the superoxide radical was suggested to be a result of both thermodynamic and kinetic factors.

An inhibitor of the copper/zinc-containing superoxide dismutases that is effective both *in vivo* and *in vitro* is diethyldithiocarbamate ion (dtc), and two papers have appeared that discuss its mode of action.[22, 23] Measurements of the optical and e.p.r. spectra, and of enzyme activity, led[22] to the suggestion that the reaction has two steps: in the first, one molecule of dtc binds to the copper and forms an intense yellow complex, but with no loss of activity of the enzyme. A second, slower, step with another molecule of dtc results in removal of copper from the active site and loss of activity; however, the copper–dtc complex remains bound to the enzyme. Additional support for the removal of copper comes from some pulse-radiolysis experiments.[23] Both dtc and penicillamine (pen) remove copper from the enzyme, but whereas the Cu–pen complex still has dismutase activity, the Cu–dtc complex does not.

The isolation of two new copper/zinc superoxide dismutases has been described; the properties of hog liver and erythrocyte enzymes have been compared to those of the bovine enzyme.[24] Two isozymes, both containing copper, were isolated from leaves of the kidney bean, *Phaseolus vulgaris*.[25] They were identified by molecular weight, their sensitivity to cyanide, and cross-reaction with antibodies against spinach enzyme.

Several new methods of assay for superoxide dismutases have been presented. Baret *et al.*[26] iodinated bovine, rat, and human enzymes to a very high specific activity (51 μCi μg^{-1} or 1700 Ci mmol^{-1}). Subsequent use of this material in an immunoassay enabled them to detect 20—80 pg of the enzyme. Details of a rapid and inexpensive assay, especially suitable for clinical work, have been

[20] G. D. Lawrence and D. T. Sawyer, *Biochemistry*, 1979, **18**, 3045.
[21] P. Wardman, *Stud. Phys. Theor. Chem.*, 1979, **6**, 189.
[22] H. P. Misra, *J. Biol. Chem.*, 1979, **254**, 11 623.
[23] E. Lengfelder, *Z. Naturforsch.*, *Teil. C*, 1979, **34**, 1292.
[24] A. Bartowiak, W. Leyko, and R. Fried, *Comp. Biochem. Physiol.*, B, 1979, **62**, 61.
[25] Y. Kono, M. Takahashi, and K. Asada, *Plant Cell Physiol.*, 1979, **20**, 1229.
[26] A. Baret, P. Michel, M. R. Imbert, J. M. Morcellet, and A. M. Michelson. *Biochem. Biophys. Res. Commun.*, 1979, **88**, 337.

published,[27] whilst Hurley and co-workers have described two electrophoretic systems for analysing this enzyme.[28, 29]

Responses of the levels of superoxide dismutase to a number of environmental factors have been investigated. Levels of the enzyme in the brain tissue of rats fell by approximately one third between 50 and 900 days post partum.[30] This decrease did not correlate with concentrations of copper in tissues, nor did administering exogenous copper have any effect. Levels of superoxide dismutase in a variety of tissues are diminished under hypoxia,[31] and the authors suggest that this may account for the increased peroxidation of lipids that occurs post-hypoxia. Whole-body X-irradiation of rats results in a superoxide dismutase which, after isolation, has only 1 % of the specific activity of the native enzyme.[32] The enzyme as isolated did not exhibit either visible or e.p.r. spectra; however, these properties, along with enzyme activity, reappear after storage. Similar results were observed when the isolated enzyme was irradiated. Enzyme levels in T and non-T lymphocytes have been determined and compared with those found in granulocytes.[33] In both T and non-T cells the levels are higher than in granulocytes. Levels of erythrocyte dismutase have been employed as an indicator of copper status in deficient and supplemented chicks.[34] The location of the gene for the copper/zinc enzyme in mouse cells was determined to be on chromosome 16 by the use of somatic cell hybrids.[35]

Manganese and Iron Superoxide Dismutases. Several new iron-containing dismutases were described in 1979. Two iron superoxide dismutases have been isolated from *Euglena gracilis* and their physical properties described.[36] These enzymes are distinctly different from the iron superoxide dismutases previously isolated from prokaryotes. In general, they seem to be much more susceptible to a wide range of chemical and physical denaturing conditions. Lengfelder and Elstner have also reported the isolation of two proteins from *E. gracilis* that exhibited superoxide dismutase activity in assays that used xanthine oxidase as a source of superoxide.[37] However, only one of the proteins showed dismutase activity with pulse-radiolytically generated superoxide, and it was characterized as an iron superoxide dismutase. The other enzyme was a ferredoxin–$NADP^+$ oxidoreductase, and its apparent superoxide dismutase activity resulted from its interfering with the one-electron reduction of oxygen by xanthine oxidase. Another iron-containing superoxide dismutase, isolated from *Anacystis nidulans*, has been described by Cséke *et al.*[38] The analysis for iron (1 atom per dimer of

[27] M. Minami and H. Yoshikawa, *Clin. Chim. Acta*, 1979, **92**, 337.
[28] B. Lönnerdal, C. L. Keen, and L. S. Hurley, *FEBS Lett.*, 1979, **108**, 51.
[29] G. De Rosa, D. S. Duncan, C. L. Keen, and L. S. Hurley, *Biochim. Biophys. Acta*, 1979, **566**, 32.
[30] H. R. Massie, V. R. Aiello, and A. A. Iodice, *Mech. Ageing Dev.*, 1979, **10**, 93
[31] V. N. Chumanov and L. F. Osinskaya, *Vopr. Med. Khim.*, 1979, **25**, 261.
[32] M. A. Symonyan and R. M. Nalbandyan, *Biochem. Biophys. Res. Commun.*, 1979, **90**, 1207.
[33] Y. Kobayashi, S. Okahata, T. Sakano, K. Tanabe, and T. Usui, *FEBS Lett.*, 1979, **98**, 391.
[34] W. J. Bettger, J. E. Savage, and B. L. O'Dell, *Nutr. Rep. Int.*, 1979, **19**, 893.
[35] U. Francke and R. T. Taggart, *Proc. Natl. Acad. Sci. USA*, 1979, **76**, 5230.
[36] S. Kanematsu ad K. Asada, *Arch. Biochem. Biophys.*, 1979, **195**, 535.
[37] E. Lengfelder and E. F. Elstner, *Z. Naturforsch., Teil. C*, 1979, **34**, 374.
[38] C. Cséke, L. I. Horváth, P. Simon, G. Borbély, L. Keszthelyi, and G. L. Farkas, *J. Biochem. (Tokyo)*, 1979, **85**, 1397.

mol. wt 37 000) was by the technique of proton-induced X-ray elemental analysis (PIXIE); e.p.r. spectroscopy showed virtually all of the iron to be high-spin iron(III), in a rhombic environment.

Reversible reconstitution of the manganese superoxide dismutase from *E. coli* was achieved with a variety of metal ions.[39] Although only manganese could restore enzymatic activity, cobalt, nickel, zinc, iron, and copper all competed with manganese and inhibited reconstitution. Sufficient quantities of the cobalt derivative could be prepared to record a visible spectrum. A number of studies of the distribution of the different metal-containing forms of superoxide dismutase were reported in 1979. The subcellular distribution of iron- and manganese-containing enzymes in blue-green algae was studied, using a variety of methods.[40] Results were consistent with the localization of the iron enzyme to the cytosol and the manganese enzyme to the thylakoids. Iron enzymes from bacteria could be identified on polyacrylamide gels,[41] and the authors claim that strains from within the same genus have enzymes with similar mobilities.

Two further studies on the induction of bacterial superoxide dismutases have appeared.[42, 43] *Bacteroides fragilis* shows a three- to five-fold increase in dismutase levels when grown in 2% oxygen as compared to those grown anaerobically.[42] Based upon its inhibition patterns, the induced enzyme was claimed to be an iron superoxide dismutase. Inhibition of the terminal oxidase in, or the addition of redox dyes to, *Escherichia coli* results in an increased superoxide flux.[43] There is also a concomitant increase in cellular levels of manganese superoxide dismutase, and this was shown to result from protein synthesis *de novo*.[43]

A cyanide-resistant (presumably manganese-containing) superoxide dismutase has been found in carp erythrocytes and accounts for about 10% of the total activity in this tissue.[44] Dismutase levels in a number of tissues, both normal and malignant, have been measured.[45]

Non-Blue Copper Oxidases.—A review on the role of copper in **dopamine-β-hydroxylase** (EC 1.14.17.1) has appeared.[46] The activity of this enzyme in the chromaffin tissue of the cod was measured under a variety of conditions.[47] The activity reached a plateau at 10 μmol l^{-1} of copper ions in the assay mixture and showed pH and temperature optima at 5.4 and 27 °C respectively. In another study of copper in dopamine-β-hydroxylase, removal of the metal ions with EDTA or bathocuproine disulphonate resulted in loss of activity.[48] Titration with copper sulphate revealed four bound copper ions per molecule of enzyme, and the kinetics of reconstitution yielded an apparent Michaelis constant

[39] D. E. Ose and I. Fridovich, *Arch. Biochem. Biophys.*, 1979, **194**, 360.
[40] S. Okada, S. Kanematsu, and K. Asada, *FEBS Lett.*, 1979, **103**, 106.
[41] S. S. Bang, M. J. Woolkalis, and P. Baumann, *Curr. Microbiol.*, 1979, **1**, 371.
[42] C. T. Privalle and E. M. Gregory, *J. Bacteriol.*, 1979, **138**, 139.
[43] H. M. Hassan and I. Fridovich, *Arch. Biochem. Biophys.*, 1979, **196**, 385.
[44] F. Mazeaud, J. Maral, and A. M. Michelson, *Biochem. Biophys. Res. Commun.*, 1979, **86**, 1161.
[45] J. N. A. Balgooy and E. Roberts, *Comp. Biochem. Physiol.*, *B*, 1979, **62**, 263.
[46] T. Skotland and T. Ljones, *Inorg. Perspect. Biol. Med.*, 1979, **2**, 151.
[47] A.-C. Jönsson and S. Nilsson, *Comp. Biochem. Physiol.*, *C*, 1979, **62**, 5.
[48] T. Skotland and T. Ljones, *Eur. J. Biochem.*, 1979, **94**, 145.

of 0.03—0.2 μmol l^{-1}.[48] Various nucleotides activate the enzyme, ADP being the most effective.[49] This activation resulted in an increase in V_{max} with no change in K_m. In a study of the oxidation of ascorbate during turnover it was proposed that a one-electron oxidation occurs and the resulting ascorbate free-radical then non-enzymatically dismutes. Parallels with the mechanism of ascorbate oxidase have been drawn.[50] The binding of dopamine to bovine plasma **amine oxidase** (EC 1.4.3.6) was investigated by studying the reaction of a chirally tritiated (at C-1 and C-2) substrate; the results were consistent with two modes of binding.[51] Using a similar approach, benzylamine oxidase was shown to remove the *Si*-hydrogen atom from benzylamine stereospecifically.[52]

Barker *et al.* have characterized the copper centres in pig plasma **benzyl-amine oxidase** by e.p.r. spectroscopy and by measuring relaxation times of protons of water.[53] This enzyme is a dimer of two identical subunits, each containing one copper ion. Simulation of the 35 GHz e.p.r. spectrum suggested that the two metal ions have different symmetries, and both are sensitive to changes in pH and ionic strength. When cyanide or azide is bound, only a single type of copper is observed, and, from the relaxation times of protons of water, these anions are claimed to bind equatorially. These data and kinetic measurements have been interpreted in terms of the oxygen molecule binding to the copper in the same environment as the anions. Inhibition of amine oxidase by β-bromoethyl-amine reveals that this is an enzyme-activated irreversible inhibitor, *i.e.* a suicide substrate. Implications for the mechanism are discussed.[54]

The binding of fluoride to **galactose oxidase** (EC 1.1.3.9) has been investigated by measurement of ^{19}F relaxation times.[55] Temperature and frequency dependencies reveal that the principal relaxation mechanism is due to unpaired electron-spin density in the fluoride-centred *p*-orbitals. Comparisons of the isotropic and anisotropic hyperfine coupling constants and the derived binding constant for fluoride with those for CuF$^+$ are consistent with axial co-ordination to the enzyme.

A review on the role of **tyrosinase** (EC 1.14.18.1) in the biosynthesis of aromatic phenols in plants has appeared.[56] The sequence of tyrosinase published in 1978 contained an error, and a correction has appeared.[57] Positions 94 and 96 are cysteine and histidine respectively, instead of the reverse, as first reported. Two papers on the inactivation of the enzyme have appeared.[58, 59] One deals with inactivation by u.v. light, and Sharma *et al.*[58] have attributed this to the destruction of aromatic amino-acid residues and a reaction at the copper centre.

[49] E. Tachikawa, T. Ohuchi, K. Morita, Y. Ishimura, M. Oka, and F. Izumi, *FEBS Lett.*, 1979, **100**, 331.
[50] T. Ljones and T. Skotland, *FEBS Lett.*, 1979, **108**, 25.
[51] M. C. Summers, R. Markovic, and J. P. Klinman, *Biochemistry*, 1979, **18**, 1969.
[52] A. R. Battersby, J. Staunton, J. Klinman, and M. C. Summers, *FEBS Lett.*, 1979, **99**, 297.
[53] R. Barker, N. Boden, G. Cayley, S. C. Charlton, R. Henson, M. C. Holmes, I. D. Kelly, and P. F. Knowles, *Biochem. J.*, 1979, **177**, 289.
[54] H. Kumagai, H. Uchida, and H. Yamada, *J. Biol. Chem.*, 1979, **254**, 10 913.
[55] R. J. Kurland and B. J. Marwedel, *J. Phys. Chem.*, 1979, **83**, 1422.
[56] V. S. Butt, *Recent Adv. Phytochem.*, 1979, **12**, 433.
[57] K. Lerch, *Proc. Natl. Acad. Sci. USA*, 1979, **76**, 6021.
[58] R. C. Sharma, R. Ali, and O. Yamamoto, *J. Radiat. Res.*, 1979, **20**, 186.
[59] M. Miranda, G. Urbani, L. Di Vito, and D. Botti, *Biochim. Biophys. Acta*, 1979, **585**, 398.

Model studies suggested that autoinactivation during the synthesis of melanin is due to the presence of indolic autoxidation products of DOPA. It was suggested that this may play a role in the regulation of the synthesis of melanin *in vivo*.[59] A paper has appeared in the Russian literature in which it is claimed that the oxidation of catechol by **tyrosinase** (EC 1.10.3.1) occurs *via* a free-radical mechanism.[60] E.p.r. spectra of substrate free-radicals and enzyme-bound copper were reported.

Multi-copper Oxidases.—The enzyme **laccase** (EC 1.10.3.2), the most extensively studied of the multi-copper oxidases, continues to attract interest. Gray and co-workers[61] have analysed the optical, c.d., and m.c.d. spectra at room temperature and the low-temperature optical spectra of fungal and tree laccases in a variety of different states. They interpreted their data in terms of a description of the type I copper centre as having a flattened tetrahedral geometry. The ground state of the copper is $^2B_2$ ($d_{x^2-y^2}$), as shown in Figure 1. Assignments

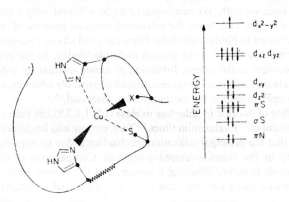

Figure 1 *The geometry and electronic energy levels of the 'blue' copper centre in multi-copper oxidases; X is methionine or a substitute*
(Reproduced by permission from *J. Am. Chem. Soc.*, 1979, **101**, 5038)

of the optical transitions of the type I copper chromophore are presented and bands due to the type II and type III copper centres described. The optical spectra of laccase in which the type I copper has been replaced by cobalt have been presented.[62] The spectra include an intense charge-transfer band, assigned to a sulphur → cobalt(II) transition. Ligand-field transitions show smaller splittings than in cobalt stellacyanin, and this has been attributed to a more nearly tetrahedral geometry in the latter. Laccase may also be prepared from which the type II copper has been selectively removed, and the properties of this form of the enzyme have been described.[63] The optical and e.p.r. spectra are analysed

[60] S. E. Bresler, E. N. Kazbekov, A. T. Sukhodolova, and V. N. Shadrin, *Biokhimiya*, 1979, **44**, 741.
[61] D. M. Dooley, J. Rawlings, J. H. Dawson, P. J. Stephens, L.-E. Andréasson, B. G. Malmström, and H. B. Gray, *J. Am. Chem. Soc.*, 1979, **101**, 5038.
[62] J. A. Larrabee and T. G. Spiro, *Biochem. Biophys. Res. Commun.*, 1979, **88**, 753.
[63] B Reinhammar and Y. Oda, *J. Inorg. Biochem.*, 1979, **11**, 115.

and its behaviour during reduction and re-oxidation is discussed in terms of a possible mechanism for the enzyme.

Binding of anions to the type II copper centre, with the copper in the type I centre reduced, has been followed, using e.p.r. spectroscopy.[64] The authors of this paper suggest that the anions bind to the copper to yield either a pseudo-tetrahedral or a five-co-ordinate complex, with the metal ion in an essentially hydrophobic environment. Measurements of the magnetic susceptibility of peroxide-treated laccase show an increase of 50% (at room temperature) as compared to the native enzyme.[65] This result has been interpreted as indicating a change in the antiferromagnetic coupling between the type III copper ions from 250—300 cm^{-1} to 120 ± 10 cm^{-1}.

The rates of reduction of the type I copper in laccase have been measured, with 25 mono- and di-substituted hydroquinones as reductants.[66] Results are consistent with the rapid formation of an enzyme–substrate complex followed by rate-limiting intramolecular electron-transfer. In general, the equilibrium constant for formation of a complex and the rate of transfer of an electron are relatively insensitive to the nature of the hydroquinone, both being controlled by protein-dependent processes. Stopped-flow optical and rapid-freeze e.p.r. spectroscopies have been used to follow the reactions of the copper centres in laccase under both anaerobic and turnover conditions.[67] The pH dependence of the rate of reduction of the type II copper ion has been attributed to the ionization of a bound water molecule; both type I and type II copper centres can donate electrons to the type III copper pair. These results have been incorporated into an expanded kinetic scheme that is shown in Figure 2. In a paper by Tarase-vich *et al.*, the cathodic reduction of oxygen at a carbon-black electrode is claimed to be catalysed by the adsorption of laccase.[68] Reduction of oxygen occurs with virtually no overpotential and the dependence on pH and on oxygen tension are consistent with Nernstian behaviour. An extracellular enzyme, isolated from *Rhizoctonia praticola*, that has phenol oxidase activity has been identified as a laccase.[69]

Two procedures for the isolation of **ascorbate oxidase** (EC 1.10.3.3) have been published.[70, 71] In one,[70] a method for the isolation of four blue copper proteins from *Cucumis sativus* is described. As well as ascorbate oxidase, stella-cyanin, plantacyanin, and plastocyanin can be prepared; Table 1 summarizes the properties of these four proteins. A modified method for isolating ascorbate oxidase from *Cucurbita pepo* var. *medullosa* has been described by Marchesini and Kroneck.[71] This preparation has a stoicheiometry of eight copper atoms per molecule and a molecular weight of 140 000, and, from a combination of metal analysis and e.p.r. spectroscopy, the distribution of the copper centres

[64] A. Desideri, L. Morpurgo, G. Rotilio, and B. Mondovì, *FEBS Lett.*, 1979, **98**, 339.

[65] O. Farver and I. Pecht, *FEBS Lett.*, 1979, **108**, 436.

[66] J. D. Clemmer, B. L. Gillard, R. A. Bartsch, and R. A. Hokserda, *Biochim. Biophys. Acta*, 1979, **568**, 307.

[67] L.-E. Andréasson and B. Reinhammar, *Biochim. Biophys. Acta*, 1979, **568**, 145.

[68] M. R. Tarasevich, A. I. Yaropolov, V. A. Bogdanovskaya, and S. D. Varfolomeev, *Bio-electrochem. Bioenerg.*, 1979, **6**, 393.

[69] J.-M. Bollag, R. D. Sjoblad, and S.-Y. Lui, *Can. J. Microbiol.*, 1979, **25**, 229.

[70] V. Ts. Aikazyan and R. M. Nalbandyan, *FEBS Lett.*, 1979, **104**, 127.

[71] A. Marchesini and P. M. H. Kroneck, *Eur. J. Biochem.*, 1979, **101**, 65.

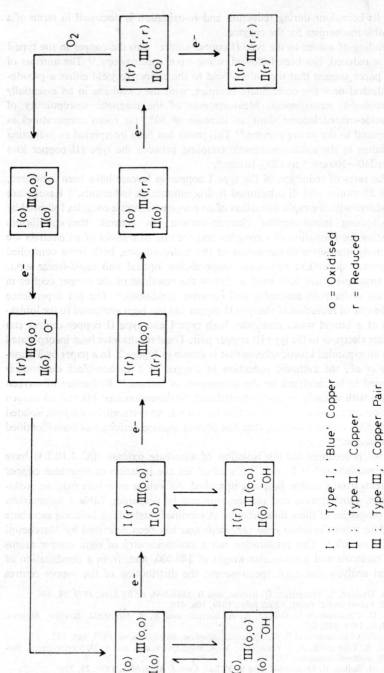

Figure 2 *A proposed mechanism for the action of laccase* (Redrawn from *Biochim. Biophys. Acta*, 1979, **568**, 145)

Table 1 *A comparison of some properties of the four copper-bearing proteins from cucumber peelings*

	Ascorbate oxidase	Plastocyanin	Stellacyanin	Plantacyanin
Mol. wt	140 000	10 000	20 000 and 40 000	9000
Copper content[a]	8	1	1	1
Isoelectric point	8.0	4.1	7.5	10.6
λ_{max}/nm[b]	610	597	605	593
A_{280}/A_{600}[c]	32	1.4	6	10
Yield/mg[d]	80—100	30—35	10—15[e]	20—25

(*a*) Content per molecule; (*b*) refers to the major visible band; (*c*) A_{600} is the absorption at the major visible band of each protein; (*d*) from 500g acetone powder; (*e*) this is the quantity obtained of both molecular forms.

was determined as three of type I, one of type II, and two of type III. Under some conditions, two of the type I copper ions may be lost. Preliminary results on the reaction of the enzyme with ascorbic and reductic acids and with hydrogen peroxide are described. The enzyme from the same source has also been subjected to *X*-ray diffraction measurements.[72] Suitable crystals of the enzyme have been grown, and they diffract to 2.5 Å; the number of molecules per unit cell was determined as two. Subjection of ascorbate oxidase to conditions that had previously been shown to remove the type II copper from laccase reduces the optical absorbance at both 750 and 330 nm.[73] Concomitant with this, the type II features disappear from the e.p.r. spectrum and the activity of the enzyme is decreased by 95%. All of these changes are reversed by addition of copper; from the stoicheiometries, the authors have suggested the same composition as in reference 71, *viz.* three type I, one type II, and two type III centres. Ascorbate oxidase exhibits the property of 'enzyme memory', *i.e.* the rate of re-oxidation of the enzyme depends upon the nature of the reductant.[74] Optical spectra of the reduced enzyme also vary, dependent upon the reductant, and the authors argue that both effects are a consequence of different (substrate-dependent) conformations of the reduced enzyme.

Descriptions of two studies of the amino-acid sequence of **caeruloplasmin** (EC 1.16.3.1), which is a multi-copper oxidase that is found in plasma, have been published.[75, 76] In one,[75] the sequences of two peptides were compared with those of the blue copper proteins azurin, plastocyanin, and stellacyanin and that of subunit II of mitochondrial cytochrome oxidase. The sequences were analysed in terms of the phylogenetic tree for the evolution of blue copper proteins that is shown in Figure 3. In contrast to this study, the amino-acid sequence of a histidine-rich (8.8%), 159-residue fragment that is obtained from treatment of human caeruloplasmin with plasmin shows no homology with

[72] R. Ladenstein, A. Marchesini, and S. Palmieri, *FEBS Lett.*, 1979, **107**, 407.
[73] L. Avigliano, A. Desideri, S. Urbanelli, B. Mondovì, and A. Marchesini, *FEBS Lett.*, 1979, **100**, 318.
[74] M. Katz and J. Westley, *J. Biol. Chem.*, 1979, **254**, 9142.
[75] L. Ryden and J.-O. Lundgren, *Biochimie*, 1979, **61**, 781.
[76] I. B. Kingston, B. L. Kingston, and F. W. Putnam, *Proc. Natl. Acad. Sci. USA*, 1979, **76**, 1668.

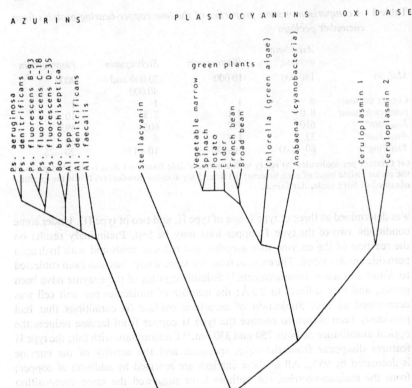

Figure 3 *A phylogenetic tree of proteins that contain the type I 'blue' copper centre. The genera from which azurins were obtained are* Pseudomonas, Bordetella, *and* Alcaligenes

(Reproduced by permission from *Biochimie*, 1979, **61**, 781)

other copper proteins.[76] However, the sequence shows multiple His-X-His regions; in view of the proximity of these to cysteine-134 and methionine-144, it has been suggested that this is a potential (blue) copper-binding region.

Several papers have appeared from a Russian group on the structure of human caeruloplasmin. The isolation of the enzyme in the presence and absence of protease inhibitors and its characterization by electrophoresis and ultracentrifugation[77] and by electron microscopy[78] have been described. The authors suggest that the intact molecule consists of a single polypeptide chain (of mol.

[77] K. A. Moshkov, Kh. M. Karimova, S. A. Neifakh, S. Lakatos, J. Haidu, P. Závodsky, T. G. Samsonidze, and N. A. Kiselev, *Bioorg. Khim.*, 1979, **5**, 395.
[78] T. G. Samsonidze, K. A. Moshkov, N. A. Kiselev, and S. A. Neifakh, *Int. J. Peptide Protein Res.*, 1979, **14**, 161.

wt 130 000) which is cleaved into six fragments by endogenous proteases. These six fragments may correspond to the six domains in the whole molecule that can be seen by electron microscopy. Further support for these ideas is presented in studies of the effect of endogenous [79] or exogenous [80] proteases. It is suggested that fragmentation by the endogenous proteases *in vivo* is prevented by the presence of inhibitors that are separated during plasma fractionation.[79]

Caeruloplasmin has been shown to inhibit a number of superoxide-dependent reactions;[81] however, the decreased production of hydrogen peroxide under these conditions implies that it does so by oxidation of the superoxide rather than by catalysing its dismutation.

The properties of the copper centres in human caeruloplasmin have been investigated by low-temperature absorption spectroscopy and by room-temperature c.d. and m.c.d. in both native and anion-bound states.[82] Optical spectra of the type I centre are consistent with a flattened tetrahedral geometry (CuN_2SS^*), and binding of an anion at the type II centre leads to changes in the type I spectra. A nine-line superhyperfine splitting in the e.p.r. spectrum of the type II copper has been interpreted in terms of four equivalent nitrogen ligands.

2 Haem-containing Oxidases and Reductases

Cytochromes *P*-450.—As in previous Reports, the cytochromes *P*-450 are discussed by source. There are no explicit references made to the other components in the cytochrome *P*-450 mono-oxygenase systems except where they are involved in the chemistry of the haemoprotein. Recent reviews to various aspects of this enzyme occur in references 83—94.

Liver Cytochromes P-450 (EC 1.14.14.1). The liver microsomal enzyme is still by far the most widely investigated of the *P*-450-type cytochromes, and several purification procedures have been described. A partially purified enzyme from human livers, both adult[95] and foetal,[96] has been isolated and shown[97] to

[79] K. A. Moshkov, S. Lakatos, J. Haidu, P. Závodsky, and S. A. Neifakh, *Eur. J. Biochem.*, 1979, **94**, 127.
[80] V. B. Vasiliev, M. M. Shavlovsky, and S. A. Neifakh, *Bioorg. Khim.*, 1979, **5**, 1045.
[81] I. M. Goldstein, H. B. Kaplan, H. S. Edelson, and G. Weissmann, *J. Biol. Chem.*, 1979, **254**, 4040.
[82] J. H. Dawson, D. M. Dooley, R. Clark, P. J. Stephens, and H. B. Gray, *J. Am. Chem. Soc.*, 1979, **101**, 5046.
[83] V. Ullrich, *Top. Curr. Chem.*, 1979, **83**, 67.
[84] J. T. Groves, *Adv. Inorg. Biochem.*, 1979, **1**, 119.
[85] B. W. Griffin, J. A. Peterson, and R. W. Estabrook, *Porphyrins*, 1979, **7**, 333.
[86] S. Ahmad, *Drug. Metab. Rev.*, 1979, **10**, 1.
[87] E. Hodgson, *Drug Metab. Rev.*, 1979, **10**, 15.
[88] F. Guengerich, *Pharmacol. Ther.*, 1979, **6**, 99.
[89] T. Kitagawa, *Shokubai*, 1979, **21**, 104.
[90] E. F. Johnson, *Rev. Biochem. Toxicol.*, 1979, **1**, 1.
[91] M. J. Coon, Y. L. Chiang, and J. S. French, *Symp. Med. Hoechst*, 1979, **14**, 201.
[92] R. Sato, Y. Imai, and H. Taniguchi, *Symp. Med. Hoechst*, 1979, **14**, 213.
[93] J. Werringloer and R. W. Estabrook, *Symp. Med. Hoechst*, 1979, **14**, 269.
[94] F. Mitani, *Mol. Cell. Biochem.*, 1979, **24**, 21.
[95] Ph. Beaune, P. Dansette, J. P. Flinois, S. Columelli, D. Mansuy, and J. R. Leroux, *Biochem. Biophys. Res. Commun.*, 1979, **88**, 826.
[96] M. Kitada and T. Kamataki, *Biochem. Pharmacol.*, 1979, **28**, 793.
[97] T. Kamataki, M. Sugiura, Y. Yamazoe, and R. Kato, *Biochem. Pharmacol.*, 1979, **28**, 1993.

contain low-spin iron. West *et al.*[98] have presented a method for the rapid purification of the enzyme from phenobarbital-treated rats, whilst the isolation of the enzyme from 3,4,5,4',5'-pentachlorobiphenyl-treated rats,[99] from pheno-barbital-treated rabbits,[100] from neonatal and foetal rats,[101] and from rabbits, with various inducers,[102] has been achieved.

The intimate involvement of the hepatic cytochromes *P*-450 in the metabolism of a whole variety of xenobiotic and endogenous molecules continues to be investigated, and Table 2[103–121, 150, 164, 195] shows some of the xenobiotic substances that are metabolized by cytochrome *P*-450. As well as those materials metabolized by the hepatic mono-oxygenase, many have been shown to alter the levels of this enzyme *in vivo*, and some of those reported in 1979 are also summarized in the Table.

One characteristic of the hepatic cytochromes *P*-450 that are induced by treatment with various compounds is that they possess different physical and catalytic properties,[90] and additional work presented in 1979 has added to our understanding of the nature of these multiple forms. Botelho *et al.*[122] have shown that the three forms of the enzyme that were isolated after treatment with different inducers have quite distinct amino-acid compositions and *N*- and *C*-terminal sequences, and they argue that this supports the view that the three proteins are produced by different genes. A similar approach has been used by Johnson *et al.*,[123] who showed that different forms of cytochrome *P*-450 yield different peptide maps after proteolysis; the authors also suggest that these maps may be used to characterize the different forms. In addition to peptide

[98] S. B. West, M.-T. Huang, G. T. Miwa, and A. Y. H. Lu, *Arch. Biochem. Biophys.*, 1979, **193**, 42.

[99] N. Ozawa, S. Yoshihara, K. Kawano, Y. Okada, and H. Yoshimura, *Biochem. Biophys. Res. Commun.*, 1979, **91**, 1140.

[100] I. I. Karuzina, G. I. Bachmanova, D. E. Mengazetdinov, K. N. Myasoedova, V. O. Zhikhareva, G. P. Kuznetsova, and A. I. Archakov, *Biokhimiya*, 1979, **44**, 1049.

[101] Th. Cresteil, J. P. Flinois, A. Pfister, and J. P. Leroux, *Biochem. Pharmacol.*, 1979, **28**, 2057.

[102] A. R. Boobis, D. S. Davis, and K. M. Lewis, *Br. J. Pharmacol.*, 1979, **66**, 427P.

[103] E. Dybing, T. Aune, and S. D. Nelson, *Biochem. Pharmacol.*, 1979, **28**, 43.

[104] E. Dybing, T. Aune, and S. D. Nelson, *Biochem. Pharmacol.*, 1979, **28**, 51.

[105] L. W. Condie and D. R. Buhler, *Biochem. Pharmacol.*, 1979, **28**, 375.

[106] K. M. Ivanetich, S. A. Lucas, and J. A. Marsh, *Biochem. Pharmacol.*, 1979, **28**, 785.

[107] D. Hultmark, K. Sundh, L. Johansson, and E. Arrhenius, *Biochem. Pharmacol.*, 1979, **28**, 1473.

[108] I. S. Owens, C. Legraverend, and O. Pelkonen, *Biochem. Pharmacol.*, 1979, **28**, 1623.

[109] H. Kurebayashi, A. Tanaka, and T. Yamaha, *Biochem. Pharmacol.*, 1979, **28**, 1719.

[110] H. Vainio and E. Elovarra, *Biochem. Pharmacol.*, 1979, **28**, 2001.

[111] W. Lenk, *Biochem. Pharmacol.*, 1979, **28**, 2149.

[112] F. P. Guengerich, *Biochem. Pharmacol.*, 1979, **28**, 2883.

[113] H. L. Gurtoo and R. P. Dahms, *Biochem. Pharmacol.*, 1979, **28**, 3441.

[114] K. A. Hassall and S. A. Addala, *Biochem. Pharmacol.*, 1979, **28**, 3199.

[115] T. Wolff, E. Deml, and H. Wanders, *Drug Metab. Dispos.*, 1979, **7**, 301.

[116] P. Hlavica and S. Huelsmann, *Biochem. J.*, 1979, **182**, 109.

[117] R. M. Facino and R. Lanzani, *Pharmacol. Res. Commun.*, 1979, **11**, 433.

[118] J. M. Guenther, J. M. Fysh, and D. W. Nebert, *Pharmacology*, 1979, **19**, 12.

[119] G. Feurerstein, H. Govrin, and J. Kapitulnik, *Biochem. Pharmacol.*, 1979, **28**, 3167.

[120] R. El Hazary and J. G. Mannering, *Mol. Pharmacol.*, 1979, **15**, 698.

[121] E. I. Ciaccio and R. J. Fruncillo, *Biochem. Pharmacol.*, 1979, **28**, 3151.

[122] L. H Botelho, D. E. Ryan, and W. Levin, *J. Biol. Chem.*, 1979, **254**, 5635.

[123] E. F. Johnson, M. C. Zounes, and U. Muller-Eberhard, *Arch. Biochem. Biophys.*, 1979, **192**, 282.

Table 2 *Various xenobiotic substances that have been shown to be metabolized by hepatic cytochromes P-450 or to alter the levels of the latter in vivo*

Xenobiotic substance	Reference
2,4-Diaminoanisole	103, 104
Hexachlorophene	105
Fluorinated anaesthetics	106
Dichloro-*p*-nitroanisole	107
3- and 9-Hydroxybenzo[*a*]pyrenes	108
Cycloalkylamines	109
Styrene oxide	110
4-Chloro-anilides	111
Aflatoxin B_1	112, 113
Dimethylated oxiran	114
Aldrin	115
NN-Dimethylaniline	116
β-Adrenergic-blocking drugs	117
2,3,7,8-Tetrachlorodibenzo-*p*-dioxin	118
Li^+	119
Interferon-inducing agents	120
Burn injury	121
1-Piperidinoanthraquinone	150
Ecdysone	164
Cholesterol	195

maps, a wide variety of physical and catalytic criteria have been used to distinguish between three highly purified cytochromes *P*-450 that were isolated from rats that had been treated with polychlorinated biphenyls.[124] The same techniques were used to compare these forms with those isolated from phenobarbital- or 3-methylcholanthrene-treated animals, and it was further shown that antibodies to each form did not cross-react with the other forms. Immunological cross-reactivity studies were also used to distinguish substrate specificities of two forms of cytochrome *P*-450 that were formed by different inducers.[125] Similarly, multiple forms of the enzyme from 3-methylcholanthrene-treated mice[126] or β-naphthoflavone-induced rabbits[127] have been characterized. Circular dichroism studies have been used by two groups to investigate different cytochromes *P*-450.[128, 129] The spectra are quite different, and respond differently to the addition of lipids or denaturants.[128] In the far-u.v., the spectra are consistent with there being 40—50% of α-helix,[129] which increases upon addition of lipid.[128] Imai has reported an extensive kinetic study of the dealkylase activity of seven forms of cytochrome *P*-450,[130] and has concluded that the characteristics of the four forms that are active in his reconstituted system are both cytochrome-

[124] D. E. Ryan, P. E. Thomas, D. Korzeniowski, and W. Levin, *J. Biol. Chem.*, 1979, **254**, 1365.
[125] R. Masuda-Mikawa, Y. Fujii-Kuriyama, M. Negishi, and Y. Tashiro, *J. Biochem. (Tokyo)*, 1979, **86**, 1383.
[126] M. Negishi and D. W. Nebert, *J. Biol. Chem.*, 1979, **254**, 11 015.
[127] P. R. McIntosh and R. B. Freedman, *FEBS Lett.*, 1979, **105**, 217.
[128] Y. L. Chiang and M. J. Coon, *Arch. Biochem. Biophys.*, 1979, **195**, 178.
[129] T. Shimizu, T. Nozawa, M. Hatano, H. Satake, Y. Imai, C. Hashimoto, and R. Sato, *Biochim. Biophys. Acta*, 1979, **579**, 122.
[130] Y. Imai, *J. Biochem. (Tokyo)*, 1979, **86**, 1697.

and substrate-dependent. In two different studies of xenobiotic metabolism[131, 132] it was demonstrated that the different forms preferentially activate specific mutagens,[131] and that the production of different warfarin metabolites is mediated by separate cytochromes *P*-450.[132] Other work on multiple forms has appeared in refs. 133 and 134.

Although treatment of animals with inducers leads to increased levels of the cytochromes *P*-450, this enzyme is also present, albeit at low levels, in uninduced animals. In this case it is also present as multiple forms, as shown by their isolation and characterization from uninduced rat livers.[135] Interestingly, both forms that were isolated metabolized RO-20458, an analogue of insect juvenile hormone. Koop and Coon have also isolated an uninduced cytochrome *P*-450 and described some of its physicochemical properties.[136] The purified enzyme has appreciable testosterone hydroxylase activity, but only half the NADH oxidase activity of inducible forms. Testosterone hydroxylase activity in rat liver microsomes was also investigated, where it was found to be mediated by three different enzymes, with distinct catalytic and regulatory properties.[137] Four cytochromes *P*-450 have been resolved in uninduced rats and the effects of different inducers on each form investigated.[138]

Although the induction of cytochromes *P*-450 has been known for many years, the molecular and cellular events involved are only just beginning to be elucidated, and the genetic regulation of one form, the aryl hydrocarbon hydroxylase, has recently been reviewed.[139] Much interest attaches to the role played by mRNA in the induction of cytochrome *P*-450, and Lechner *et al.*[140] have shown, using a heterologous cell-free system, that total hepatic mRNA yields a cytochrome *P*-450 peptide subunit upon translation. Prior treatment with phenobarbital yielded larger amounts of cytochrome *P*-450 peptide *in vitro*, concomitant with increased enzyme activity *in vivo*. Dubois and Waterman[141] have followed the time course of appearance of mRNA that codes for the cytochrome, together with measurements of the enzyme levels, following a single dose of phenobarbital. The latter lagged behind the former by 30 hours, showing that synthesis of the messenger is not rate-limiting. Protein synthesis by polysomes *in vitro* is elevated if prior administration of phenobarbital has occurred.[142] The quantity of material that is active against cytochrome *P*-450 is also elevated, all increases occurring within 24 hours. Utilizing antibodies against pheno-

131 R. L. Norman, U. Muller-Eberhard, and E. F. Johnson, *Biochem. Biophys. Res. Commun.*, 1979, **89**, 195.
132 M. J. Fasco, L. J. Piper, and L. S. Kaminsky, *Biochem. Pharmacol.*, 1979, **28**, 97.
133 N. Ozawa, S. Yoshihara, and H. Yoshimura, *J. Pharmacobio-Dynam.*, 1979, **2**, 309.
134 E. F. Johnson, G. E. Schwab, and U. Muller-Eberhard, *Mol. Pharmacol.*, 1979, **15**, 708.
135 M. Agosin, A. Morello, R. White, Y. Repetto, and J. Pedemonte, *J. Biol. Chem.*, 1979, **254**, 9915.
136 D. R. Koop and M. J. Coon, *Biochem. Biophys. Res. Commun.*, 1979, **91**, 1075.
137 M. Warner and A. H. Neims, *Drug. Metab. Dispos.*, 1979, **7**, 188.
138 K. T. Shwerick and A. H. Neims, *Drug. Metab. Dispos.*, 1979, **7**, 290.
139 D. W. Nebert and N. M. Jensen, *Crit. Rev. Biochem.*, 1979 **6**, 401.
140 M. C. Lechner, M. T. Freire, and B. Groner, *Biochem. Biophys. Res. Commun.*, 1979, **90**, 531.
141 R. Dubois and M. R. Waterman, *Biochem. Biophys. Res. Commun.*, 1979, **90**, 150.
142 R. A. Colbert, E. Bresnick, W. Levin, D. E. Ryan, and P. E. Thomas, *Biochem. Biophys. Res. Commun.*, 1979, **91**, 886.

barbital-induced cytochrome *P*-450, Bhat and Padmanaban[143] have shown that both *in vivo* and in a homologous system *in vitro*, cytochrome *P*-450 is synthesized as the native protein and not as a precursor. The same result was obtained with isolated mRNA in a heterologous cell-free system; treatment with the inducer increased the messenger activity that could be measured in the heterologous system. Many of the studies of the biosynthesis of the cytochrome rely on immunochemical methods of detection, and Thomas *et al.*[144] have described the preparation of monospecific antibodies against phenobarbital- and 3-methylcholanthrene-induced enzymes. One interesting difference between these two inducers is that whereas the former can act without prior metabolism, for the latter it is a metabolite that initiates the synthesis of protein.[145]

The site of synthesis and the subsequent fate of cytochrome *P*-450 have been investigated by Rabin and co-workers.[146-148] They have shown that there is no preference for the biosynthesis of the enzyme on either the heavy or the light rough endoplasmic reticulum, as measured by the incorporation of radioactive amino-acids into enzyme and total protein.[146] Enzyme that has been synthesized by the rough endoplasmic reticulum is to a substantial extent (40%) resistant to proteases prior to treatment with detergent,[147] and the authors have interpreted this in terms of translocation of the protein across the membrane during synthesis. This work was extended in a further publication[148] to reveal that newly synthesized enzyme is translocated into the intracisternal space and then passed through other components of the endoplasmic reticulum before insertion at its ultimate membrane location. Fujii-Kuriyama *et al.*[149] have shown that the main site of synthesis is on tightly bound ribosomes, whence it is incorporated into the rough endoplasmic reticulum before being evenly distributed between the rough and the smooth endoplasmic reticula.

Liver microsomal cytochrome *P*-450 exists *in vivo* as a tightly bound membrane-associated protein, and it may be isolated by treatment with detergent. An understanding of the chemistry of the physiological processes that are catalysed by the enzyme therefore requires consideration of the effects of lipids on the detergent-solubilized form. The association of the enzyme with its reductase in a variety of partially or completely disassembled states and the behaviour of the reconstituted system led Blancke *et al.*[151] to conclude that the re-assembly of the cytochrome/reductase system in the presence of lipids is a thermodynamically favoured reaction. In a similar fashion, the gel filtration of detergent-

[143] K. S. Bhat and G. Padmanaban, *Arch. Biochem. Biophys.*, 1979, **198**, 110.
[144] P. E. Thomas, D. Korzeniowski, D. Ryan, and W. Levin, *Arch. Biochem. Biophys.*, 1979, **192**, 524.
[145] I. B. Tsyrlov, N. E. Polyakova, O. A. Gromova, N. B. Rivkino, and V. L. Lyakhovich, *Biochem. Pharmacol.*, 1979, **28**, 1473.
[146] M. B. Cooper, J. A. Craft, M. R. Estall, and B. R. Rabin, *Biochem. Biophys. Res. Commun.*, 1979, **91**, 95.
[147] J. A. Craft, M. B. Cooper, M. R. Estall, and B. R. Rabin, *FEBS Lett.*, 1979, **98**, 403.
[148] J. A. Craft, M. B. Cooper, M. R. Estall, D. E. Rees, and B. R. Rabin, *Eur. J. Biochem.*, 1979, **96**, 379.
[149] Y. Fujii-Kuriyama, M. Negishi, R. Mikawa, and Y. Tashiro, *J. Cell Biol.*, 1979, **81**, 510.
[150] L. M. Vainer, V. I. Yamkovi, I. B. Tsyrlov, L. N. Pospelova, and V. V. Lyakhovich, *Biokhimiya*, 1979, **44**, 634.
[151] J. Blalcke, K. Rohde, J. Behlke, G.-R. Jänig, D. Pfeil, and K. Ruckpaul, *Acta Biol. Med. Ger.*, 1979, **38**, 399.

solubilized microsomes on Sephadex LH-20 results in the formation of 'microsome-like particles' that have 80% of the benzopyrene hydroxylase activity of intact microsomes.[152] The authors attribute this to the self-assembly of protein and lipid in the course of the filtration. Difference spectroscopy has been used to study the interaction of purified cytochrome *P*-450 with phospholipids directly.[153]

Physical and catalytic properties of cytochrome *P*-450 incorporated into phospholipid have been described[154] and compared to those of the enzyme in microsomes. Studies of highly purified cytochrome *P*-450 adsorbed on to polyamide may also be relevant to the lipid-associated enzyme.[155] The effect of phospholipids on the cumyl-hydroperoxide-supported *N*-demethylase activity of cytochrome *P*-450 has also been investigated.[156] Binding of the 'type I' substrates to soluble enzyme is worse than to lipid-associated forms, and there are differences in the *N*-demethylase and hydroxylase activities.[157] Although the properties of reconstituted cytochrome *P*-450 are closer to those of the microsomal system than the soluble enzyme, there are definite differences that are dependent upon the amount of lipid. In vesicles (excess lipid) and microsomes, hydroxylase kinetics were sigmoidal, whilst in reconstituted systems with smaller quantities of lipid they were hyperbolic.[158] This may be due to a protein–protein co-operativity in the former case. In addition to studies in which lipid was added to soluble cytochrome *P*-450, an investigation of the effects of lipid depletion on mono-oxygenase activity in lyophilized microsomes has been carried out.[159] Bachmanova *et al.*[160] have compared the inhibitory effect of copper–tyrosine complexes on the activity of microsomal, reconstituted, and soluble preparations.

The hydrodynamic properties of cytochrome *P*-450$_{LM}$ in the presence of the non-ionic detergent Triton N-101 have been measured.[161] Analysis of the results shows that the sedimentation behaviour depends upon the ratio of detergent to enzyme, and that at least fifteen detergent molecules must bind before changes are elicited. Richter *et al.* have measured the rotational diffusion of cytochrome *P*-450 in the microsomal membrane,[162] and the orientation of the haem group with respect to the membrane has been determined by Rich *et al.*[163] These authors measured the e.p.r. spectra of this haem protein and of cytochrome b_5 in oriented

152 V. M. Mishin, A. Yu. Grishanova, and V. V. Lyakhovich, *FEBS Lett.*, 1979, **104**, 300.
153 A. A. Akhrem, P. A. Kiselev, M. A. Kisel, S. A. Usanov, and D. I. Metelitsa, *Dokl. Akad. Nauk SSSR*, 1979, **245**, 234.
154 B. Boesterling, A. Stier, A. G. Hildebrandt, J. H. Dawson, and J. R. Trudell, *Mol. Pharmacol.*, 1979, **16**, 332.
155 A. N. Eremin, S. A. Usanov, D. I. Metelitsa, and A. A. Akhrem, *Vestsi Akad. Navuk B. SSR, Ser. Khim. Navuk*, 1979, No. 3, p. 86.
156 A. A. Akhrem, M. A. Kisel, S. A. Usanov, and D. I. Metelitsa, *Dokl. Akad. Nauk B. SSR*, 1979, **23**, 276.
157 A. Archakov, G. Bachmanova, I. Kurzina, and D. Mengazetdinov, *Acta Biol. Med. Ger.*, 1979, **38**, 299.
158 M. Ingleman-Sundberg, I. Johansson, and A. Hansson, *Acta Biol. Med. Ger.*, 1979, **38**, 379.
159 M. Vore and E. Soliven, *Biochem. Pharmacol.*, 1979, **28**, 3659.
160 G. I. Bachmanova, I. I. Karuzina, D. A. Mengazetdinov, A. I. Archakov, I. V. Galushchenko, and V. V. Obraztsov, *Biokhimiya*, 1979, **44**, 1361.
161 J. H. Behlke, G.-R. Jänig, and D. Pfeil, *Acta Biol. Med. Ger.*, 1979, **38**, 389.
162 C. Richter, K. H. Winterhalter, and R. J. Cherry, *FEBS Lett.*, 1979, **102**, 151.
163 P. R Rich, D. M. Tiede, and W. D. Bonner, Jnr., *Biochim. Biophys. Acta*, 1979, **546**, 307.

multilayers, and concluded that, for the mono-oxygenase, the plane of the haem moiety is parallel to the membrane. They also investigated the relaxation behaviour of low-spin cytochrome *P*-450 in the presence of both reduced and oxidized cytochrome b_5; the lack of any difference is consistent with there being no appreciable interaction between these two proteins in the membrane.[163] The membrane-binding properties also tend to result in self-aggregation of cytochrome *P*-450,[164] and analytical centrifugation has been used to study the effect of substrates, phospholipid, and reductase on this aggregation.[165] Thermodynamic data are consistent with hydrophobic interactions mediating the association; the activity of the enzyme appears to be relatively insensitive to the state of aggregation.[165]

Two papers[166, 167] have appeared which describe attempts at isoelectric focusing of hepatic cytochrome *P*-450. Although the enzyme can be electrofocused by using a mixture of non-ionic and zwitterionic detergents, this treatment leads to loss of some of the haem and also conversion into the *P*-420 form.[166] Guengerich has shown that cytochrome *P*-450 that is judged as homogenous by several criteria gives a multiple banding pattern when electrofocused,[167] and suggests that this may be due to the binding of carrier ampholytes.

A key event in the metabolic activation of oxygen is the provision of reducing equivalents, and for the hepatic enzyme these are supplied by NADPH *via* a flavoprotein, *i.e.* cytochrome *P*-450 reductase.[168] Blanck *et al.*[169] have investigated the rates of reduction of cytochrome *P*-450 in microsomal, reconstituted, and solubilized preparations, and they have shown that it is biphasic in the first two whereas the fast phase is lost in the latter. Biphasic kinetics have also been demonstrated by Coon and co-workers,[170, 171] who have shown that the rate-constants for the two phases are independent of the NADPH concentrations over a 5000-fold range, and of reductase to mono-oxygenase ratio over a 10-fold range.[170] In a second publication they interpret the observed kinetics in terms of an enzyme–enzyme complex that is formed prior to the rate-determining step, and the reduction of the cytochrome by the semiquinone form of the flavin being dependent on the presence of NADPH.[171] These authors have attributed the biphasic kinetics to a property of the reductase rather than of the cytochrome. Biphasic kinetics have also been observed by Taniguchi *et al.*,[172] working with purified enzymes incorporated into phosphatidylcholine vesicles; however, they interpret dependencies on the reductase-to-cytochrome and protein-to-lipid ratios in terms of a mechanism involving random planar diffusion of the proteins in the vesicle membrane. A model system for the observed

[164] S. L. Smith, W. E. Bollenbacher, D. Y. Cooper, H. Schleyer, J. J. Wielgus, and L. I. Gilbert, *Mol. Cell. Endocrinol.*, 1979, **15**, 111.

[165] F. P. Guengerich and L. A. Holladay, *Biochemistry*, 1979, **18**, 5442.

[166] L. M. Hjelmeland, D. W. Nebert, and A. Chrambach, *Anal. Biochem.*, 1979, **95**, 201.

[167] F. P. Guengerich, *Biochim. Biophys. Acta*, 1979, **577**, 132.

[168] M. J. Coon and R. E. White, in 'Metal Ion Activation of Dioxygen', ed. T. G. Spiro, Wiley-Interscience, New York, 1980, p. 73.

[169] J. Blanck, J. H. Behlke, G.-R. Jänig, D. Pfeil, and K. Ruckpaul, *Acta Biol. Med. Ger.*, 1979, **38**, 11.

[170] K. P. Vatsis, D. D. Oprian, and M. J. Coon, *Acta Biol. Med. Ger.*, 1979, **38**, 459.

[171] D. D. Oprian, K. P. Vatsis, and M. J. Coon, *J. Biol. Chem.*, 1979, **254**, 8895.

[172] H. Taniguchi, Y. Imai, T. Iyanagi, and R. Sato, *Biochim. Biophys. Acta*, 1979, **550**, 341.

kinetic behaviour in the reduction of solubilized cytochrome P-450 has been described.[173] In this case, neither the formation of a complex between reductase and cytochrome nor preceding diffusional steps are rate-limiting. Jänig *et al.*[174] have described the reductase activity of immobilized reductase/cytochrome P-450, in addition to the benzphetamine demethylase activity. Miwa *et al.*[175] have analysed the same reaction in a reconstituted system, with purified components, and concluded that a catalytically active complex of reductase and cytochrome is involved. They have calculated a dissociation constant of 50 nmol l^{-1} for this complex.

In the preceding work, kinetic evidence has been presented for the binding of reductase to cytochrome. Black *et al.*[176] have possibly isolated the domain of the reductase that is involved in the formation of a complex. These workers treated detergent-solubilized reductase with trypsin and showed the loss of a hydrophobic peptide of mol. wt 6100 from the N-terminus of the enzyme. This peptide is a potent inhibitor of cytochrome P-450$_{LM2}$ (80% at 3 μg ml^{-1}) in a reconstituted system.

As a contrast to the analysis presented above, Archakov *et al.*[177] claim that aerobically, though not anaerobically, electron flow from flavoprotein to cytochrome is mediated by the superoxide radical anion. This group has also attempted the electrochemical reduction of microsomal cytochrome P-450 but could only detect the reduced P-420 form at a potential of -1.5 V.[178]

An important factor that governs the binding and reactivity of oxygen in the cytochrome P-450 type of mono-oxygenase is the spin state of the iron. One method of probing the spin states is to measure the temperature dependence of optical spectra.[179,180] Sligar *et al.*[179] have studied the effect of binding a substrate on redox potential and on temperature-dependent spin-equilibria of the partially purified enzyme, and they concluded that the redox potential of the iron is controlled by its spin state. They discuss their model in the context of the control of electron flow *in vivo*. The same group have also used the temperature dependence of the optical spectra to measure the thermodynamic parameters of the high-spin–low-spin equilibrium in microsomal and solubilized preparations.[180] After solubilization, the enzyme is essentially low-spin, as compared to approximately equal quantities of the forms of both spin states in the microsomal enzyme. By way of a contrast, the 3-methylcholanthrene-induced enzyme in rabbits has optical and e.p.r. spectra that are consistent with a temperature-independent spin equilibrium.[181] However, here again about half (55%) is

[173] J. Blanck, K. Rohde, and K. Ruckpaul, *Acta Biol. Med. Ger.*, 1979, **38**, 23.
[174] G.-R. Jänig, D. Pfeil, and K. Ruckpaul, *Acta Biol. Med. Ger.*, 1979, **38**, 409.
[175] G. T. Miwa, S. B. West, M.-T. Huang, and A. Y. H. Lu, *J. Biol. Chem.*, 1979, **254**, 5695.
[176] S. D. Black, J. S. French, C. H. Williams, Jnr., and M. J. Coon, *Biochem. Biophys. Res. Commun.*, 1979, **91**, 1528.
[177] A. I. Archakov, G. I. Bachmanova, M. V. Izotov, and G. P. Kuznetsova, *Biokhimiya*, 1979, **44**, 2026.
[178] B. A. Kuznetsov, N. M. Mestechkina, M. V. Izotov, I. I. Karuzina, A. V. Karyakin, and A. I. Archakov, *Biokhimiya*, 1979, **44**, 1569.
[179] S. G. Sligar, D. L. Cinti, G. G. Gibson, and J. B. Schenkman, *Biochem. Biophys. Res. Commun.*, 1979, **90**, 925.
[180] D. L. Cinti, S. G. Sligar, G. G. Gibson, and J. B. Schenkman, *Biochemistry*, 1979, **18**, 36.
[181] J. Friedrich, G. Butschak, G. Scheunig, O. Ristau, H. Rein, K. Ruckpaul, and G. Smettan, *Acta Biol. Med. Ger.*, 1979, **38**, 207.

high-spin. Further support for microsomal cytochrome *P*-450 existing as a mixture of high- and low-spin forms has been presented by Rein *et al.*[182] and by Werringloer *et al.*[183] The previous authors have also analysed the thermo-dynamics of the spin equilibrium, and concluded that the change in spin state is coupled to the breaking of about fifteen haem–protein contacts.[184] They do not invoke the loss of an axial ligand.

Binding of a substrate by liver microsomal cytochromes *P*-450 has been studied by a number of methods in attempts to elucidate the various factors involved. Šipal *et al.*[185] have correlated the 'lipophilicity' of a large number of hydro-carbons with their spectrally determined dissociation constants. When a sub-strate exhibits multiple types of binding, the determination of the dissociation constants can be difficult. However, a method has been described for the spectro-scopic determination of the binding constants;[186] those for butan-1-ol show a sex dependence in mice.[187] The factors governing the binding of the drug iopronic acid have also been analysed.[188]

An alternative approach to the investigation of binding of substrates is through nuclear relaxation measurements for protons of a solvent.[189] Analysis of the temperature- and the frequency-dependence of the spin–lattice relaxation times in a number of microsomal and solubilized preparations has allowed the authors to suggest the relative accessibilities of water to the haem iron. These correlate with substrate specificity. It is suggested that the latter is controlled by the 'tightness' of the haem crevice.[189] Magnetic resonance methods have also been used to study the binding of substrates by employing spin-labelled alkylamines and isocyanides.[190, 191] Both the room-temperature e.p.r. spectrum of the spin label and the low-temperature e.p.r. spectrum of the haem iron are consistent with there being a strong magnetic interaction between the two centres. Considerable flexibility in reactivity of the bound substrate is shown by the hydroxylation of (*R*)- and (*S*)-warfarins, where the stereo- and regio-selectivity of the reaction depends on a variety of factors.[192] Similarly, hydroxylation of the four cyclohexane derivatives shown in Figure 4 led to a variety of products.[193] In addition, the authors observed a large intramolecular isotope effect (11.9)

[182] H. Rein, O. Ristau, R. Misselwitz, E. Buder, and K. Ruckpaul, *Acta Biol. Med. Ger.*, 1979, **38**, 187.

[183] J. Werringloer, S. Kawano, and R. Estabrook, *Acta Biol. Med. Ger.*, 1979, **38**, 163.

[184] O. Ristau, H. Rein, S. Greschner, G.-R. Jänig, and K. Ruckpaul, *Acta Biol. Med. Ger.*, 1979, **38**, 177.

[185] Z. Šípal, P. Azenbacher, Z. Putz, and O. Křivanová, *Acta Biol. Med. Ger.*, 1979, **38**, 483.

[186] A. P. van den Berg, J. Noordhoek, and E. Koopman-Kool, *Biochem. Pharmacol.*, 1979, **28**, 37.

[187] A. P. van den Berg, J. Noordhoek, and E. Koopman-Kool, *Biochem. Pharmacol.*, 1979, **28**, 31.

[188] R. Maefei Facino, E. Raffaeli, and D. Pitre, *Pharmacol. Res., Commun.*, 1979, **11**, 105.

[189] S. Maričić, B. Benko, G. Butschak, G. Scheunig, G.-R. Jänig, H. Rein, and K. Ruckpaul, *Acta Biol. Med. Ger.*, 1979, **38**, 217.

[190] J. Pirrwitz, G. Lassmann, H. Rein, G.-R. Jänig, S. Pečar, and K. Ruckpaul, *Acta Biol. Med. Ger.*, 1979, **38**, 235.

[191] J. Pirrwitz, H. Rein, G. Lassmann, G.-R. Jänig, S. Pečar, and K. Ruckpaul, *FEBS Lett.*, 1979, **101**, 195.

[192] L. Kaminsky, M. J. Fasco, and F. P. Guengerich, *J. Biol. Chem.*, 1979, **254**, 9657.

[193] R. E. White, J. T. Groves, and G. A. McClusky, *Acta Biol. Med. Ger.*, 1979, **38**, 675.

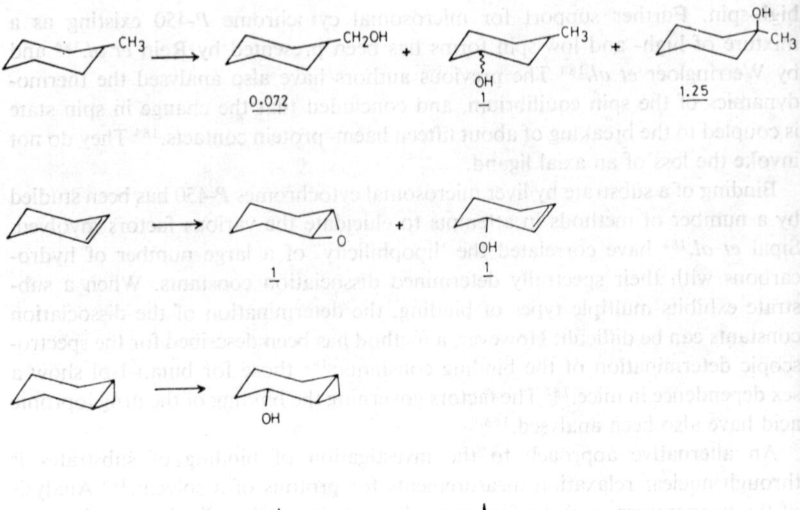

Figure 4 *Substrates and products from the action of liver microsomal cytochrome P-450_{LM2} on cyclohexane derivatives*
(Reproduced by permission from *Acta Biol. Med. Ger.*, 1979, **38**, 675)

with trinorbornane which they interpreted as reflecting initial abstraction of a hydrogen atom.

However, not all metabolic reactions that are dependent on cytochrome *P*-450 need involve the binding of a substrate. Cederbaum *et al.* claim that the oxidation of ethanol in microsomes is mediated, in part, by hydroxyl radicals that are generated by the mono-oxygenase system.[194] In addition to hydroxylation reactions, microsomal cytochromes *P*-450 will also catalyse *O*-dealkylation and epoxidation, and Mansour *et al.*[196] have investigated the metabolism of 4-methoxy-β-chlorostyrene, which is a molecule with sites available for all three reactions. Further complications in the kinetic behaviour, at least in the dealkylase reaction, are provided by the observation of biphasic Eadie–Hofstee plots; these have been interpreted as reflecting the existence of at least two kinetically different forms of the enzyme.[197] The possibility of so called 'non-equilibrium forms' of the enzyme has been probed by reducing the iron centre,

194 A. I. Cederbaum, G. Miwa, C. Cohen, and A. Y. H. Lu, *Biochem. Biophys. Res. Commun.*, 1979, **91**, 747.
195 R. Hansson and K. Wikvall, *Eur. J. Biochem.*, 1979, **93**, 419.
196 B. Mansour, V. Ullrich, and K. Pfleger, *Biochem. Pharmacol.*, 1979, **28**, 2321.
197 A. R. Boobis, M. J. Brodie, D. S. Davies, G. C. Kahn, and C. Whyte, *Br. J. Pharmacol.*, 1979, **66**, 428P.

at 77 K, with either hydrated electrons[198] or γ-rays.[199] These reduced forms have been characterized by a variety of spectroscopic methods.

Perhaps the most interesting and least understood step in the mono-oxygenase cycle is the last one, *i.e.* of hydroxylation of the substrate, and a number of ingenious methods have been applied to try and unravel it. Brown *et al.*[200] have explored the use of perfluorocarbons as oxygen reservoirs, but they discovered that, at concentrations that do not inhibit the enzyme, the capacity for oxygen is limited. Optical spectroscopic properties of the oxyferro-form of liver microsomal cytochrome *P*-450 have been measured by using fluid mixtures at sub-zero temperatures.[201] Although this form is stable under the conditions reported, it rapidly autoxidizes if the temperature or the concentration of protons is increased. Kinetic measurements at sub-zero temperatures have also shown that the dealkylation of 7-ethoxycoumarin is coupled in a single turnover of an intermediate of fully reduced enzyme and oxygen.[202]

Although the physiological action of cytochrome *P*-450 involves dioxygen, it may be driven *in vitro* by other molecules, acting as donors of oxygen atoms. One of these is iodosylbenzene, and a number of derivatives were synthesized and tested for their ability to support microsomal hydroxylation of steroids. The most efficient derivative was diacetoxyiodo-2-nitrobenzene, which was several hundred times as effective as NADPH in the hydroxylation of androstenedione.[203] The authors therefore suggest that the actual hydroxylation step is not rate-limiting. Peroxides may also support reactions that are dependent on cytochrome *P*-450, and Mohr *et al.*[204] have compared the reaction of this enzyme and of horseradish peroxidase with hydrogen peroxide. The authors argue that, although the cytochrome forms an adduct similar to compound I of the peroxidase, there is nothing analogous to compound II; and that this may be the reason for the difference in catalytic properties. In another publication the same group describe their investigation of electrochemical methods of driving the cytochrome, and they claim that electrodically generated hydrogen peroxide is more active than added hydrogen peroxide.[205]

When either cumene hydroperoxide or chloroperbenzoic acid is used as an oxygen donor, the kinetic data are consistent with their acting as one-electron acceptors.[206] Stopped-flow spectrophotometry suggests that there is an initial enzyme–oxidant complex, followed by transfer of an oxygen atom to yield the hydroxylating species.

The role of cytochrome *P*-450 in drug metabolism and in the activation of

[198] S. Greschner, R. M. Davydov, G.-R. Jänig, K. Ruckpaul, and L. R. Blumenfeld, *Acta Biol. Med. Ger.*, 1979, **38**, 443.

[199] R. M. Davydov, S. Greschner, and K. Ruckpaul, *Mol. Biol.*, 1979, **13**, 1397.

[200] N. A. Brown, K. J. Netter, and J. W. Bridges, *Biochem. Pharmacol.*, 1979, **28**, 2850.

[201] C. Bonfils, P. Debey, and P. Maurel, *Biochem. Biophys. Res. Commun.*, 1979, **88**, 1301.

[202] K. K. Andersson, P. Debey, and C. Balny, *FEBS Lett.*, 1979, **102**, 117.

[203] J.-A. Gustafsson, L. Rondahl, and J. Bergman, *Biochemistry*, 1979, **18**, 865.

[204] P. Mohr, M. Kühn, E. Wesuls, R. Renneberg, and F. Scheller, *Acta Biol. Med. Ger.*, 1979, **38**, 495.

[205] F. Scheller, R. Renneberg, W. Schwartze, G. Strnad, K. Pommerening, H.-J. Prümke, and P. Mohr, *Acta Biol. Med. Ger.*, 1979, **38**, 503.

[206] M. J. Coon, R. C. Blake, II, D. D. Oprian, and D. P. Ballou, *Acta Biol. Med. Ger.*, 1979, **38**, 449.

carcinogens has prompted much research into the inhibition or inactivation of this enzyme. Table 3 lists[207-234] some of the compounds that are known to inhibit or to inactivate cytochrome *P*-450 *in vivo* or *in vitro*. In the case of many of the compounds in the Table, the mechanism of inactivation has not been fully worked out, although it is likely that the proximate agents are metabolic products of the added compound; that is, they act as suicide substrates. This seems to have been shown in the case of cyclopropylamines.[227]

The mechanism of action of two inhibitors, *i.e.* copper(II) complexes and allyl-(isopropyl)acetamide, has been worked out in detail; they provide a fascinating contrast. It was originally thought that lipophilic copper(II) complexes, usually with amino-acids as ligands, act by catalysing the breakdown of the oxy-form of the cytochrome. However, Estabrook and co-workers[208, 209] have measured the rate of production of hydrogen peroxide as well as of the metabolism of substrate, and they concluded that the copper complexes inhibit the cytochrome *P*-450 reductase. Allyl(isopropyl)acetamide has long been known to act by degrading and removing the haem from the cytochrome, and has been used to induce experimental porphyrias in animals, a process that can be reversed by

207 V. V. Lyakhovich, I. B. Tsyrlov, N. M. Mishin, L. M. Weiner, G. Rumyantseva, S. I. Eremeko, and V. G. Budker, *Acta Biol. Med. Ger.*, 1979, **38**, 201.
208 R. W. Estabrook, S. Kawano, J. Werringloer, H. Kuthan, H. Tsuji, H. Graf, and V. Ullrich, *Acta Biol. Med. Ger.*, 1979, **38**, 423.
209 J. Werringloer, S. Kawano, N. Chacos, and R. W. Estabrook, *J. Biol. Chem.*, 1979, **254**, 11 839.
210 I. I. Karuzina, G. I. Bachmanova, G. P. Kuznetsova, M. V. Izotov, and A. I. Archakov, *Biokhimiya*, 1979, **44**, 1796.
211 V. P. Kurchenko, S. A. Usanov, and D. I. Metelitsa, *React. Kinet. Catal. Lett.*, 1979, **12**, 31.
212 H. G. Shertzer and G. S. Duthu, *Biochem. Pharmacol.*, 1979, **28**, 873.
213 A. Tunek and F. Oesch, *Biochem. Pharmacol.*, 1979, **28**, 3425.
214 I. B. Tsylov and V. V. Lyakhovich, *Biokhimiya*, 1979, **44**, 1172.
215 J. L. Kraus and J. J. Yaounac, *Pharmacol. Res. Commun.*, 1979, **11**, 597.
216 G. K. Gourlay and B. H. Stock, *Biochem. Pharmacol.*, 1979, **28**, 1421.
217 M. A. Correia, G. C. Farrell, R. Schmid, P. R. Ortiz de Montellano, G. S. Yost, and B. A. Mico, *J. Biol Chem.*, 1979, **254**, 15.
218 G. C. Farrell, R. Schmid, K. L. Kunze, and P. R. Ortiz de Montellano, *Biochem. Biophys. Res. Commun.*, 1979, **89**, 456.
219 P. R. Ortiz de Montellano, G. S. Yost, B. A. Mico, S. E. Dinizo, M. A. Correia, and H. Kumbara, *Arch. Biochem. Biophys.*, 1979, **197**, 524.
220 F. de Matteis and L. Cantoni, *Biochem. J.*, 1979, **183**, 99.
221 T. Shimada and R. Sato, *Biochem. Pharmacol.*, 1979, **28**, 1777.
222 G. R. Peterson, R. M. Hostetler, T. Lehman, and H. P. Covault, *Biochem. Pharmacol.*, 1979, **28**, 1783.
223 G. J. Mannering, *Mol. Pharmacol.*, 1979, **15**, 410.
224 D. I. Metelitsa, A. A. Akhrem, A. N. Erjomin, M. A. Kissel, and S. A. Usanov, *Acta Biol. Med. Ger.*, 1979, **38**, 511.
225 P. E. B. Reilly and D. E. Ivey, *FEBS Lett.*, 1979, **97**, 141.
226 P. R. Ortiz de Montellano, K. L. Kunze, G. S. Yost, and B. A. Mico, *Proc. Natl. Acad. Sci. USA*, 1979, **76**, 746.
227 R. P. Hanzlik, V. Kishore, and R. Tullman, *J. Med. Chem.*, 1979, **22**, 759.
228 J. M. Patel, *Toxicol. Appl. Pharmacol.*, 1979, **48**, 337.
229 M. Dickins, C. R. Elcombe, S. J. Moloney, K. J. Netter, and J. W. Bridges, *Biochem. Pharmacol.*, 1979, **28**, 231.
230 T. R. Fennell, M. Dickins, and J. W. Bridges, *Biochem. Pharmacol.*, 1979, **28**, 1427.
231 M. Hirata, B. Lindeke, and S. Orrenius, *Biochem. Pharmacol.*, 1979, **28**, 479.
232 B. Lindke, U. Paulsen, and E. Anderson, *Biochem. Pharmacol.*, 1979, **28**, 3629.
233 A. R. Dahl and E. Hodgson, *Chem.-Biol. Interact.*, 1979, **27**, 163.
234 S. T. Omaye and J. D. Turnbull, *Biochem. Pharmacol.*, 1979, **28**, 3651.

Table 3 *Compounds that are known to inhibit or to inactivate cytochrome P-450 activity either* in vivo *or* in vitro

Inhibitor/Inactivator	Reference
Copper(II) complexes	207—211
Nitrite ion	212
Metyrapone	213, 214
SKF 525 A	215
Dihydroxyacetophenone	216
Allyl(isopropyl)acetamide	217—220
1,1,1-Trichloropropene 2,3-oxide	221
Propoxyphene	222
Fatty-acyl-CoA	223
Cumene hydroperoxide	224
Chloramphenicol	225
Ethynyl sterols	226
Cyclopropylamines	227
Phthalaldehyde	228
Isosafrole metabolites	229, 230
Norbenzphetamine metabolites	231
Methamphetamine metabolites	232
Dioxolans	233
Ascorbate deficiency	234
Spironolactone	267

administering exogenous haemin.[218] Ortiz de Montellano *et al.*[219] have shown that the degraded haem, the so-called 'green pigment', is a covalent adduct of the drug with the porphyrin, probably an *N*-alkylated derivative,[220] and they have postulated a mechanism of action that involves prior metabolism by cytochrome *P*-450. A similar process may also be involved in the mode of action of ethynyl sterols.[226]

Carbon monoxide is also an inhibitor of cytochrome *P*-450, and it was the unusual wavelength of the Soret band of the adduct that gave this protein its name. Böhm *et al.*[235] have presented data on the infrared spectra of various cytochromes *P*-450 and have shown that the appearance of the bands reflects both the source and the inducer of the enzyme. Electronic spectra of the complexes of carbon monoxide and of oxygen with ferrous cytochrome *P*-450 and with haemoglobin suggest that, in the former, the spectral properties are very sensitive to the distance between iron and the axial ligand.[236] On the basis of calculations, it is claimed that, on conversion into the inactive *P*-420 form, this distance increases by 0.2 Å.

Although the bulk of hepatic cytochrome *P*-450 is found in the microsomal fraction, quantities are also detected in other organelles. Rat liver nuclei contain approximately 3% of the microsomal content of this enzyme, with a relative enrichment of the nuclear membrane.[237] Thomas *et al.*[238] have further shown

[235] S. Böhm, H. Rein, G. Butschak, G. Scheunig, B. Billwitz, and K. Ruckpaul, *Acta Biol. Med. Ger.*, 1979, **38**, 249.

[236] Chr. Jung, J. Friederich, and O. Ristau, *Acta Biol. Med. Ger.*, 1979, **38**, 363.

[237] H. Mukhtar, T. H. Elmamlouk, and J. R. Bend, *Arch. Biochem. Biophys.*, 1979, **192**, 10.

[238] P. E. Thomas, D. Korzeniowski, E. Bresnick, W. A. Bornstein, C. B. Kasper, W. E. Fahl, C. R. Jefcoate, and W. Levin, *Arch. Biochem. Biophys.*, 1979, **192**, 22.

that the nuclear mono-oxygenase is immunochemically identical to the microsomal cytochrome *P*-448.[239] A steroid hydroxylase that appears to be similar to the adrenal enzymes can be isolated from liver mitochondria (see below). This steroid *C*-26-hydroxylase requires an iron–sulphur protein for activity, and the turnover of the solubilized system is higher (5—15 times) than for intact mitochondria,[240] suggesting that there is rate-limiting transport of steroid into the mitochondria. Vitamin D_3 25-hydroxylase activity is also exhibited by a *P*-450-type cytochrome that has been isolated from rat liver mitochondria in a reconstituted system with bovine adrenal ferredoxin and ferredoxin reductase.[241]

In addition to studies *in vivo* with whole animals and investigations *in vitro* employing microsomal, solubilized, or purified preparations, results have been reported when cultured hepatocytes or other cell cultures were used. Treatment of hepatocytes with either phenobarbital, *β*-naphthoflavone, or 3-methylcholanthrene certainly induces cytochrome *P*-450, and, as in animals, different forms are induced by different materials.[242] However, Fahl *et al.*[243] have concluded that the control of the synthesis of this mono-oxygenase is substantially different in cultured cells compared to intact rat livers. Differences also appear to occur in the mechanism of demethylation of aminopyrine, as the kinetics in hepatocytes appear to describe two separate reactions, as compared to only one *in vivo*.[244] The factors that affect the expression of mono-oxygenase activity in cultured cells are still being investigated, and it has been shown that, although insulin is not required for induction of the enzyme, both dexamethasone and tri-iodothyronine resulted in an increased inducibility.[245] It has also been shown that the decrease in enzyme levels in hepatocyte cultures is related to ascorbate deficiency.[246] In addition to hepatocytes, fibroblast cultures have been used to study the metabolism of benzo[*a*]pyrene.[247]

Adrenal Cytochromes P-*450*. In contrast to the liver microsomal cytochrome *P*-450, the mono-oxygenase system that is located in the mitochondria of the adrenal gland is largely involved with the metabolism of endogenous compounds, largely steroids.[94, 248]

Akhrem *et al.*[249] have described a method for isolating the three components of the cholesterol hydroxylase system, making extensive use of affinity chromatography. They have also described some of the basic physical and chemical properties of their preparation. The same group have also used a similar approach to

239 E. Bresnick, D. Boraker, B. Hassuk, W. Levin, and P. E. Thomas, *Mol. Pharmacol.*, 1979, **16**, 324.
240 J. I. Pederson, I. Björkhem, and J. Gustafsson, *J. Biol. Chem.*, 1979, **254**, 6464.
241 J. I. Pederson, I. Holmberg, and I. Björkhem, *FEBS Lett.*, 1979, **98**, 394.
242 F. R. Althaus, J. F. Sinclair, P. Sinclair, and U. A. Meyer, *J. Biol. Chem.*, 1979, **254**, 2148.
243 W. E. Fahl, G. Michalopoulos, G. L. Sattler, C. R. Jefcoate, and H. C. Pitot, *Arch. Biochem. Biophys.*, 1979, **192**, 61.
244 D. J. Stewart and T. Inaba, *Biochem. Pharmacol.*, 1979, **28**, 461.
245 J. F. Sinclair, P. R. Sinclair, and H. L. Bonkowsky, *Biochem. Biophys. Res. Commun.*, 1979, **86**, 710.
246 D. M. Bissell and P. S. Guzelian, *Arch. Biochem. Biophys.*, 1979, **192**, 569.
247 E. B. Gehly, W. E. Fahl, C. R. Jefcoate, and C. Heidelberger, *J. Biol. Chem.*, 1979, **254**, 5041.
248 E. R. Simpson, *Mol Cell. Endocrinol.*, 1979, **13**, 213.
249 A. A. Akhrem, V. N. Lapko, A. G. Lapko, V. M. Shumatov, and V. L. Chashchin, *Acta Biol. Med. Ger.*, 1979, **38**, 257.

isolate the 11β-hydroxylase (EC 1.14.15.4) by affinity chromatography on adreno-doxin-Sepharose.[250] Purification of the 11β-hydroxylase is also described by Rydström *et al.*,[251] who show that the active species is a monomer, of relative molecular mass 48 000, provided that sufficient Triton X-100 is present.

A structural analysis of the enzyme from adrenal mitochondria that cleaves the side-chain of cholesterol reveals that it is a single polypeptide chain[252] that appears to consist of two domains.[253] Interestingly, if these domains are cleaved by mild trypsinolysis, then the larger fragment both complexes to adrenodoxin and is enzymatically active in a reconstituted system.[253]

One similarity that the adrenal cytochromes *P*-450 do share with the liver enzymes is that they are bound to membranes, and it has been shown that lipids that are bound to the solubilized enzyme are remarkably similar in composition to those found in the mitochondria.[254] The location of the enzymes on the inner mitochondrial membrane is thought to be on the matrix side, as both the activity for cleavage of the side-chain of cholesterol and the 11β-hydroxylase activity are more susceptible to the influence of trypsin in disrupted as compared to intact membranes.[255] Further evidence for the membrane-binding nature for the enzyme that cleaves side-chains comes from the observation that it will spontaneously associate with lipid vesicles that contain cholesterol.[256] The same is not true for adrenodoxin or for adrenodoxin reductase, and spectrophoto-metric and activity titrations support the idea that the formation of tertiary and quaternary complexes is not necessary for activity of the enzyme.

Binding of adrenodoxin to the enzyme is claimed to depend upon the presence of both cholesterol and detergent (Emulgen 913), and the latter also substantially enhances the activity of the enzyme.[257] In the same paper it is shown that the addition of detergent to the cholesterol-bound enzyme results in a conversion from high-spin to low-spin which is reversed when adrenodoxin is bound. Another non-ionic detergent, Triton X-100, also stimulates the cleavage of the side-chain. However, for a series of non-ionic detergents, the degree of stimula-tion did not correlate with either critical micellar concentration or hydrophil–lipophil balance.[258] Incorporation of highly purified, soluble, side-chain-cleavage cytochrome *P*-450 into synthetic lipid vesicles increased both the apparent V_{max} and the stability of the enzyme, and induced a corresponding change in spin state (from high to low).[259] The vesicular enzyme is freely accessible, as it reacts

[250] A. A. Akhrem, S. P. Martsev, and V. L. Chashchin, *Bioorg. Khim.*, 1979, **5**, 786.
[251] J. Rydström, M. Ingleman-Sundberg, J. Montelius, and J.-Å. Gustafsson, *Acta Biol. Med. Ger.*, 1979, **38**, 275.
[252] A. A. Akhrem, V. N. Lapko, A. G. Lapko, V. M. Shkumatov, and V. L. Chashchin, *Bioorg. Khim.*, 1979, **5**, 1201.
[253] A. A. Akhrem, V. I. Vasilevski, V. M. Shkumatov, and V. L. Chashchin, *Bioorg. Khim.*, 1979, **5**, 789.
[254] P. F. Hall, M. Watanuki, J. Degroot, and G. Rouser, *Lipids*, 1979, **14**, 148.
[255] P. F. Churchill and T. Kimura, *J. Biol. Chem.*, 1979, **254**, 10 443.
[256] D. W. Seybert, J. R. Lancaster, Jr., J. D. Lambeth, and H. Kamin, *J. Biol. Chem.*, 1979, **254**, 12 088.
[257] T. Kido, M. Arakawa, and T. Kimura, *J. Biol. Chem.*, 1979, **254**, 8377.
[258] S. Nakajin, Y. Ishii, M. Shinoda, and M. Shikita, *Biochem. Biophys. Res. Commun.*, 1979, **87**, 524.
[259] P. F. Hall, M. Watanuki, and B. A. Hamkalo, *J. Biol. Chem.*, 1979, **254**, 547.

with dithionite, cholesterol, adrenodoxin, and antibody when these materials are added to the aqueous phase.

There is considerable kinetic evidence supporting the view that adrenodoxin forms a complex with the enzyme that cleaves side-chains, and this is supported by two independent experiments.[260, 261] Miura *et al.*[260] have spin-labelled adrenodoxin and examined its e.p.r. spectrum at room temperature in the presence and absence of the haemoprotein. At low ionic strength, an 'immobilized' component is observed in the presence of cytochrome *P*-450, and titration experiments reveal that it forms a 1:1 complex. The same enzyme also binds tightly to adrenodoxin-Sepharose.[261]

The side-chain-cleavage and 11β-hydroxylase cytochromes *P*-450 have quite distinct enzymatic activities; they also appear to possess different environments of the haem moiety, at least as reflected by their e.p.r. spectra.[262] Ligand-field parameters for high- and low-spin ferric forms have been calculated for both enzymes from the *g* values. Electron paramagnetic resonance, in addition to optical spectroscopy and equilibrium dialysis, has been used to study the binding of various steroids to the side-chain-cleavage enzyme.[263] As isolated, the enzyme contains two moles of bound cholesterol per mole of haem; by using different freezing techniques, the authors have shown that, unless the changes in spin-state occur very rapidly (in < 5 ms), the low-temperature e.p.r. spectra reflect the spin state at room temperature. In contrast to the conclusions presented in reference 174, for the enzyme associated with vesicles, Kido and Kimura[264] claim spectrophotometric evidence for the formation of binary and tertiary complexes of the three proteins of the side-chain-cleavage system only in the presence of both substrate (cholesterol) and lipid (or detergent). Spectroscopic dissociation constants are presented and the authors discuss their results in terms of the control of steroidogenesis in adrenocortical mitochondria.

An alternative approach to reconstitution has been adopted by Narasimhulu,[265] who has depleted adrenal microsomal membranes of lipid and observed a decrease in the dissociation constant of the substrate, implying that, *in vivo*, the membrane may control the binding of substrate, perhaps by affecting the conformation of the enzyme. Interestingly, it has been observed that the mitochondrial activities of cytochromes *P*-450 increase in sodium-depleted rats, and spectrophotometric measurements indicate that this may be due to enhanced binding of substrate.[266]

The last compound shown in Table 3 destroys the steroid 17α-hydroxylase (EC 1.14.99.9) activity in adrenal mitochondria by acting as a suicide substrate,[267] and a sulphur atom at the 7α-position of spironolactone is essential for this process. Finally in this section, the kinetics of recombination of carbon monoxide

260 R. Miura, T. Sugiyama, and T. Yamano, *J. Biochem.* (*Tokyo*), 1979, **85**, 1107.
261 A. A. Akhrem, V. M. Shkumatov, and V. L. Chashchin, *Dokl. Akad. Nauk SSSR*, 1979, **245**, 1490.
262 S. Kominami, H. Ochi, and S. Takemori, *Biochim. Biophys. Acta*, 1979, **577**, 170.
263 N. R. Orme-Johnson, D. R. Light, R. W. White-Stevens, and W. H. Orme-Johnson, *J. Biol. Chem.*, 1979, **254**, 2103.
264 T. Kido and T. Kimura, *J. Biol. Chem.*, 1979, **254**, 11 806.
265 S. Narasimhulu, *Biochim. Biophys. Acta*, 1979, **556**, 457.
266 R. E. Kramer, S. Gallant, and A. C. Brownie, *J. Biol. Chem.*, 1979, **254**, 3953.
267 R. H. Menard, T. M. Guenther, H. Kon, and J. R. Gillette, *J. Biol. Chem.*, 1979, **254**, 1726.

with the enzyme that cleaves side-chains have been shown to be a multi-phasic or both high- and low-spin forms.[268]

Cytochromes P-450 from Other (Non-Bacterial) Sources. No other, non-bacterial, cytochrome P-450 has been studied to the extent of either the liver or adrenal enzymes. This section summarizes papers published on the enzymes from other sources.

Two forms of pulmonary microsomal cytochrome P-450 have been isolated, and shown to produce different products of hydroxylation of benzo[a]pyrene and to differ in their susceptibility to inhibition by α-naphthoflavone.[269] These two pulmonary enzymes are also immunologically distinct, although one of the isoenzymes is apparently identical to the phenobarbital-induced liver enzyme.[270] In perfused rat lung, the rate of demethylation of p-nitroanisole is dependent upon the oxygen tension, and it is reduced by 50% when the oxygen content of the gas is 0.3 mmHg.[271] 4-Methylbenzaldehyde acts as a suicide substrate for the pulmonary enzyme, inhibiting it by 50%, with concomitant loss of the haem. Interestingly, however, one product of the metabolism of 4-methylbenzaldehyde also protects the enzyme.[272]

A cytochrome P-450 has been isolated from rabbit intestinal mucosa and shown to hydroxylate fatty acids at the ω- and (ω − 1)-positions.[273] A comparison has also been carried out between intestinal and liver microsomal mono-oxygenases, acting on a variety of substrates.[274] Preliminary evidence for the occurrence of cytochrome P-450 in rat mammary glands has been presented[275] and metabolic profiles for the hydroxylation of benzo[a]pyrene in rat and mouse ovaries have been compared.[276] Testicular steroid 17,20-lyase in rats can be inhibited by either steroids or metyrapone, and although the binding of different steroid inhibitors is mutually exclusive, metyrapone and steroid inhibitors can bind simultaneously.[277]

Spectrophotometric evidence for the presence of cytochrome P-450 in bovine brain mitochondria and the isolation of an iron–sulphur protein, from the same source, that can substitute for adrenodoxin argue for the presence of a hydroxylase system in this organ.[278] Different cytochromes P-450 appear to be located in different regions of the kidney, and can be distinguished by a variety of criteria. These isoenzymes are also different from the liver enzyme.[279] Benzo[a]pyrene

[268] A. A. Akhrem, B. M. Dzhagarov, V. Rumas, V. M. Timinski, V. M. Shkumatov, and V. L. Chashchin, *Dokl. Akad. Nauk SSSR*, 1979, **244**, 1256.
[269] C. R. Wolf, B. R. Smith, L. M. Ball, C. Serabjit-Singh, J. R. Bend, and R. M. Philpot, *J. Biol. Chem.*, 1979, **254**, 3658.
[270] C. J. Serabjit-Singh, C. R. Wolf, and R. M. Philpot, *J. Biol. Chem.*, 1979, **254**, 9901.
[271] A. N. Fisher, N. Itakura, C. Dodia, and R. G. Thurman, *J. Clin. Invest.*, 1979, **64**, 770.
[272] J. M. Patel, C. R. Wolf, and R. M. Philpot, *Biochem. Pharmacol.*, 1979, **28**, 2031.
[273] K. Ichihara, I. Yamakawa, E. Kusnose, and M. Kusnose, *J. Biochem. (Tokyo)*, 1979 **86**, 139.
[274] R. S. Shirkey, J. Chakraborty, and J. W. Bridges, *Biochem. Pharmacol.*, 1979, **28**, 2835.
[275] L. E. Rikans, D. D. Gibson, and P. B. McCay, *Biochem. Pharmacol.*, 1979, **28**, 3039.
[276] D. R. Mattison, D. M. West, and R. H. Menard, *Biochem. Pharmacol.*, 1979, **28**, 2101.
[277] G. Betz and P. Tsai, *J. Steroid Biochem.*, 1979, **10**, 393.
[278] H. Oftebro, F. C. Størmer, and J. I. Pederson, *J. Biol. Chem.*, 1979, **254**, 4331.
[279] H. J. Armbrecht, L. S. Birnbaum, T. V. Zenser, M. B. Mattammal, and B. B. Davis, *Arch. Biochem. Biophys.*, 1979, **197**, 277.

hydroxylase activity in rabbit aorta[280] and in the fish *Stenotomus versicolor* has been reported,[281] whilst cytochrome *P*-450 can be induced by manganese(II) ions, ethanol, phenobarbital, and herbicides in higher plants.[282]

Mono-oxygenases of the cytochrome *P*-450 type have been detected in yeasts, and the partial purification of an enzyme from *Candida tropicalis* has been described.[283] This protein has fatty-acid ω-hydroxylase activity when reconstituted with NADPH and the corresponding reductase. Another species of *Candida*, *C. guilliermondii*, also has an ω-hydroxylase, although in this case the substrates are C_{10}—C_{18} alkanes.[284] Benzo[*a*]pyrene hydroxylase activity has been detected in *Saccharomyces cerevisiae*.[285]

Bacterial Cytochromes P-450. The camphor 5-exohydroxylase (EC 1.14.15.1) from *Pseudomonas putida* is the best studied of the bacterial cytochromes *P*-450 and provides the bulk of the references in this section. However, a *P*-450-type cytochrome has been detected in extracts of the luminescent bacterium *Photobacterium fischeri*.[286] Another bacterial cytochrome *P*-450 is the steroid 11β-hydroxylase from *Bacillus megaterium*, and the enzyme from this source has been purified to homogeneity.[287] It is a three-component system, with the electron flow from reduced pyridine nucleotide to cytochrome being *via* a flavoprotein and an iron–sulphur protein; some of the components of the bacterial system will substitute for those of the bovine adrenal steroid hydroxylase. The same group has reported the isolation of progesterone 15β-hydroxylase from *B megaterium*[288] and has characterized the ferredoxin component. They have also compared the effectiveness of the proteins in mixed mono-oxygenase systems with adrenal or hepatic cytochromes *P*-450.

Cytochrome *P*-450 from *Pseudomonas putida* has provided a fine 'model' for investigations of the other mono-oxygenases of this type, and 1979 provided further publications on its mechanism. In the mono-oxygenation reaction, the uptake of protons occurs, and these seem to modulate the nature of the haem pocket; Sligar and Gunsalus have studied the effect of pH on the redox potential high-spin–low-spin equilibrium, and the rate of reduction by putidaredoxin.[28] They discuss the proton dependence of the free-energy profile and suggest a possible coupling of the transfer of protons and of electrons. Titrimetric and spectrophotometric determinations have revealed that uptake of protons does not occur when either dioxygen or carbon monoxide bind to the reduced cyto-

[280] J. A. Bond, C. J. Omiecinski, and M. R. Juchau, *Biochem. Pharmacol.*, 1979, **28**, 305.
[281] J. J. Stegeman and R. L. Binder, *Biochem. Pharmacol.*, 1979, **28**, 1623.
[282] D. Reichhart, J.-P. Salaün, I. Benveniste, and F. Durst, *Arch. Biochem. Biophys.*, 1979, **196**, 301.
[283] J. C. Bertrand, M. Gilewicz, H. Bazin, M. Zachek, and E. Azoulay, *FEBS Lett.*, 1979, **105**, 143.
[284] H.-G. Müller, W.-H. Schunck, P. Riege, and H. Honeck, *Acta Biol. Med. Ger.*, 1979, **38**, 345.
[285] L. F. J. Woods and A. Wiseman, *Biochem. Soc. Trans.*, 1979, **7**, 124.
[286] A. D. Ismailov, N. S. Egorov, and V. S. Danilov, *Dokl. Akad. Nauk SSSR*, 1979, **249**, 482.
[287] A. Berg, M. Ingleman-Sundberg, and J.-Å. Gustafsson, *Acta Biol. Med. Ger.*, 1979, **38**, 333.
[288] A. Berg, M. Ingleman-Sundberg, and J.-Å. Gustafsson, *J. Biol. Chem.*, 1979, **254**, 5264.
[289] S. G. Sligar and I. C. Gunsalus, *Biochemistry*, 1979, **18**, 2290.

chrome,[290] and the results are consistent with retention of the axial mercaptide ligand. Lange and Debey[291] and Lange et al.[292, 293] have used cryo-enzymological methods to provide a detailed kinetic and thermodynamic analysis of the haem crevice. A local negative electrostatic potential results in the pH in the active site being lower than that in the medium,[293] and this effect is modulated by cations, especially potassium.[291] These changes in turn affect the spin-state equilibrium, and the authors are able to account for all of their static and dynamic data by Scheme 1.[292]

$$LS \cdot H_{in}{}^{n+} \underset{}{\overset{[1]}{\rightleftharpoons}} LS + n\,H^{+}$$

$$[2] \downarrow$$

$$LS' \overset{[3]}{\rightleftharpoons} HS'$$

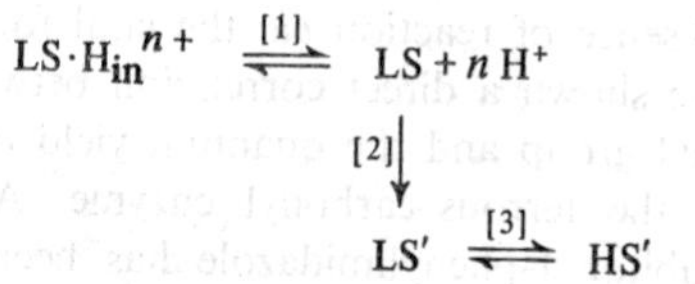

LS = low-spin ferric cytochrome P-450; LS' = a different conformation of LS
HS' = high-spin ferric cytochrome P-450
reaction [2] is rate-determining and reactions [1] and [3] are fast, reaction [3] possibly being a thermal equilibrium

Scheme 1

The nature of the axial ligands to the iron atom in the various forms of the cytochrome P-450 of $Ps.\ putida$ is still not unambiguously decided. Philson et al.[294] have measured the relaxation times for protons of water in ferric substrate-free and substrate-bound enzymes; on the basis of their analysis of temperature and frequency dependence, they speculate that water may be the sixth ligand. One approach to trying to elucidate the nature of the axial ligands is through model studies. Berzinis and Traylor[295] have compared ^{13}C and ^{1}H n.m.r. spectra of mercaptide–haem–CO complexes and suggest that two diagnostic features of axial ligation of mercaptide are (i) a large upfield shift of the ^{13}CO resonance (at 197.0 p.p.m.) compared with the shift for ligation of imidazole at 204.7 p.p.m.), and (ii) the shift to -3 p.p.m. for the protons that are α to the mercaptide ion. Optical and e.p.r. spectra of model thiolate–protoporphyrin-IX complexes with a variety of sixth ligands have led Ullrich et al.[296] to conclude that, in the enzyme, the sixth ligand is a hydroxyl oxygen, whilst Traylor and Mincey have mimicked the hydrophobic nature of the active site by incorporating their model complexes into micelles of cetyltrimethylammonium bromide.[297] Peisach et al.[298] have measured ^{14}N electron–nuclear double-resonance spectra in

[0] D. Dolphin, B. R. James, and H. C. Welborn, $Biochem.\ Biophys.\ Res.\ Commun.$, 1979, 88, 415.
[1] R. Lange and P. Debey, $Eur.\ J.\ Biochem.$, 1979, 94, 485.
[2] R. Lange, G. Hui Bon Hoa, P. Debey, and I. C. Gunsalus, $Eur.\ J.\ Biochem.$, 1979, 94, 491.
[3] R. Lange, G. Hui Bon Hoa, P. Debey, and I. C. Gunsalus, $Acta\ Biol.\ Med.\ Ger.$, 1979, 38, 143.
[4] S. B. Philson, P. G. Debrunner, P. G. Schmidt, and I. C. Gunsalus, $J.\ Biol.\ Chem.$, 1979, 254, 10 173.
[5] A. P. Berzinis and T. G. Traylor, $Biochem.\ Biophys.\ Res.\ Commun.$, 1979, 87, 229.
[6] V. Ullrich, H. Sakurai, and H. H. Ruf, $Acta\ Biol.\ Med.\ Ger.$, 1979, 38, 287.
[7] T. G. Traylor and T. Mincey, $Acta\ Biol.\ Med.\ Ger.$, 1979, 38, 351.
[8] J. Peisach, W. B. Mims, and J. L. Davis, $J.\ Biol.\ Chem.$, 1979, 254, 12 379.

cytochrome *P*-450 and in a variety of low-spin haems and haemoproteins that include nitrogen- and sulphur-containing axial ligands. On the basis of their data they suggest that an imidazole–haem–mercaptide complex is not a good model for the enzyme.

Peterson and Mock[299] have used rapid reaction techniques to show that complexes form between the cytochrome and putidaredoxin, and they have concluded that transfer of electrons within this complex is rate-limiting (33 s^{-1}).

Several physicochemical techniques have been used to probe the haem site of the enzyme; product analyses of deuteriated camphor and *peri*-cyclocamphanone are consistent with the absence of reaction *via* the enol form of the substrate.[300] Shimada *et al.*[301] have shown a direct correlation between the stretching frequency of the carbonyl group and the quantum yield of photodissociation of carbon monoxide in the ferrous–carbonyl enzyme. An azido photoaffinity derivative of the inhibitor 1-phenylimidazole has been synthesized;[302] after photolysis, it formed a singly labelled derivative in which the label was located near to the haem. Pulse radiolysis has been used to reduce the enzyme,[303] and whereas several pyridinyl radicals reduced the iron, neither CO_2^- nor O_2^- was effective. Hydrated electrons reacted at sites other than the metal ion. The reactivity of camphor with several radical species was also studied by pulse radiolysis and the results were discussed in terms of possible enzymatic reactions.[304]

Other Haemoprotein Oxidoreductases.—*Cytochrome* b_5. Cytochrome b_5 is usually isolated from liver microsomes or erythrocytes, where it occurs in greatest abundance; it has, however, recently been detected in the sarcoplasmic reticulum of rabbit slow muscle,[305] and purified from the larval midgut of the southern army worm.[306] The microsomal enzyme appears to have multiple functions, some of which may be closely involved with cytochrome *P*-450 (see p. 195); for example, Sugiyama *et al.*[307] have used cytochrome b_5 on an affinity column to isolate a cytochrome *P*-450 *p*-nitroanisole demethylase. Reconstitution requires NADH, NADH–cytochrome b_5 reductase, and cytochrome b_5 as well as the components of the cytochrome *P*-450 system. The authors suggest that cytochrome b_5 transfers an electron to cytochrome *P*-450. It has also been shown that antibodies to cytochrome b_5 inhibit both the NADH- and the NADPH-supported microsomal oxidation of 7-ethoxycoumarin and benzo[*a*] pyrene.[308] A role for the cytochrome in the elongation of fatty acids in liver

[299] J. A. Peterson and D. M. Mock, *Acta Biol. Med. Ger.*, 1979, **38**, 153.

[300] S. Banerjee and G. Dec, *Biochem. Biophys. Res. Commun.*, 1979, **88**, 833.

[301] H. Shimada, T. Iizuka, R. Ueno, and Y. Ishimura, *FEBS Lett.*, 1979, **98**, 290.

[302] R. A. Swanson and K. M. Dus, *J. Biol. Chem.*, 1979, **254**, 7238.

[303] P. Debey, E. J. Land, R. Santus, and A. J. Swallow, *Biochem. Biophys. Res. Commun.*, 1979, **86**, 953.

[304] E. J. Land and A. J. Swallow, *J. Chem. Soc., Faraday Trans. 1*, 1979, **75**, 1849.

[305] G. Salviati, R. Betto, S. Salvatori, and A. Margreth, *Biochim. Biophys. Acta*, 1979, **57**, 280.

[306] K. A. McFadden, D. L. Crankshaw, H. K. Hetnarski, and C. F. Wilkinson, *Insect Biochem.*, 1979, **9**, 301.

[307] T. Sugiyama, N. Miki, and T. Yamano, *Biochem. Biophys. Res. Commun.*, 1979, **90**, 71.

[308] M. Noshiro, N. Harada, and T. Omura, *Biochem. Biophys. Res. Commun.*, 1979, **91**, 20.

microsomes has been postulated,[309] whilst Ohba *et al.*[310] have shown that it functions in the desaturation of fatty acids in yeast microsomes. Abe and Sugita[311] have characterized a human erythrocyte cytochrome b_5 and measured the rate of its reduction by the NADH-dependent reductase and that of its reduction of methaemoglobin. The values that they obtained are consistent with the protein functioning physiologically as a reductant of methaemoglobin.

The (non-physiological) electron-transfer reaction between cytochromes b_5 and c has been investigated as a function of ionic strength.[312] A protein–protein complex, formed by complementary charge interactions between lysine residues on the cytochrome c and aspartate and glutamate residues on the cytochrome b_5, is suggested, and hydrodynamic evidence is also presented to support the occurrence of such a complex. Chemical modification studies have been used to probe the interaction of cytochrome b_5 with its reductase, and the results implicate the carboxyl groups of the aspartic acid residues 47, 48, and 52 and a haem propionate.[313] Modification with methylamine (charge destroyed) results in the same apparent V_{max} and a larger apparent K_m, whilst modification with taurine (charge displaced) leaves the apparent K_m unaffected but reduces the apparent V_{max}. In a detergent or phospholipid vesicle system, cytochrome P-450 reductase from liver microsomes can rapidly reduce cytochrome b_5 in an NADPH-dependent reaction; and in a reconstituted system cytochrome P-450 reductase is almost as effective as cytochrome b_5 reductase in the desaturation of stearoyl-CoA.[314]

Microsomal cytochrome b_5 is a membrane-bound protein, and several studies have been aimed at elucidating its interactions with lipids. Fluorescent transfer of energy from tryptophan-109 to an acceptor on a lipid vesicle has been measured with both holo-cytochrome b_5 and with the membrane-binding segment, and in both cases the results are consistent with a distance between tryptophan and the surface of approximately 21 Å.[315] The fluorescence of tryptophan has also been employed to measure the kinetics of binding of protein to phosphatidylcholine vesicles; the enhancement in fluorescence is biphasic.[316] The authors interpreted the first (fast) phase as the association of monomeric cytochrome with the vesicles; the second slower phase is due to (rate-limiting) dissociation of cytochrome octamers prior to rapid binding of monomers. In a further study of the effect of cholesterol on the binding of cytochrome b_5 to phosphatidylcholine vesicles, Tajima and Sato[317] have shown that cholesterol inhibits the binding of cytochrome when there are limiting quantities of vesicles. Cholesterol has no effect if the vesicles are present in large excess over the cytochrome.

[09] S. R. Keyes, J. A. Alfano, I. Jansson, and D. L. Cinti, *J. Biol. Chem.*, 1979, **254**, 7778.
[10] M. Ohba, R. Sato, Y. Yoshida, C. Bieglmayer, and H. Ruis, *Biochim. Biophys. Acta*, 1979, **572**, 352.
[11] K. Abe and Y. Sugita, *Eur. J. Biochem.*, 1979, **101**, 423.
[12] J. Stonehauser, J. B. Williams, and F. Millett, *Biochemistry*, 1979, **18**, 5422.
[13] H. A. Dailey and P. Strittmatter, *J. Biol. Chem.*, 1979, **254**, 5388.
[14] H. G. Enoch and P. Strittmatter, *J. Biol. Chem.*, 1979, **254**, 8976.
[15] P. J. Fleming, D. E. Koppel, A. L. Y. Lau, and P. Strittmatter, *Biochemistry*, 1979, **18**, 5458.
[16] T. L. Leto and P. W. Holloway, *J. Biol. Chem.*, 1979, **254**, 5015.
[17] S. Tajima and R. Sato, *Biochim. Biophys. Acta*, 1979, **550**, 357.

The transfer of cytochrome b_5 between unilamellar vesicles has been demonstrated, although the protein could not be transferred from microsomal membranes to artificial acceptors.[318] Release of the the haem peptide by carboxypeptidase-1 only occurred from vesicle-bound protein, and the authors concluded that the binding to artificial and natural membranes is different. A rhodium(III)-substituted form of cytochrome b_5 was prepared by Vaz *et al.*[319] and used to measure the rates of rotational diffusion in artificial membranes under a variety of conditions. Further confirmation that the *C*-terminal portion of the molecule is the membrane-binding segment has come from work with arylazido-containing lipids.[320] Incorporation into egg lecithin vesicles that contained the protein and subsequent photolysis resulted in the labelling of the *C*-terminal portion of the cytochrome. Amino-acid sequence analysis of the membrane-binding segment of rabbit cytochrome b_5 shows a substantial homology with the corresponding portion of the molecule from other species.[321] The amino-acid sequence of calf liver cytochrome b_5 has been compared with the haem domains of sulphite oxidase from chicken liver and with yeast cytochrome b_2.[322] By considering the sequence similarities and the crystal structure of cytochrome b_5, the author postulate a common fold and then go on to discuss the evolutionary implications for this group of proteins. A review of the *X*-ray work on cytochrome b_5 has appeared.[323]

Plant Peroxidases. Most of this section deals with horseradish peroxidase (HRP), which is the most readily available and best studied of all of the peroxidases (EC 1.11.1.7). A recent review on peroxidases, catalases, and chloroperoxidases has appeared.[324]

Peroxidase assays have been described, using iodide as the electron donor and following the reaction by calorimetric or potentiometric methods,[325] while a second study has compared *o*-dianisidine, guiacol, and pyrogallol as donors in the peroxidase reaction.[326] Detailed mechanistic studies of the aerobic oxidation of indoleacetic acid by two isozymes of HRP have been carried out by Nakajima and Yamazaki;[327] their mechanism involves the intermediate formation of a hydroperoxide of indoleacetic acid, and is shown in Scheme 2. Indoleacetic acid is also peroxidized by HRP (isoenzyme C) in the presence of hydrogen peroxide, and the pH dependence is consistent with tighter binding at lower pH.[328] HRP, like cytochrome *P*-450, is able to oxidize benzo[*a*]pyrene to compounds that bind to DNA, and it has been postulated that this occurs *via* one-electron oxidation by HRP/hydrogen peroxide to form a radical cation

318 H. G. Enoch, P. J. Fleming, and P. Strittmatter, *J. Biol. Chem.*, 1979, **254**, 6483.
319 W. L. C. Vaz, R. H. Austin, and H. Vogel, *Biophys. J.*, 1979, **26**, 415.
320 R. Bisson, C. Montecucco, and R. A. Capaldi, *FEBS Lett.*, 1979, **106**, 317.
321 K. Kondo, S. Tajima, R. Sato, and K. Narita, *J. Biochem. (Tokyo)*, 1979, **86**, 1119.
322 B. Guiard and F. Lederer, *J. Mol. Biol.*, 1979, **135**, 639.
323 F. S. Mathews, E. W. Czerwinski, and P. Argos, *Porphyrins*, 1979, **7**, 107.
324 W. D. Hewson and L. P. Hager, *Porphyrins*, 1979, **7**, 295.
325 J. K. Grimes and K. R. Lockhart, *Anal. Chim. Acta*, 1979, **106**, 251.
326 M. Marshall and G. W. Chism, *J. Food Sci.*, 1979, **44**, 942.
327 R. Nakajima and I. Yamazaki, *J. Biol. Chem.*, 1979, **254**, 872.
328 A. Chappet, M. Deschamps-Mudry, and D. Job, *Can. J. Bot.*, 1979, **57**, 1078.

$$\text{HRP} + \text{R'OOH} \longrightarrow \text{HRP[I]} + \text{R'OH} \qquad k = 2 \times 10^6 \, \text{l} \, \text{mol}^{-1} \, \text{s}^{-1}$$

$$\text{HRP[I]} + \text{RH} \longrightarrow \text{HRP[II]} + \text{R}^{\cdot} \qquad \text{very fast}$$

$$\text{HRP[II]} + \text{RH} \longrightarrow \text{HRP} + \text{R}^{\cdot} \qquad k = 1.2 \times 10^4 \, \text{l} \, \text{mol}^{-1} \, \text{s}^{-1}$$

$$\text{R}^{\cdot} + \text{O}_2 \longrightarrow \text{R'OO}^{\cdot} + \text{CO}_2$$

$$\text{R'OO}^{\cdot} + \text{RH} \longrightarrow \text{R'OOH} + \text{R}^{\cdot}$$

$$\text{R'OOH} \longrightarrow [\text{X}] \longrightarrow \text{3-methyleneoxindole}$$

(RH = indoleacetic acid; R'H = skatole)

Scheme 2

with the charge localized at C-6 of the hydrocarbon.[329] Aerobic oxidation of ascorbate is catalysed by HRP in a reaction that is inhibited by azide, enhanced by Cu^{2+}, Ca^{2+}, and Fe^{3+}, and unaffected by Mn^{2+}.[330] HRP will also catalyse the aerobic oxidation of phenylhydrazine, although in this case the enzyme is simultaneously inhibited,[331] the process apparently involving the generation of free radicals and their subsequent reaction with the haem.

The classical peroxidase mechanism involves initial oxidation followed by subsequent one-electron reductions,[332] as shown in Scheme 3. However, it has

$$\text{peroxidase} + \text{H}_2\text{O}_2 \longrightarrow \text{compound I}$$

$$\text{compound I} + \text{AH}_2 \longrightarrow \text{compound II} + \text{AH}^{\cdot}$$

$$\text{compound II} + \text{AH}_2 \longrightarrow \text{peroxidase} + \text{AH}^{\cdot}$$

$$2\text{AH}^{\cdot} \longrightarrow \text{A} + \text{AH}_2$$

Scheme 3

recently become apparent that not all substrates are peroxidized in this fashion, and Claiborne and Fridovich, in two papers,[333, 334] have described their investigation of the peroxidation of *o*-dianisidine. They have used chemical trapping and spectroscopic methods to demonstrate that the reaction is a direct two-electron process.[333] The transient coloration produced during turnover is assigned to a charge-transfer complex of the product quinonedi-imine with the starting material. In their second paper[334] they report changes in the spectra of the enzyme that are quite unlike compounds I and II, and suggest that the observed intermediate may be a substrate radical–enzyme complex. Rate-limiting breakdown of this complex is enhanced by the addition of nucleophilic nitrogenous bases. Aerobic oxidation of NADH by HRP is also unusual in that it shows both periodic and non-periodic behaviour when the system is continuously supplied with reactants.[335] In addition to the catalytic intermediates of peroxidase,

[329] E. G. Rogan, P. A. Katomski, R. Wroth, and E. L. Cavalieri, *J. Biol. Chem.*, 1979, **254**, 7055.

[330] N. T. Bakardjieva, *Dokl. Bolg. Akad. Nauk*, 1979, **32**, 675.

[331] O. V. Lebedeva, N. N. Ugarova, and I. V. Berezin, *Biokhimiya*, 1979, **44**, 2235.

[332] B. Chance, H. Sies, and A. Boveris, *Physiol. Rev.*, 1979, **59**, 527.

[333] A. Claiborne and I. Fridovich, *Biochemistry*, 1979, **18**, 2324.

[334] A. Claiborne and I. Fridovich, *Biochemistry*, 1979, **18**, 2329.

[335] L. F. Olsen, *Z. Naturforsch.*, *Teil. A*, 1979, **34**, 1544.

i.e. compounds I and II, there is also compound III (or oxyperoxidase), formed by the reaction of ferrous peroxidase with oxygen. γ-Irradiation of this complex and oxymyoglobin at low temperatures gives adducts with $[Fe^{II}O_2]^-$ centres, and the optical and e.p.r. spectra have been compared.[336] Nitric oxide forms an adduct with both native, iron(III), and ferro-peroxidase, and the e.p.r. spectrum of the latter has been compared with that of the former after γ-irradiation.[337] Differences have been attributed to different conformations.

Two papers on the reactivity of HRP in mixed organic–aqueous solvents were published during 1979; in one,[338] dimethyl sulphoxide was the co-solvent, and a variety of pH-dependent processes were studied in different mixtures. All of the results were consistent with the hypothesis that hydrogen peroxide binds to the enzyme as its conjugate base, HO_2^-. When methanol is the co-solvent, the rate of oxidation of 2,3-dimethyl-1,4-naphthoquinol-1-dimethylphosphate showed a marked dependence upon the dielectric constant of the medium.[339]

Another non-classical pathway of action of peroxidase is in the generation of products in excited electronic states. Aerobic oxidation of malonaldehyde results in the emission of light with λ_{max} of 700 nm, attributed to singlet oxygen,[340] whilst oxidation of NADH and acetoacetate by HRP is catalysed by eosin, and light is emitted that has a wavelength of 550 nm.[341] Triplet acetone is produced during the aerobic oxidation of isobutanal by HRP, and the quenching of this excited state by a variety of collisional agents has been used to suggest a possible mechanism of formation.[342] Light is also emitted during the peroxidation of phenols, and the intensity is dependent on the amount of enzyme present, providing the basis for a sensitive assay.[343]

Alkyl hydroperoxides react with HRP (isoenzyme A_2) to form compound I, and the rate of reactivity appears to correlate best with the pK_a.[344] Many reducing agents will react with compound I to yield compound II. However, it has only recently been shown to be possible to oxidize compound II to compound I. Nadezhdin and Dunford[345] have achieved this with periodate ($k = 10^{-2}$ l mol^{-1} s^{-1}) or with oxidizing free-radicals ($CO_3^{\cdot-}$, $Cl_2^{\cdot-}$, $Br_2^{\cdot-}$, or $SO_4^{\cdot-}$) that were generated photolytically. Hewson and Hager[346, 347] could effect the same transformation with chlorite in addition to the oxidation of the native enzyme to compound II with the same reagent. These authors demonstrated that chlorination of monochlorodimedone by HRP and chlorite involves the enzymatic

336   Z. Gasyna, *FEBS Lett.*, 1979, **106**, 213.
337   Z. Gasyna, *Stud. Biophys.*, 1979, **76**, 77.
338   P. A. Adams, D. A. Baldwin, G. S. Collier, and J. M. Pratt, *Biochem. J.*, 1979, **179**, 273.
339   G. B. Sergeev, V. A. Batyuk, B. M. Sergeev, T. Saprykina, and G. V. Yamshchikova, *Vestn. Mosk. Univ., Ser. 2 Khim.*, 1979, **20**, 140.
340   C. Vidigal-Martinelli, K. Zinner, B. Kachar, N. Dúran, and G. Cilento, *FEBS Lett.*, 1979, **108**, 266.
341   B. Kachar, K. Zinner, C. C. C. Vidiga, Y. Shimizu, and G. Cilento, *Arch. Biochem. Biophys.*, 1979, **195**, 245.
342   E. J. H. Bechara, O. M. M. Faria Oliveira, N. Dúran, R. Casadei de Baptista, and G. Cilento, *Photochem. Photobiol.*, 1979, **30**, 101.
343   M. Halmann, B. Velan, T. Sery, and H. Schupper, *Photochem. Photobiol.*, 1979, **30**, 165.
344   K. G. Paul, P. I. Ohlsson, and S. Wold, *Acta Chem. Scand., Ser. B*, 1979, **33**, 747.
345   A. Nadezhdin and H. B. Dunford, *Can. J. Biochem.*, 1979, **57**, 1080.
346   W. D. Hewson and L. P. Hager, *J. Biol. Chem.*, 1979, **254**, 3175.
347   W. D. Hewson and L. P. Hager, *J. Biol. Chem.*, 1979, **254**, 3182.

disproportionation of chlorite to chlorine dioxide and the reaction of the latter with monochlorodimedone:[346]

$$5HClO_2 \longrightarrow 4ClO_2 + Cl^- + 2H_2O + H^+$$

$$ClO_2 + \text{monochlorodimedone} \longrightarrow \text{dichlorodimedone}$$

In their second paper they show that the reactivity of compound II is controlled by a group with a pK_a of 8.6, the base form being relatively unreactive.[347] The authors use this information to choose appropriate conditions to effect the transformations:

$$ClO_2^- + \text{peroxidase} \longrightarrow \text{compound II} + [ClO]$$

$$[ClO] + \text{compound II} \longrightarrow \text{compound I} + HOCl$$

Coupled proton transfers occur during peroxidase reactions, and interest has centred around the nature of the acid/base group(s) involved. Dunford and Araiso[348] have speculated that aspartate-43 may be important in this respect and suggest that metmyoglobin does not form a compound I with hydrogen peroxide because the corresponding residue is glycine. The same authors have reconstituted apoperoxidase with 2,4-dimethyldeuterohaemin and its mono- and di-methyl esters;[349] the rates of formation of compound I are essentially the same in all three forms ($\sim 10^7$ l mol^{-1} s^{-1}). This tends to rule out the haem propionates as being involved. HRP is inactivated by photolysis, and the mechanism has been proposed to be *via* excitation of an electron from the porphyrin and subsequent transfer from an amino-acid residue.[350] Several chemical modifications of HRP have been reported; the reactions of the lysine residues with a variety of acid anhydrides and with picrylsulphonic acid give several products.[351] The thermal stability of the products depended upon the extent rather than the nature of modification, and the increased stability of the modified forms at low temperatures ($< 75\,°C$) has been attributed to a decrease in conformational mobility around the haem. HRP may also be modified at carboxy- or amino-groups to give products with unaltered spectroscopic or enzymatic properties but with different isoelectric points.[352] N-Bromosuccinimide modifies HRP at both pH 4.1 and pH 7; however, only in the latter case is there retention of haem and of enzyme activity.[353]

The redox potentials for the compound I/compound II and compound II/native enzyme couples have been determined for HRP isoenzymes A_2 and C.[354] In the neutral and acidic pH regions, the measured potentials are the same, and the previous conclusion that compound I is a better oxidant than compound II arises from kinetic rather than thermodynamic restraints. In the pH range investigated, the redox potentials can be fitted to the equation:

$$E'_0 = E_0 + 0.058 \log(K_r[H^+] + [H^+]^2)/(K_0 + [H^+])$$

[348] H. B. Dunford and T. Araiso, *Biochem. Biophys. Res. Commun.*, 1979, **89**, 764.
[349] T. Araiso and H. B. Dunford, *Biochem. Biophys. Res. Commun.*, 1979, **90**, 520.
[350] A. D. Nadezhdin and H. B. Dunford, *Photochem. Photobiol.*, 1979, **29**, 899.
[351] N. N. Ugarova, G. D. Rozhkova, and I. V. Berezin, *Biochim. Biophys. Acta*, 1979, **570**, 31.
[352] H. G. Rennke and M. A. Venkatachalam, *J. Histochem. Cytochem.*, 1979, **27**, 1352.
[353] A. P. Savitskii, E. V. Plotitsyna, and N. N. Ugarova, *Bioorg. Khim.*, 1979, **5**, 1390.
[354] Y. Hayashi and I. Yamazaki, *J. Biol. Chem.*, 1979, **254**, 9101.

where K_r and K_o are proton dissociation constants for the reduced and oxidized components.

Many spectroscopic methods have been used to study peroxidase; binding of inhibitor and of substrate, for example, has been investigated by both e.p.r.[355] and n.m.r.[356] In the first case, a spin-labelled hydroxamic acid was synthesized and its spectrum at room temperature, in the presence of enzyme, is consistent with its being strongly immobilized. An analysis of the dipolar interaction between the nitroxide and iron centres set a lower limit of 22 Å on the distance between them.[355] High-resolution ^{1}H n.m.r. of HRP and its complexes with indolepropionic acid, p-cresol, and benzhydroxamic acid revealed that, whilst the first two were in fast exchange, the last compound was in slow exchange on the n.m.r. timescale.[356] By contrast, with the cyano-HRP all three molecules are in slow exchange; a study of pH and temperature dependencies of the dissociation constants led to the model for substrate binding that is shown in Figure 5. Proton n.m.r. has also been used to follow the binding of calcium to the native enzyme and to compound I; in both cases there are changes in the chemical shifts of peripheral methyl groups of the haem when calcium is bound

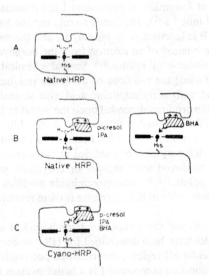

Figure 5 *Postulated modes of binding of the aromatic donor molecules indolepropionic acid (IPA), benzhydroxamic acid (BHA), and p-cresol to horseradish peroxidase (HRP), where A shows the structure of the native enzyme, B shows the binding of p-cresol and IPA to a sterically hindered site at the distal side of haem and of BHA to two sites, one of these being the same as for p-cresol and IPA and the other at the co-ordination position of the iron of haem, and C shows how all three substrates bind to the cyanide complex of HRP*

(Reproduced by permission from *J. Biol. Chem.*, 1979, **254**, 2814)

[355] G .Rakhit and C. F. Chignell, *Biochim. Biophys. Acta*, 1979, **580**, 108.
[356] I. Morishima and S. Ogawa, *J. Biol., Chem.*, 1979, **254**, 2814.

at a stoicheiometry of 1:1.[357] Decay of compound I is also much faster in the absence of Ca^{2+}. La Mar and de Ropp[358] have assigned the resonance of the proximal N^1-proton and HRP and metmyoglobin (96 and 100 p.p.m. respectively), and have shown that in the former case it does not exchange in D_2O, arguing for a much more hydrophobic environment. In addition, when indole-propionic acid is added there is no change in the position of resonance, suggesting that the substrate is bound on the distal side.

Spiro et al.[359] have compared the resonance Raman spectra of a number of high-spin ferric porphyrins and of HRP and have concluded that the latter is five-co-ordinate, while Mineev et al.[360] have presented an analysis of the m.c.d. of nitrosyl ferroperoxidase. The electronic states of compounds I and II have been investigated by Mössbauer and e.p.r. spectroscopies; for compound I, both methods are consistent with an Fe^{4+} ($S=1$) centre that is weakly coupled to a radical ($S'=\frac{1}{2}$).[361] Rapid-passage conditions have to be used to obtain the e.p.r. spectrum, which absorbs over a field of >0.2 T (at 9.2 GHz); integration of the signal amounts to 0.7 spins per haem moiety. Mössbauer spectra of compound II as a function of temperature and applied field can be accounted for by an Fe^{4+} ($S=1$) spin Hamiltonian.[362] Electron paramagnetic resonance spectroscopy has also been used to study the native enzyme, and the spectra of HRP in vivo, in an extract and in two purified isoenzymes, have been compared.[363] All of the samples had the characteristic high-spin ($g\approx6$) and quantum-mechanical mixed-spin ($g\approx5$) signals, and the authors concluded that the two existed in chemical equilibrium. Oxygen-17 superhyperfine broadening in the e.p.r. spectrum of the benzhydroxamic acid–HRP complex ($A=1.3$ G) is consistent with the co-ordination of water or a hydroxyl group to the iron.[364]

The pH dependence of absorption and fluorescence spectra of protoporphyrin-IX in the free porphryin and in the HRP-bound states has been discussed in terms of the environment of the active site and the possible requirements for the sixth ligand.[365] A second paper from the same group[366] presents evidence that a group with a pK_a of 4.3 considerably influences the mobility of the active site; they discuss the consequences of this for the mechanism.

A complete amino-acid sequence for HRP isoenzyme C has been determined. The single polypeptide chain is 308 amino-acid residues long and is blocked at the N-terminus by cyclization of glutamate-1 to a pyrrolidonecarboxylyl group.[367]

[57] S. Ogawa, Y. Shiro, and I. Morishima, *Biochem. Biophys. Res. Commun.*, 1979, **90**, 674.
[58] G. N. La Mar and J. S. de Ropp, *Biochem. Biophys. Res. Commun.*, 1979, **90**, 36.
[59] T. G. Spiro, J. D. Strong, and P. Stein, *J. Am. Chem. Soc.*, 1979, **101**, 2648.
[60] A. P. Mineev, Yu. A. Sharonov, and V. A. Figlovskii, *Int. J. Quantum Chem.*, 1979, **16**, 883.
[61] C. E. Schulz, P. W. Devaney, H. Winkler, P. G. Debrunner, N. Doan, R. Chiang, R. Rutter, and L. P. Hager, *FEBS Lett.*, 1979, **103**, 102.
[62] C. Schulz, R. Chiang, and P. G. Debrunner, *J. Phys. Colloq. (Orsay, Fr.)*, 1979, 534.
[63] M. M. Maltempo, P.-I. Ohlsson, K.-G. Paul, L. Petersson, and A. Ehrenberg, *Biochemistry*, 1979, **18**, 2935.
[64] R. K. Gupta, A. S. Mildvan, and G. R. Schonbaum, *Biochem. Biophys. Res. Commun.*, 1979, **89**, 1334.
[65] A. P. Savitskii, N. N. Ugarova, and I. V. Berezin, *Bioorg. Khim.*, 1979, **5**, 1210.
[66] A. P. Savitskii, N. N. Ugarova, and I. V. Berezin, *Bioorg. Khim.*, 1979, **5**, 1843.
[67] K. G. Welinder, *Eur. J. Biochem.*, 1979, **96**, 483.

Other Peroxidases. **Cytochrome *c* peroxidase** (EC 1.11.1.5) is most commonly isolated from yeast, where it is synthesized as a larger precursor protein, transported into the inner membrane space, and proteolytically processed to its mature form.[368] Resonance Raman spectroscopy of ferrous and ferric cytochrome *c* peroxidase and their various liganded forms, as well as of compound I, has been used by Sievers *et al.* to assign spin states.[369] They suggest that the native enzyme is a five-co-ordinate high-spin ferric compound whilst compound I contains low-spin iron(IV). Hoffman *et al.*[370] have interpreted the e.p.r. and ENDOR spectra of yeast cytochrome *c* peroxidase compound I as containing the second oxidizing equivalent as a sulphur-centred free-radical. The authors suggest that it might be a dimeric thioether radical cation from two methionine residues. Two spectral forms of the native enzyme have been detected in acetate buffer, and they appear to be in equilibrium.[371] The two forms have absorbance maxima at 645 and 620 nm; only the former reacts with dithionite or hydrogen peroxide, and it is the only species that is present in phosphate buffer. Both states are believed to have anions bound to them. A paper has appeared in which the preparation of heavy-metal derivatives is described.[372]

Cytochrome *c* peroxidase may also be isolated from species of the bacterial genus *Pseudomonas*, and a purification scheme for this enzyme from *Ps. denitrificans* has been presented.[373] The isolation of potential substrates, *e.g.* azurin and cytochromes *c*-551 and c_4, and preliminary kinetic data have also been described. In the peroxidase of *Ps. aeruginosa*, the two haems appear to be inequivalent, as judged by kinetic, spectral, and ligand-binding studies.[374, 375] One haem appears to be high-spin and one low-spin;[374] only one binds carbon monoxide, and in doing so it undergoes a change in spin state.[375]

Chloroperoxidase (EC 1.11.1.10) is in many ways an atypical member of the peroxidase family, showing many similarities to cytochrome *P*-450. A resonance Raman study of the enzyme bears out the similarities with cytochrome *P*-450$_{cam}$, and has been used to investigate the different spin states.[376] The reactivity of chloroperoxidase is also quite distinct; the peroxidation of indole yields oxindole as the major product[377] whilst this compound is not detected at all after HRP-catalysed peroxidation. 2-Methylindole was not affected, and acted as a good inhibitor, whilst the 1-methyl isomer was both a poor substrate and a weak inhibitor. Anilines that are substituted at the 4-position are only oxidized to the corresponding nitroso-compound by chloroperoxidase and hydrogen peroxide, and the turnover number decreases with increasing bulk of the 4-substituent.[378]

[368] M. L. Maccecchini, Y. Rudin, and G. Schatz, *J. Biol. Chem.*, 1979, **254**, 7468.
[369] G. Sievers, K. Osterlund, and N. Ellfolk, *Biochim. Biophys. Acta*, 1979, **581**, 1.
[370] B. M. Hoffman, J. E. Roberts, T. G. Brown, C. H. Kang, and E. Margoliash, *Proc. Natl. Acad. Sci. USA*, 1979, **76**, 6132.
[371] R. A. Mathews and J. B. Wittenberg, *J. Biol. Chem.*, 1979, **254**, 5991.
[372] B. Eriksson, B. Ersson, P. Kierkegaard, L. O. Larsson, U. Skoglund, M. Ludwig, and T. Yonetani, *J. Mol. Biol.*, 1979, **127**, 225.
[373] A. F. W. Coulson and R. I. C. Oliver, *Biochem. J.*, 1979, **181**, 159.
[374] M. Rönnberg and N. Ellfolk, *Biochim. Biophys. Acta*, 1979, **581**, 325.
[375] M. Rönnberg, N. Ellfolk, and R. Soininen, *Biochim. Biophys. Acta*, 1979, **578**, 392.
[376] R. D. Remba, P. M. Champion, D. B. Fitchen, R. Chiang, and L. O. Hager, *Biochemistry* 1979, **18**, 2280.
[377] M. D. Corbett and B. R. Chipko, *Biochem. J.*, 1979, **183**, 269.
[378] M. D. Corbett, D. G. Baden, and B. R. Chipko, *Bioorg. Chem.*, 1979, **8**, 91.

Myeloperoxidase (MPO) is most commonly found in white blood cells, and some properties of the bovine enzyme have been described;[379] the enzyme is claimed to retain 86% of its activity at 80 °C. MPO, in the presence of hydrogen peroxide, is able to oxidize chloride, probably to hypochlorite; in the presence of organic compounds, a number of further reactions can occur. Glycine yields mono- and di-chloroamines, and the latter decompose to HCN and ClCN;[380] however, cyanogen chloride can also be produced enzymatically by the MPO-catalysed chlorination of HCN. Inhibition of α-1-proteinase inhibitor by MPO/H_2O_2/Cl^- is apparently due to the oxidation of an essential methionine residue to a sulphoxide,[381] and the enhanced proteolytic activity of neutrophils may be due to this reaction. Iron(II) MPO reacts with hydrogen cyanide to yield two cyano-complexes, the one that is formed first being converted into the second, more stable, form *via* a change of conformation of the enzyme.[382] The author suggests that a conformational, kinetic stabilization of the high-spin form of the native enzyme may be of importance during catalysis by the enzyme. Two forms of MPO that interconvert with a pK_a of 4.6 have been described, and they can be distinguished by their opposing affinities for chloride and hydrogen peroxide.[383]

Thyroid peroxidase has been purified to 80—95% and to a very high specific activity (~ 800 μmol of guiacol oxidized per minute per milligram of protein), and some of the basic physical, chemical, and kinetic properties have been described.[384] Thiourea is an inhibitor of thyroid peroxidase both *in vivo* and *in vitro*, and Davidson *et al.*[385] have postulated that this compound reacts with the iodinating intermediate in enzyme turnover and is oxidized to formamidine; the latter decomposes to cyanamide, which is the proximate inactivating material.

Lactoperoxidase is also effective in iodinating proteins, and the mechanism of two other anti-thyroid drugs has been probed with this enzyme.[386] Both methyl-thiouracil and methylmercaptoimidazole inhibit iodination of *N*-acetyltyrosylamide at high pH, and the binding of iodide prevents this to some extent. Pommier and Cahnmann[387] have presented kinetic and spectrophotometric evidence for the binding of thiol to lactoperoxidase in a reaction that is potentiated by the simultaneous presence of di-iodotyrosine. A reciprocal effect was seen with the effect of thiols on di-iodotyrosine, and the authors have exploited this observation to develop an isolation procedure for the enzyme from milk, based upon a di-iodotyrosine affinity column. The haem of lactoperoxidase is claimed to be protoporphyrin-IX, although it is only released after proteolysis,

[379] O. Yu. Yankovsky, V. N. Koryakov, and S. N. Lyzlova, *Ukr. Biokhim. Zh.*, 1979, **51**, 492.

[380] J. M. Zgliczýnski and T. Stelmaszýnska, *Biochim. Biophys. Acta*, 1979, **567**, 309.

[381] N. R. Matheson, P. S. Wong, and J. Travis, *Biochem. Biophys. Res. Commun.*, 1979, **88**, 402.

[382] J. E. Harrison, *J. Biol. Chem.*, 1979, **254**, 4536.

[383] J. M. Zgliczýnski, *Przegl. Lek.*, 1979, **36**, 331.

[384] A. B. Rawitch, A. Taurog, S. B. Chernoff, and M. L. Dorris, *Arch. Biochem. Biophys.*, 1979, **194**, 244.

[385] B. Davidson, M. Soodak, H. V. Strout, J. T. Neary, C. Nakamura, and F. Maloof, *Endocrinology*, 1979, **104**, 919.

[386] J. L. Michot, J. Nunez, M. L. Johnson, G. Irace, and H. Edelhoch, *J. Biol. Chem.*, 1979, **254**, 2205.

[387] J. Pommier and H. J. Cahnmann, *J. Biol. Chem.*, 1979, **254**, 3006.

and there is always some loss due to degradation.[388] A binding site in a crevice deep in the molecule is claimed to account for the difficulty of removing the haem, and the unusual pyridine haemochrome is attributed to retention of one of the axial ligands of the protein.

Kinetic behaviour and spectroscopic properties of lactoperoxidase and intestinal peroxidase are similar for compounds I, II, and III; the two enzymes are, however, different in their reaction with benzhydroxamic acid.[389] The dissociation constants for the complex show haem-linked ionizations of 3.5 and 4.75 respectively, and the inhibitor is much more effective with the intestinal enzyme. A new spectral intermediate was detected in the reaction of the inhibitor with compound II of this enzyme.

In addition to the above relatively well characterized enzymes, peroxidase activity is present in many tissues; a recent isolation of two isoenzymes of human foetal peroxidase has been described.[390] Oestradiol was covalently bound to the enzyme in the presence of hydrogen peroxide. A microsomal peroxide that reduces prostaglandin hydroperoxides has been characterized and a number of observations have implicated the production of free radicals during turnover.[391] The authors suggest that radical-induced auto-inactivation of the enzyme may play a role in the regulation of prostaglandin biosynthesis. Finally, the induction of lysis of erythrocytes by 2-methylpropanol has been attributed to the production of triplet acetone from this compound by an erythrocyte peroxidase.[392]

Glutathione peroxidase (EC 1.11.1.9) from erythrocytes has had its structure determined by X-ray crystallography to a resolution of 2.8 Å,[393] which is sufficient to allow a tentative amino-acid sequence to be proposed. The single selenium atom (as selenocysteine) is located on the surface of the molecule in a largely hydrophobic environment, and difference Fourier maps between oxidized and reduced forms suggest that the former has one or more oxygen atoms attached to the selenium. An interesting topological similarity between the structure of part of the peroxidase and thioredoxin and rhodanese was noted by the authors, and they point out that the selenium atom occupies the same relative position as the disulphide group in thioredoxin and the persulphide group in rhodanese. There may in fact be several redox forms of the active site of glutathione peroxidase, as it shows sigmoidal kinetics.[394] Pre-incubation with reduced glutathione abolishes a lag in the activity, and this is not due to aggregation of the enzyme; a possible model of enzyme action is presented to account for the observed behaviour. A kinetic analysis of the rates of oxidation of linoleic acid and cumene hydroperoxides was made, using integrated rate equations; from the rate-constants, a half-life of 10^{-4} min was calculated for lipid hydroperoxides *in vivo*.[395]

[388] G. Sievers, *Biochim. Biophys. Acta*, 1979, **579**, 181.
[389] S. Kimura and I. Yamazaki, *Arch. Biochem. Biophys.*, 1979, **198**, 580.
[390] L. Dimitrijevic, C. Aussel, A. Muchielli, and R. Masseyeff, *Biochimie*, 1979, **61**, 535.
[391] R. W. Egan, P. H. Gale, and F. A. Kuehl, Jr., *J. Biol. Chem.*, 1979, **254**, 3295.
[392] N. Dúran, Y. Makita, and L. H. Innocentini, *Biochem. Biophys. Res. Commun.*, 1979, **88**, 642.
[393] R. Ladenstein, O. Epp, K. Bartels, A. Jones, R. Huber, and A. Wendel, *J. Mol. Biol.*, 1979, **134**, 199.
[394] A. G. Splittgerber and A. L. Tappel, *J. Biol. Chem.*, 1979, **254**, 9807.
[395] J. W. Forstrom, F. H. Stults, and A. L. Tappel, *Arch. Biochem. Biophys.*, 1979, **193**, 51.

Several compounds are inhibitors of the enzyme; diethyldithiocarbamate ion inhibits both lung and liver glutathione peroxidases *in vivo*,[396] and heavy metals are also effective.[397] In the latter case, cadmium is the best, and it is competitive with reduced glutathione and non-competitive with the peroxide, the degree of inhibition by those metal ions that were studied being correlated with softness and size of the ion.[397] Both the kinetic patterns and the formation of mixed disulphides between glutathione and thiols are consistent with the enzyme only being specific for the first donor substrate.[398] It was also suggested that, in the absence of reduced glutathione, the selenium may become over-oxidized. Glutathione peroxidase has been detected in fish, frogs, salamanders, turtles, crayfish, and snails, but not in insects, earthworms, plants, or micro-organisms,[399] and a fluorometric assay for the enzyme has been used to measure its levels in human platelets.[400] Human placental glutathione peroxidase has been purified and seems to be a typical member of this class of enzymes; it is inhibited by cyanide, with concomitant loss of selenium, in a process that is protected against by thiols.[401]

Catalases (EC 1.11.1.6). The isolation of a catalase from nitrate-induced *Neurospora crassa* has been described by Jacob and Orme-Johnson.[402] The enzyme is rather different from the mammalian forms; it has a molecular weight of 320 000, is composed of four subunits, and has an activity of 9—12 kilounits mg^{-1}. The pyridine haemochrome is, however, quite unlike that from mammalian catalases or any other known haemoprotein; radioimmunoassay shows that this form accounts for 80% of the nitrate-induced catalase activity in this organism. Further characterization of the haem by e.p.r., electronic spectroscopy, and chemical methods led the same authors to conclude that it is a polar chlorin with four carboxyl groups.[403] Spectroscopically, the haem resembles bacterial chlorin a$_2$. As isolated, the enzyme appears to exist in multiple high-spin forms. *Escherichia coli* possesses two catalases, and one of these has been purified to homogeneity;[404] in addition to its catalatic activity it will peroxidize *o*-dianisidine and a variety of other donors. *o*-Dianisidine is a powerful competitive inhibitor of the catalatic activity and a suicide substrate during peroxidatic turnover.

Several structural studies of catalase have been described; preliminary *X*-ray crystallographic data for the enzyme from *Micrococcus luteus* reveal that the four subunits are identical.[405] The crystal structure of another catalase, that from *Penicillium vitale*, has been determined to a resolution of 6 Å, and it also shows

[396] B. D. Goldstein, M. G. Rozen, J. C. Quintavalla, and M. A. Amoruso, *Biochem. Pharmacol.*, 1979, **28**, 27.
[397] A. G. Splittgerber and A. L. Tappel, *Arch. Biochem. Biophys.*, 1979, **197**, 534.
[398] J. W. Forstrom and A. L. Tappel, *J. Biol. Chem.*, 1979, **254**, 2888.
[399] J. Smith and A. Shrift, *Comp. Biochem. Physiol., B*, 1979, **63**, 39.
[400] R. Martinez, J.-M. Launay, and C. Dreux, *Anal. Biochem.*, 1979, **98**, 154.
[401] Y. C. Awasthi, D. D. Dao, A. K. Lal, and S. K. Srivastava, *Biochem. J.*, 1979, **177**, 471.
[402] G. S. Jacob and W. H. Orme-Johnson, *Biochemistry*, 1979, **18**, 2967.
[403] G. S. Jacob and W. H. Orme-Johnson, *Biochemistry*, 1979, **18**, 2975.
[404] A. Claiborne and I. Fridovich, *J. Biol. Chem.*, 1979, **254**, 4245.
[405] A. L. Marie, F. Parak, and W. Hoppe, *J. Mol. Biol.*, 1979, **129**, 675.

the presence of four identical subunits,[406] the overall molecular dimensions being $70 \times 90 \times 120$ Å. The possible location of the haem groups between subunits is reported. Electron-microscopic image-reconstruction techniques have also been used by the Russian group to investigate tubular beef heart catalase;[407] these results are also consistent with there being four subunits per molecule and the overall dimensions being $69 \times 87 \times 92$ Å.

Optical absorption and c.d. spectra of native human catalase and of its adducts with F^-, CN^-, and N_3^- have been reported by Palcic and Dunford;[408] they have also described the e.p.r. spectra of the native and cyanide-bound forms. Some pH-dependent changes in the spectra were interpreted in terms of dissociation of the subunits and loss of the haem. A change in the Mössbauer spectrum of the catalase of *M. luteus* is observed at pH < 7.[409] C.d., m.c.d., and absorption spectra have also been used to study the anion complexes of bovine liver catalase; the m.c.d. results are consistent with the presence of maximal amounts of high-spin iron in the fluoride adduct and maximal amounts of low-spin iron in the cyanide adduct.[410] All of the other anions (CNO^-, CNS^-, and N_3^-) yield mixed-spin species. A change in the m.c.d. spectrum is seen upon lowering the pH from 6.9 to 5.0. Abe *et al.*[411] have reported the pH-dependence of beef liver catalase, and noted the existence of two activity-linked pK_a values of 5 and 5.9 for the binding of hydrogen peroxide.

The biosynthesis of catalase has attracted considerable interest and has been reviewed by Ruis.[412] Catalase T of *Saccharomyces cerevisiae* can be synthesized *in vitro* in either the wheat-germ or the reticulocyte cell-free system, using either total RNA or poly(A)-RNA.[413] Rat liver catalase has also been synthesized in the same two cell-free translation systems. Although the rabbit reticulocyte lysate requires the addition of haemin,[414] the translation could be inhibited by the addition of a haemin-regulated inhibitor (HRI); from the patterns of inhibition and relief, Scheme 4 was suggested.

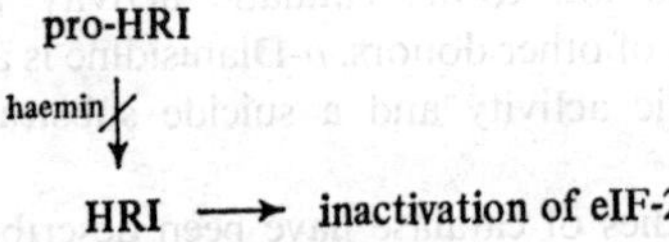

pro-HRI

haemin $\downarrow$

HRI $\longrightarrow$ inactivation of eIF-2

Scheme 4

The effect of intraperitoneal injections of haemin in rats has been investigated, and a single dose caused a transient decrease in catalase levels,[415] possibly due

[406] B. K. Vainshtein, F. R. Melik-Adamyan, V. V. Barynin, A. A. Vagin, and V. Yu. Nekrasov. *Dokl. Akad. Nauk SSSR*, 1979, **246**, 220.

[407] V. V. Barynin, B. K. Vainshtein, O. N. Zograf, and S. Ya. Karpukhina, *Mol. Biol.* (*Moscow*), 1979, **13**, 1189.

[408] M. Palcic and H. B. Dunford, *Can. J. Biochem.*, 1979, **57**, 321.

[409] F. Parak, D. Bade, and A. L. Marie, *J. Phys. Colloq.* (*Orsay, Fr.*), 1979, 528.

[410] W. R. Browett and M. J. Stillman, *Biochim. Biophys. Acta*, 1979, **577**, 291.

[411] K. Abe, N. Makino, and F. K. Anan, *J. Biochem.* (*Tokyo*), 1979, **85**, 473.

[412] H. Ruis, *Can. J. Biochem.*, 1979, **57**, 1122.

[413] G. Ammerer and H. Ruis, *FEBS Lett.*, 1979, **99**, 242.

[414] J. Sakamoto and T. Higashi, *J. Biochem.* (*Tokyo*), 1979, **85**, 389.

[415] M. Hamato and T. Higashi, *J. Biochem.* (*Tokyo*), 1979, **85**, 397.

to increased degradation rather than impaired synthesis. Repeated administration of haemin caused a considerable decrease; there was, however, only a slight impairment of catalase-synthesizing activity in cell-free systems. Rat liver peroxisomal catalase has a half-life of 39 h *in vivo*.[416] Berte and Sels[417] have isolated a mutant (*kat 80*) of *S. cerevisiae* that displays catalase activity even when grown anaerobically; they have interpreted their work in terms of continued synthesis of apo-catalase-T during anaerobic growth and of aerobic activation by insertion of haem.

Other Oxidoreductases. **Sulphite reductase** (EC 1.8.99.1) has been isolated from *Thiobacillus denitrificans*[418] and from *Chromatium vinosum*;[419] the two enzymes appear to be very similar, containing both sirohaem and iron–sulphur prosthetic groups. The enzymes are also very similar to the dissimilatory sulphite reductases from *Desulphovibrio* species, and the authors suggest that the physiological role for the enzyme of *T. denitrificans* may be oxidation of sulphide.[418] However, both enzymes reduce sulphite to sulphide, thiosulphate, and trithionate. Two e.p.r. studies of sulphite reductases from various species of sulphate-reducing bacteria have been presented; Hall *et al.*[420] have assigned the iron of the sirohaem as high-spin ferric in the isolated enzyme. The enzyme is slowly reduced by dithionite to reveal a signal from the iron–sulphur cluster; an intermediate free-radical was also observed during the course of the reaction. Redox titrations yielded redox potentials of -310 mV (sirohaem) and -560 mV (iron–sulphur cluster), and the enzyme can also reduce nitrate to nitric oxide. Liu *et al.*[421] have reported the e.p.r. spectra of sulphite reductases from *Desulphovibrio gigas* (desulphoviridin), *D. sulphuricans* (desulphorubidin), and *Desulfotomaculum ruminis* (P-582) in both oxidized and reduced forms. They also note that the enzyme is only slowly reduced by dithionite alone but rapidly reduced in the presence of dithionite and methyl viologen. The authors concluded that the three enzymes have similar functional properties.

Flavocytochromes can be isolated from a variety of sources, and the protein from *Chromatium vinosum* will oxidize sulphide with either horse or yeast cytochrome *c* as the electron acceptor, although not with the cytochromes *c*-553 or c_1 from this organism.[422] The same protein will also reduce sulphur, using viologens as the electron donor. Subunit structures and the isolation of cytochrome and flavin moieties of this protein and a similar one from *C. limicola* have been described.[423] Another flavocytochrome has been isolated from *Alcaligenes eutrophus*; it contains one protoporphyrin-IX and one FAD group in the molecule (mol. wt 43 000), and the protein has NADH–cytochrome *c* reductase activity and binds oxygen at the haem.[424] The authors speculate that

[416] H. Hayashi, *Biochim. Biophys. Acta*, 1979, **585**, 220.
[417] C. Berte and A. Sels, *Mol. Gen. Genet.*, 1979, **172**, 45.
[418] M. Schedel and H. G. Trüper, *Biochim. Biophys. Acta*, 1979, **568**, 454.
[419] M. Schedel, M. Vanselow, and H. G. Trüper, *Arch. Microbiol.*, 1979, **121**, 29.
[420] M. H. Hall, R. H. Prince, and R. Cammack, *Biochim. Biophys. Acta*, 1979, **581**, 27.
[421] C. L. Liu, D. V. Der Vartanian, and H. D. Peck, Jr., *Biochem. Biophys. Res. Commun.*, 1979, **91**, 962.
[422] Y. Fukumori and T. Yamanaka, *J. Biochem. (Tokyo)*, 1979, **85**, 1405.
[423] T. Yamanaka, Y. Fukumori, and K. Okunuki, *Anal. Biochem.*, 1979, **95**, 209.
[424] I. Probst, G. Wolf, and H. G. Schlegel, *Biochim. Biophys. Acta*, 1979, **576**, 471.

its role may either be (i) a terminal oxidase, (ii) a hydroxylase, or (iii) an oxygen store.

Cytochrome b_2 (Lactate–cytochrome c reductase) (EC 1.1.2.3) is the best studied of the flavocytochromes, and Pompon and Lederer[425] have described their results on the activity of the yeast enzyme that has been reconstituted with 5-deazaflavin. This analogue is slowly reduced by lactate and oxidized by pyruvate, although no reduction of the haem takes place; sulphite binds to this analogue, but the inhibitor 2-hydroxy-3-butynoic acid does not form an adduct. Comparisons with the lactate oxidase from *Mycobacterium smegmatis* have been drawn. Cytochrome b_2 is composed of four subunits, and a rotation-function analysis of X-ray diffraction data has shown these to be equivalent.[426] The enzyme has also been adsorbed on to 'organic metals' that are held in contact with a platinum electrode; the current that was observed during cyclic voltammetry was a function of the concentration of lactate present in the medium.[427] The current showed saturation behaviour with a K_m of 2 mmol l^{-1}.

Indoleamine-2,3-dioxygenase (EC 1.13.11.11) has been investigated by Taniguchi *et al.*[428] and the rates of the individual steps in the catalytic reaction have been determined (Figure 6). A consideration of the relative rates has led the authors to conclude that the superoxide radical anion is a substrate when all of the enzyme is present in the ferric form. The inhibition of rat liver trypto-

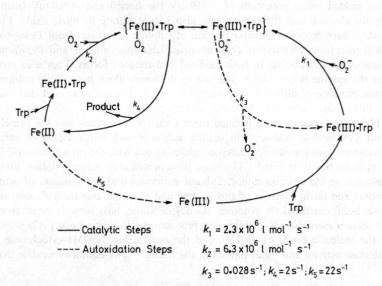

— Catalytic Steps $\quad k_1 = 2.3 \times 10^6$ l mol^{-1} s^{-1}

--- Autoxidation Steps $\quad k_2 = 6.3 \times 10^6$ l mol^{-1} s^{-1}

$$k_3 = 0.028\, s^{-1};\ k_4 = 2\, s^{-1};\ k_5 = 22\, s^{-1}$$

Figure 6 *Catalytic cycles of indoleamine dioxygenase*
(Redrawn from *J. Biol. Chem.*, 1979, **254**, 3288)

425 D. Pompon and F. Lederer, *Eur. J. Biochem.*, 1979, **96**, 571.
426 P. H. Bethge and F. S. Mathews, *Acta Crystallogr.*, *Sect. A*, 1979, **35**, 412.
427 J. Kulys and G. Svirmickas, *Dokl. Akad. Nauk SSSR*, 1979, **245**, 137.
428 T. Taniguchi, M. Sono, F. Hirata, O. Hayaishi, M. Tamura, K. Hayashi, T. Iizuka, and Y. Ishimura, *J. Biol. Chem.*, 1979, **254**, 3288.

phan oxygenase by 4-hydroxypyrazole *in vitro* has been studied to try and elucidate its mode of action as an inhibitor *in vivo*. At low concentrations (<0.1
mmol l^{-1}) it in fact activates, and at higher concentrations it inhibits competitively with tryptophan ($K_i = 30$ μmol l^{-1}).[429]

An **aldehyde oxidase** has been purified from *Methylomonas methylovora* which
oxidizes a wide range of aldehydes in the presence of an artificial electron-
acceptor (phenazine methosulphate). The enzyme contains a single haem per
unit of mol. wt 43 000 and is inhibited by sulphydryl reagents and by copper(I)
and iron(II).[430]

Peroxygenase is the name given to an enzyme from pea seed microsomes that
catalyses the reaction:

$$ROOH + SH \longrightarrow ROH + SOH$$

The enzyme has been partially purified and some of its basic physical and chemical properties have been described.[431]

3 Metallo-oxidoreductases of the Respiratory Chain

Complex I (EC 1.6.99.2).—Mitochondrial NADH dehydrogenase has been
purified from larvae and embryos of *Drosophila hydei* and some of its basic
physical and enzymatic properties have been described.[432] It has been shown to
be one of the 'heat shock' polypeptides. Partial purification of a bacterial NADH–
ubiquinone reductase from the slightly halophilic organism *Vibrio alginolyticus*
has been achieved, and it was shown to be the site of Na$^+$-dependent respiratory
stimulation.[433] *Micrococcus lysodeikticus* contains a membranous NADH
dehydrogenase that is inhibited by antibodies to the enzyme. Interestingly, in
this bacterium the cytoplasmic activity is also inhibited by the same antibodies.[434]

Treatment of complex I of beef heart with a chaotropic agent such as sodium
perchlorate resolves the NADH dehydrogenase activity; the protein has three
subunits and flavin:iron:sulphur ratios of 1:5—6:5—6.[435] The donor activities
for NADH and NADPH with a variety of acceptors were measured and the
authors found that both these and the transhydrogenase activity were enhanced
in the presence of guanidine derivatives, implying an important role for charge
neutralization. Heron *et al.*[436] have used one- and two-dimensional electrophoresis to resolve 26 polypeptides in purified complex I. They have determined
the distribution of these between the resolved iron- and flavin-binding proteins
and have proposed a model for the complex of a hydrophilic core of high-
molecular-weight peptides that contains the flavin and one or more of the
iron–sulphur binding subunits. This core is surrounded by hydrophobic peptides of lower molecular weight that interact with the membrane. Further light

[429] H. Rouach, C. Ribiere, J. Nordmann, and R. Nordmann, *FEBS Lett.*, 1979, **101**, 149.
[430] R. N. Patel, C. T. Hou, and A. Felix, *Arch. Microbiol.*, 1979, **122**, 241.
[431] A. Ishimaru, *J. Biol. Chem.*, 1979, **254**, 8427.
[432] L. Hermans, *Biochim. Biophys. Acta*, 1979, **567**, 125.
[433] T. Unemoto and M. Hayashi, *J. Biochem. (Tokyo)*, 1979, **85**, 1461.
[434] E. F. Kharatyan, I. G. Zhukova, G. K. Degteva, N. F. Sepeton, V. M. Temnova, D. A.
Turgenbaeva, and D. N. Ostrovskij, *Biokhimiya*, 1979, **44**, 2187.
[435] Y. M. Galante and Y. Hatefi, *Arch. Biochem. Biophys.*, 1979, **192**, 559.
[436] C. Heron, S. Smith, and C. I. Ragan, *Biochem. J.*, 1979, **181**, 435.

has been shed upon the interaction between complex I (and II) and the mito-chondrial membrane by replacing the naturally occurring lipid with 1,2-ditetra-decanoyl-*sn*-glycero-3-phosphocholine.[437] The effects of temperature and lipid:ubiquinone:protein ratios are reported and discussed in terms of a model for protein–membrane interactions. Electron paramagnetic resonance spectro-scopy has been applied to study the stoicheiometry of the iron–sulphur centres of complex I in submitochondrial particles (SMPs).[438] The results from ten different preparations are consistent with the individual centres 1a and 1b being present at a concentration of one quarter of that of centre 2.

Chen and Guillory[439] have employed the photo-affinity label arylazido-β-alanyl-NAD$^+$ to probe the enzyme activities of complex I. In the dark, this compound competitively inhibits ubiquinone reductase activity with respect to both quinone and NADH; from its inhibitory action on the other activities of complex I and from the effects of photolysis the authors concluded that (i) both NADH and NADPH donate electrons to the respiratory chain at a common site, (ii) the ubiquinone reductase and transhydrogenase activities are at different sites, and (iii) the two transhydrogenase activities are also at distinct sites. Other inhibitors of complex I that have been described are the metal chelators 4,7-diphenyl-1,10-phenanthroline and 1,10-phenanthroline[440] and the hormone luliberin.[441] The latter possibly acts *via* activation of the mitochondrial phospho-lipase. The use of polymeric viologens as acceptors for NADH dehydrogenase has also been described.[442]

Complex II (EC 1.3.99.1).—Reviews on iron–sulphur flavoprotein dehydro-genases[443] and succinate dehydrogenase[444] have appeared. The nature of the iron–sulphur centres in succinate dehydrogenase was determined by Coles *et al.*,[445] using two types of core-extrusion experiment. In the first the extruded cores were characterized by ^{19}F n.m.r. of the fluorinated thiol; in the second, e.p.r. spectroscopy of reconstituted apo-iron–sulphur proteins was employed. The methods agree, both indicating two Fe_2S_2 and one Fe_4S_4 centre per flavin. Salerno *et al.*[446] have presented e.p.r. data supporting the idea that the two binuclear centres (S-1 and S-2) are spin-coupled, with a calculated separation of 10 Å. The centres, although having similar e.p.r. spectra, possess quite different mid-point potentials, and the role of the two groups and their relationship to the high-potential centre (S-3) are discussed. Further support for a magnetic interaction between S-1 and S-2 has come from e.p.r. studies of plant mito-chondria, using power-saturation measurements.[447] Evidence was also provided

[437] C. Heron, M. G. Gore, and C. I. Ragan, *Biochem. J.*, 1979, **178**, 415.
[438] S. P. J. Albracht, F. J. Leeuwerik, and B. van Swol, *FEBS Lett.*, 1979, **104**, 197.
[439] S. Chen and R. J. Guillory, *J. Biol. Chem.*, 1979, **254**, 7220.
[440] I. B. Minkov and L. M. Maglova, *Biokhimiya*, 1979, **44**, 832.
[441] G. Ya. Bakalkin, I. P. Krasinskaya, E. N. Komissarova, L. S. Yaguzhinskij, and V. A. Isachenkov, *Biokhimiya*, 1979, **44**, 1353.
[442] Ya. A. Aleksandrovskij, A. A. Sukhno, and Yu. V. Rodionov, *Biokhimiya*, 1979, **44**, 2130.
[443] T. Ohnishi, *Membr. Proteins*, 1979, **2**, 1.
[444] B. L. Trumpower and A. G. Katki, *Membr. Proteins*, 1979, **2**, 89.
[445] C. J. Coles, R. H. Holm, D. M. Kurtz, Jr, W. H. Orme-Johnson, J. Rawlings, T. P. Singer, and G. B. Wong, *Proc. Natl. Acad. Sci. USA*, 1979, **76**, 3805.
[446] J. C. Salerno, J. Lim, T. E. King, H. Blum, and T. Ohnishi, *J. Biol. Chem.*, 1979, **254**, 4828.
[447] H. Rupp and A. L. Moore, *Biochim. Biophys. Acta*, 1979, **548**, 16.

for a magnetic interaction between S-3 and an interacting pair of ubisemiquinone radicals.

The isolation of complex II (and complexes III and IV) from *Neurospora crassa* in a reconstitutively active form has been described by Weiss and Kolb.[448] Weiss and Wingfield[449] have investigated the kinetic behaviour of these preparations in detergent micelles in three different types of assay, and concluded that, when the substrate has to transfer from micelle to micelle prior to reaction, then the rate-limiting step in detergents with small head-groups is transfer. On the other hand, with a detergent with a large head-group, the enzyme reactions were rate-limiting. In reactions between micelle-bound complex II and complex III, no protein–micelle–protein–micelle reactions occurred, but rather the ubiquinol shuttled between the two enzymes; comparisons between the micellar and membranous enzymes have been made on the basis of these results. Effects of lipids on the behaviour of succinate–cytochrome *c* reductase have been probed in *Drosophila melanogaster*, using a mutant that shows a break in the Arrhenius plot at a much higher temperature than the wild-type enzyme.[450] However, by lipid re-binding experiments between native and mutant enzymes it was shown that it is the protein that is responsible for the behaviour, and not the lipid. Solubilization of the succinate dehydrogenase from *Micrococcus lysodeikticus* with the non-ionic detergent Triton X-100 only releases 66% of the enzyme that is detected by immunoelectrophoresis; the remainder can be extracted by treatment with sodium dodecylsulphate.[451]

The inhibition of complex II can be effected by a number of compounds, such as phenols[452] or Pd^{2+}, the latter showing an apparent K_i of 8 μmol l^{-1}.[453] Thenoyltrifluoroacetone (TTFA) is a classical inhibitor of complex II; Trumpower and Simmons[454] have concluded that it inhibits by binding to a single site, which they interpret as the S-3 centre, where reduction of ubisemiquinone to ubiquinol occurs. In the presence of antimycin, TTFA is a much less effective inhibitor; this may be due to an increased superoxide-dependent reduction of cytochrome *c*, the former being generated from enhanced autoxidation of ubisemiquinone. 3-Nitropropionate is a suicide inhibitor of succinate dehydrogenase; consistent with this, the dehydrogenation product, 3-nitroacrylate, is a rapid and effective inhibitor.[455] The reaction between 3-nitroacrylate and an essential thiol group on the enzyme is shown in Scheme 5. An effective artificial electron acceptor for succinate dehydrogenase is Wurster's blue (the radical cation of tetramethylphenylenediamine), which appears to react at the level of ubisemiquinone.[456] Re-oxidation of the enzyme occurs prior to release of the product (fumarate). Absorbance and c.d. spectral changes during the active-inactive transition of succinate dehydrogenase by oxaloacetate have been

[448] H. Weiss and H. J. Kolb, *Eur. J. Biochem.*, 1979, **99**, 131.
[449] H. Weiss and P. Wingfield, *Eur. J. Biochem.*, 1979, **99**, 151.
[450] L. Sondergaard, *Biochim. Biophys. Acta*, 1979, **557**, 208.
[451] P. Owen and H. Doherty, *J. Bacteriol.*, 1979, **140**, 881.
[452] V. A. Kostyrko and L. S. Yaguzhinskij, *Biokhimiya*, 1979, **44**, 1884.
[453] T. Z. Liu, S. D. Lee, and R. S. Bhatnagar, *Toxicol. Lett.*, 1979, **4**, 469.
[454] B. L. Trumpower and Z. Simmons, *J. Biol. Chem.*, 1979, **254**, 4608.
[455] C. J. Coles, D. E. Edmondson, and T. P. Singer, *J. Biol. Chem.* 1979, **254**, 5161.
[456] A. D. Vinogradov, V. G. Grivennikova, and E. V. Gavrikova, *Biochim. Biophys. Acta*, 1979, **545**, 141.

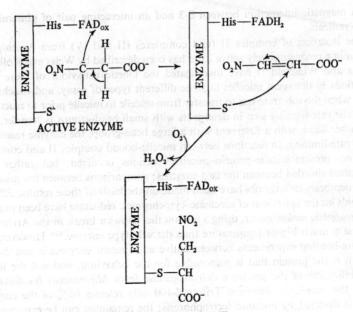

Scheme 5

interpreted in terms of alterations of both flavin and iron–sulphur centres.[457] Oxaloacetate-induced conformational changes also shift the mid-point for flavin from $+6$ to -186 mV. Reduction of cytochrome c by succinate has been reported to yield a stoicheiometry of four protons translocated for every two electrons transferred.[458]

Complex III (EC 1.6.99.3).—The stoicheiometry of the redox centre of complex III has been determined by e.p.r. and diffuse reflectance spectroscopies; it is 1:1:1:1:2:2 for iron–sulphur centre I:iron–sulphur centre II:cytochrome b-558:cytochrome b-556:cytochrome b-562:cytochrome c_1.[459] In the same paper the authors report that the e.p.r. parameters of the iron–sulphur centre I are dependent on the overall redox state of the complex and that cytochrome b-562 is probably two types of cytochrome, with different midpoint potentials. Two types of cytochrome c_1 may also be present, as Holmquist and Moise[460] have described the resolution of two fractions of material that contain cytochrome c_1 from Triton X-100 extracts. The exact subunit stoicheiometry of complex III is still not certain, and Augen *et al.*[461] claim that the so-called 'core protein I' is an artefact, arising from the aggregation of subunits of lower molecular weight that occurs at high temperatures.

[457] M. Gutman, F. Bonomi, S. Pagani, and P. Cerletti, *FEBS Lett.*, 1979, **104**, 371.
[458] A. Alexandre and A. L. Lehninger, *J. Biol. Chem.*, 1979, **254**, 11 555.
[459] S. de Vries, S. P. J. Albracht, and F. J. Leeuwerik, *Biochim. Biophys. Acta*, 1979, **546**, 316.
[460] R. Holmquist and H. Moise, *Biochimie*, 1979, **61**, 697.
[461] J. Augen, S. Power, and G. Palmer, *Biochem. Biophys. Res. Commun.*, 1979, **86**, 271.

Complex III, like the other enzymes of the respiratory chain, is membrane-bound, and membranous crystals of the protein from *Neurospora crassa* have been studied by electron microscopy; using image-reconstruction methods, the molecule appears to be rod-shaped, with dimensions that are consistent with its spanning the membrane.[462] The orientation of the metal centres in the enzyme with respect to the membrane has been studied by e.p.r.[463, 464] and by optical[464] spectroscopy of oriented multilayers. The e.p.r. measurements for the two-iron–two-sulphur (Rieske's) centre are consistent with the iron–iron vector lying in the plane of the membrane,[463, 464] whilst the ubisemiquinone pair is approximately normal to the membrane.[463] Both the b and the c_1 cytochromes also appear to be oriented at right angles to the membrane.[464] The effect of lipids on the redox potentials of the cytochromes b in a variety of preparations has been investigated.[465]

'Oxidation factor', which is required for the transfer of electrons from ubiquinol to complex III, has been identified as an iron–sulphur-containing subunit of the latter, and has been prepared from resolved complex III in a reconstitutively active form.[466] The interaction domain on cytochrome c with complex III has been defined by using singly modified cytochromes c; all of the modified proteins showed a higher apparent K_m and a greater sensitivity to ionic strength than the native cytochrome c.[467] As the interaction domain for the reductase broadly overlaps that of the oxidase, the authors suggest that cytochrome c has a mobility that enables it to shuttle electrons from the former to the latter. Bosshard *et al.*[468] have studied the binding of cytochrome c to complex III, and the isolated b- and c_1-containing subunits, by gel filtration and protection from chemical modification. The results are completely consistent with the cytochrome c binding to the c_1-containing subunit, and a dissociation constant of less than 0.1 μmol l^{-1} was calculated. Although cytochrome c is known to transfer electrons from the reductase to the oxidase, Chiang and King[469] have described the preparation of a 1:1 reductase–oxidase complex and have presented some of the optical and sedimentation properties. A 1:1:1 complex with cytochrome c has also been prepared, and possible roles for a reductase–oxidase complex *in vivo* are discussed.

Circular dichroism spectra of reduced cytochrome bc_1 in the presence of various compounds that uncouple electron flow from phosphorylation show the presence of haem–haem interactions between the b-type cytochromes.[470] The authors discuss their results in terms of a possible conformational mechanism for synthesis of ATP. Reduced complex III has been spin-labelled, and the resulting e.p.r. spectrum reveals both tightly and weakly immobilized labels; after

[462] P. Wingfield, T. Arad, K. Leonard, and H. Weiss, *Nature (London)*, 1979, **280**, 696.
[463] J. C. Salarno, H. Blum, and T. Ohnishi, *Biochim. Biophys. Acta*, 1979, **547**, 270.
[464] M. Erecińska and D. F. Wilson, *Arch. Biochem. Biophys.*, 1979, **192**, 80.
[465] C.-A. Yu, L. Yu, and T. E. King, *Arch. Biochem. Biophys.*, 1979, **198**, 314.
[466] B. L. Trumpower and C. A. Edwards, *FEBS Lett.*, 1979, **100**, 13.
[467] S. H. Speck, S. Ferguson-Miller, N. Osheroff, and E. Margoliash, *Proc. Natl. Acad. Sci. USA*, 1979, **76**, 155.
[468] H. R. Bosshard, Z. Zürrer, H. Schägger, and G. von Jagow, *Biochem. Biophys. Res. Commun.*, 1979, **89**, 250.
[469] K-L. Chiang and T. E. King, *J. Biol. Chem.*, 1979, **254**, 1845.
[470] J. Reed, T. A. Reed, and B. Bess, *Proc. Natl. Acad. Sci. USA*, 1979, **76**, 1045.

oxidation, the amount of weakly immobilized label increases.[471] This implies that a more mobile structure exists for the oxidized form; an increase in total concentration of spin after oxidation is also consistent with a magnetic interaction of the spin-label with one of the redox centres in the enzyme.

The subunit of complex III that binds cytochrome b is one of a small number of mitochondrially synthesized proteins, and as such it has been subjected to detailed biosynthetic studies. Alexander *et al.*[472] have mapped the *cob* gene in *Saccharomyces cerevesiae* mutants that have defects in this region; using genetic analysis and restriction endonuclease digestions, they have shown it to contain 3100 base pairs, and, as well as a lack of apocytochrome b, the mutants also showed regulatory effects on cytochrome oxidase. In a second paper, the same group[473] show that the structural gene for apocytochrome b is dispersed over several regions of the *cob* gene. In similar studies, Solioz and Schatz[474] showed that mutations in the intron regions of the *cob* gene cause the accumulation of polypeptides that are immunologically related to cytochrome b, but some are considerably larger. By contrast, mutations in the exon regions yield mainly cytochrome b fragments. In a more detailed study, Haid *et al.*[475] have mapped the *cob* region and found at least five clusters of mutations; three of these are probably coding regions for the cytochrome. Subunit V of complex III is synthesized cytoplasmically in *S. cerevisiae*; translation of total RNA in a cell-free system and pulse-chase labelling in intact yeast spheroplasts suggest that the subunit is synthesized as a larger precursor and then processed either during or subsequent to entry into the mitochondrion.[476] Beattie *et al.*[477] have shown the presence of two mitochondrially synthesized proteins that cross-react with anti-cytochrome b antibody, although they are claimed not to be precursors to the cytochrome. A review on cytochrome bc_1 has appeared.[478]

Cytochrome Oxidase.—Also known as complex IV in the mitochondrial respiratory chain, cytochrome oxidases are found in bacteria in addition to the mitochondria of higher organisms. The bacterial cytochrome oxidases may also react with other terminal electron-acceptors, and are discussed below (see p. 247).

Mitochondrial Cytochrome Oxidase (EC 1.9.3.1). Several reviews have appeared on cytochrome oxidase,[479-484] and they provide an up-to-date account of this

471 U. Dasgupta, D. C. Wharton, and J. S. Rieske, *J. Bioenerg. Biomembr.*, 1979, **11**, 79.
472 N. J. Alexander, R. D. Vincent, P. S. Perlman, D. H. Miller, D. K. Hanson, and H. R. Mahler, *J. Biol. Chem.*, 1979, **254**, 2471.
473 D. K. Hanson, D. H. Miller, H. R. Mahler, N. J. Alexander, and P. S. Perlman, *J. Biol. Chem.*, 1979, **254**, 2480.
474 M. Solioz and G. Schatz, *J. Biol. Chem.*, 1979, **254**, 9331.
475 A. Haid, R. J. Schweyen, H. Bechmann, F. Kaudewitz, M. Solioz, and G. Schatz, *Eur. J. Biochem.*, 1979, **94**, 451.
476 C. Côté, M. Solioz, and G. Schatz, *J. Biol. Chem.*, 1979, **254**, 1437.
477 D. S. Beattie, Y.-S. Chen, L. Clejan, and F.-E. H. Lin, *Biochemistry*, 1979, **18**, 2400.
478 W. A. Cramer, J. Whitmarsh, and P. Horton, *Porphyrins*, 1979, **7**, 71.
479 D. F. Wilson and M. Erecińska, *Porphyrins*, 1979, **7**, 1.
480 B. G. Malmström, *Biochim. Biophys. Acta*, 1979, **549**, 281.
481 'Developments in Biochemistry', ed. T. E. King, Y. Orii, B. Chance, and K. Okunuki, Elsevier, New York, 1979, Vol. 5.
482 F. W. Cope, *Physiol. Chem. Phys.*, 1979, **11**, 261.
483 R. A. Capaldi, *Membr. Proteins*, 1979, **2**, 201.
484 M. A. Kulish and A. F. Mironov, *Bioorg. Khim.*, 1979, **5**, 965.

enzyme. Purification of the enzyme from rat liver[485, 486] and from adrenocortical mitochondria[487] has been described, although most of the current work is still carried out with beef heart or yeast enzymes.

The subunit composition of cytochrome oxidase is still not unequivocally established, and electrophoresis on polyacrylamide gel is most often used to try and establish this; Penttilä *et al.*[488] have described a two-dimensional system, and claim that there are six subunits in the bovine enzyme. Subunit III appears not to be necessary for the oxidation of cytochrome *c*, and the authors stress the importance of carefully characterizing preparations, because their analytical system detected appreciable quantities of aggregates, impurities, and partial breakdown products. A different two-dimensional gel system has been described by Freedman *et al.*,[489] who claim that it avoids migration of free haem *a*; their results are consistent with the latter compound being associated with a complex of subunits I, II, and IV. The same group has also described a very-high-resolution one-dimensional gel system, although the plot of log (molecular weight) *versus* mobility for a number of protein standards is sigmoidal, and subunits I, III, and V show anomalous mobilities as a function of acrylamide concentration ('Ferguson Plot').[490] However, when molecular weight is plotted against retardation coefficient, a linear relationship holds, and molecular weights of the subunits can be determined to $\pm 7\%$. A similar detergent/urea gel electrophoresis method has been described by Verhuel *et al.*[491] This system did not prove to be suitable for scaling up to preparative use, however, and an alternative detergent/gel-permeation chromatographic method was employed. Subunits could be prepared from 350 mg of oxidase, although the separation was critically dependent upon the amount of lipid and of cholate in the preparation. The authors present some preliminary properties of the six major subunits isolated.

Extraction of cytochrome oxidase with 10% formic acid resolves it into two fractions; one is insoluble, and contains the three large hydrophobic subunits (I–III), whilst the soluble portion contains subunits IV—VII.[492a] Haem *a* appears to be largely associated with the former and copper with the latter fractions. Mixing of the two fractions reconstitutes most of the original oxidase activity. Subunit I has been resolved into smaller fractions by oxidation with performic acid;[492b] one of these is formed on partial oxidation, and is electrophoretically identical to subunit III. If oxidized cytochrome oxidase is treated with trypsin or chymotrypsin, about 45% of the protein can be digested away with relatively little change in spectroscopic properties; a decrease in activity of about 80% occurs, although there is no change in the capacity for binding

[485] R. J. Rascatti and P. Parsons, *J. Biol. Chem.*, 1979, **254**, 1586.
[486] T. Nagasawa, H. Nagasawa-Fujimori, and P. C. Heinrich, *Eur. J. Biochem.*, 1979, **94**, 31.
[487] S. S. Mardanyan and R. M. Nalbandyan, *Biokhimiya*, 1979, **44**, 1203.
[488] T. Penttilä, M. Saraste, and M. Wikström, *FEBS Lett.*, 1979, **101**, 295.
[489] J. A. Freedman, R. P. Tracy, and S. H. P. Chan, *J. Biol. Chem.*, 1979, **254**, 4305.
[490] R. P. Tracy and S. H. P. Chan, *Biochim. Biophys. Acta*, 1979, **576**, 109.
[491] F. E. A. M. Verhuel, J. C. P. Boonman, J. W. Draijer, A. O. Muijsers, D. Borden, G. E. Tarr, and E. Margoliash, *Biochim. Biophys. Acta*, 1979, **548**, 397.
[492] (*a*) M. Fry, *Biochem. Biophys. Res., Commun.*, 1979, **90**, 1119; (*b*) M. Fry, G. A. Blondin, and D. E. Green, *ibid.*, 1979, **91**, 192.

cytochrome c.[493] Analysis of the residual protein suggests that it is largely composed of two large hydrophobic subunits, I and II, and the authors suggest that these bind both the prosthetic group and cytochrome c. Fry and Green[494] have incorporated subunit I into lipid vesicles, which then transport ions in either direction in the presence of a salt gradient. These results are consistent with electron-transfer and ion-transport roles for the different subunits.

Amino-acid sequence data have been obtained for subunits II,[495] IV,[496, 497] V,[498] and VII[499, 500] of beef-heart and for subunit II[501] of *N. crassa* cytochrome oxidase. The fungal and beef-heart subunits II have the *N*-terminal sequence homologous, although only the latter has an *N*-terminal *N*-formylmethionine residue;[495] however, it is suggested that the enzyme from *N. crassa* is synthesized as a precursor that has an *N*-formylmethionine terminal residue but which is rapidly cleaved to the mature enzyme.[501] Subunit II of beef heart has a region (residues 56—227) that shows substantial homology with the small 'blue' copper proteins and includes a putative copper-binding site.[495] The subunit VII of beef heart cytochrome oxidase has a single methionine residue, and the sequence in this region is very similar to that around the ligand methionine in *c*-type cytochromes, although the characteristic -Cys-X-X-Cys- sequence of the latter is not present.[500] Whether this is due to a different mode of binding of haem *a* or whether haem *a* does not bind to this subunit is not known. Subunit V of cytochrome oxidase has also been claimed to be a haem-binding subunit, and it shows about 20% homology with the *β*-subunit of haemoglobin.[498]

Cytochrome oxidase exists in a pH-dependent monomer–dimer equilibrium, the proportion of the former increasing with increasing pH; a corresponding loss in the spectral contribution of cytochrome *a* was also observed.[502] However, the authors concluded that the spectral properties of monomeric and dimeric forms are intrinsic, and not directly perturbed by changes in pH. Image-reconstruction of electron micrographs of cytochrome oxidase in two-dimensional crystals in the presence of deoxycholate has led to the suggested structure shown in Figure 7;[503] the oxidase is present as monomers, with both sides exposed to solvent.

The interaction of cytochrome oxidase with lipids is crucial for its biological function, and has been extensively investigated by a wide array of physical and chemical methods. Radioactive photoaffinity-labelled lipids and radioactive membrane-impermeant compounds have been used to try and determine the

[493] J. C. P. Boonman, G. G. M. van Beek, A. O. Muijsers, and B. F. van Gelder, *Mol Cell. Biochem.*, 1979, **26**, 183.
[494] M. Fry and D. E. Green, *Proc. Natl. Acad. Sci. USA*, 1979, **76**, 2664.
[495] G. J. Steffens and G. Buse, *Hoppe-Seyler's Z. Physiol. Chem.*, 1979, **360**, 613.
[496] R. Sacher, G. Buse, and G. J. Steffens, *Hoppe-Seyler's Z. Physiol. Chem.*, 1979, **360**, 1377.
[497] R. Sacher, G. J. Steffens, and G. Buse, *Hoppe-Seyler's Z. Physiol. Chem.*, 1979, **360**, 1385.
[498] M. Tanaka, M. Haniu, K. T. Yasunobu, C.-A. Yu, L. Yu, Y.-H. Wei, and T. E. King, *J. Biol. Chem.*, 1979, **254**, 3879.
[499] G. C. M. Steffans, G. J. Steffans, G. Buse, L. Witte, and H. Nau, *Hoppe-Seyler's Z. Physiol. Chem.*, 1979, **360**, 1633.
[500] G. C. M. Steffans, G. J. Steffans, and G. Buse, *Hoppe-Seyler's Z. Physiol. Chem.*, 1979, **360**, 1641.
[501] W. Machleidt and S. Werner, *FEBS Lett.*, 1979, **107**, 327.
[502] H. E. Auer, M. Sun, and M. Greulick, *Physiol. Chem. Phys.*, 1979, **11**, 9.
[503] S. D. Fuller, R. A. Capaldi, and R. Henderson, *J. Mol. Biol.*, 1979, **134**, 305.

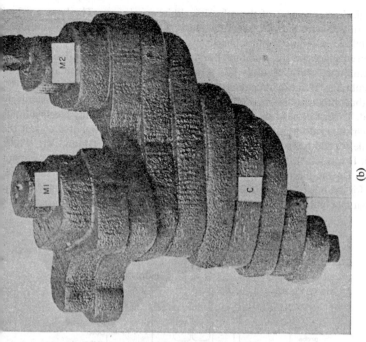

(a)

(b)

Figure 7 *The structure of mitochondrial cytochrome oxidase, based upon electron microscopy image-reconstruction. (a) Several unit cells of the crystal. (b) The orientation of the matrix (M1 and M2) and cytoplasmic (C) domains* (Reproduced by permission from *J. Mol. Biol*, 1979, **134**, 305)

topological arrangement of the subunits within membranes. The latter method has been used with diazobenzenesulphonic acid and subunit-specific antibodies in solubilized, sub-mitochondrial particle and whole mitochondrial preparations; it has been suggested that subunits I and VI are buried, subunits II and III span the membrane, whilst subunits IV, V, and VII extend from the matrix side of the membrane.[504] In experiments with azidophosphatidylcholine, Bisson *et al.*[505a] found that, after irradiation, subunits I, III, VI, and VII were the most extensively labelled. This work has been extended by employing photoaffinity-labelled lipids in which the azido-group is at either the polar or the hydrophobic end of the molecule.[505b] Both resulted in heavy labelling of subunits I, III, and VII, whilst subunit II was labelled to a lesser extent, IV only poorly, and V and VI were not labelled. On the basis of the difference between these two probes, a schematic model for the positioning of the cytochrome molecule has been presented, and it is shown in Figure 8. Cerletti and Schatz[506] have used a similar approach with

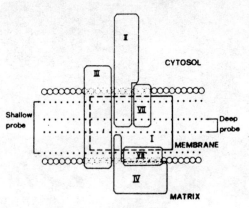

Figure 8 *A schematic model of the arrangement of subunits of cytochrome oxidase in the lipid bilayer. The locations of subunits V and VI are extremely uncertain, and are thus not included*
(Reproduced by permission from *J. Biol. Chem.*, 1979, **254**, 9962)

two hydrophobic aryl azides, and they concluded that, for both mitochondrial and vesicle-bound oxidase, subunits I, II, III, VII, and possibly V were the major membrane-associated portions of the molecule. A comparison of these results shows that although subunits I and III are clearly closely associated with the membrane, the situation with the remaining subunits is far from unequivocal.

The orientation of the haem groups in the enzyme with respect to the membrane has been probed by using oriented multilayers and measuring the angular dependence of the e.p.r. spectrum.[507,508] Using the nitric oxide complex, Barlow and

[504] B. Ludwig, N. W. Downer, and R. A. Capaldi, *Biochemistry*, 1979, **18**, 1401.
[505] (*a*) R. Bisson, C. Montecucco, H. Gutweniger, and A. Azzi, *Biochem. Soc. Trans.*, 1979, **7**, 156; (*b*) R. Bisson, C. Montecucco, H. Gutweniger, and A. Azzi, *J. Biol. Chem.*, 1979, **254**, 9962.
[506] N. Cerletti and G. Schatz, *J. Biol. Chem.*, 1979, **254**, 7746.
[507] C. Barlow and M. Erecińska, *FEBS Lett.*, 1979, **98**, 9.
[508] M. Erecińska, D. F. Wilson, and J. K. Blaise, *Biochim. Biophys. Acta*, 1979, **18**, 3257.

Erecińska[507] determined an Fe–N–O angle of 135° normal to the plane of the membrane. Erecińska *et al.*[508] have also performed similar experiments with native cytochrome oxidase from pigeon breast, and they concluded that the angular dependencies of the x, y, and z components of the g tensor for low-spin ferric were consistent with the haem being normal to the plane of the membrane in all the systems studied. However, the g_x and g_y axes appeared to be differently oriented in oxidized and in ligand-bound (N_3^- or S^{2-}) half-reduced enzymes, and they concluded that this was because the respective e.p.r. signals belonged to the two different haems.

In addition to these structural studies, a considerable amount of work has been carried out on the dynamics of protein–lipid interactions, using either n.m.r. or e.p.r. methods. Deuterium n.m.r. spectra of [^{2}H]lecithins as a function of temperature and added cytochrome oxidase are consistent with an increased disordering of the protein-associated (boundary) lipid near the head-group; above the gel–liquid-crystal transition temperature, the boundary and bulk lipid are in fast exchange on the ^{2}H n.m.r. timescale ($\tau_M \lesssim 10^{-3}$ s).[509] The apparent conflict between this model and that derived from spin-label e.p.r. measurements is due to the quite different timescales that the two techniques sense. In a second paper on ^{2}H n.m.r. quadrupolar splittings and the anisotropies of ^{31}P n.m.r. chemical shifts of 1-(5,5-dideuteriopalmitoyl)-2-oleyl-*sn*-glycero-3-phospho-choline in a mixture of the lipid with cytochrome oxidase the authors showed that the spectroscopic parameters depended on the amount of residual cholate in the enzyme preparation.[510] As the quantity of cholate is reduced, the spectra change from those characteristic of an ordered system to those very like the unperturbed lipid bilayer. These results are discussed in terms of the authors' model for protein–lipid interactions. Falk and Karlsson[511] have observed the ^{1}H n.m.r. spectra of egg phosphatidylcholine in the presence of cytochrome oxidase, and interpreted their results in terms of an immobilized layer of oxidase-associated lipid. Titrations yield 0.7 mg of immobilized lipid per mg of oxidase, which corresponds to three molecular layers of lipid.

Knowles *et al.*[512] have followed the e.p.r. spectrum of spin-labelled phospha-tidylcholine as a function of added cytochrome oxidase from *S. cerevisiae* and concluded that 55 ± 5 lipid molecules are immobilized per protein molecule in the first shell. Beyond this, a second shell is also affected, and then a third less strongly so; possibly a further two or three shells of lipid are weakly per-turbed. These results correlate with the dependence of enzyme activity on the lipid/protein ratio. Preferential binding of spin-labelled phosphatidylcholine to solubilized cytochrome oxidase has been observed by Benga *et al.*,[513] although the authors caution that some contribution to 'free' spin-label signal may arise from molecules in the aqueous phase. The interaction of spin-labelled fatty acid and phospholipid with cytochrome oxidase has been used to show, in the

[509] S. Y. Kang, H. S. Gutowsky, J. C. Hsung, R. Jacobs, T. E. King, D. Rice, and E. Old-field, *Biochemistry*, 1979, **18**, 3257.
[510] D. M. Rice, J. C. Hsung, T. E. King, and E. Oldfield, *Biochemistry*, 1979, **18**, 5885.
[511] K. E. Falk and B. Karlsson, *FEBS Lett.*, 1979, **98**, 25.
[512] P. F. Knowles, A. Watts, and D. Marsh, *Biochemistry*, 1979, **18**, 4480.
[513] G. Benga, O. Popescu, and V. Pop, *Rev. Roum. Biochem.*, 1979, **16**, 175.

latter case, that some mobility is still present even at the level of boundary lipid.[514] Differential scanning calorimetry and e.p.r. of spin-labelled lipids have been used to try and account for breaks in Arrhenius plots for both membrane-bound and reconstituted cytochrome oxidase.[515] Although the breaks did not correspond to phase changes of the lipid, as monitored by the former method, they did correspond to changes in mobility of the lipid as detected by e.p.r. The authors attribute the results to temperature-dependent changes in specific lipid–protein interactions that are induced by changes in equilibrium between immobilized and bulk lipid pools. Yoshida *et al.*[516] have also measured breaks in Arrhenius plots for both soluble and lipid-vesicle-bound cytochrome oxidase; the break temperature ($\sim 26\,^\circ$C) did not correspond with a phase-transition temperature of the lipid. This led the authors to ascribe the change in kinetics to a conformational change between 'hot' and 'cold' forms of the enzyme itself.

On a macroscopic scale, the lateral translational diffusion of cytochrome oxidase in the mitochondrial inner membrane was studied by a combination of monospecific antibody labelling, freeze-fracture microscopy, and temperature-induced phase transitions of the lipid.[517] The results were consistent with the lateral diffusion of cytochrome oxidase occurring either independently or in conjunction with that of other integral membrane proteins.

The biosynthesis of cytochrome oxidase has been extensively studied in *Saccharomyces cerevisiae*, where a large number of *mit⁻* mutants that lack the enzyme are known. The *oxi 1* mutants lack subunit II, and physical mapping of the mitochondrial DNA with restriction endonucleases as well as genetic analysis by recombination with *rho⁻* (*petite*) mutants have located a fragment consisting of 2400 base-pairs that contains at least part of the *oxi 1* locus.[518] In the case of one mutant, the site of mutation could be further located as a region of 75 base-pairs in the centre of this fragment. Another *oxi 1* mutant has been shown to be due to a single deletion in the middle of this locus, and the sequence of the DNA in a fragment of 9000 base-pairs around this deletion was determined.[519] The predicted amino-acid sequence for subunit II was 38% homologous to that of the bovine subunit, although it was noted that a TGA stop codon appeared to be translated as tryptophan. Coruzzi and Tzagoloff[520] have also reported the DNA sequence for the gene for subunit II, and they noted the consistency of the amino-acid sequence that it generated with the composition of the subunit, as well as its substantial homology with the bovine subunit.

Another *mit⁻* mutant of *S. cerevisiae*, namely *oxi 3*, has been analysed by immunoprecipitation and gel electrophoresis of mitochondrial homogenates; comparison with the wild type shows an absence of subunit I.[521] Restriction endonuclease mapping of *oxi 3* mutants has shown that two of these mutants

[514] G. Benga, T. Porumb, and P. T. Frangopol, *Cell Biol. Int. Rep.*, 1979, **3**, 651.
[515] A. S. Denes and N. Z. Stanacev, *Can. J. Biochem.*, 1979, **57**, 238.
[516] S. Yoshida, Y. Orii, S. Kawato, and A. Ikegami, *J. Biochem. (Tokyo)*, 1979, **86**, 1443.
[517] M. Höchli and C. R. Hackenbrock, *Proc. Natl. Acad. Sci. USA*, 1979, **76**, 1236.
[518] T. D. Fox, *J. Mol. Biol.*, 1979, **130**, 63.
[519] T. D. Fox, *Proc. Natl. Acad. Sci. USA*, 1979, **76**, 6534.
[520] G. Coruzzi and A. Tzagoloff, *J. Biol. Chem.*, 1979, **254**, 9324.
[521] E. Keyhani, *Biochem. Biophys. Res. Commun.*, 1979, **89**, 1212.

have large deletions (7500 and 5400 base-pairs) in their DNA, and furthermore that the corresponding RNAs were also missing.[522] A protein, labelled Pr_{IV-VII}, that stimulates the synthesis of the mitochondrially coded subunits *in vitro*, was identified by immunoprecipitation and peptide mapping as corresponding to subunits IV–VII of cytochrome oxidase.[523] This polyprotein has a relative molecular mass that is some 8000 units larger than the sum of the component molecular weights, and pulse labelling has revealed that it is transported to the inner mitochondrial membrane, where it is processed by proteolysis.[524] Two intermediates in this processing were detected. The role of oxygen in the biosynthesis of yeast cytochrome oxidase has been investigated, and, although not required for protein synthesis, oxygen seems to be involved in the assembly of subunits I and II with VI and VII, both in whole cells and in isolated mitochondria.[525] This assembly does not require either cytoplasmic or mitochondrial synthesis of protein.

Cytochrome-oxidase-deficient mutants of *Neurospora crassa* are also known, and immunological methods have shown that at least seven of the eight genes identified as being involved in the biosynthesis of cytochrome oxidase have a regulatory function.[526] In three mutants, a larger version of subunit I is produced, implying a deficiency in processing of the precursor. Further support for the production of a large precursor to subunit I comes from a study where antibody against this subunit recognizes a translation product in an oxidase-deficient mutant that is larger than wild-type subunit I.[527] Induction of oxidase in this mutant can be achieved by treatment with antimycin and is paralleled by a decrease in the amount of anti-(subunit I)-precipitable material. The analysis of temperature-sensitive nuclear mutants of *N. crassa* revealed that the enzyme did not differ from the wild type in either its thermal stability or the composition of the subunits, though no assembly of the subunits occurred in the mutants at the higher temperature.[528] Rascati and Parsons[529] have shown that three subunits of cytochrome oxidase can be synthesized *in vitro* in isolated rat liver mitochondria, and thus do not require the presence of other cellular constituents.

The nature of the haem and copper centres in cytochrome oxidase continues to be of interest, and descriptions of three resonance Raman studies on the enzyme have been published. Because excitation at 441.6 nm (the Soret band of ferrous oxidase) causes photoreduction, Babcock and Salmeen[530] used a flow cell to study native and inhibitor-bound forms; they concluded that the spectrum is mainly due to low-spin haem *a* in the ferric form. Their results are discussed in terms of haem–haem interactions, mainly *via* hydrogen-bonding or conjugation of the formyl group of haem *a*. By contrast, Adar and Erecińska[531] have made use of the photoreduction that is caused by excitation at 441.6 nm

[522] R. Morimoto, A. Lewin, and M. Rabinowitz, *Mol. Gen. Genet.*, 1979, **170**, 1.
[523] R. O. Poyton and E. McKemmie, *J. Biol. Chem.*, 1979, **254**, 6763.
[524] R. O. Poyton and E. McKemmie, *J. Biol. Chem.*, 1979, **254**, 6772.
[525] G. Woodrow and G. Schatz, *J. Biol. Chem.*, 1979, **254**, 6088.
[526] H. Bertrand and S. Werner, *Eur. J. Biochem.*, 1979, **98**, 9.
[527] S. Werner and H. Bertrand, *Eur. J. Biochem.*, 1979, **99**, 463.
[528] F. E. Nargang, H. Bertrand, and S. Werner, *Eur. J. Biochem.*, 1979, **102**, 297.
[529] R. J. Rascati and P. Parsons, *J. Biol. Chem.*, 1979, **254**, 1594.
[530] G. T. Babcock and I. Salmeen, *Biochemistry*, 1979, **18**, 2493.
[531] F. Adar and M. Erecińska, *Biochemistry*, 1979, **18**, 1825.

9

to carry out reductive titrations of whole mitochondrial cytochrome oxidase; they interpreted their results in terms of three states of reduction, and assigned a band at 1609—1623 cm^{-1} as indicating that there are interacting haem moieties. In addition, they suggest that a band at 216 cm^{-1} may indicate a haem–copper interaction, and note that this band is perturbed by exogenous ligands. Excitation in the 600 nm region of the cytochrome oxidase does not lead to photoreduction, and a resonance Raman study of the native, cyanide-bound, and reduced forms has appeared.[532] The spectrum appears to be relatively insensitive to the oxidation state of the protein or to the spin state of the haem a_3; the first observation implies that the absorption at 600 nm does not contain a contribution from copper, as the resonance Raman spectra of type I copper centres are dramatically affected by reduction.

Stevens *et al.*[533] have measured the superhyperfine splittings in the e.p.r. spectra of ferrocytochrome oxidase with ^{14}NO or ^{15}NO present, and their results are consistent with NO being an axial ligand and a second, endogenous, axial nitrogen ligand also being present. They discuss the possibility of the endogenous ligand being a histidine residue. Electron paramagnetic resonance spectroscopy has also been used to study the effect of incorporation of the enzyme into lipid vesicles;[534] the dependence of the high-spin ($g \approx 6$) haem signal on pH was interpreted in terms of a high- to low-spin transition with increasing pH, possibly due to deprotonation of an axial water molecule. Both the e.p.r. spectrum and the activity of the enzyme are consistent with vesicular oxidase experiencing a pH that is one unit lower than that of the bulk medium. In contrast to the reduced enzyme,[533] nitric oxide appears to bind to the copper that is associated with haem a_3 in the oxidized oxidase,[535] and in doing so it yields a species with an e.p.r. signal at $g \approx 6$, presumed to arise from the haem a_3 as the magnetic coupling to the copper is broken. Addition of azide to the ferricytochrome oxidase–NO complex leads to reduction of haem a_3 and the formation of a nitric oxide bridge between copper and haem a_3; the e.p.r. spectrum of this species shows a half-field transition of $g \approx 4$ and a copper hyperfine splitting of 0.020 cm^{-1}. This last observation is consistent with a type II environment for the copper that is associated with haem a_3.

Near-infrared and *X*-ray absorption-edge spectroscopies have been used to characterize the copper centres in resting, mixed-valence state, and CO-reduced forms of cytochrome oxidase.[536] Comparisons with various models, including the blue copper protein stellacyanin, have led the authors to conclude that the copper that is associated with haem a_3 is 'ionic', type I, or 'blue-like', whilst the other copper centre is more covalent. Evidence against the contribution of a type I copper to the 600 nm absorption of oxidized cytochrome oxidase is

[532] D. F. Bocian, A. T. Lemley, N. O. Peterson, G. W. Brudvig, and S. I. Chan, *Biochemistry*, 1979, **18**, 4396.
[533] T. H. Stevens, D. F. Bocian, and S. I. Chan, *FEBS Lett.*, 1979, **97**, 314.
[534] B. Lanne, B. G. Malmström, and T. Vänngård, *Biochim. Biophys. Acta*, 1979, **545**, 205.
[535] T. H. Stevens, G. W. Brudvig, D. F. Bocian, and S. I. Chan, *Proc. Natl. Acad. Sci. USA*, 1979, **76**, 3320.
[536] L. Powers, W. E. Blumberg, B. Chance, C. H. Barlow, J. S. Leigh, J. Smith, T. Yonetani, S. Vik, and J. Peisach, *Biochim. Biophys. Acta*, 1979, **546**, 520.

claimed by Brudvig and Chan,[537] who observed no effect upon addition of either Ag^I or Hg^{II} to the enzyme. Low-frequency e.p.r. spectroscopy (2—4 GHz) has been used to resolve previously undetected hyperfine structure in the 'e.p.r.-visible' copper of cytochrome oxidase;[538] magnetic interaction with another paramagnetic centre in the molecule may account for the observed spectrum. The interaction of this so-called 'visible copper' with a variety of anions has been investigated and it was concluded[539] that the anions could be divided into three groups, *i.e.* (i) those that cause reduction of the copper (CN^-, CNS^-, and S^{2-}), (ii) those that modify its environment (N_3^- and NO_2^-), and (iii) those that have little or no effect (NO_3^- and halides).

Cyanide binds to haem a_3 in a pH-dependent process, and the rate of this reaction has been interpreted in terms of the protonation of an axial haem ligand, induced by a transmembrane potential.[540] The authors suggest that protonation of haem a_3 *in vivo* may be a stage in the formation of a proton gradient. Haems a and a_3 show pH-dependent mid-point potentials in the native enzyme. This result may be explained by the high affinity of the half-reduced form (a_3^{3+}, a^{2+}) for azide.[541] Possible candidates for the ligand that mediates the strong anti-ferromagnetic coupling ($-J > 200$ cm^{-1}) between haem a_3 and copper have been considered by Reed and Landrum.[542] They reject the proposal of a bridging imidazolate and suggest instead a μ-oxo bridge, by comparison with the properties of known μ-oxo-bridged species; their suggested catalytic cycle is shown in Scheme 6.

The rates of reaction of cytochrome oxidase with ferrocytochrome c under a wide variety of conditions have been measured spectrophotometrically or polarographically, and in certain cases up to a 30-fold difference was observed.[543] A correlation between rate of consumption of oxygen and quantity of cytochrome c in the oxidized form in the aerobic steady state was interpreted in terms of the formation of an especially reactive cytochrome c–cytochrome oxidase complex under certain conditions. Dethmer *et al.*[544] have investigated the binding and reactivity of various c-type cytochromes with beef heart and yeast cytochrome oxidases, and from their data they have concluded that other factors, in addition to charge interactions, are important in determining the rates of cross-reactions. When cytochrome oxidase is incorporated into vesicles, approximately 50% of the enzyme molecules are oriented so that they can react with external cytochrome c, the turnover rates being of the order of 300 s^{-1}. During the oxidation of cytochrome c, the internal pH of the vesicles rises when measured by incorporating an indicator.[545] Reversed electron flow, *i.e.* from cytochrome

[537] G. W. Brudvig and S. I. Chan, *FEBS Lett.*, 1979, **106**, 139.
[538] W. Froncisz, C. P. Scholes, J. S. Hyde, Y.-H. Wei, T. E. King, R. W. Shaw, and H Beinert, *J. Biol. Chem.*, 1979, **254**, 7482.
[539] K. A. Markosyan, G. G. Pogosyan, N. A. Paityan, and R. M. Nalbandyan, *Biokhimiya*, 1979, **44**, 844.
[540] I. M. Andreev, V. Yu. Artsatbanov, A. A. Konstantinov, and V. P. Skulachev, *Dokl. Akad. Nauk SSSR*, 1979, **244**, 1013.
[541] J. G. Lindsay, *Biochem. Soc. Trans.*, 1979, **7**, 741.
[542] C. A. Reed and J. T. Landrum, *FEBS Lett.*, 1979, **106**, 265.
[543] L. Smith, H. C. Davies, and M. E. Nava, *Biochemistry*, 1979, **18**, 3140.
[544] J. K. Dethmer, S. Ferguson-Miller, and E. Margoliash, *J. Biol. Chem.*, 1979, **254**, 11 973.
[545] J. M. Wrigglesworth and P. Nicholls, *Biochim. Biophys. Acta*, 1979, **547**, 36.

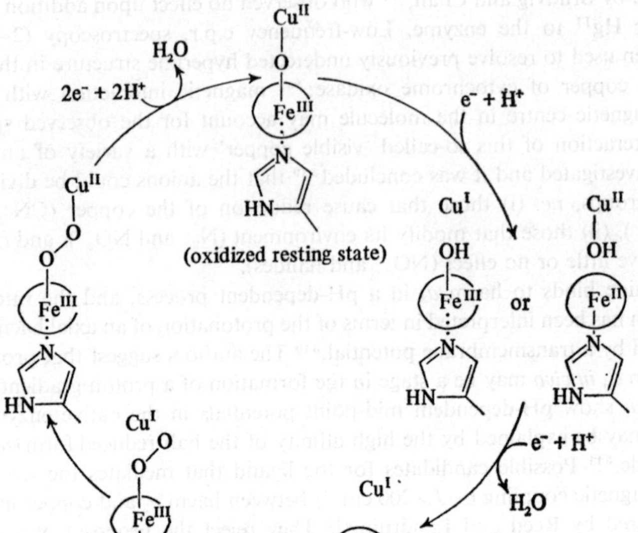

The involvement of a μ-oxo-bridged species in the catalytic cycle of cytochrome oxidase. The ellipse represents haem a_3

Scheme 6

oxidase to cytochrome c, can be achieved by making advantage of the apparent lowering of the mid-point potential of haem a_3 under appropriate conditions (photodissociation of the CO complex in the presence of formate). Thus the photolysis of mixed-valence-state carbonmonoxy-cytochrome oxidase, $[a^{3+}, a_3^{2+}CO]$, in the presence of formate and cytochrome c results in the reduction of both the haem a and the cytochrome c.[546] Spectroscopic properties of a non-equilibrium form of cytochrome oxidase that was prepared by irradiation with thermalized electrons at 77 K have been reported.[547] Kinetic data have long supported the concept of low- and high-affinity sites for cytochrome c on its oxidase; recent results by Krab and Slater[548] on the stimulation of ferrocyanide-dependent consumption of oxygen by sub-stoicheiometric amounts of cytochrome c or excess clupein are consistent with proteins binding at the high-affinity site and regulating the access of electron-donors at the low-affinity site.

The reaction of cytochrome oxidase with oxygen continues to elicit interest, and the use of low-temperature methods has resulted in the elucidation of further

[546] R. Boelens and R. Weaver, *Biochim. Biophys. Acta*, 1979, **547**, 296.
[547] R. M. Davydov, R. Vilu, A. A. Kondrashim, S. N. Magonov, and L. A. Blyumenfel'd, *Biofizika*, 1979, **24**, 565.
[548] K. Krab and E. C. Slater, *Biochim. Biophys. Acta*, 1979, **547**, 58.

details of the reaction. Clore and Chance[549] have investigated the reaction of membrane-bound oxidase with oxygen at low temperatures; from changes in the α, Soret, and near-i.r. bands, the authors have suggested possible structures for the intermediates in the reaction. To provide complementary information to previous optical studies, Karlsson *et al.*[550] have followed the reaction of solubilized cytochrome oxidase with oxygen at 173 K by e.p.r. spectroscopy. Their results were interpreted in terms of the four intermediate branching schemes derived from previous optical measurements (Scheme 7). The reaction

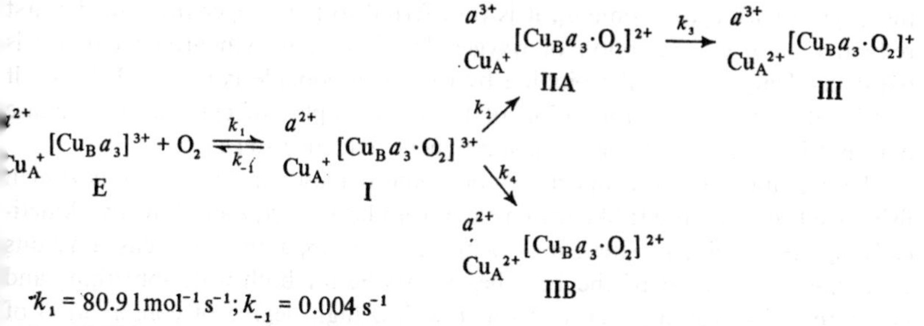

$$\underset{E}{Cu_A^+[Cu_B a_3]^{3+} + O_2} \underset{k_{-1}}{\overset{k_1}{\rightleftharpoons}} \underset{I}{\overset{a^{2+}}{Cu_A^+[Cu_B a_3 \cdot O_2]^{3+}}}$$

$$\overset{k_2}{\nearrow} \quad \underset{IIA}{\overset{a^{3+}}{Cu_A^+[Cu_B a_3 \cdot O_2]^{2+}}} \overset{k_3}{\longrightarrow} \underset{III}{\overset{a^{3+}}{Cu_A^{2+}[Cu_B a_3 \cdot O_2]^{+}}}$$

$$\overset{k_4}{\searrow} \quad \underset{IIB}{\overset{a^{2+}}{Cu_A^{2+}[Cu_B a_3 \cdot O_2]^{2+}}}$$

$k_1 = 80.9 \, l \, mol^{-1} \, s^{-1}; k_{-1} = 0.004 \, s^{-1}$
$k_2 = 0.005 \, 61 \, s^{-1}; k_3 = 0.000 \, 988 \, s^{-1}; k_4 = 0.008 \, 05 \, s^{-1}$

Scheme 7

of cytochrome oxidase with oxygen is often initiated by photolysis of the fully reduced or mixed-valence-state complexes with carbon monoxide. Chance *et al.*[551] have compared the optical spectra of the low-temperature intermediates of these two forms during reaction with oxygen, and they concluded that the spectroscopic (and hence presumably the chemical) properties differ appreciably. The properties of the hydrophobic active-site pocket have been investigated by following changes in the absorbance at 666 nm in the presence of carbon monoxide or oxygen, as a function of temperature.[552] In both native and solubilized enzymes, in the fully reduced or mixed-valence state, the hydrophobic active-site pocket appears to undergo temperature-dependent conformational changes. Finally, a note of warning has been sounded about possible artefacts in low-temperature studies; vitrification–devitrification phase changes in the glasses will cause changes in scattering of light, and hence in the apparent absorbance.[553]

An e.p.r. spectrum of a low-temperature intermediate in the reaction of cytochrome oxidase with oxygen that has g values of 5, 1.78, and 1.79 has been described.[554] The temporal course of the appearance of this spectrum and of a characteristic optical spectrum has led the authors to assign it to the 'oxygenated' oxidase. The kinetic properties of resting and the so-called 'pulsed' cytochrome

[549] G. M. Clore and E. M. Chance, *Biochem. J.*, 1979, **177**, 613.
[550] B. Karlsson, L.-E. Andréasson, R. Aasa, B. G. Malmström, and G. M. Clore, *Acta Chem. Scand., Ser. B*, 1979, **33**, 615.
[551] B. Chance, C. Saronio, and J. S. Leigh, *Biochem. J.*, 1979, **177**, 931.
[552] M. Denis and G. M. Clore, *Biochim. Biophys. Acta*, 1979, **545**, 483.
[553] K. de Fonseka and B. Chance, *Biochem. J.*, 1979, **183**, 375.
[554] R. W. Shaw, R. E. Hansen, and H. Beinert, *Biochim. Biophys. Acta*, 1979, **548**, 386.

oxidases have been compared; the 'pulsed' oxidase contains one molecule of oxygen per cytochrome oxidase and contains four oxidizing equivalents.[555] Analysis of the kinetic data suggest that the 'pulsed' oxidase (i) approaches the steady state more rapidly, (ii) maintains a higher steady-state level of ferricytochrome c, and (iii) exhausts the oxygen more rapidly. A possible role for the 'pulsed' oxidase *in vivo* is discussed in the light of these results. Nicholls[556] has followed the optical changes that occur during the formation of an aerobic carbon-monoxide complex of reduced cytochrome oxidase and its subsequent reactions with inhibitory ligands. It is suggested that, after initial association of the oxygen with cytochrome a_3, it is transferred to the copper prior to the first reductive step. Consumption of oxygen by liposomal cytochrome oxidase is pH-dependent, whereas the binding by carbon monoxide is not.[557] This result implies that, whilst the formation of the oxy-complex of cytochrome oxidase does not involve proton transfers, subsequent reductive steps do.

The activities of mitochondrial cytochrome oxidases from some Amazonian fish have been measured; like the enzyme from beef heart, they show two kinetically significant binding sites for cytochrome c.[558] Apparent Michaelis constants were similar to those of the beef heart enzyme for both water-breathing and facultative air-breathing fish, consistent with a high degree of conservation of the enzyme. Activities of cytochrome oxidase in mitochondria from ascites tumours, host liver, and normal rat liver were found to differ in their response to pH, to calcium ions, to detergents, to chelating agents, and to uncouplers.[559]

A comprehensive review on the role of cytochrome oxidase as a proton pump has appeared,[560] and the advantages and limitations of each of the experimental approaches in studying this function are discussed. The same authors have also presented a mathematical model that relates proton translation to redox processes in membrane-bound cytochrome oxidase.[561] This model successfully predicts a stoicheiometry of one proton ejected per electron transferred, and the thermodynamic consequences of proton pumping are elaborated. Further experiments that have been interpreted in terms of a proton-pumping mechanism have come from Sigel and Carafoli,[562] who measured the stoicheiometry of cytochrome oxidase that had been incorporated into lipid vesicles; they determined that two charges (K^+) are transported for each electron passed, in the presence of valinomycin. Under controlled turnover conditions of cytochrome oxidase in lipid vesicles, the external medium became acidified in a process that is dependent on the ratio of protein to lipid and which can be prevented by the addition of protonophores.[563] A stoicheiometry of 0.9 protons excluded per molecule of ferrocytochrome c oxidized was determined.

555 M. Brunori, A. Colosimo, G. Rainori, M. T. Wilson, and E. Antonini, *J. Biol. Chem.* 1979, **254**, 10769.
556 P. Nicholls, *Biochem. J.*, 1979, **183**, 519.
557 L. Chr. Petersen, *Biochim. Biophys. Acta*, 1979, **548**, 636.
558 M. T. Wilson, J. Bonaventura, and M. Brunori, *Comp. Biochem. Physiol., A*, 1979, **624**, 245.
559 C. Mihai and G. Dinescu-Romalo, *Rev. Roum. Biochim.*, 1979, **16**, 49.
560 M. Wikström and K. Krab, *Biochim. Biophys. Acta*, 1979, **549**, 177.
561 K. Krab and M. Wikström, *Biochim. Biophys. Acta*, 1979, **548**, 1.
562 E. Sigel and E. Carafoli, *J. Biol. Chem.*, 1979, **254**, 10572.
563 R. P. Casey, J. B. Chappell, and A. Azzi, *Biochem. J.*, 1979, **182**, 149.

Proton pumping in whole mitochondria or in reconstituted lipid vesicles can be inhibited by the addition of dicyclohexylcarbodi-imide, although the oxidation of cytochrome c is little affected.[564] In this respect the oxidase is analogous to the di-imide-sensitive ATPase proton pumps. Sorgato and Ferguson[565] have summarized the arguments for a proton-pumping role for cytochrome oxidase; however, Lorusso *et al.*[566] dispute a proton-pumping function for cytochrome oxidase. These authors discuss the mechanism of generation of a transmembrane proton gradient and conclude that it is due to the consumption of protons during reduction of oxygen rather than to membrane translocation. The effect of pH on redox potentials, *i.e.* the redox Bohr effect, has been determined for membrane-bound cytochrome oxidase by simultaneous pH and absorbance measurements.[567] A new method for preparing planar synthetic membranes that incorporate cytochrome oxidase has been described;[568] in the presence of reduced cytochrome c, a transmembrane potential of 60 mV is generated. The addition of cyanide collapses this potential.

Microbial Terminal Oxidases. The bacterial terminal oxidases are a diverse group of enzymes, and in some cases may operate physiologically with compounds other than oxygen as the terminal electron-acceptor. This is certainly the case with the soluble cytochrome cd_1 (EC 1.9.3.2) from *Pseudomonas aeruginosa* which reduces nitrite to nitric oxide. Silvestrini *et al.*[569] have reported a re-evaluation of some structural and steady-state kinetic properties of the enzyme; they have shown that the product, nitric oxide, is an inhibitor of the enzyme and has a much higher affinity for the d_1 haem than for the c haem. This differential binding becomes more marked with increasing pH. In the reaction of this enzyme with oxygen, the kinetic behaviour in the presence and absence of ethyl hydroperoxide supports a two-pathway mechanism, with a high apparent Michaelis constant and maximal velocity at high oxygen tensions.[570] It is claimed that two oxygen molecules bind to the enzymes and are reduced to hydroperoxide, and there then follows a catalatic reaction *in situ* in the high-oxygen-tension pathway. In the low-oxygen-tension pathway, a direct four-electron reduction of hydroperoxide to water occurs.

Preliminary X-ray crystallographic data for the enzyme have been obtained and the unit-cell dimensions are $122.8 \times 87.2 \times 73.4$ Å; the two subunits are crystallographically identical and there is one per asymmetric unit.[571] The spin states of the haems in oxidized and reduced forms, with and without cyanide present, have been measured by variable-temperature m.c.d.[572] In all cases,

[564] R. P. Casey, M. Thelen, and A. Azzi, *Biochem. Biophys. Res. Commun.*, 1979, **87**, 1044.

[565] M. C. Sorgato and S. J. Ferguson, *Biochem. Soc. Trans.*, 1979, **7**, 219.

[566] M. Lorusso, F. Capuano, D. Boffoli, R. Stefanelli, and S. Papa, *Biochem. J.*, 1979, **182**, 133.

[567] S. Papa, F. Guerrieri, and G. Izzo, *FEBS Lett.*, 1979, **105**, 213.

[568] R. H. Tredgold and M. Elgamal, *Biochim. Biophys. Acta*, 1979, **555**, 381.

[569] M. C. Silvestrini, A. Colosimo, M. Brunori, T. A. Walsh, D. Barber, and C. Greenwood, *Biochem. J.*, 1979, **183**, 701.

[570] W. J. Ingledew and M. Saraste, *Biochem. Soc. Trans.*, 1979, **7**, 166.

[571] T. Takano, R. E. Dickerson, S. A. Schichman, and T. E. Meyer, *J. Mol. Biol.*, 1979, **133**, 185.

[572] T. A. Walsh, M. K. Johnson, C. Greenwood, D. Barber, J. P. Springall, and A. J. Thomson, *Biochem. J.*, 1979, **177**, 29.

both haems are low-spin except for the haem d_1 in the reduced enzyme, where it is high-spin. Spectrophotometric redox titrations of the enzyme have been fitted to a high-potential c haem and a d_1 haem of lower potential ($\Delta E_m = 59$ mV), with a negative co-operativity (27 mV) between the c haems in different subunits and a positive co-operativity (27 mV) between the d_1 haems.[573] Although no c–d_1 haem–haem interaction need be invoked to fit the data, a spectroscopic perturbation of a haem in one subunit by the same haem in the second subunit is required. The consequences of these results for the mechanism of the enzyme are discussed.

A similar cytochrome cd_1 to that from *Ps. aeruginosa* can be isolated from *Paracoccus denitrificans*, and the rates of consumption of oxygen and of oxidation of ferrocytochrome c are consistent with this enzyme effecting a direct four-electron reduction of oxygen to water.[574] The lack of effect of superoxide dismutase, catalase, or peroxidase on these rates is also in agreement with the absence of products arising from one- or two-electron reduction. In a second kinetic study of this enzyme, the time courses for reduction of oxygen or of nitrite and for oxidation of cytochrome c-550 were measured and an integrated rate equation was derived.[575] Various possible mechanisms for the enzyme are discussed in the light of the kinetic data. A cytochrome cd has been purified from *Thiobacillus denitrificans*; in the presence of nitrite and a reductant, the e.p.r. spectrum of an enzyme–nitric oxide complex is observed.[576] The authors suggest that both c and d haems bind nitric oxide.

Some bacterial terminal oxidases possess properties very similar to those of the mitochondrial cytochrome oxidase, and the isolation of an enzyme that contains haem aa_3 and copper from the thermophilic bacterium PS3 has been described.[577] The enzyme was active with artificial electron-donors and with yeast cytochrome c, and it could generate a membrane potential when incorporated into vesicles. Another similarity to eukaryotic cytochrome oxidase occurs with the haem-a-containing terminal oxidases from *Thiobacillus novellus* and *Nitrobacter agilis*; their estimated minimal molecular weights are 55 000 and 66 000 respectively, and they each yield subunits with molecular weights of 32 000 and 33 000 and of 40 000 and 27 000.[578] The authors point out the similarity in size of these subunits to the mitochondrially synthesized subunits (I—III) of eukaryotic cytochrome oxidase.

The behaviour of the terminal oxidase of *Azotobacter vinelandii* in whole cells and in subcellular fractions has been investigated, using 3,3'-diaminobenzidine as an electron-donor,[579] and it was shown to be sensitive to cyanide, azide, and hydroxylamine. Mid-point potentials for the cytochrome o of terminal oxidase and an associated cytochrome c_4 from *A. vinelandii* have been determined

573 Y. Blatt and I. Pecht, *Biochemistry*, 1979, **18**, 2917.
574 R. Timkovich and M. K. Robinson, *Biochem. Biophys. Res. Commun.*, 1979, **88**, 649.
575 M. K. Robinson, K. Martinkus, P. J. Kennelly, and R. Timkovich, *Biochemistry*, 1979, **18**, 3921.
576 J. Legall, W. J. Payne, T. V. Morgan, and D. Dervartanian, *Biochem. Biophys. Res. Commun.*, 1979, **87**, 355.
577 N. Sone, T. Ohyama, and Y. Nagawa, *FEBS Lett.*, 1979, **106**, 39.
578 T. Yamanaka, K. Fujii, and Y. Kamita, *J. Biochem. (Tokyo)*, 1979, **86**, 821.
579 P. Jurtshuk, D. N. McQuitty, and W. H. Riley, *Curr. Microbiol.*, 1979, **2**, 349.

n mixtures of the two components.[580] The values obtained were -18 and $+260$ mV respectively, and the former was roughly independent of pH; some of the physiological consequences of these values were discussed. Cytochrome o is also found in *Vitreoscilla*, and spectrophotometric titrations of this enzyme with oxygen have resulted in the identification of several intermediates in the reaction.[581] Its reaction with oxygen has been proposed to occur *via* a complex between an oxygen molecule and two cytochrome molecules, which then reacts with a further molecule of oxygen to yield an oxygenated cytochrome, hydrogen peroxide, and an oxidized cytochrome molecule.

Low-temperature techniques have been employed to study the reaction of cytochrome o of *E. coli* with oxygen in a fashion analogous to that employed with mitochondrial cytochrome oxidase.[582, 583] After photolysis of the reduced carbon-monoxide complex, the first spectroscopic intermediate has characteristics consistent with its formulation as an $[Fe^{2+} \cdot O_2]$ complex.[582] Subsequent reactions give rise to kinetics very similar to those observed with the mammalian oxidase, and similar spectroscopic intermediates can be detected.[583] Rotational mobilities of cytochrome o of *E. coli* and of cytochrome a_1 of *Thiobacillus ferro-oxidans* in the cytoplasmic membrane were determined by polarized laser flash linear dichroism of the reduced carbon-monoxide complexes.[584] The results of these measurements and electron microscopy are consistent with there being a relatively fluid cytoplasmic membrane in *E. coli* and a much more rigid one in *T. ferro-oxidans*.

A terminal oxidase has been isolated and purified to homogeneity from *Photobacterium phosphoreum*; on the basis of optical spectra and determination of the pyridine ferrohaemochrome, it was identified as a cytochrome bd.[585] In the presence of ascorbate and a mediator, an apparent Michaelis constant of less than 10 μmol l^{-1} and a maximum velocity of 4×10^3 moles per mole of enzyme per minute were determined for consumption of oxygen. Finally, two papers have appeared that describe the characterization of the respiratory chain of the protozoan *Paramecium tetraurelia*.[586, 587]

4 Non-Haem Iron-containing Oxidoreductases

Lipoxygenase (EC 1.13.11.12).—The mechanism of action of soybean lipoxygenase has recently been reviewed,[588] and a description of the isolation of four isozymes from pea seeds and some of their physicochemical properties have been presented.[589] Two isozymes of the wheat-germ enzyme have been prepared and

[580] T. Yang, D. O'Keefe, and B. Chance, *Biochem. J.*, 1979, **181**, 763.
[581] B. Tyree and D. A. Webster, *J. Biol. Chem.*, 1979, **254**, 176.
[582] R. K. Poole, A. J. Waring, and B. Chance, *FEBS Lett.*, 1979, **101**, 56.
[583] R. K. Poole, A. J. Waring, and B. Chance, *Biochem. J.*, 1979, **184**, 379.
[584] P. B. Garland, M. T. Davison, and C. H. Moore, *Biochem. Soc. Trans.*, 1979, **7**, 1112.
[585] H. Watanabe, Y. Kamita, T. Nakamura, A. Takimoto, and T. Yamanaka, *Biochim. Biophys. Acta*, 1979, **547**, 70.
[586] J. Doussière, A. Sainsard-Chanet, and P. V. Vignais, *Biochim. Biophys. Acta*, 1979, **548**, 224.
[587] J. Doussière, A. Sainsard-Chanet, and P. V. Vignais, *Biochim. Biophys. Acta*, 1979, **548**, 236.
[588] J. F. G. Vliegenthart, G. A. Veldink, and J. Boldingh, *J. Agric. Food Chem.*, 1979, **27**, 623.
[589] S. Yoon and B. P. Klein, *J. Agric. Food Chem.*, 1979, **27**, 955.

an activator has been isolated from the same source.[590] One of the isozymes could be further separated into two distinct forms, and the kinetic properties of both the isozymes and the activator have been described. The existence of multiple forms of lipoxygenase was further illustrated by Aoshima,[591] who resolved electrophoretically homogenous lipoxygenase 1 into five distinct bands of activity by high-performance liquid chromatography. Each peak had a different specific activity, and the author discusses the advantages and disadvantages of the technique.

Three forms of the lipoxygenase molecule have been implicated in catalysis, and Spaapen *et al.*[592] have analysed the c.d. spectra of each form; each shows a quite distinct spectrum in the visible region but they are essentially identical in the ultraviolet. An analysis of the far-u.v. c.d. reveals 34% of α-structure, 27% of β-structure, and 39% of random coil. The e.p.r. spectrum of the nitric-oxide complex of the enzyme is claimed to be due to two species.[593] One, with g values of around 4, has been analysed in terms of an $S=\frac{3}{2}$ spin state from antiferromagnetic coupling of a high-spin ferrous nucleus with the nitric oxide. The other species, with g values less than 2, has also been attributed to a spin-coupled system with the zero-field splitting much larger than the exchange-coupling constant.

A modified Clark-type oxygen electrode has been described for use in lipoxy-genase assays;[594] the modifications lead to improved accuracy and sensitivity. The interaction of 13-hydroperoxylinoleic acid and lipoxygenase has been investigated by stopped-flow absorbance and fluorescence measurements,[595] and the results obtained were interpreted in terms of the series of reactions shown in Scheme 8. Soybean lipoxygenase-2 shows a much lower pH optimum

$$E \cdot Fe^{II} + ROOH \longrightarrow E \cdot Fe^{II} \cdot ROOH \xrightarrow{\ OH^-\ } E \cdot Fe^{III} \xrightarrow{\ ROOH\ } E \cdot Fe^{III} \cdot ROOH$$
(native)   (yellow)   (purple)
                                                                        +
                                                                      RO·

Scheme 8

than isozyme-1; however, this is apparently due to substrate inhibition at higher values of pH.[596] At pH 9 this isozyme produces a novel product in the 9-L hydroperoxylinoleic acid, and the similarities to prostaglandin synthetase are discussed. Although linoleic acid is the physiological substrate, lipoxygenase will react with certain furans and will catalyse the transformation shown in Scheme 9. Both singlet dioxygen and *m*-chloroperoxybenzoic acid also catalyse this reaction, and the authors discuss the mechanism in terms of an enzyme-

590 J. M. Wallace and E. L. Wheeler, *Phytochemistry*, 1979, **18**, 389.
591 H. Aoshima, *Anal. Biochem.*, 1979, **95**, 371.
592 L. J. M. Spaapen, G. A. Veldink, T. J. Liefkens, J. F. G. Vliegenthart, and C. M. Kay, *Biochim. Biophys. Acta*, 1979, **574**, 301.
593 J. C. Salarno and J. N. Siedow, *Biochim. Biophys. Acta*, 1979, **579**, 246.
594 H. W. Cook, G. Ford, and W. E. M. Lands, *Anal. Biochem.*, 1979, **96**, 341.
595 H. Aoshima, T. Kajiwara, A. Hatanaka, and H. Nakatani, *Agric. Biol. Chem.*, 1979, **43**, 167.
596 C. P. A. van Os, G. P. M. Rijke-Schilder, and J. F. G. Vliegenthart, *Biochim. Biophys. Acta*, 1979, **575**, 479.

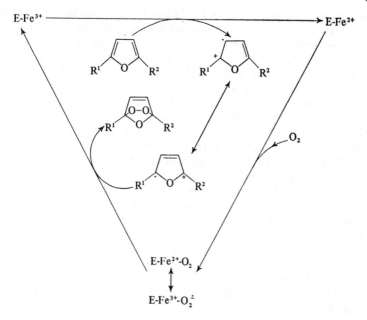

Scheme 9

superoxide complex and furan radical intermediates.[597] van Os *et al.*[598] have described a method for determining the enantiomeric composition of the products from the action of lipoxygenase on linoleic acid. They employed a series of chemical steps to convert the hydroperoxy products into α-trifluoromethyl-phenylacetate esters and then measured their ^{1}H n.m.r. spectra in the presence of the shift reagents Eu(fod)$_3$ or Pr(fod)$_3$.

During turnover of lipoxygenase there is a co-oxidation of β-carotene; soybean lipoxygenase-2 has been resolved into two fractions, both of which effect this co-oxidation in a reaction that is stimulated by the addition of 13-hydroperoxylinoleic acid.[599] A test paper for lipoxygenase has been described that is impregnated with *NN*-dimethyl-*p*-phenylenediamine and linoleic acid.[600] Silymarin has been shown to inhibit the enzyme.[601]

Lipoxygenase activity has been detected in mammalian as well as plant tissues, and the purification of the enzyme of rabbit reticulocytes has been described by Rapoport *et al.*[602] The enzyme has a molecular weight of 78 000 and contains 2 gram atoms of iron, and it is responsible for the 'respiratory protein inhibition' activity in cell lysates. During peroxidation of lipids it is autoinactivated, and

[597] R. F. Boyer, D. Litts, J. Kostishak, R. C. Wijesundera, and F. D. Gunstone, *Chem. Phys. Lipids*, 1979, **25**, 237.
[598] C. P. A. van Os, M. Vente, and J. F. G. Vliegenthart, *Biochim. Biophys. Acta*, 1979, **574**, 103.
[599] W. Grosch and G. Laskawy, *Biochim. Biophys. Acta*, 1979, **575**, 439.
[600] B. O. Lumen and S. J. Kazeniac, *Anal. Biochem.*, 1979, **99**, 118.
[601] F. Fiebrich and H. Kock, *Experientia*, 1979, **35**, 1548.
[602] S. M. Rapoport, T. Schewe, R. Wiesner, W. Halangk, P. Ludwig, M. Janicke-Höhne, C. Tannert, C. Hiebsch, and D. Klatt, *Eur. J. Biochem.*, 1979, **96**, 545.

measurement of the amount of malondialdehyde that is produced has revealed a higher activity of the enzyme with mitochondrial lipids than with lipids of the plasma membrane. The authors discuss the role of the enzyme in the maturation of erythrocytes. Rat testis lipoxygenase acts on linoleic acid to yield 33% and 67% respectively of the 9- and 13-hydroperoxy-derivatives.[603] However, additional polar products are also produced, and the major one of these has been identified. The nature of the polar product indicates that a hydroperoxyisomerase is possibly present in the purified lipoxygenase.

Although nearly all of the lipoxygenase enzymes that are known contain non-haem iron, that isolated from *Fusarium oxysporum* is a haemoprotein. In a study of the binding of substrates, the iron was shown to undergo a change in spin state from low- to high-spin, as judged from the e.p.r. spectrum.[604] The same change in spin state is observed at pH 12, although in this case the lines are narrower, and this may be related to the higher activity at this pH. Both catalase and superoxide dismutase inhibit the enzyme of *F. oxysporum*, although scavengers of hydroxyl radical have no effect; low concentrations of hydrogen peroxide (1—10 mmol l⁻¹) activate the enzyme.[605] These results suggest that superoxide and hydrogen peroxide are produced endogenously and are essential intermediates in the reaction. During storage of the enzyme of *F. oxysporum*, heterogeneity develops; this has been shown to be due to partial degradation of the enzyme.[606]

Hydrogenase.—Hydrogenases are widely distributed enzymes, and their physiological role may be either in the evolution of hydrogen (EC 1.18.3.1) or in its utilization (EC 1.12.1.2 and EC 1.12.2.1); reviews on metabolism of hydrogen in photosynthetic organisms[607] and on the properties and possible applications of hydrogenase enzymes[608] have appeared. The enzyme may occur as either a soluble or a membrane-bound form, and the purification of both types from *Alcaligenes eutrophus* has been described;[609,610] properties of these and other purified hydrogenases are collected in Table 4.[609-619] Although not purified, two types of hydrogenase have been found in bacteroids in the root nodules of soybean and lupin; only one of these types was stimulated by ATP, and they

[603] S. Grossman, I. Shahin, and B. Sredni, *Biochim. Biophys. Acta*, 1979, **572**, 293.
[604] Y. Matsuda, T. Beppu, and K. Arima, *Biochem. Biophys. Res. Commun.*, 1979, **86**, 319.
[605] Y. Matsuda, T. Beppu, and K. Arima, *Agric. Biol. Chem.*, 1979, **43**, 1179.
[606] Y. Matsuda, T. Beppu, and K. Arima, *Agric. Biol. Chem.*, 1979, **43**, 189.
[607] A. Ben-Amotz, *Encycl. Plant Physiol.* (*New Ser.*), 1979, **6**, 497.
[608] A. I. Krasna, *Enzyme Microb. Technol.*, 1979, **1**, 165.
[609] K. Schneider, R. Cammack, H. G. Schlegel, and D. O. Hall, *Biochim. Biophys. Acta*, 1979, **578**, 445.
[610] B. Schink and H. G. Schlegel, *Biochim. Biophys. Acta*, 1979, **567**, 315.
[611] E. E. Pinchukova, S. D. Varfolomeev, and E. N. Kondrat'eva, *Biokhimiya*, 1979, **44**, 605.
[612] M. W. W. Adams and D. O. Hall, *Biochem. J.*, 1979, **183**, 11.
[613] M. W. W. Adams and D. O. Hall, *Arch. Biochem. Biophys.*, 1979, **195**, 288.
[614] G. S. Schoenmaker, L. F. Oltmann, and A. H. Stouthamer, *Biochim. Biophys. Acta*, 1979, **567**, 511.
[615] M. J. Llama, J. L. Serra, K. K. Rao, and D. O. Hall, *FEBS Lett.*, 1979, **98**, 342.
[616] C. van Dijk, S. G. Mayhew, H. J. Grande, and C. Veeger, *Eur. J. Biochem.*, 1979, **102**, 317.
[617] E. Sim and P. M. Vignais, *Biochim. Biophys. Acta*, 1979, **570**, 43.
[618] E. Sim and R. B. Sim, *Eur. J. Biochem.*, 1979, **97**, 119.
[619] D. J. Arp and R. H. Burris, *Biochim. Biophys. Acta*, 1979, **570**, 221.

Table 4 *Properties of some hydrogenases that were purified in 1979*

Source[a]	Nature	Mol. wt	Prosthetic groups	Notes	Ref.
A. eutrophus	Soluble	—	$2 \times [Fe_4S_4^*]$, $2 \times [Fe_2S_2^*]$, FAD	Values of E_i are -455 mV (Fe_4 centre) and -325 mV (Fe_2 centre); there is spin coupling between iron clusters	609
A. eutrophus	Soluble	200 000 ($2 \times 65\,000 + 2 \times 30\,000$)	—	Inactivation and kinetic data for enzyme when purified and *in vivo*	611
A. eutrophus	Membranous	98 000 ($67\,000 + 31\,000$)	—	No FAD	610
E. coli	Membranous	113 000 ($2 \times 56\,000$)	13Fe, 12S*	Kinetic and inhibition data presented	612
R. rubrum	Membranous	66 000	4Fe, 4S*	E.p.r. spectrum of HiPIP	613
P. mirabilis	Membranous	205 000 ($2 \times 63\,000 + 2 \times 33\,000$)	24Fe, 24S*	E.p.r. spectrum of HiPIP	614
S. maxima	Soluble	56 000	—	Inhibited by CO	615
M. elsdenii	Soluble	50 000	12Fe, 12S*	Detailed kinetic analysis	616
P. denitrificans	Membranous	63 000	—	Forms tetramers in detergents and aggregates when detergent is removed	617, 618
R. japonicum	Membranous	63 000	12Fe	Kinetic data are consistent with re-cycling of H_2 that is generated by nitrogenase	619

(a) A. eutrophus is *Alcaligenes eutrophus*, E. coli is *Escherichia coli*, R. rubrum is *Rhodospirillum rubrum*, P. mirabilis is *Proteus mirabilis*, S. maxima is *Spirulina maxima*, M. elsdenii is *Megasphaera elsdenii*. P. denitrificans is *Paracoccus denitrificans*, and R. japonicum is *Rhizobium japonicum*.

also differed in size and in their inhibition by CO.[620] Particulate and soluble hydrogenase enzymes have also been found in *Methanobacterium* strain GR2.[621] The blue-green alga *Anacystis nidulans* has a membranous hydrogenase during both aerobic and anaerobic growth. In the former case it appears to be involved in electron flow from hydrogen to oxygen and concomitant oxidative phosphorylation,[622] whilst anaerobically it seems to catalyse the light-driven electron flow from hydrogen to ferredoxin for subsequent photoreduction of carbon dioxide.[623]

Core-extrusion experiments have been carried out on the hydrogenase from *Desulfovibrio vulgaris* and, with the results of inhibition of the enzyme by mercury(II), they are consistent with the presence of $Fe_4S_4^*$ clusters.[624] The preparation and the reconstitution of the enzyme from the same source have been described; however, the reconstituted material has only 1% of the activity of the native enzyme.[625]

A commonly used artificial electron-donor for hydrogenase is methyl viologen, and Gluck *et al.*[626] have compared the activities of the hydrogenase of *Clostridium pasteurianum* and that of *Desulfovibrio sulphuricans* with both methyl viologen and with polymeric viologens. The binding and maximal rates of evolution of hydrogen in this system are discussed in terms of potential for solar production of hydrogen. Methyl viologen has also been compared with cytochrome c_3 in a hydrogen-uptake assay with *D. vulgaris* hydrogenase;[627] from an analysis of the kinetics in the presence and absence of inhibitors, it was concluded that both electron-acceptors are bound at the same site on the enzyme. The reaction between cytochrome c_3 and hydrogenase has also been studied for the proteins in the dry state, and Kimura *et al.*[628] found that dried hydrogenase is capable of catalysing the interconversion of ortho- and para-hydrogen, hydrogen–deuterium exchange, and reversible reduction of the cytochrome. The authors also demonstrated exchange between gaseous hydrogen and protons that are bound to the cytochrome. Photo-evolution of hydrogen from water has been achieved by coupling the photosensitized reduction of methyl viologen by triphenylamine to the reaction of hydrogenase.[629] In a further characterization of hydrogenase from *D. vulgaris*, Okura *et al.*[630] have proposed a mechanism that involves double binding of substrate on the basis of kinetic data.

A common property of hydrogenases is their sensitivity towards dioxygen; even those that are 'oxygen-stable' are only so in the resting state, and they tend

[620] T. Suzuki and Y. Maruyama, *Agric. Biol. Chem.*, 1979, **43**, 1833.
[621] R. C. McKellar and G. D. Sprott, *J. Bacteriol.*, 1979, **139**, 231.
[622] G. A. Peschek, *Biochim. Biophys. Acta*, 1979, **548**, 203.
[623] G. A. Peschek, *Biochim. Biophys. Acta*, 1979, **548**, 187.
[624] I. Okura, K. I. Nakamura, and S. Nakamura, *J. Mol. Catal.*, 1979, **6**, 307.
[625] I. Okura, K. I. Nakamura, and S. Nakamura, *J. Mol. Catal.*, 1979, **6**, 299.
[626] B. R. Glick, W. G. Martin, J. J. Giroux, and R. E. Williams, *Can. J. Biochem.*, 1979, **57**, 1093.
[627] I. Okura, S. Nakamura, and K. I. Nakamura, *J. Mol. Catal.*, 1979, **5**, 315.
[628] K. Kimura, A. Suzuki, H. Inokuchi, and T. Yagi, *Biochim. Biophys. Acta*, 1979, **567**, 96.
[629] I. Okura and N. K. Im-Thuan, *J. Mol. Catal.*, 1979, **6**, 449.
[630] I. Okura, K. I. Nakamura, and S. Nakamura, *J. Mol. Catal.*, 1979, **6**, 311.

to be inactivated during aerobic turnover. Klibanov *et al.*[631] have shown that the stability of the hydrogenase of *C. pasteurianum* towards oxygen could be considerably enhanced by the presence of agents that chelate metal ions, such as EDTA, sulphosalicylic acid, or Chelex 100. They suggest that protection is afforded by removal of the metal ions that accelerate the reaction of dioxygen with thiol groups. Scavengers for superoxide or hydroxyl radicals or for hydrogen peroxide were ineffective in protecting against oxidative inactivation.

Other Non-Haem Iron-containing Oxidoreductases.—A method has been described for the purification of substantial quantities of **toluene dioxygenase** from *Pseudomonas putida*, using affinity chromatography.[632] The enzyme has a molecular weight of 151 000 and contains two iron atoms and two acid-labile sulphide groups per molecule; in the presence of NADH, a flavoprotein reductase, and a ferredoxin, the enzyme catalyses the reaction that is shown in Scheme 10.

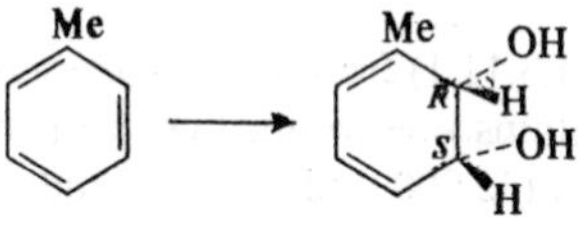

Scheme 10

Benzene dioxygenase (EC 1.14.12.3) can also be isolated from *Ps. putida*, and an improved purification method has been described for two iron–sulphur components.[633] Protein B has one Fe_2S_2* centre whilst protein A1 has two Fe_2S_2* centres; the respective molecular weights are 12 000 and 215 300. The electron flow is from NADH to a flavoprotein to protein B to protein A, which binds oxygen and substrate. Iron(II) is also required for activity. Methane monooxygenase from *Methylococcus capsulatus* (Bath) is another multi-component enzyme system, and the purification of component C (a metallo-flavoprotein) has been described.[634] Chemical analysis and core-extrusion measurements show the presence of one FAD and one Fe_2S_2* centre per molecule. The kinetic properties of the protein as an NADH acceptor–reductase are described. The isolation of **protocatechuate-2,3-dioxygenase** from *Bacillus macerans* has been described, and evidence for the transient accumulation of 5-carboxy-2-hydroxymuconic semialdehyde is presented.[635]

Optical spectroscopy, e.p.r. spectroscopy, and reaction with *o*-phenanthroline and subsequent reconstitution support the view that the iron in **anthranilate hydroxylase** (EC 1.14.16.3) is in the ferrous state. When anthranilate was added, a high-spin ferric e.p.r. signal appeared.[636]

[631] A. M. Klibanov, N. O. Kaplan, and M. D. Kamen, *Biochim. Biophys. Acta*, 1979, **547**, 411.
[632] V. Subramanian, T.-N. Liu, W. K. Yeh, and D. T. Gibson, *Biochem. Biophys. Res. Commun.*, 1979, **91**, 1131.
[633] S. E. Crutcher and P. J. Geary, *Biochem. J.*, 1979, **177**, 393.
[634] J. Colby and H. Dalton, *Biochem. J.*, 1979, **177**, 903.
[635] R. L. Crawford, J. W. Bromley, and P. E. Perkins-Olson, *Appl. Environ. Microbiol.*, 1979, **37**, 614.
[636] V. Subramanian, M. Sugmaran, and C. S. Vaidyanathan, *Indian J. Biochem. Biophys.*, 1979, **16**, 370.

Amongst the most studied of the non-haem iron-containing dioxygenases is **protocatechuate-3,4-dioxygenase** (EC 1.13.11.3), and Bull *et al.*[637] have reported the resonance Raman spectrum of this enzyme from *Ps. putida* in the oxidized and reduced forms. Resonance-enhanced bands have been assigned to tyrosine ligands and compared to those previously observed in the enzyme from *Ps. aeruginosa* and in iron transferrin. The occurrence of additional low-frequency bands suggests that there are extra ligands. Keyes *et al.*[638] have obtained resonance Raman spectra of the stable enzyme–substrate–oxygen complex, using a computer-controlled spectrometer.[639] Resonance-enhanced tyrosine bands are shifted ~ 10 cm^{-1} to lower frequencies compared to the native and binary (enzyme–substrate) forms. Binding of dioxygen is suggested to be distal from the iron, and the shift has been assigned to conformational changes at the iron centre. Kohlmiller and Howard[640,641] have reported the complete amino-acid sequence of the 200 residues in the α subunit of protocatechuate-3,4-dioxygenase.

The **pyrocatechase** (EC 1.3.11.1) from *Pseudomonas arvilla* C-1 has a molecular weight of 63 000 and contains one iron(III) centre per molecule; the enzyme is composed of two subunits (32 000 and 30 000), and their amino-acid compositions and *N*-terminal sequences have been reported.[642] Que and Heistand[643] have reported the resonance Raman spectra of pyrocatechase, and concluded that iron–tyrosine ligation is also present in this enzyme. Binding of a substrate changes the spectrum, and comparison with iron–catechol complexes suggests that the substrate binds as the dianion form.

The **dimethylamine dehydrogenase** from *Hyphomicrobium X* contains four iron and four acid-labile sulphide groups, as well as 6-*S*-cysteinesulphonyl-FMN; in this respect it resembles **trimethylamine dehydrogenase** (EC 1.5.99.7) from *bacterium* W_3A_1.[644] Phenylhydrazine acts on the latter enzyme as a typical suicide substrate in the presence of an electron-acceptor;[645] the product of dehydrogenation, *i.e.* phenyldiazene, inhibits the enzyme much more rapidly, and does not require the presence of an electron-acceptor. Analysis of the flavin of the inhibited enzyme shows that it has been arylated at C-4a.

Spinach **nitrite reductase** contains one $Fe_4S_4^*$ centre and one sirohaem group per molecule; as isolated, the sirohaem is high-spin ferric. However, it is likely that the enzyme is devoid of some of the sirohaem after isolation.[646] The $Fe_4S_4^*$ centre is at an unusually low potential, although the binding of carbon monoxide to the sirohaem results in a shift to higher potentials. Kinetic studies show that

637 C. Bull, D. P. Ballou, and I. Salmeen, *Biochem. Biophys. Res. Commun.*, 1979, **87**, 836.
638 W. E. Keyes, T. M. Loehr, M. L. Taylor, and J. S. Loehr, *Biochem. Biophys. Res. Commun.*, 1979, **89**, 420.
639 T. M. Loehr, W. E. Keyes, and P. A. Pincus, *Anal. Biochem.*, 1979, **96**, 456.
640 N. A. Kohlmiller and J. B. Howard, *J. Biol. Chem.*, 1979, **254**, 7302.
641 N. A. Kohlmiller and J. B. Howard, *J. Biol. Chem.*, 1979, **254**, 7309.
642 C. Nakai, H. Kagamiyama, Y. Saeki, and M. Nozaki, *Arch. Biochem. Biophys.*, 1979, **195**, 12.
643 L. Que, Jr, and R. H. Heistand, *J. Am. Chem. Soc.*, 1979, **101**, 2219.
644 D. J. Steenkamp, *Biochem. Biophys. Res. Commun.*, 1979, **88**, 244.
645 J. Nagy, W. C. Kenney, and T. P. Singer, *J. Biol. Chem.*, 1979, **254**, 2684.
646 J. R. Lancaster, J. M. Vega, H. Kamin, N. R. Orme-Johnson, W. H. Orme-Johnson, R. J. Krueger, and L. M. Sigel, *J. Biol. Chem.*, 1979, **254**, 1268.

the iron–sulphur centre is re-oxidized rapidly enough ($k = 100$ s^{-1}) by nitrite to be involved in catalysis.

5 Molybdenum-containing Oxidoreductases

Xanthine Oxidase and Xanthine Dehydrogenase.—These two enzymes are closely related, and are distinguished by the nature of the terminal electron-acceptors; the oxidase (EC 1.2.3.2) preferentially uses oxygen and the dehydrogenase (EC 1.2.1.37) uses NAD$^+$.

Xanthine dehydrogenase has been purified and characterized from *Pseudomonas synxantha* A3,[647] from *Streptomyces cyanogenus*,[648] and from locusts.[649] The enzyme from the *Pseudomonas* sp. only uses NAD$^+$, and not NADP$^+$, as an electron-acceptor and it is inhibited by Hg^{II}, Cu^{II}, Ag^{II}, cyanide, and *p*-chloromercuribenzenesulphonate. Large quantities of iron (14 mole per mole) can be detected in the enzyme from *S. cyanogenus*, and also FAD and acid-labile sulphide (2 mole per mole); little molybdenum was found. The enzyme is also unusual in its substrate specificity, oxidizing hypoxanthine faster than xanthine and also using guanine as a substrate. By contrast, the locust enzyme appears to be a normal representative of this class of protein, and the levels of enzyme during development have been measured.[650]

The appearance of xanthine oxidase in serum is an indicator of certain types of hepatitis, and McHale *et al.*[651] have described a fluorometric enzyme assay that is applicable to serum samples. Heinz *et al.*[652] have also developed an assay for serum xanthine oxidase in which the coupled system that is shown in Scheme 11 is employed and the NAD(P)H is monitored spectrophotometrically.

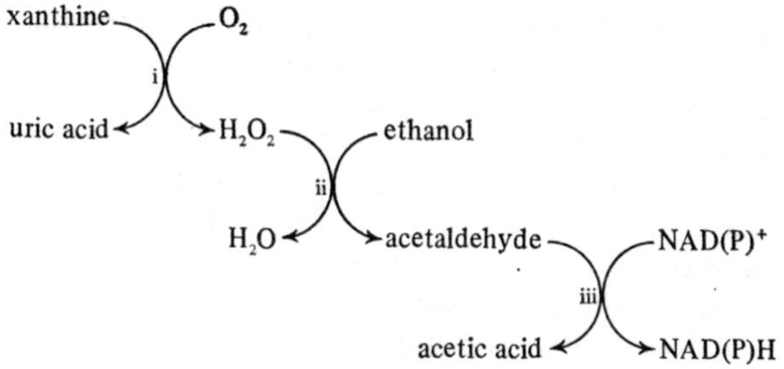

Reagents: i, xanthine oxidase; ii, catalase; iii, aldehyde dehydrogenase

Scheme 11

Another spectrophotometric assay for rat liver xanthine dehydrogenase, xanthine oxidase, and the intermediate form in the conversion of the former into the latter has been utilized by Kamiński and Jezewska[653] to follow the modification

[647] T. Sakai and H.-K. Jun, *Agric. Biol. Chem.*, 1979, **43**, 753.
[648] T. Ohe and Y. Watanabe, *J. Biochem (Tokyo)*, 1979, **86**, 45.
[649] T. J. Hayden and E. J. Duke, *Insect Biochem.*, 1979, **9**, 583.
[650] T. J. Hayden and E. J. Duke, *Insect Biochem.*, 1979, **9**, 589.
[651] A. McHale, H. Grimes, and M. P. Coughlan, *Int. J. Biochem.*, 1979, **10**, 317.
[652] F. Heinz, S. Reckel, and J. R. Kalden, *Enzyme*, 1979, **24**, 239.
[653] Z. W. Kamiński and M. M. Jezewska, *Biochem. J.*, 1979, **181**, 177.

from dehydrogenase to oxidase in the presence of copper(II). Fresh preparations of the enzyme contained no oxidase form at all.

The *N*-methyl- and *NN'*-dimethyl-derivatives of allopurinol (an inhibitor of xanthine oxidase), as well as the 4-thioxo-homologues, have been tested as inhibitors of the enzyme.[654] Only the 7-methyl-substituted derivative and its 4-thioxo-homologue were effective, and both were also oxidized, suggesting that, for reaction to occur at C-6, both N-1 and N-2 of the pyrazole ring must be bound to the enzyme, causing tautomerization so that the double-bond is at the 5,6 position in the pyrimidine ring. *p*-Chloromercuribenzoate reacts with xanthine oxidase and apparently forms two complexes with the enzyme; the first is formed rapidly and reversibly, competitive with substrate. In a second, slower reaction a sulphydryl group is irreversibly modified, leading to inactivation.[655] The inhibition of xanthine oxidase by a series of substituted 1,2,3-triazoles has been investigated; K_i values lie in the range 200—2000 μmol l^{-1}.[656] Autoinactivation of xanthine oxidase during turnover of acetaldehyde has been investigated by Lynch and Fridovich.[657] The effects of catalase, superoxide dismutase, and the variation of oxygen tension, and also the protection that is afforded by histidine, have led the authors to conclude that the inactivation is due to $^1O_2$ or something similar. Aerobic turnover of xanthine dehydrogenase also results in inactivation, and hydrogen peroxide, which is produced by the reaction of the reduced enzyme, has been shown to be the proximate cause.[658] Hydrogen peroxide appears to react much more rapidly with the reduced than with the oxidized enzyme, and the authors have presented evidence to support the hypothesis that the cyanolysable sulphur atom at the molybdenum centre is the site of reaction.

In the presence of an electron-donor, xanthine oxidase catalyses the *cis*↔*trans* isomerization of 3-substituted nitrofurans at a much greater rate than it reduces them.[659] This has led Tatsumi *et al.*[659] to postulate a mechanism based upon radical anion intermediates, and some support is provided by pulse-radiolysis experiments with the nitrofurans. Reduction of xanthine dehydrogenase by both xanthine and NADH has been followed spectrophotometrically, and whilst the degree of reduction by xanthine is dependent upon the content of functional active sites, the same is not true in the case of NADH.[660] The authors also show that the persistence of a flavin semiquinone at high substrate concentrations and long reaction times distinguishes the dehydrogenase from the oxidase. In the reactions of xanthine oxidase, dehydrogenase, and aldehyde oxidase (EC 1.2.3.1) with different oxidizing substrates, various of the redox centres in the enzyme may react as shown in Table 5.[661] Cytochrome *c* and nitroblue tetrazolium may be reduced by superoxide or directly by the flavin or iron–sulphur centres,

[654] F. Bergmann, A. Frank, and H. Govrin, *Biochim. Biophys. Acta*, 1979, **571**, 215.
[655] A. I. Kozachenko, L. G. Nagler, and L. S. Vartanyan, *Biokhimiya*, 1979, **44**, 1401.
[656] A. Lucacchini, D. Segnini, A. Da Settimo, and O. Livi, *Ital. J. Biochem.*, 1979, **28**, 194.
[657] R. E. Lynch and I. Fridovich, *Biochim. Biophys. Acta*, 1979, **571**, 195.
[658] S. L. Betcher-Lange, M. P. Coughlan, and K. V. Rajagopalan, *J. Biol. Chem.*, 1979, **254**, 8825.
[659] K. Tatsumi, N. Koga, S. Kitamura, H. Yoshimura, P. Wardman, and Y. Kato, *Biochim. Biophys. Acta*, 1979, **567**, 75.
[660] I. Ní Fhaolain and M. P. Coughlan, *Proc. R. Ir. Acad., Sect. B*, 1979, **79**, 111.
[661] M. P. Coughlan and I. Ní Fhaolain, *Proc. R. Ir. Acad., Sect. B*, 1979, **79**, 169.

Table 5 *The interaction of oxidizing substrates with the redox centres of xanthine oxidase, dehydrogenase, or aldehyde oxidase*[661]

Redox centre	Oxidizing substrates
Molybdenum	Dichlorophenolindophenol, trinitrobenzenesulphonate, and (possibly) phenazine methosulphate
Flavin (FAD)	Oxygen, NAD^+
Iron–sulphur	Methylene blue, ferricyanide, and (possibly) phenazine methosulphate

depending upon both the enzyme and the conditions. Bray *et al.*[662] have described the mechanism for the action of xanthine oxidase, shown in Scheme 12. The reaction involves nucleophilic attack at C-8 by a group on the enzyme, coupled to transfer of hydride to a terminal sulphide ligand on the molybdenum.

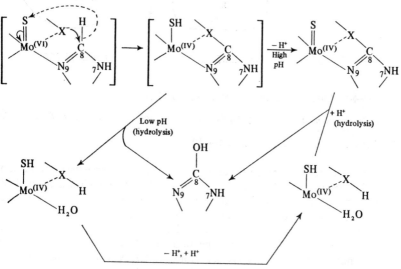

Proposed intermediates that give rise to the e.p.r. spectra in the reaction of xanthine oxidase with xanthine

Scheme 12

Several papers published in 1979 have yielded further insight into the nature of both the enzyme and the prosthetic groups of the xanthine-oxidizing enzymes. The action of various proteases on xanthine dehydrogenase suggests that the molecule is composed of three domains, and treatment with subtilisin yields a fragment of mol. wt 65 000 (*cf.* the mol. wt of 150 000 for the native enzyme) that is devoid of flavin but which contains both the iron–sulphur and the molybdenum centres;[663] the fragment does not catalyse xanthine-dependent reduction

[662] R. C. Bray, S. Gutteridge, D. A. Stotter, and S. J. Tanner, *Biochem. J.*, 1979, **177**, 357.
[663] M. P. Coughlan, S. L. Betcher-Lange, and K. V. Rajagopalan, *J. Biol. Chem.*, 1979, **254**, 10 694.

of NAD⁺, although dichlorophenolindophenol is still reduced (*cf.* Table 5). A possible model for the structure of the whole molecule has been presented on the basis of these results, and it is shown in Figure 9. Core extrusion of the iron-sulphur centres of xanthine oxidase and quantitation by ^{19}F n.m.r. of trifluoroacetylated thiol has shown all of the iron to be present as $[Fe_2S_2]$ groups (four per molecule, two per flavin).[664]

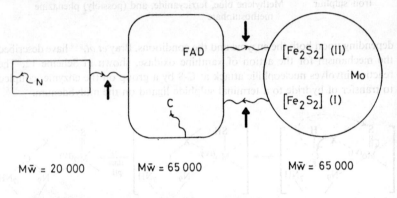

M$\bar{w}$ = 20 000   M$\bar{w}$ = 65 000   M$\bar{w}$ = 65 000

⟶   Cleavage Sites of Protease Action

Figure 9 *The domain structure of sulphite oxidase, based upon the results of proteolytic cleavage*
(Redrawn from *J. Biol. Chem.*, 1979, **254**, 10 694)

Two main spectroscopic methods have been used to observe the molybdenum atom in various forms of the enzyme, these being the established method of e.p.r. and recently the method of extended *X*-ray absorption fine structure (EXAFS). Results from the latter technique, along with *X*-ray absorption-edge measurements, on oxidized and reduced xanthine oxidase have been interpreted by Tullius *et al.*[665] in terms of a co-ordination environment of the molybdenum of two terminal oxygen atoms at 1.71 Å, two sulphur atoms at 2.54 Å, and one sulphur atom at 2.84 Å. The authors note that the EXAFS spectra are quite different from those previously obtained from the molybdenum protein of nitrogenase, and they mention the possibility of a heterogeneous molybdenum centre in xanthine oxidase, due to the presence of the desulpho-form. Bordas *et al.*,[666] in their EXAFS analysis of xanthine oxidase, have attempted to subtract the component due to the desulpho-enzyme, and they concluded that the molybdenum is in a quite different environment in the two forms. Comparison with some model compounds suggests a MoO_2S_2 centre for the desulpho-enzyme, and the cyanolysable sulphur atom is proposed to be co-ordinated to the molybdenum with a Mo–S bond length of ~2.3 Å. Berg *et*

[664] G. B. Wong, D. M. Kurtz, Jr., R. H. Holm, L. E. Mortenson, and R. G. Upchurch. *J. Am. Chem. Soc.*, 1979, **101**, 3078.
[665] T. D. Tullius, D. M. Kurtz, Jr., S. D. Conradson, and K. O. Hodgson, *J. Am. Chem. Soc.*, 1979, **101**, 2776.
[666] J. Bordas, R. C. Bray, C. D. Garner, S. Gutteridge, and S. S. Hasnain, *J. Inorg. Biochem.*, 1979, **11**, 181.

al.[667] have compared the metal–ligand bond lengths from X-ray crystallography and EXAFS measurements in two molybdenum complexes of the form MoO_2L, where L is $(SCH_2CH_2)_2NCH_2CH_2X$ and X is either SMe or NMe_2. The results compare favourably, and the authors discuss the application of the complex where X is SMe to model the active sites of xanthine oxidase and sulphite oxidase.

Electron paramagnetic resonance spectra of the molybdenum(v) states of xanthine oxidase and sulphite oxidase in the presence of $H_2^{17}O$ show super-hyperfine broadening ($A \approx 1$ mT), owing to a rapidly exchanging oxygen atom (of a water molecule) that is co-ordinated to the molybdenum.[668] The effect was most pronounced in sulphite oxidase and differed between the high- and low-pH forms. Superhyperfine coupling has also been observed in the e.p.r. spectrum of one of the intermediates of xanthine oxidase (the rapid signal) generated in ^{17}O-enriched water.[669] A preliminary attempt to simulate the spectrum yields one component of $A = 1.6$ mT, which is consistent with co-ordination of oxygen. Coffman and Buettner[670] have presented a general treatment of spin–spin dipolar coupling between paramagnets and have applied their analysis to the interaction of the molybdenum and iron–sulphur centres in xanthine oxidase. They arrived at an estimate of 8—14 Å for the distance between the two groups.

Aldehyde Oxidase, Sulphite Oxidase, Nitrate Reductase, and Formate Dehydrogenase.—Despite their diverse catalytic functions, these four enzymes show many similarities both between themselves and with the xanthine-oxidizing enzymes, and appear to be quite distinct from nitrogenase.

The molybdenum centre in sulphite oxidase (EC 1.8.3.1) has been probed by e.p.r. spectroscopy[668] and by X-ray absorption edge and EXAFS.[671] Spectra in the latter case were obtained for oxidation states v and iv, and analysis of the data and consideration of the known crystal structures of molybdenum complexes led the authors to propose the structures (1)—(3) for the active sites. The ligand

(1)   (2)   (3)

X is probably either nitrogen or oxygen. An amino-acid sequence of the haem-binding domain of chicken liver sulphite oxidase has been determined,[672] and the 97-residue fragment shows substantial homology ($\sim 27\%$ of identical residues) with microsomal cytochrome b_5, suggesting a common evolutionary origin for the two proteins. The authors are able to fit the sequence of sulphite

[667] J. M. Berg, K. O. Hodgson, S. P. Cramer, J. L. Corbin, A. Elsberry, N. Pariyadath, and E. I. Steifel, *J. Am. Chem. Soc.*, 1979, **101**, 2774.
[668] S. P. Cramer, J. L. Johnson, K. V. Rajagopalan, and T. N. Sorrell, *Biochem. Biophys. Res. Commun.*, 1979, **91**, 434.
[669] S. Gutteridge, J. P. G. Malthouse, and R. C. Bray, *J. Inorg. Biochem.*, 1979, **11**, 355.
[670] R. E. Coffman and G. R. Buettner, *J. Phys. Chem.*, 1979, **83**, 2392.
[671] S. P. Cramer, H. B. Gray, and K. V. Rajagopalan, *J. Am. Chem. Soc.*, 1979, **101**, 2772.
[672] B. Guiard and F. Lederer, *Eur. J. Biochem.*, 1979, **100**, 441.

oxidase to the crystal structure of cytochrome b_5 with relatively few modifications of the backbone.

A physiological role for aldehyde oxidase (EC 1.2.3.1) has long been a problem; this may now be resolved, due to the work of Brandänge and Lindblom.[673] They have shown that the enzyme catalyses the oxidation of the intermediate iminium ion that is formed during the metabolism of nicotine. The low K_m (2 μmol l^{-1}) for this substrate and the ubiquity of iminium ions during the microsomal metabolism of nitrogen-containing heterocycles have led the authors to propose that this enzyme may be an iminium oxidase. Aldehyde oxidase is inhibited by 6,6'-azopurine,[674] which also acts as a substrate, being oxidized to 8,8'-dioxo-6,6'-azopurine.

Several enzymes with nitrate reductase activity are catalogued in the latest recommendations (1978) of the Nomenclature Committee of the International Union of Biochemistry;[675] this section is concerned only with those that contain molybdenum, namely the bacterial dissimilatory (respiratory) enzyme (EC 1.7.99.4) and the assimilatory enzyme (EC 1.6.6.3) that is found in both bacteria and plants. The isolation and characterization of several nitrate reductases were reported in 1979, and details are collected together in Table 6.[676-680] Immunological studies on the enzyme from *Bacillus licheniformis* have shown that neither the subunits nor the molybdenum cofactor contains any antigenic determinant in common.[676] A possible structure for the trimeric assimilatory

Table 6 *Nitrate reductases whose isolation was reported in 1979*

Organism[a]	Mol. wt (Subunits)	Cofactor content	Notes	Ref.
B. licheniformis	193 000 (150 000 + 57 000)	7Fe, 7S, and 1Mo	Does not cross-react with anti-sera to enzymes of E. coli or of Klebsiella aerogenes	676
C. vulgaris	270 000 (3 × 90 000)	1 FAD, 1 Haem, and 1 Mo per subunit	—	677
S. aureus	—	—	—	678
A. nidulans	—	—	—	679
R. capsulata AD2	170 000 (2 × 85 000)	1.5 Haem and 0.8 Mo	—	680

(a) *B. licheniformis* is *Bacillus licheniformis*, *C. vulgaris* is *Chlorella vulgaris*, *S. aureus* is *Staphylococcus aureus*, *A. nidulans* is *Aspergillus nidulans*, and *R. capsulata* AD2 is *Rhodopseudomonas capsulata* AD2.

673 S. Brandänge and L. Lindblom, *Biochem. Biophys. Res. Commun.*, 1979, **91**, 991.
674 A. L. Jadhav, K. G. Bhansali, and J. R. Davis, *J. Pharm. Sci.*, 1979, **68**, 1202.
675 'Enzyme Nomenclature (1978)', International Union of Biochemistry, Academic Press, 1979.
676 J. Van't Riet, F. B. Wientjes, J. Van Doorn, and R. J. Planta, *Biochim. Biophys. Acta*, 1979, **576**, 347.
677 L. Giri and C. S. Ramadoss, *J. Biol. Chem.*, 1979, **254**, 11 703.
678 K. A. Burke and J. Lascelles, *J. Bacteriol.*, 1979, **139**, 120.
679 R. J. Downey and F. X. Steiner, *J. Bacteriol.*, 1979, **137**, 105.
680 K. Alef and J.-H. Klemme, *Z. Naturforsch., Teil. C*, 1979, **34**, 33.

enzyme from *Chlorella vulgaris* has been derived from electron microscopy,[677] and it is consistent with hydrodynamic data. An extremely sensitive immuno-assay for nitrate reductase from *Neurospora crassa* has been described and used to measure enzyme levels independently of the expression of activity.[681] The method can also be used to investigate the partial electron-transfer steps and to probe the nature of the *nit* mutants in *N. crassa*.

Fido *et al.*[682] have measured the mid-point potential (-60 mV) of the haem group of the spinach enzyme, and have also presented low-temperature optical spectra. Mid-point potentials for the enzyme from *Escherichia coli* K12 have also been determined, using e.p.r. spectroscopy to monitor the molybdenum(v) and iron–sulphur contents.[683] The results obtained are $+220$ mV ($Mo^{VI/V}$) and $+180$ mV ($Mo^{V/IV}$) for the molybdenum centre and $+50$ mV and $+80$ mV for iron–sulphur centres in the reduced and oxidized forms of the enzyme, respectively. Hewitt *et al.*[684] have presented a short discussion of the properties of, and a possible mechanism for the action of, eukaryotic nitrate reductase.

Solubilization and purification of a membranous formate dehydrogenase (EC 1.2.1.2) from *Vibrio succinogenes* have been achieved and some of its basic properties determined.[685] The enzyme is a dimer, with a molecular weight of 263 000, and it contains one molybdenum and 18 or 19 iron and acid-labile sulphide groups per subunit. Variable substoicheiometric amounts of *b*- and *c*-type cytochromes were also present. Electron flow from formate to mena-quinone was suggested to be *via* first the formate dehydrogenase and then a *b*-type cytochrome. Formate dehydrogenase from *Methanococcus vannielii* contains selenium in addition to molybdenum and iron, and Jones *et al.*[686] have demonstrated that this element is present as a selenocysteine residue.

Nitrogenase.—Several reviews[687–691] and a book[692] on nitrogenase (EC 1.18.2.1) and nitrogen fixation appeared in 1979. A successful cell-free preparation of the enzyme from a symbiotic actinomycete from the root nodules of alder has been described, and the activity of the enzyme from this source appears to be very similar to those of enzymes of other nitrogen-fixing organisms.[693] Nitrogenase has also been purified from *Anabaena cylindrica*;[694] during the course of this purification the authors used an assay for the iron–molybdenum enzyme that is based upon extracts from cells grown in the presence of ammonia and

[681] N. A. Amy and R. H. Garrett, *Anal. Biochem.*, 1979, **95**, 97.
[682] R. J. Fido, E. J. Hewitt, B. A. Notton, O. T. G. Jones, and A. Nasrulhaq-Boyce, *FEBS Lett.*, 1979, **99**, 180.
[683] S. P. Vincent, *Biochem. J.*, 1979, **177**, 757.
[684] E. J. Hewitt, B. A. Notton, and C. D. Garner, *Biochem. Soc. Trans.*, 1979, **7**, 629.
[685] A. Kröger, E. Winkler, A. Innerhofer, H. Hackenberg, and H. Schägger, *Eur. J. Biochem.*, 1979, **94**, 465.
[686] J. B. Jones, G. L. Dilworth, and T. C. Stadtman, *Arch. Biochem. Biophys.*, 1979, **195**, 255.
[687] L. E. Mortenson and R. N. F. Thorneley, *Annu. Rev. Biochem.*, 1979, **48**, 387.
[688] T. L. Jones, *FEBS Lett.*, 1979, **98**, 1.
[689] B. Quebedeaux, *Encycl. Plant Physiol. (New Ser.)*, 1979, **6**, 472.
[690] R. N. F. Thorneley, D. J. Lowe, R. R. Eady, and R. W. Miller, *Biochem. Soc. Trans.*, 1979, **7**, 633.
[691] R. L. Richards, *Chem. Soc. Spec. Publ.*, 1979, **36**, 258.
[692] F. Bottomley and R. C. Burns, 'Treatise on Dinitrogen Fixation', J. Wiley, New York, 1979.
[693] D. R. Benson, D. J. Arp, and R. H. Burris, *Science*, 1979, **205**, 688.
[694] P. C. Hallenbeck, P. J. Kostel, and J. R. Beneman, *Eur. J. Biochem.*, 1979, **98**, 275.

tungsten. It was suggested that part of the protection mechanism against oxygen in this aerobic organism is due to a barrier to diffusion that is presented by the cell wall of the heterocyst. The nitrogenase of *Klebsiella pneumoniae* has been purified at $-20\,^{\circ}C$; the treatment has little effect on yields or on specific activities, and causes only a slight improvement in sensitivity to oxygen, as compared to conventional methods.[695]

Purified components of nitrogenase are extremely labile towards oxygen, and much interest exists in the protective mechanisms *in vivo*; a diffusion barrier in *A. cylindrica* has already been mentioned,[694] and in *Anabaena* sp. CA a mutant has been isolated that has an oxygen-sensitive nitrogenase.[696] This mutation appears to be in the protection mechanism rather than in the enzyme, and therefore provides a useful tool for further study. In *Azotobacter chroococcum* the stability of the nitrogenase towards oxygen appears to be related to a protective iron–sulphur protein that associates in a 1:1 ratio with the other two components of nitrogenase.[697] Addition of this protective protein to the purified enzyme restores tolerance to oxygen. Finally, if the oxygen stress becomes too severe, *A. chroococcum* responds by repressing the synthesis of both nitrogenase and its possible electron-donor flavodoxin.[698]

The factors that control the expression of nitrogenase activity *in vivo* are of considerable interest, and a number of recent results have appeared in this area. Meyer and Vignais[699] have studied the effect of two inhibitors of glutamine synthetase in *Rhodopseudomonas capsulata* and have shown that they cause derepression of the biosynthesis of nitrogenase; however, both compounds also inhibit nitrogenase activity *in vivo*. In free-living *Rhizobium japonicum*, pulse-labelling experiments have shown that enzyme levels rise to a maximum after three days of growth in nitrogen-starved conditions.[700] Addition of oxygen reduced the enzyme levels to nearly zero, whilst nitrate, although suppressing nitrogenase activity, did not affect the amount of enzyme that was detected by immunoprecipitation. *Rhodospirillum rubrum* synthesizes two forms of nitrogenase, depending upon the growth conditions.[701] Under nitrogen-starved conditions, the A-form is expressed; this form requires a membrane component, ATP, and a bivalent metal ion for activity, whilst the R-form (which is synthesized in nitrogen-supplemented media) does not. The differences appear to be in the nature of the iron enzyme, and conversion of A- into R-form occurs rapidly (~ 10 min) upon adding fixed nitrogen to the culture. Symbiotic nitrogen-fixing organisms have an additional problem in co-ordinating the synthesis of nitrogenase with the plant's synthesis of the oxygen-binding protein leghaemoglobin. In *Pisum sativum*, inoculation with *Rhizobium leguminosarum* leads to the prior synthesis of leghaemoglobin, followed by that of the nitrogenase

[695] A. D. Tsopanakis, R. C. Bray, and B. E. Smith, *Biochem. J.*, 1979, **181**, 509.
[696] J. F. Grillo, P. J. Bottomley, C. van Baalen, and F. R. Tabita, *Biochem. Biophys. Res. Commun.*, 1979, **89**, 685.
[697] R. L. Robson, *Biochem. J.*, 1979, **181**, 569.
[698] R. L. Robson, *FEMS Microbiol. Lett.*, 1979, **5**, 259.
[699] J. Meyer and P. M. Vignais, *Biochem. Biophys. Res. Commun.*, 1979, **89**, 353.
[700] D. B. Scott, H. Hennecke, and S. T. Lim, *Biochim. Biophys. Acta*, 1979, **565**, 365.
[701] R. P. Carithers, D. C. Yoch, and D. I. Arnon, *J. Bacteriol.*, 1979, **137**, 779.

proteins.[702] However, the two components of nitrogenase are not synthesized co-ordinately; the iron–molybdenum protein precedes the iron protein in appearance.

As mentioned previously, one form of nitrogenase from *R. rubrum* requires a membrane component as well as a metal ion and ATP for activity, and this membrane component has been shown to relieve ammonium-ion-dependent deactivation and to abolish the lag in evolution of hydrogen and reduction of acetylene.[703] The authors suggest that the membrane component acts upon the iron protein. This is consistent with the results of Ludden and Burris,[704] who showed that an adenine-like moiety is removed from the iron protein during activation. The authors have suggested a role for the metal ion and ATP, and proposed a mechanism for the control of the interconversion of the different forms.

Nitrogenase proteins have provided a fertile ground for spectroscopists, and recent studies have involved Mössbauer,[705, 706] electronic absorption,[707] n.m.r.,[708, 709] fluorescence,[710] and EXAFS[711] measurements. Mössbauer studies of the $S = \frac{3}{2}$ centres (M-centres) in the iron–molybdenum protein have revealed a spin-coupled system containing six iron atoms, in good agreement with previous e.p.r. studies. In this work, a resolution-enhancement technique is described[705] and used to demonstrate the so-called spectral component S. Absorption, c.d., and m.c.d. spectra of both nitrogenase components from *Azotobacter vinelandii* and *Clostridium pasteurianum* have been reported and shown to be very similar for the corresponding components from different organisms.[707] Spectra of the iron–molybdenum protein in different oxidation states cannot be accounted for purely in terms of iron–sulphur clusters, whereas those of the iron protein are as expected for [Fe$_4$S$_4$] centres. The c.d. spectrum of the iron protein shows a substantial perturbation by magnesium-ATP. Binding of magnesium to the iron protein has been measured by ^{25}Mg n.m.r.,[708] and the dependence of the relaxation time on the concentrations of protein and of iron is consistent with the metal ion being in fast exchange on the ^{25}Mg n.m.r. timescale. Core-extrusion methods and subsequent characterization of the iron–sulphur complexes by ^{19}F n.m.r. of fluorinated thiols identified the iron–sulphur centres in the iron–molybdenum proteins as of the [Fe$_4$S$_4$] type.[709] Under the conditions used, no reaction with the isolated iron–molybdenum cofactor occurred. Mitsova *et al.*[710] have labelled a surface sulphydryl group of nitrogenase with

[702] T. Bisseling, R. C. Van Den Bos, M. W. Weststrate, M. J. J. Hakkaart, and A. Van Kammen, *Biochim. Biophys. Acta*, 1979, **562**, 515.

[703] S. Nordlund and U. Eriksson, *Biochim. Biophys. Acta*, 1979, **547**, 429.

[704] P. W. Ludden and R. H. Burris, *Proc. Natl. Acad. Sci. USA*, 1979, **76**, 6201.

[705] B. H. Huynh, E. Münck, and W. H. Orme-Johnson, *Biochim. Biophys. Acta*, 1979, **576**, 192.

[706] B. H. Huynh, E. Münck, and W. H. Orme-Johnson, *J. Phys. Colloq. (Orsay, Fr.)*, 1979, 526.

[707] P. J. Stephens, C. E. McKenna, B. E. Smith, H. T. Nguyen, M.-C. McKenna, A. J. Thomson, F. Devlin, and J. B. Jones, *Proc. Natl. Acad. Sci. USA*, 1979, **76**, 2585.

[708] E. O. Bishop, S. J. Kimber, B. E. Smith, and P. J. Beynon, *FEBS Lett.*, 1979, **101**, 31.

[709] D. M. Kurtz, Jr., R. S. McMillan, B. K. Burgess, L. E. Mortenson, and R. H. Holm, *Proc. Natl. Acad Sci USA*, 1979, **76**, 4986.

[710] I. Z. Mitsova, V. I. Evreinov, and R. I. Gvozdev, *Mol. Biol. (Moscow)*, 1979, **13**, 1044.

[711] B.-K. Teo and B. A. Averill, *Biochem. Biophys. Res. Commun.*, 1979, **88**, 1454.

a fluorescent probe and measured the fluorescence yields of various forms of the enzyme; both iron–molybdenum and the iron proteins (before and after destruction of the iron–sulphur centres). From their data they determined the distances from the probe to iron–sulphur clusters as 12—14 Å for the iron–molybdenum protein and 18—20 Å for the iron protein; they also calculated the distance between the nearest iron–sulphur clusters of the two proteins as less than 14—16 Å. A new model for the iron–molybdenum cofactor has been proposed,[711] based upon a re-interpretation of EXAFS data. The structure (4) is consistent with other chemical and spectroscopic data, and a possible mechanism for the enzyme is discussed in the light of this model.

(4)

A number of new kinetic studies of nitrogenase activity have been published; Hageman and Burris[712] measured reduction of the iron–molybdenum protein by e.p.r. and showed it to be biphasic, the fast phase occurring over the timescale of the observed lag in evolution of hydrogen. The authors argue that the evolution of hydrogen is supported by two electrons per e.p.r.-detectable centre, and that the two nitrogenase proteins dissociate after each transfer of an electron. The hydrogen-evolution activities of nitrogenase and of hydrogenase occur together in some blue-green algae; whilst low concentrations of viologens inhibit the activity of the former *in vivo*, they still support the latter, enabling the two contributions to be separated.[713] Inhibition of nitrogenase by carbon monoxide has been investigated by Davis *et al.*,[714] and it was found that the type of inhibition depended upon the substrate. Thus for reduction of nitrogen, inhibition is non-competitive, whilst for reduction of acetylene the kinetic patterns are complex, due, in part, to a contribution from inhibition by substrate. Electron paramagnetic resonance spectra for the inhibited enzyme with different isotopic substitutions are consistent with the postulate that carbon monoxide binds to iron in the iron–molybdenum protein. The authors suggest that the $[Fe_4S_4]$ cluster is at the -1 oxidation level at high partial pressures of carbon monoxide whilst at low partial pressures it is at the -3 level. A paper has appeared in the Russian literature[715] on the kinetics of nitrogenase and the

[712] R. V. Hageman and R. H. Burris, *J. Biol. Chem.*, 1979, **254**, 11 189.

[713] A. Daday, G. R. Lambert, and G. D. Smith, *Biochem. J.*, 1979, **177**, 139.

[714] L. C. Davis, M. T. Henzl, R. H. Burris, and W. H. Orme-Johnson, *Biochemistry*, 1979, **18**, 4860.

[715] V. V. Kochetov, V. R. Linde, and G. I. Likhtenstein, *Mol. Biol.* (*Moscow*), 1979, **13**, 402.

analysis of the data by graph theory; it is suggested that inhibitors and substrates other than nitrogen interact with oxidized or half-reduced enzyme whilst nitrogen interacts with the fully reduced enzyme.

ATP is required for nitrogenase activity, and the enzyme from *A. vinelandii* will catalyse exchange of ^{18}O between water and phosphate in an ATP-stimulated reaction.[716] Reduction of the enzyme increases the rate of hydrolysis of ATP and decreases the exchange rate, whilst *o*-phenanthroline inhibits both processes. Exchange of oxygen between water and the terminal phosphate of ATP is also observed during turnover,[717] and the results have been interpreted in terms of the mechanism of hydrolysis of ATP coupled to changes in redox potential.

McKenna and co-workers[718-720] have exploited the use of cyclopropene as a substrate for nitrogenase in a number of ways. Reduction of cyclopropene yields both cyclopropane and propene, and kinetic parameters and product ratios are very similar for whole cells, extracts, or various degrees of purified enzymes; this has led the authors to propose the use of this substrate for comparing the activities of the enzyme *in vivo* and *in vitro*.[718] In contrast to the enzyme, isolated iron–molybdenum cofactor only reduces cyclopropene to cyclopropane with borohydride.[719] In view of this, the authors urge caution in drawing analogies between the mode of action of the enzyme and the cofactor; they also note the difference in half-life for reconstitutive activity (< 1 minute) and acetylene-reducing activity (~ 1 hour) of the isolated cofactor during oxidative inactivation. The stereochemistries of the products of action of nitrogenase on the three isomeric C_3H_4 hydrocarbons in D_2O have been determined,[720] and they are listed in Table 7. Restraints upon the mechanism of action of nitrogenase that are consistent with these products are discussed.

Table 7 *The products of the action of nitrogenase on allene, methylacetylene, and cyclopropene in* D_2O

Substrate	Product(s)
Allene	[2,3-$^2H_2$]propene
Methylacetylene	*cis*- and *trans*-[1,2-$^2H_2$]propene
Cyclopropene	*cis*-[1,2-$^2H_2$]cyclopropane, [1,3-$^2H_2$]propene, and [2,3-$^2H_2$]propene

Oxidative inactivation of the iron–molybdenum protein from the nitrogenase of *C. pasteurianum* is potential-dependent, only occurring at potentials above $+350$ mV; a potential- and temperature-dependent lag is observed in the rate of inactivation, which is accompanied by the loss of iron, molybdenum, and acid-labile sulphide groups.[721] Reconstitution of the enzyme with these species yields a product with an e.p.r. spectrum that is characteristic of the aerially inactivated material.

[716] E. G. Skvortsevich, L. A. Sytsova, A. M. Uzenskaya, N. I. Tertyshnaya, and N. S. Panteleeva, *Dokl. Akad. Nauk SSSR*, 1979, **244**, 241.
[717] R. I. Gvozdev, A. P. Sadkov, and I. Mitsova, *Biokhimiya*, 1979, **44**, 2145.
[718] C. E. McKenna and C. W. Huang, *Nature (London)*, 1979, **280**, 609.
[719] C. E. McKenna, J. B. Jones, H. Eran, and C. W. Huang, *Nature (London)*, 1979, **280**, 611.
[720] C. E. McKenna, M.-C. McKenna, and C. W. Huang, *Proc. Natl. Acad. Sci. USA*, 1979, **76**, 4773.
[721] C. Gomez-Moreno and B. Ke, *Mol. Cell. Biochem.*, 1979, **26**, 111.

6

Zinc Metalloenzymes

A. GALDES

Introduction.—Zinc is one of the most abundant trace metals, and the essential role of Zn^{II} in the maintenance of life processes is now firmly established. It is known that Zn^{II} exerts its biological influence mainly by participating as an essential cofactor to a large number of proteins, and this article reviews the publications concerning zinc-requiring enzymes that appeared during 1979.

It is remarkable that, in spite of intensive research, the precise function of zinc in metalloenzymes is still not fully understood. A survey of the enzymes that are reviewed below reveals that in almost every case zinc enzymes act on C—O groups, either as hydratases (*e.g.* carbonic anhydrase) or hydrolyases (*e.g.* peptidases, β-lactamase II), or by aiding the oxidoreduction (*e.g.* alcohol dehydrogenase) and rearrangement (*e.g.* aldolases) of such groups. From the earliest days of the study of zinc metalloenzymes, two types of mechanisms whereby the Zn^{II} can effect these reactions have been proposed. One, termed the zinc-carbonyl mechanism, proposes that the substrates bind directly to the Zn^{II} through the oxygen of the C—O group, and the function of the Zn^{II} is then envisaged as being the polarization of the group, thereby facilitating nucleophilic attack at the carbon (*e.g.* carbonic anhydrase, peptidases) or attack by the bound oxygen atom (*e.g.* nucleotidyl polymerases). The second, termed the zinc-hydroxide mechanism, proposes that the Zn^{II} mediates its function through a water molecule, bound to the metal ion. According to this hypothesis, the function of the Zn^{II} is to lower the pK_a of the bound water molecule from ~ 14 to ~ 7. The resultant metal-bound hydroxide ion can then attack the carbon atom of the susceptible C—O group (*e.g.* carbonic anhydrase, peptidases), or acts as a proton acceptor and ionizes the oxygen atom to O^- (*e.g.* alcohol dehydrogenase). It is of course possible that the Zn^{II} can act in either manner, depending on the enzyme in question. The zinc-carbonyl mechanism has generally found more support than the zinc-hydroxide mechanism, and at one time or another has been proposed to be operating for all the major zinc metalloenzymes, including carbonic anhydrase, carboxypeptidase, alcohol dehydrogenase, and aldolase. The most definitive example for this type of mechanism was thought to be supplied by carboxypeptidase A, where X-ray diffraction studies have shown that the pseudo-substrate Gly-Try is bound to the active-site Zn^{II} through the peptide carbonyl group. Recently, however, the validity of this X-ray study to the mechanism of the enzyme has been questioned, and in 1979 it was claimed in one publication, for the first time since the diffraction results appeared over ten years ago, that the mechanism of the enzyme proceeds *via* a zinc-hydroxide mechanism. To date, the best case for a zinc-hydroxide mechanism can be made for carbonic anhydrase; for this enzyme, a large amount of experimental evidence obtained during the past decade, including the work reviewed below, supports such a mechanism. However, even for carbonic anhydrase, this mechanism has not yet found universal support, and there is some evidence that the C—O group of the substrate is bound, albeit at a distant fifth co-ordination site, to the metal during catalysis.

The latest proposed mechanisms for several zinc-containing metalloenzymes combine elements from both types of mechanism by suggesting that the substrate binds to the enzyme *via* the C—O group, but that in the process the metal-bound water molecule is *not* displaced, so that the reaction proceeds *via* a five-co-ordinate intermediate. This hybrid mechanism is discussed below in greater detail for alcohol dehydrogenase, and it will be interesting to see if it will stand the test of time better than the two previous postulates.

Carbonic Anhydrase.—*X*-Ray diffraction studies[1] on carbonic anhydrase have suggested that an ordered array of eight water molecules, presumed to be of great functional significance, is present in the active site of the enzyme; this suggestion has been questioned by some authors.[2] In a recent theoretical study, a Monte-Carlo simulation of the structure of water in the active site of carbonic anhydrase was performed.[3] It was concluded that three water molecules are present in the first solvation sphere of the Zn^{II} ion and that another four water molecules are present in the active site of the enzyme. Hence this study suggests that the active cleft of carbonic anhydrase contains an ordered array of seven water molecules and that the Zn^{II} ion is six-co-ordinate; this latter conclusion is in conflict with previous experimental studies on this enzyme, which indicate that the Zn^{II} is only four- or five-co-ordinate. An *ab initio* SCF study on the interaction of Zn^{II} with H_2O, NH_3, and imidazolate has been carried out, in an attempt to understand the role of the Zn^{II} in carbonic anhydrase.[4] This study confirms earlier results obtained by the same group,[5] and indicates that the deprotonation of a metal-bound water molecule is favoured over that of a histidine ligand, so that the active nucleophile in the reaction of carbonic anhydrase is suggested to be a metal-bound hydroxide ion. The SCF computations have also been extended to a model of the active site of the enzyme that involves a glutamate–threonine–water hydrogen-bonded chain.[6] The results suggest that, when bound to Zn^{II}, the chain is more stable in the ionized form Glu-Thr-OH⁻, and this lends further support for a mechanism of action involving a metal-bound hydroxide ion. Another theoretical study focuses on the binding of substrate to carbonic anhydrase.[7] This study uses the additional stabilization of the transition-state complex as a qualitative measure of the effect of the active-site residues on the hydration reaction of CO_2.

[1] K. K. Kannan, B. Notstrand, K. Fridberg, S. Lovgren, A. Ohlsson, and M. Petef, *Proc. Natl. Acad. Sci. USA*, 1975, **72**, 51.
[2] Y. Pocker and S. Sarkanen, *Adv. Enzymol.*, 1978, **47**, 149.
[3] E. Clementi, G. Corongiu, B. Jonsson, and S. Romano, *FEBS Lett.*, 1979, **100**, 313.
[4] A. Pullman and D. Demoulin, *Int. J. Quantum Chem.*, 1979, **16**, 641.
[5] D. Demoulin, A. Pullman, and B. Sarkar, *J. Am. Chem. Soc.*, 1977, **99**, 8498.
[6] D. Demoulin and A. Pullman, *Jerusalem Symp. Quantum. Chem. Biochem.*, 1979, **12**, 51.
[7] A. Sawaryn and W. A. Sokolski, *Int. J. Quantum. Chem.*, 1979, **16**, 293.

A recent report re-evaluates the spin–lattice relaxation rate of solvent water protons in solutions of Co^{II} carbonic anhydrase.[8] The paramagnetic relaxivity is responsive to the ionization of two groups with pK_a values of 6.1 and 8.5. The first of these pK_a values had previously been observed,[9] and it is usually ascribed to the metal-bound solvent molecule. The second pK_a value is novel and, since there are no suitable amino-acid residues present in the active site, the authors suggest that it may be due to bicarbonate ion that is bound to the enzyme.

In addition to the hydration of carbon dioxide, carbonic anhydrase can catalyse the reversible hydration of aldehydes;[10, 11] this reaction has, however, been largely neglected. The interaction between hydrated trichloroacetaldehyde and Co^{II} bovine carbonic anhydrase has been investigated in a recent paper.[12] It was concluded that chloral is bound directly to the metal ion, the apparent affinity constant being pH-dependent.

A study of the electronic spectra of Co^{II} bovine carbonic anhydrase in un-buffered and HEPES-buffered solutions indicates that HEPES buffer does not bind to the enzyme, and that the spectra are not affected by the ionic strength of the solution.[13] Cobalt(II) bovine carbonic anhydrase has been oxidized to the Co^{III} enzyme with H_2O_2.[14] The oxidation was inhibited by specific inhibitors of carbonic anhydrase. As expected, the Co^{III} complex is diamagnetic and possesses an octahedral symmetry. It is not, however, totally inert towards substitution, and cyanide or azide ions bind to the Co^{III} enzyme. The Co^{III} can be reduced by dithionite or borohydride to give the fully active Co^{II} enzyme. The unfolding of Co^{II} carbonic anhydrase B in guanidine hydrochloride has been followed by a variety of spectroscopic techniques.[15] Separable stages of unfolding were observed, with the loss of the conformation of the active site occurring in the early stages. The unfolded protein cannot be re-folded to form the active enzyme.

A number of other metal derivatives of carbonic anhydrase have been prepared and utilized as probes of the active site of the enzyme. Oxovanadium(IV) com-bines with apocarbonic anhydrase in aqueous solution to yield the corresponding metal-bound derivative.[16] Indium(III)-111 bovine carbonic anhydrase has been studied by perturbed directional correlations of γ-rays.[17] This study focused on the conformational change of the enzyme that is induced by guanidine hydrochloride. The inhibition of carbonic anhydrase by a variety of heavy-metal ions has been investigated.[18] The order of effectiveness for inhibition was $Cu^{II} > Cd^{II} > Zn^{II} > Ag^I > Pb^{II}$, with Hg^{II} and Co^{II} being weakly activatory. The ^{13}C n.m.r. spectra of CO_2 and HCO_3^- in the presence of Cu^{II} carbonic

8 J. W. Wells, S. I. Kandal, and S. H. Koenig, *Biochemistry*, 1979, **18**, 1989.
9 M. E. Fabry, S. H. Koenig and W. E. Schillinger, *J. Biol. Chem.*, 1970, **245**, 4256.
10 Y. Pocker and J. E. Meany, *Biochemistry*, 1965, **4**, 2535.
11 Y. Pocker and J. E. Meany, *J. Am. Chem. Soc.*, 1965, **87**, 1809.
12 I. Bertini, E. Borghi, G. Canti, and C. Luchinat, *J. Inorg. Biochem.*, 1979, **11**, 49.
13 D. Barzi, I. Bertini, C. Luchinat, and A. Scozzafava, *Inorg. Chim. Acta*, 1979, **36**, L431.
14 H. Shinar and G. Navon, *Eur. J. Biochem.*, 1979, **93**, 313.
15 L. F. McCoy and K.-P. Wong, *Biopolymers*, 1979, **18**, 2893.
16 I. Bertini, G. Canti, C. Luchinat, and A. Scozzafava, *Inorg. Chim. Acta*, 1979, **36**, 9.
17 G. R. Demille, K. Larlee, D. L. Livesey, and K. Mailer, *Chem. Phys. Lett.*, 1979, **64**, 534.
18 K. Pfaff and M. Kujawa, *Nahrung*, 1979, **23**, 91.

anhydrase have been interpreted[19] as indicating that HCO_3^- is directly bound to the metal ion, and the CO_2 is bound in the active site of the enzyme, though not directly to the copper(II).

Another [13]C n.m.r. study investigates the CO_2/HCO_3^- exchange that is catalysed by native human carbonic anhydrases C.[20] The rate of exchange shows no deuterium isotope effect, and is not affected by the presence of buffer ions. The maximal rate of exchange is larger than the maximal turnover rate. It was concluded that the proton-transfer process which is rate-limiting in the turnover reaction is not directly involved in the CO_2/HCO_3^- exchange step. This would be consistent with a nucleophilic mechanism whereby a metal-co-ordinated OH^- group reacts with the CO_2 to give HCO_3^- ion. Also consistent with this mechanism is a study of the anionic inhibition of human carbonic anhydrases B and C.[21] This study shows that anions compete with OH^- (in the hydration reaction) and with HCO_3^- (in dehydration) for the active site of carbonic anhydrase, as expected from the above mechanism. When CO_2 is the substrate, inhibition by the anions appears to be mixed. The order of inhibition by anions is $F^- \ll$ acetate, NO_3^-, $Cl^- < Br^- < I^-$, $ClO_4^- < CNS^- \ll CNO^-$ for both isoenzymes, with the affinity of each anion for the B isoenzyme being generally much greater than that for the C isoenzyme. The only exception to the latter observation is ClO_4^-, which does not discriminate between the B and C forms. This order of inhibition by anions corresponds to the lyotropic series, which is related to the size of anions, and their effects upon the ordered structure of water.

The rate constants for the exchange of [18]O between CO_2 and H_2O and between [12]CO_2 and [13]CO_2 in the presence of human carbonic anhydrase C have been measured.[22] It was concluded that an internal transfer of protons between two groups in the enzyme occurs, and that the rate of this transfer becomes rate-limiting at high buffer concentrations. The pK_a of the groups was estimated to be approximately 6.8, and the rate constant for the proton transfer was calculated to be 3.5×10^6 s^{-1}.

The kinetics of the absoption of CO_2 into CO_3^{2-}/HCO_3^- solutions, when catalysed by carbonic anhydrase, have been evaluated.[23] The process was first-order, with a rate constant of 0.9 l mol^{-1} s^{-1} and an activation energy of 9 kcal mol^{-1}.

The inhibition of carbonic anhydrases by sulphonamides is well documented.[2] These inhibitors bind to the metal ion of the active site as $-SO_2NH^-$.[24] An apparent exception to this was thought to be neoprontosil (4'-sulphamylphenyl-2-azo-7-acetamido-1-hydroxynaphthalene-3,6-disulphonate), which, on the basis of resonance Raman (RR) spectroscopy, was postulated to bind in the proton-ated form ($-SO_2NH_2$).[25] However, a re-evaluation of the binding of neopronto-

[19] I. Bertini, E. Borghi, and C. Luchinat, *J. Am. Chem. Soc.*, 1979, **101**, 7069.
[20] I. Simonsson, B. H. Jonsson, and S. Lindskog, *Eur. J. Biochem.*, 1979, **93**, 409.
[21] T. H. Maren and E. O. Couto, *Arch. Biochem. Biophys.*, 1979, **196**, 501.
[22] D. N. Silverman, C. K. Tu, S. Lindskog, and G. C. Wynns, *J. Am. Chem. Soc.*, 1979, **101**, 6734.
[23] E. Apler, M. Lohse, and W.-D. Deckwer, *Chem.-Ing.-Tech.*, 1979, **51**, 980.
[24] K. Kumar, R. W. King, and P. R. Carey, *Biochemistry*, 1976, **15**, 2195.
[25] R. L. Petersen, T.-Y. Li, J. T. McFarland, and K. L. Watters, *Biochemistry*, 1977, **16**, 726.

sil to carbonic anhydrase does not substantiate this conclusion.[26] This latest study demonstrates that neoprontosil has two coupled ionizable groups, —OH and —SO_2NH_2, in the pH range 10.5—11.5. The close proximity of the microscopic pK values for these two groups precludes the spectroscopic characterization of the separate —SO_2NH_2/O⁻ and —SO_2NH^-/OH species, and hence no conclusion can be drawn on the ionization state of the drug when it is bound to carbonic anhydrase.

It has been known for some time that the toxic gas carbonyl sulphide (COS), which is produced during the metabolism of CS_2 and which is a by-product of some industrial processes, is metabolized to carbon dioxide.[27] In a recent publication it is suggested that this reaction is mediated by carbonic anhydrase.[28] According to this scheme, metabolism of COS proceeds as shown in Scheme 1 Hence the toxicity of COS may be due to the H_2S that is released.

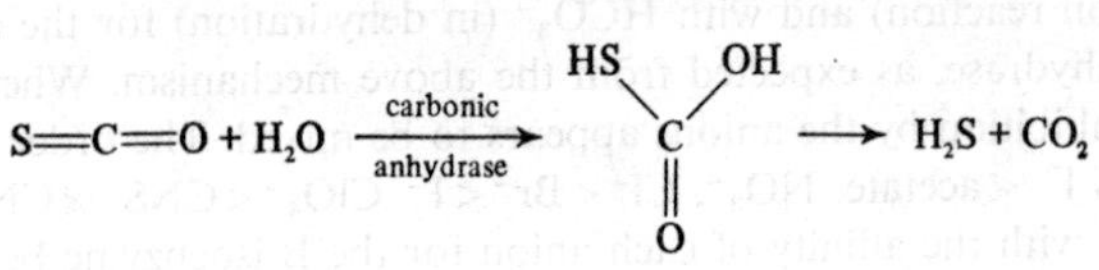

$$S{=}C{=}O + H_2O \xrightarrow[\text{anhydrase}]{\text{carbonic}} \underset{\underset{O}{\|}}{\overset{HS\diagdown \diagup OH}{C}} \longrightarrow H_2S + CO_2$$

Scheme 1

Carbonic anhydrase undergoes reversible denaturation on lyophilization This process has been followed by tracer studies, using $^{65}Zn^{II}$.[29] The recombination of Zn^{II} with the active site of the enzyme was found to account for only a small part of the re-activation process.

Several new methods for the detection of carbonic anhydrase in biological materials have been reported. A radioimmunosorbent technique for the assay of the B- and C-type isoenzymes in human tissue fluids is claimed to have a sensitivity of 0.1 ng per ml of fluid.[30] A specific method for the detection of carbonic anhydrase in tissues, based on its reaction with dimethylamino-naphthalene-5-sulphonamide to yield a highly fluorescent complex, has been described.[31] Other new methods for the localization of carbonic anhydrase involve immunohistochemical techniques, utilizing indirect fluorescent antibodies and immunoperoxidase.[32]

The characterization of carbonic anhydrase from a number of novel sources has been reported. A carbonic anhydrase that was isolated from human ciliary processes has been identified as the C isoenzyme.[33] It was concluded from this study that the differences that were observed in the reduction of intra-ocular pressure in a group of patients with glaucoma who were treated with systemic carbonic anhydrase inhibitors cannot be attributed to variations in the isoenzyme

26 P. R. Carey and R. W. King, *Biochemistry*, 1979, **18**, 2834.
27 R. Dalvi, A. Hunter, and R. Neal, *Chem.-Biol. Interact.*, 1975, **10**, 347.
28 C. P. Chengelis and R. A. Neal, *Biochem. Biophys. Res. Commun.*, 1979, **90**, 993.
29 O. De Jesus, E. D. Handel, and H. J. Ache, *Radiochem. Radioanal. Lett.*, 1979, **41**, 133.
30 T. Waahlstrand, K. G. Knuuttila, and P. J. Wistrand, *Scand. J. Clin. Lab. Invest.*, 1979, **39**, 503.
31 C. Pochhammer, P. Dietsch, and P. R. Siegmund *J. Histochem. Cytochem.*, 1979, **27**, 1103
32 T. Kumpulainen, *Histochemistry*, 1979, **62**, 271.
33 P. C. Dobbs, D. L. Epstein, and P. J. Anderson, *Invest. Ophthalmol. Visual Sci.*, 1979 **18**, 867.

composition of the ciliary processes. Human carbonic anhydrase III from skeletal muscle has been purified and characterized.[34] This isoenzyme is less effective as a hydratase and esterase than the erythrocyte isoenzymes, but is similar to these isoenzymes in molecular size (mol. wt 28 000). The limiting viscosities of the isoenzymes from human, bovine, and ovine erythrocytes have been compared.[35] The data suggest that all the isoenzymes possess stable conformations. A comparative study of the effects of chloride on a number of mammalian and insect carbonic anhydrases has been published.[36] A new type of carbonic anhydrase has been found in ovine parotid gland.[37] This enzyme has a molecular weight of 238 000 and shows no immunological cross-reactivity with the erythrocyte enzyme. This is the first report of a high-molecular-weight carbonic anhydrase that has been isolated from a mammalian source; however, bacteria and plants typically contain oligomeric carbonic anhydrases with molecular weights in this range. The isoenzyme composition of ovine erythrocyte carbonic anhydrase has also been studied.[38] Carbonic anhydrases B and C have been isolated from the erythrocytes of the turtle *Malaclemys terrapin centrata*.[39] The isoenzymes have molecular weights of 28 000 and 30 000, respectively, and both contain 1 mole of Zn^{II} per mole of protein. The effects of chelating agents and zinc on the carbonic anhydrase from *Tradescantia fluminensis* have been investigated.[40]

Plant carbonic anhydrases are in many respects similar to the mammalian erythrocyte enzymes, suggesting a common catalytic mechanism. A striking difference between the enzymes from these two sources is that the plant enzymes generally contain a large number (approx. seven per subunit) of cysteine residues. In a recent study, the chemical modification of spinach carbonic anhydrase by thiol reagents was investigated.[41] Three cysteine residues in the active site, whose modification abolishes catalysis, were found, but none of these residues was considered to be involved in catalysis. It has been reported that the fixation of CO_2 by ribulose-1,5-bisphosphatase carboxylase at low CO_2 concentrations is greatly enhanced by carbonic anhydrase.[42] It is known that carbonic anhydrase tends to accumulate in plant tissue under conditions of low CO_2 concentrations, and hence the above observation indicates that the accumulation of the enzyme under these conditions is required if a high rate of fixation of CO_2 is to be maintained.

Carboxypeptidase.—Although chelating agents such as 1,10-phenanthroline and 8-hydroxyquinoline-5-sulphonate have been used to remove the metal ion from Zn^{II} metalloenzyme for a large number of years, kinetic

[34] N. Carter, S. Jeffery, A. Shiels, Y. Edwards, T. Tipler, and D. Hopkinson, *Biochem. Genet.*, 1979, **17**, 837.

[35] M. Bouthier, J.-M. Gulian, and B. Mallet, *Biochimie*, 1979, **61**, 1161.

[36] J. W. Johnston and A. M. Jungreis, *Comp. Biochem. Physiol.*, *B*, 1979, **62**, 465.

[37] R. T. Fernley, R. D. Wright, and J. P. Coghlan, *FEBS Lett.*, 1979, **105**, 299.

[38] B. Mallet, J. M. Gulian, M. Sciaky, G. Laurent, and M. Charrel, *Biochim. Biophys. Acta*, 1979, **576**, 290.

[39] G. E. Hall and R. Schraer, *Comp. Biochem. Physiol.*, *B*, 1979, **63**, 561.

[40] A. V. Kositsyn and T. I. Igoshina, *Fiziol. Rast.* (*Moscow*), 1979, **26**, 81.

[41] D. L. Cybulsky, A. Nagy, S. I. Kandal, M. Kandal, and G. Gornall, *J. Biol. Chem.*, 1979, **254**, 2032.

[42] Y. Shiraiwa and S. Miyachi, *FEBS Lett.*, 1979, **106**, 243.

studies on the interaction of such reagents with metalloproteins are almost non-existent. It is therefore very interesting to see that a recent publication reports the interaction of chelating agents with Zn^{II} carboxypeptidase A.[43] Polydentate ligands such as EDTA, nitrilotriacetic acid, and *trans*-1,2-cyclo-hexanediaminetetra-acetate ion do not accelerate the rate at which Zn^{II} dissociates from the enzyme; these reagents inhibit the enzyme purely by scavenging the free metal ion from solution, and thereby inducing the dissociation of the holo-protein. On the other hand, bi- and ter-dentate ligands are able to remove the Zn^{II} directly from the protein; the rate for this process is first-order in both the enzyme and the ligand. It has been proposed that, for this latter type of ligand, the reaction proceeds *via* the ternary ligand–Zn^{II}–carboxypeptidase complex, which then dissociates into the Zn^{II}–ligand complex and the apoprotein. The results for a variety of ligands suggest that, in the ternary complex, terdentate ligands bind to the metal ion through only two donor groups. The reaction of 1,10-phenanthroline with the holoenzyme was studied over the pH range 6—9, and the results suggest that the removal of the metal ions by the ligand proceeds *via* a proton-assisted pathway, whereby protonation of the ternary complex precedes the dissociation of the metal–ligand complex from the protein. In a further publication the same author reports that amino-acids catalyse the reaction in which Zn^{II} is transferred from carboxypeptidase A to EDTA.[44] This catalysis proceeds *via* the formation of a ternary complex. In contrast to the other amino-acids, L- and D-phenylalanine and L-tyrosine act as inhibitors of the dissociation of Zn^{II}. This inhibition is probably due to the binding of these aromatic amino-acids to a hydrophobic pocket in the active site.

The interaction of the pseudo-substrate Gly-L-Tyr with carboxypeptidase A, as revealed by *X*-ray crystal analysis, has served as a basis for the mechanism of this enzyme.[45] Since this model is derived from studies in the solid state, its relevance to the mechanism of the enzyme in solution has been questioned.[46] In a recent report, the interaction of Gly-L-Tyr with Mn^{II} carboxypeptidase A in solution has been investigated by ^1H n.m.r. spectroscopy.[47] This study allowed the conclusion to be drawn that the modes of binding of Gly-L-Tyr to the Mn^{II} enzyme in solution are different from that observed with the crystalline Zn^{II} enzyme. In particular, the carbonyl group of the substrate does not co-ordinate tightly to the metal ion in solution, and the binding of the substrate does not exclude the co-ordination of water to the metal ion. This situation contrasts markedly with that seen in the crystalline state, and reinforces the doubts on the validity of the *X*-ray model for the mechanism of carboxypeptidase A.

PAC (perturbed angular correlation of γ-rays) spectroscopy has been used to investigate the structure of the active site of $^{111}Cd^{II}$ carboxypeptidase A.[48]

[43] E. J. Billo, *J. Inorg. Biochem.*, 1979, **10**, 331.
[44] E. J. Billo, *J. Inorg. Biochem.*, 1979, **11**, 339.
[45] W. N. Lipscomb, *Chem. Soc. Rev.*, 1972, **1**, 319.
[46] C. A. Spilburg, J. L. Bethune, and B. L. Vallee, *Biochemistry*, 1977, **16**, 1142.
[47] A. Nakano, M. Tasumi, K. Fujiwara, K. Fuwa, and T. Miyazawa, *J. Biochem.* (*Tokyo*), 1979, **85**, 1001.
[48] R. Bauer, C. Christensen, J. T. Johansen, J. L. Bethune, and B. L. Vallee, *Biochem. Biophys. Res. Commun.*, 1979, **90**, 679.

The enzyme was found to exist in one predominant state in the crystal state, whereas several different conformations were present in solution. It has been suggested that the metal possesses a distorted co-ordination geometry in the crystals, and that this form is in equilibrium with a penta-co-ordinate species in solution. A theoretical study of the active site of carboxypeptidase A, which involves a computer-implemented optimization procedure based on minimal steric difference and the inhibition constants of thirteen carboxylic acids, has been reported.[49] The computed map of the active site was found to predict correctly the relative rates of release of hydrophobic and charged amino-acids from the *N*-terminus of peptides, but not that of polar amino-acids, which had the potential of forming hydrogen bonds.

A group of potent metal-co-ordinating inhibitors of carboxypeptidase A and other Zn^{II} proteases have been synthesized and characterized.[50] These inhibitors are either thiols (R—SH) or phosphoramidates, $R—NHP(OPh)O_2^-$. They incorporate the anions $—S^-$ or $—PO^-$, which bind to the metal ions in the active site, and substrate-like moieties —R, which are either amino-acids or peptides; these latter groups interact with the substrate-recognition sites on the enzymes. These inhibitors can be extremely potent; thus *N*-(2-mercaptoacetyl)-D-phenylalanine inhibits carboxypeptidase A with a K_I of 0.22 μmol l^{-1}. The binding of the thiol inhibitors to the metal ion in the active site of the Co^{II}-substituted enzymes is accompanied by the appearance of an intense $S \rightarrow Co^{II}$ charge-transfer band near 340 nm (Figure 1). Analysis of the absorption, c.d., and m.c.d. spectra of Co^{II} indicates that the metal ion is four-co-ordinate in the enzyme–inhibitor complex and that the thiol group is in the inner co-ordination sphere of the metal ion. Two other groups have independently reported the use of analogues of thiol substrates[51] and of phosphoramidates[52] as inhibitors of carboxypeptidases. Some of these inhibitors are extraordinarily potent; thus 2-benzyl-3-mercaptopropanoic acid has a K_I of 11 nmol l^{-1} for carboxypeptidase A, and 2-mercaptomethyl-5-guanidinopentanoic acid has a K_I of 0.4 nmol l^{-1} for carboxypeptidase B.[51]

The effects of the inhibitor 3-phenylpropanoate ion on the hydrolysis of dipeptides by carboxypeptidase A have been investigated.[53] At non-activating concentrations of substrate, 3-phenylpropanoate acts as a mixed inhibitor of the hydrolysis of Cbz-Gly-Phe and as a non-competitive inhibitor of hydrolysis of Bz-Gly-Phe. At activating substrate concentrations, the inhibition is mixed towards both substrates, but it appears to be mostly competitive.

Another kinetic study of carboxypeptidase utilized the emerging technique of cryoenzymology to investigate the mechanism of action of this enzyme.[54] This study focuses on the hydrolysis of the ester *O*-(*trans-p*-chlorocinnamoyl)-ʟ-β-phenyl-lactate by the Zn^{II} and Co^{II} enzymes in ethylene glycol–methanol–

[49] Z. Simon, S. Holban, and I. Motoc, *Rev. Roum. Biochim.*, 1979, **16**, 141.
[50] B. Holmquist and B. L. Vallee, *Proc. Natl. Acad. Sci. USA*, 1979, **76**, 6216.
[51] M. A. Ondetti, M. E. Condon, J. Reid, E. F. Sabo, H. S. Cheung, and D. W. Cushman, *Biochemistry*, 1979, **18**, 1427.
[52] C.-M. Kam, N. Nishino, and J. C. Powers, *Biochemistry*, 1979, **18**, 3032.
[53] R. Poorman, M. Kuo, D. I. Johnson, S. Lin, and J. F. Sebastian, *Can. J. Biochem.*, 1979, **57**, 357.
[54] M. W. Makinen, L. C. Kuo, J. J. Dymowski, and J. Shams, *J. Biol. Chem.*, 1979, **254**, 356.

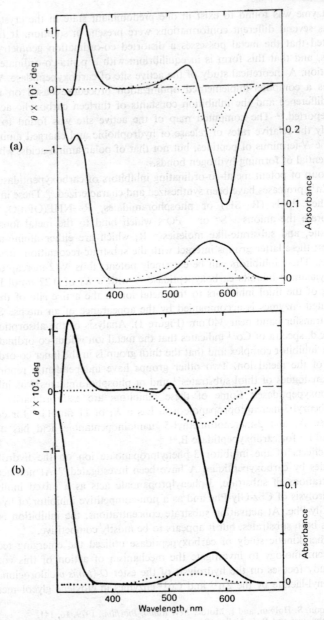

Figure 1 *The absorption (lower curves) and m.c.d. spectra (upper curves) of (a) Co^{II}-carboxypeptidase A (2.8 × 10^{-4} mol l^{-1}) in 1M-NaCl, 50mM-HEPES, at pH 7.5 and (b) Co^{II}-thermolysin (4.2 × 10^{-4} mol l^{-1}) in 50mM-HEPES, 10mM-Ca^{II} 2.5M-NaBr, at pH 7.5: spectra were collected in the absence (· · · · ·) or presence (———) of 1 equivalent of HSAc-D-Phe and HSAc-L-Phe-L-Ala, respectively* (Reproduced by permission from *Proc. Natl. Acad. Sci. USA*, 1979, **76**, 6216)

water over the temperature range 25 to $-45\,^{\circ}\text{C}$. On going from an aqueous to a mixed organic solvent, the K_M for this ester is negligibly affected, but k_{cat} drops twenty-fold. In the temperature range -25 to $-45\,^{\circ}\text{C}$, with enzyme in excess, a biphasic reaction is observed for hydrolysis of a substrate. In Arrhenius plots, the rate-constant for the slower process extrapolates to k_{cat} at $25\,^{\circ}\text{C}$ for both enzymes in aqueous solution, indicating that the same catalytic rate-limiting step is observed. It is suggested that this slow process involves the deacylation of a mixed acyl-enzyme anhydride intermediate. The pH and temperature dependence of the rate-constant for this slow step suggest that the ionization of the metal-bound water molecule is necessary for the deacylation of the mixed anhydride mechanism. It was concluded, from the work on this ester, that the hydrolysis of both ester and peptide proceeds *via* the attack of a metal-bound hydroxide group on an acyl intermediate (Scheme 2). This conclusion goes against much previous experimental work on this enzyme, including the X-ray structural analysis[55] and kinetic studies.[56, 57] In a further publica-

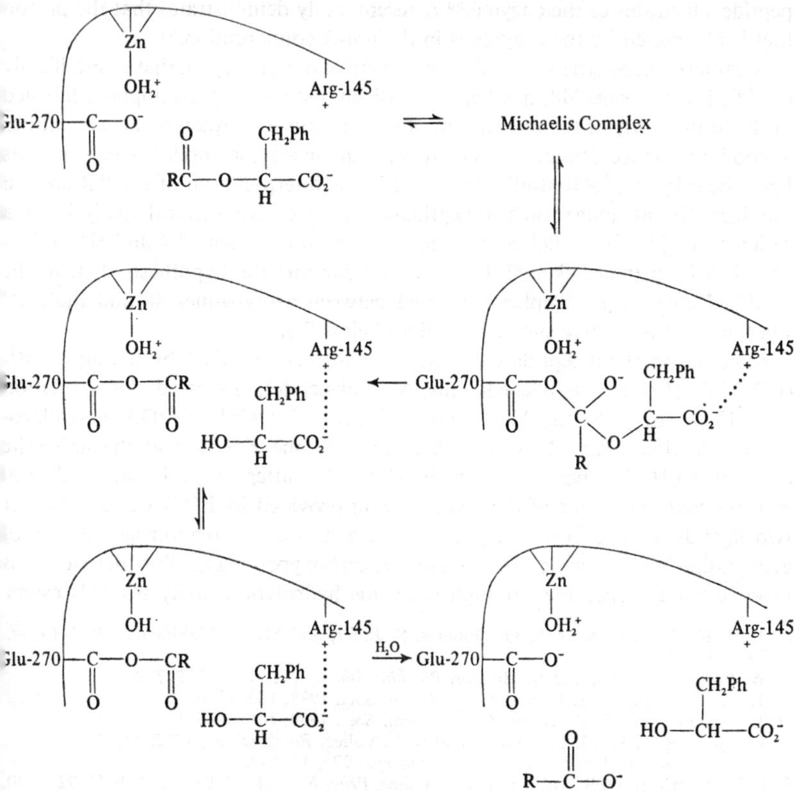

A schematic representation of the mechanism of hydrolysis of esters by carboxypeptidase A

Scheme 2

[55] J. A. Hartsuck and W. N. Lipscomb, in 'The Enzymes', ed. P. D. Boyer, Academic Press, New York, 1971, Vol. 3, p. 1.
[56] D. S. Auld and B. L. Vallee, *Biochemistry*, 1971, **10**, 2892.
[57] D. S. Auld and B. Holmquist, *Biochemistry*, 1974, **13**, 4355.

tion,[58] this same group reports the synthesis of the spin-label ester 3-(2,2,5,5-tetramethylpyrroline-1-oxyl)propen-2-oyl-L-β-phenyl-lactic acid, which is a substrate of carboxypeptidase A. It is suggested that this ester will prove useful in the characterization of intermediates of enzymic reactions by a combination of e.p.r. and cryoenzymologic techniques.

The reaction of dimethyl-(2-hydroxy-5-nitrobenzyl)sulphonium chloride with carboxypeptidase A at pH 7.5 results in the modification of one enzymatic residue.[59] Analysis of the resulting derivative indicates that the residue that is modified is not tryptophan (the amino-acid residue that is characteristically the site of modification by this reagent in proteins) but the *N*-terminal amino-group. It is postulated that this probably reflects the location of this amino-group with respect to the tryptophan residues in the tertiary structure of the enzyme.

In Volume 2 it was reported that carboxypeptidase A catalyses the stereospecific hydrogen–deuterium exchange at the methylene group of (−)-2-benzyl-3-(*p*-methoxybenzoyl)propionic acid, which is a ketonic analogue of ester and peptide substrates of the enzyme.[60] A recent study demonstrates that the proton that is abstracted by the enzyme is in the *pro-R* configuration.[61]

Arsanilazocarboxypeptidase A, the derivative of the enzyme that is specifically modified at tyrosine-248, has been extensively used as a chromophore-labelled probe in the study of the mechanism of the enzyme.[62-65] Azotyrosine constitutes a good resonance Raman (RR) probe, and hence the modified enzyme has been the subject of RR studies.[66] To aid in the interpretation of the RR spectra obtained for arsanilazocarboxypeptidase, an extensive spectral analysis of a series of azophenol model compounds and several of their ^{15}N and ^{2}H derivatives has been undertaken.[67] These studies support the hypothesis that, in the modified enzyme, a complex is formed between azotyrosine-248 and the Zn^{II} ion that is in the active site at pH values below 8.5.

Arsanilazocarboxypeptidase A has been further modified by adding [Co^{II}-(EDDA)] (EDDA = ethylenediamine-*NN'*-diacetate), followed by oxidation with H_2O_2, to yield the highly unusual derivative Co^{III}(EDDA)-arsanilazocarboxypeptidase A.[68] This derivative contains one Zn^{II} ion at the active site and one Co^{III} ion bound to tyrosine-248. The latter ion is in an octahedral environment, with four of the ligands being provided by EDDA and the other two ligands arising from the phenolic oxygen and the azo-nitrogen atoms of azotyrosine-248 (Scheme 3). Unlike azocarboxypeptidase, this derivative is inactive towards peptides, though it retains hydrolytic activity towards esters

[58] T. R. Koch, L. C. Kuo, E. G. Douglas, S. Jaffer, and M. W. Makinen, *J. Biol. Chem.* 1979, **254**, 12 310.
[59] W. Liu, N. Y. Licia, and R. Horton, *Biochim. Biophys. Acta*, 1979, **577**, 22.
[60] T. Sugimoto and E. T. Kaiser, *J. Am. Chem. Soc.*, 1978, **100**, 7750.
[61] T. Sugimoto and E. T. Kaiser, *J. Am. Chem. Soc.*, 1979, **101**, 3946.
[62] J. T. Johanson, D. M. Livingston, and B. L. Vallee, *Biochemistry*, 1972, **11**, 2584.
[63] J. T. Johanson and B. L. Vallee, *Biochemistry*, 1975, **14**, 649.
[64] L. W. Harrison, D. S. Auld, and B. L. Vallee, *Proc. Natl. Acad. Sci. USA*, 1975, **72**, 3930.
[65] L. W. Harrison, D. S. Auld, and B. L. Vallee, *Proc. Natl. Acad. Sci. USA*, 1975, **72**, 4356.
[66] R. K. Scheule, H. E. Van Wart, B. L. Vallee, and H. A. Scheraga, *Proc. Natl. Acad. Sci. USA*, 1977, **74**, 3273.
[67] R. K. Scheule, H. E. Van Wart, B. O. Zweifel, B. L. Vallee, and H. A. Scheraga, *J. Inorg. Biochem.*, 1979, **11**, 283.
[68] M. S. Urdea and J. I. Legg, *Biochemistry*, 1979, **18**, 4984.

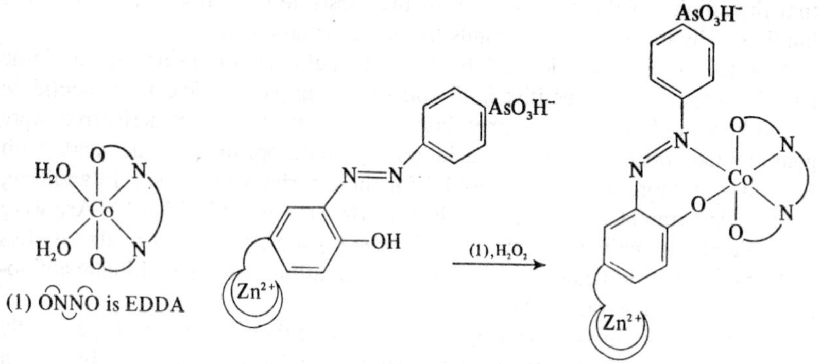

Scheme 3

It has been postulated, on the basis of these results, that tyrosine-248 is involved in the mechanism of the hydrolysis of peptides (but not of esters) by this enzyme. A further publication describes how it has been shown that peptides are excellent inhibitors of the esterase activity of this derivative.[69] This study shows that short peptides are non-competitive inhibitors of the hydrolysis of esters by Co^{III}(EDDA)-azocarboxypeptidase A, implying that the binding positions for esters and peptides are not identical. In contrast, long peptides are competitive inhibitors of the hydrolysis of esters. These results are consistent with earlier suggestions that multiple, overlapping, binding sites for substrates and inhibitors are present at the active site of carboxypeptidase A.[70, 71] Acetylation of two tyrosine residues (Tyr-248 and Tyr-198) in the active site of the native enzyme[72] produces a peptidase-inactive derivative that is distinctly different from Co^{III}-(EDDA)-azocarboxypeptidase A. In contrast with the native and Co^{III}(EDDA)-modified enzymes, peptides and esters appear to occupy a common binding position in the acetylated derivative. Peptides also bind much more poorly to the acetylated enzyme than to the native or Co^{III}(EDDA)-modified enzymes. Similar effects on the binding of peptides can be brought about by acetylation of the Co^{III}(EDDA)-modified enzyme. Therefore it appears that the acetylation of Tyr-198 results in a change in the binding position of peptides. Quantitative affinity-chromatography studies on carboxypeptidase B have indicated that multiple binding sites are also present in the active site of this enzyme.[73] Kinetic studies on bovine carboxypeptidase B, utilizing the substrates Cbz-Gly-Phe and Bz-Gly-Arg, and the effects of the inhibitors β-phenylpropionic acid and ε-aminocaproic acid provide further evidence for multiple binding sites in the active site of this enzyme.[74] It was concluded from this study that a secondary binding site, specific for hydrophobic compounds, is present in the enzyme, and

[69] M. S. Urdea and J. I. Legg, *J. Biol. Chem.*, 1979, **254**, 11 868.
[70] B. L. Vallee, J. F. Riordan, J. L. Bethune, T. L. Coombs, D. S. Auld, and M. Sokolovsky, *Biochemistry*, 1968, **7**, 354.
[71] G. M. Alter and B. L. Vallee, *Biochemistry*, 1978, **17**, 2212.
[72] R. T. Simpson, J. F. Riordan, and B. L. Vallee, *Biochemistry*, 1963, **2**, 616.
[73] J. Danner, J. E. Somerville, J. Turner, and B. M. Dunn, *Biochemistry*, 1979, **18**, 3039
[74] H. Akanuma and M. Yamasaki, *J. Biochem. (Tokyo)*, 1979, **85**, 775.

that the affinity of the primary site for the substrate is greatly enhanced by the binding of hydrophobic compounds to this secondary site.

A series of dicarboxylic acid by-product analogues of lysine and arginine have been tested as competitive inhibitors of human pancreatic carboxypeptidase B and human plasma carboxypeptidase N.[75] The most effective derivatives were guanidinoethylmercaptosuccinic acid and guanidinopropylsuccinic acid, with K_I values of approximately 1 μmol l^{-1} for these carboxypeptidases. In addition, the kinetic parameters for the hydrolysis of Bz-Ala-Lys and of Bz-Ala-Arg were determined. The enhanced affinity of these substrates, compared to Bz-Gly-Lys and Bz-Gly-Arg, indicates a significant participation of the penultimate amino-acid residue in substrate catalysis.

A single-step affinity-chromatography procedure that is capable of specifically isolating and resolving pancreatic carboxypeptidases A and B has been reported.[76] The affinity ligand used in these studies was [p-(N-caproylamino)-benzyl]succinic acid-Sepharose 4B.

The isolation of a carboxypeptidase from the crayfish *Astacus fluviatilis* has been reported.[77] The isolation procedure included affinity chromatography on a column of potato carboxypeptidase inhibitor that was covalently linked to Sepharose. The pure enzyme resembled bovine carboxypeptidase B in specificity, and was found to be a zinc metalloenzyme. The amino-acid sequence of the first nineteen N-terminal residues showed a significant homology with that of pancreatic carboxypeptidases A and B. A carboxypeptidase-A-like enzyme has also been purified from the limpet *Patella vulgata*.[78] The enzyme is a single polypeptide chain with a molecular weight of 40 000, and it exhibits both peptidase and esterase activities. Inhibition studies suggest that a metal ion is required for activity.

Two independent groups have reported the isolation of a carboxypeptidase from the micro-organism *Streptomyces griseus*.[79, 80] The enzyme has a molecular weight of approximately 40 000 and contains 1 gram atom of Zn^{II} per mole. In marked contrast to any hitherto known metallocarboxypeptidase, the enzyme of *S. griseus* hydrolyses both basic and neutral C-terminal peptide and ester substrates, thus combining the dual specificities of carboxypeptidases A and B. Chemical modification studies suggest that tyrosyl, arginyl, and glutamyl residues are required for catalytic activity;[79] these residues are known to be essential for the activity of bovine carboxypeptidase A. In addition, the carboxy-peptidase of *S. griseus* contains 2 gram atoms of tightly bound Ca^{II} ions per mole; these appear to stabilize the protein, in conjunction with two disulphide bridges. This is the first carboxypeptidase that has been found to contain structural Ca^{II} ions, though such Ca^{II} is known to be present in bacterial neutral endoproteases. Therefore it has been postulated[79] that this carboxypeptidase may represent the evolutionary intermediate between endopeptidases and carboxypeptidases.

75 T. J. McKay, A. W. Phelan, and T. H. Plummer, *Arch. Biochem. Biophys.*, 1979, **197**, 487.
76 T. J. Bazzone, M. Sokolovsky, L. B. Cueni, and B. L. Vallee, *Biochemistry*, 1979, **18**, 4362.
77 R. Zwilling, F. Jakob, H. Bauer, and H. Neurath, *Eur. J. Biochem.*, 1979, **94**, 223.
78 G. M. Hass, *Arch. Biochem. Biophys.*, 1979, **198**, 247.
79 K. Breddam, T. J. Bazzone, B. Holmquist, and B. L. Vallee, *Biochemistry*, 1979, **18** 1563
80 Y. Narahashi and K. Yoda, *J. Biochem. (Tokyo)*, 1979, **86**, 683.

Neutral Proteases.—Benzylsuccinate ion, which is a competitive inhibitor of carboxypeptidase A, has been shown to inhibit thermolysin.[81] The mode of binding of this by-product analogue (so called because it is postulated to resemble the two products of the enzymatic reaction[82]) to crystalline thermolysin has been determined to a resolution of 2.3 Å by X-ray crystallography. The inhibitor binds to the enzyme in a manner similar to that of dipeptide inhibitors, with one carboxyl oxygen displacing a water molecule from the Zn^{II}, the other carboxyl group interacting with Arg-203 and Asn-112, and with the benzyl group occupying a hydrophobic pocket. The results are consistent with a general-base mechanism of hydrolysis of peptides by thermolysin, with Glu-143 acting as a general or a specific base.

One of the most interesting reactions observed with thermolysin is its super-activation (as happens for other neutral proteases) by treatment with N-hydroxy-succinimide esters of acyl derivatives of amino-acids.[83] In a recent study[84] it was shown that this superactivation is due to the modification of tyrosine-110. This residue is known to be close to the substrate-binding site of the enzyme.

Thermolysin can be split into three fragments by treatment with CNBr. Surprisingly, c.d. spectroscopy and immunological techniques indicate that two of these fragments retain stable configurations in solution, and the tertiary structures of these fragments appear to be similar to those of the equivalent domains in the native enzyme.[85]

Thiol and phosphoramide analogues of substrates (which, as indicated above, have been found to be potent inhibitors of carboxypeptidases) also competitively inhibit neutral proteases.[50,52] Thus, HS-Ac-L-Phe-L-Ala has a K_i value of approximately 1 μmol l^{-1} towards thermolysin and the neutral protease of *Bacillus cereus*, phenylphosphoryl-L-Phe-DL-Trp has a K_i value of 0.3 μmol l^{-1} towards these two enzymes,[50] and phosphoryl-Leu-Trp has a K_i of 15 nmol l^{-1} for thermolysin.[52] Derivatives of hydroxamic acid were also found to be potent inhibitors.[86] Based on these results, an affinity resin for neutral proteases, comprising a hydroxamic acid derivative attached to agarose *via* an aminopropyl spacer group, has been prepared.

The isolation of new neutral proteases from *Bacillus thuringiensis*,[87] from *B. subtilis*,[88] from *Escherichia coli*,[89] and from *Serratia marcescens*[90] has been reported. All of these enzymes appear to be Zn^{II} metalloenzymes, their molecular weights being, respectively, 50 000, 28 000, 110 000, and 50 000; in addition to Zn^{II}, the enzyme of *S. marcescens* contains 7 gram atoms of Ca^{II} per mole.

Zinc(II)-requiring neutral endopeptidases are classically found in microbes,

[81] M. C. Bolognesi and B. W. Matthews, *J. Biol. Chem.*, 1979, **254**, 634.
[82] L. D. Byers and R. Wolfenden, *Biochemistry*, 1973, **12**, 2070.
[83] S. Blumberg and B. L. Vallee, *Biochemistry*, 1975, **14**, 2410.
[84] S. Blumberg, *Biochemistry*, 1979, **18**, 2815.
[85] C. Vita, A. Fontana, J. R. Seeman, and I. M. Chaiken, *Biochemistry*, 1979, **18**, 3023.
[86] N. Nishino and J. C. Powers, *Biochemistry*, 1979, **18**, 4340.
[87] M. Kucera, and N. M. Barbashova, *Prikl. Biokhim. Mikrobiol.*, 1979, **15**, 707.
[88] E. Savickaite, P. Petniunas, D. Sauklyte, I. Remeikaite, and A. Bukauskiene, *Metab. Ego Regul. Biol. Akt. Veshchestvami Vil'nyus*, 1979, 388 (*Chem. Abstr.*, 1979, **91**, 119 296).
[89] Y.-S. E. Cheng and D. Zipsar, *J. Biol. Chem.*, 1979, **254**, 4698.
[90] D. Lyerly and A. Kreger, *Infect. Immun.*, 1979, **24**, 411.

but a recent report details the isolation of such a protease from carp muscle.[91] Although Zn^{II} was not unequivocally shown to be present in this protein, the enzyme is inhibited by metal-chelating agents but not by di-isopropyl phosphoro-fluoridate (a reagent which characteristically inhibits serine proteases), indicating that this enzyme is indeed a novel metallo-endopeptidase.

Aminopeptidases.—The quaternary structure of leucine aminopeptidase from bovine lens has been investigated by electron microscopy.[92] The predominant images observed in the micrographs of negatively stained single molecules were triangles, many of which had a denser inset. Based on these studies, it was postulated that the quaternary structure of this hexameric enzyme consists of six asymmetrical bilobal subunits, arranged such that the main lobes are eclipsed and the inner lobes are staggered. This overall arrangement is similar to that proposed in earlier studies.[93]

Limited digestion of bovine lens aminopeptidase by trypsin results in the specific cleaving of only one bond per subunit.[94] Despite the splitting of this bond, the oligomeric structure of the enzyme remains intact, and the cleaved enzyme retains all of its catalytic properties, including activation by Mn^{II}. It has been postulated that the cleaved bond is on the surface of the molecule. Other enzymes, such as chymotrypsin, plasmin, and thrombin, fail to split the native enzyme. Dissociation of the trypsin-cleaved enzyme into its subunits reveals that each subunit is split into two fragments, with molecular weights of 7000 and 37 000. The bond that is cleaved by trypsin appears to be Arg-Lys. Determination of the N-terminal residues of the fragments reveals that the smaller fragment represents the N-terminal part, and the other fragment represents the C-terminal part, of the subunit. A series of papers concerning the properties of bovine lens leucine aminopeptidase has appeared.[95-98] These reports deal with kinetic and chemical-modification studies on the enzyme, and indicate both that substrates are bound by hydrophobic interactions at two sub-sites on the enzyme[97] and that the Zn^{II} in the active site is bound by histidyl and cystyl residues.[96]

The incorporation or loss of ^{18}O during the transpeptidation reaction Leu-NH$_2$ → Leu-Leu-NH$_2$, as catalysed by leucine aminopeptidase, has been measured.[99] Mass spectral analysis of the product showed that 50% of the label was incorporated from $H_2{}^{18}O$ or was lost from [^{18}O]leucinamide; 50% of the label was also incorporated into unreacted leucinamide. These results strongly suggest that an acyl-enzyme intermediate is not formed under these conditions for this mechanism precludes the incorporation of the label into the transpeptida

91 Y. Makinodan, M. Hirotsuka, and S. Ikeda, *J. Food Sci.*, 1979, **44**, 1110.
92 A. Taylor, F. H. Carpenter, and A. Wlodawer, *J. Ultrastruct. Res.*, 1979, **68**, 92.
93 N. A. Kiselev, V. Ya. Stel'Mashchuk, V. L. Tsuprun, M. Ludewig, and H. Hanson, *J. Mol. Biol.*, 1977, **115**, 33.
94 L. Van Loon-Klaassen, H. Th. Cuypers, and H. Bloemendal, *FEBS Lett.*, 1979, **107**, 366
95 J. Lasch, *Ophthalmic Res.*, 1979, **11**, 372.
96 M. Mueller-Frohne, *Ophthalmic Res.*, 1979, **11**, 377.
97 S. Fittkau, G. Kaemmerer, and W. Damerau, *Ophthalmic Res.*, 1979, **11**, 381.
98 K. P. Klante, S. Mqotsi, and S. Fittkau, *Ophthalmic Res.*, 1979, **11**, 386.
99 V. K. Antonov, L. P. Yavashev, L. I. Volkova, V. L. Sadovskaya, and L. M. Ginodman, *Bioorg. Khim.*, 1979, **5**, 1427

tion product. Therefore it appears likely that the reaction proceeds *via* a general base catalysis mechanism.

A new colorimetric method for the determination of serum leucine aminopeptidase has been proposed.[100] This method depends on the release of aromatic amines from substrates that contain primary aromatic amines, and their quantitation by reaction with pentacyanoamminoferroate to yield a highly coloured product (measured at 700 nm). An automated assay method for serum leucine aminopeptidase has also been described.[101]

A new aminotripeptidase has been isolated from the cytosolic fraction of rabbit intestinal mucosa.[102] This enzyme is a monomeric glycoprotein with a molecular weight of 50 000 and contains 1 gram ion of Zn^{II} per mole. Its specificity is directed primarily at tripeptides with proline at the *N*-terminus, but it will also hydrolyse other tripeptides, provided that the *N*-terminal position is not a charged residue and that the second position is not occupied by proline. Dipeptides and tetrapeptides are not hydrolysed by this enzyme. A novel aminopeptidase has also been isolated from the extracellular (culture) fluids of *Bacillus subtilis*.[103] This aminopeptidase is monomeric, with a molecular weight of 46 500, and it contains 1 gram ion of Zn^{II} per mole. The enzyme can be activated by Co^{II}; this process involves the binding of 1 mole of Co^{II} per mole of enzyme, without the displacement of the Zn^{II}. Zinc(II) competes with Co^{II} for this second site, but does not activate the enzyme. The constants for the dissociation of Zn^{II} and Co^{II} from this second site are 28 μmol l^{-1} and 1.25 mmol l^{-1}, respectively. Metal ions other than Co^{II} do not seem to activate the enzyme. The kinetic parameters for the hydrolysis of different L-α-amino-acid β-naphthylamides by this aminopeptidase have been measured.[104] For the native enzyme the order of specificity towards the *N*-terminal amino-acid is Arg > Met > Trp > Lys > Leu; this order is changed to Lys > Arg > Met > Trp > Leu in the presence of Co^{II}. In all cases, Co^{II} increases the values of k_{cat}/K_M for the hydrolysis. The activation of the aminopeptidase from *Aspergillus oryzae* has also been investigated.[105] Again, Co^{II} was found to be a specific activator, being much more effective than either Mg^{II} or Mn^{II}.

Two new extracellular aminopeptidases have been isolated from filtrates of *Keratinomyces ajelloi*.[106] Their molecular weights are 27 000 and 23 000, and both enzymes are inhibited by metal-chelating agents. An enkephalin-degrading aminopeptidase has been isolated from rat brain extracts.[107] The enzyme requires a free amino-group on the substrate and will hydrolyse basic and neutral, but not acidic, amino-acids. It will cleave the *N*-terminal tyrosine residue from leucine-enkephalin. The enzyme is dimeric and has a molecular weight of 100 000.

[100] S. Minato, *Clin. Chim. Acta*, 1979, **92**, 249.
[101] U. Lippi, R. Pavan, C. Crosta, V. Cagnin, and M. Zaninotto, *Laboratorio* (*Milan*), 1979, **6**, 127.
[102] C. Doumeng and S. Maroux, *Biochem. J.*, 1979, **177**, 801.
[103] F. W. Wagner, L. E. Ray, M. A. Ajjabnoor, P. E. Ziemba, and R. L. Hall, *Arch. Biochem. Biophys.*, 1979, **197**, 63.
[104] M. A. Ajjabnoor and F. W. Wagner, *Arch. Biochem. Biophys.*, 1979, **197**, 73.
[105] M. V. Kolodzeiskaya, S. V. Verbilenko, and L. A. Konoplich, *Ukr. Biokhim. Zh.*, 1979, **51**, No. 1, p. 56.
[106] P. Ruffin, E. Van Brussel, J. Biguet, and G. Biserte, *Biochimie*, 1979, **61**, 495.
[107] H. P. Schnebli, M. A. Phillipps, and R. K. Barclay, *Biochim. Biophys. Acta*, 1979, **569**, 89.

It is inactivated by metal-chelating agents, and the addition of Zn^{II} and Co^{II} restores the enzymatic activity.

Angiotensin-converting Enzyme.—Several recent reports deal with the development of inhibitors to angiotensin-converting enzyme. One of these involves the use of mercaptan and phosphoramidate analogues of the normal substrates.[50] Some of these inhibitors are extremely potent; for example, HS-Ac-D-Phe has a K_i value of 75 nmol l^{-1} for the enzyme. Another group reports the use of D-3-mercapto-2-methylpropanoyl-L-proline.[108] This compound has a K_i of 1.7 nmol l^{-1}, and its use in the chronic therapy of human hypersensitive disease is advocated. The other reports deal with peptide inhibitors of angiotensin-converting enzyme. One of the best known of these inhibitors is a snake venom peptide, < Glu-Trp-Pro-Arg-Pro-Gln-Ile-Pro-Pro, often referred to as SQ 20,881 or BPP9a. Recently, a new series of potent and specific orally active inhibitors of angiotensin-converting enzyme, based on these natural peptide inhibitors, have been developed.[109] From these studies, a catalytic mechanism for this enzyme has been proposed; this mechanism has been drawn in analogy with that postulated for the action of carboxypeptidase A. A series of analogues of BPP9a, containing L-3,4-dehydroproline (Δ^3Pro) at positions 3, 4, 8, or 9, have been synthesized.[109] All of these analogues were found to be more potent than BPP9a itself, and have I_{50} values ranging from 0.6 mmol l^{-1} to 0.2 nmol l^{-1}, compared to 25 mmol l^{-1} for BPP9a. The digestion of gelatin with bacterial collagenase results in the production of inhibitors of angiotensin-converting enzyme,[110] and nine peptide inhibitors were isolated from the digests. In addition to being inhibitory, six of these peptides were substrates for the enzyme. The enzyme releases the dipeptide alanylhydroxyproline from one of the peptides, and is strongly inhibited by this dipeptide. Human seminal angiotensin-converting enzyme has also been found to be inhibited by several peptides, and it is likely that this enzyme is homologous to the pulmonary enzyme.[111]

A recent report suggests that the dipeptidyl carboxypeptidase (enkephalinase) which is responsible for inactivating methionine-enkaphalin is, in fact, angiotensin-converting enzyme.[112] Other reports dispute this contention, and claim that the two enzymatic activities are distinguishable[113] and even separable.[114] The separation of the two enzymatic activities was achieved by ion-exchange chromatography on DEAE-cellulose.

The stereospecificity of angiotensin-converting enzyme has been investigated through the use of synthetic substrates.[115] The enzyme showed a high stereospecificity for an amino-acyl residue in position 3 from the C-terminus, and had an absolute requirement for an L configuration at this position. An L-amino-acid

108 D. W. Cushman, H. S. Cheung, E. F. Sabo, B. Rubin, and M. A. Ondetti, *Fed. Proc.*, 1979, **38**, 2778.
109 G. H. Fisher and J. W. Ryan, *FEBS Lett.*, 1979, **107**, 273.
110 G. Oshima, H. Shimabukuro, and K. Nagasawa, *Biochim. Biophys. Acta*, 1979, **566**, 128.
111 D. Depierre, J. P. Bargetzi, and M. Roth, *Enzyme*, 1979, **24**, 362.
112 M. Benuck and N. Marks, *Biochem. Biophys. Res. Commun.*, 1979, **88**, 215.
113 J. P. Swerts, B. Malfroy, and J. C. Schwartz, *Eur. J. Pharmacol.*, 1979, **53**, 209.
114 A. Arregui, C.-M. Lee, P. C. Emson, and L. L. Iversen, *Eur. J. Pharmacol.*, 1979, **59**, 141.
115 G. Oshima and K. Nagasawa, *J. Biochem.* (Tokyo), 1979, **86**, 1719.

in positions 1 or 2 (from the *C*-terminus) increased, but was not essential for, activity, whereas the stereochemistry at position 4 had little effect.

A comparison of the active sites of carboxypeptidase, thermolysin, and angiotensin-converting enzyme suggests that these three enzymes share very similar catalytic mechanisms.[116] The main difference appears to be that an essential lysyl residue is present in the active site of angiotensin-converting enzyme, but not in that of the other enzymes. This residue appears to be involved in the activation of this enzyme by univalent anions.

A recent re-investigation of angiotensin-converting enzyme from rabbit lung is in agreement with previous studies on this enzyme.[117]

An immunological study of the distribution of angiotensin-converting enzyme in organs of the rat indicates that two distinct forms of the enzyme are present, one of which is present in lung, serum, brain, spleen, and kidney and the other in the testis and epididymis.[118]

Several new methods of assay for angiotensin-converting enzyme have been reported. One of these utilizes furanacryloyl (FA) tripeptides (conforming to the known substrate specificity of the enzyme) for the continuous spectrophotometric assay of this peptidase in the visible region.[119] This assay is based on a blue shift of the absorption spectrum that occurs upon the hydrolysis of these substrates. The most suitable of these chromatophoric substrates for routine assay is FA-L-Phe-Gly-Gly.

Another spectrophotometric assay is based on the quantitation of the hippuric acid that is released from hippuryl-L-histidyl-L-leucine.[120] This method is fairly laborious, as it involves the extraction of the hippuric acid before it can be estimated. An easier assay, which depends on the fluorescence enhancement that is observed on hydrolysis of the substrate *o*-aminobenzoylglycyl-*p*-nitro-L-Phe-L-Pro, has also been described.[121] The fourth assay uses the substrate hippuryl-L-His-L-Leu and the quantitation of the hippuric acid end-product by high-performance liquid chromatography.[122]

Collagenases.—A new collagenase has been isolated from bovine dental pulp.[123] In contrast to other collagenases isolated fron oral tissues, this collagenase was present solely in the latent form. It could be activated by trypsin and 4-aminophenylmercuric chloride; the activated enzyme has a molecular weight of 45 000 and is inhibited by EDTA. Pig synovium, in tissue culture, has also been shown to secrete a specific collagenase in a latent form.[124] This enzyme has been purified to apparent homogeneity. Another novel collagenase has been purified from the serum-free culture medium of epidermoid carcinoma of rat prostate.[125] The

[116] P. Bünning, B. Holmquist, and J. F. Riordan, *Colloq. Ges. Biol. Chem.*, 1979, **30**, 269.
[117] N. Yoshida, *Jpn. Circ. J.*, 1979, **43**, 55.
[118] R. Polsky-Cynkin and B. L. Fanburg, *Int. J. Biochem.*, 1979, **10**, 669.
[119] B. Holmquist, P. Bünning, and J. F. Riordan, *Anal. Biochem.*, 1979, **95**, 540.
[120] A. Le Treut, H. Couliou, M. Delbary, J. J. Larzul, B. De Labarthe, and J. Y. Le Gall, *Clin. Chim. Acta*, 1979, **98**, 1.
[121] A. Carmel, S. Ehrlich-Rogozinsky, and A. Yaron, *Clin. Chim. Acta*, 1979, **93**, 215.
[122] S. G. Chiknas, *Clin. Chem.*, 1979, **25**, 1259.
[123] J. Kishi, K. Iijima, and T. Hayakawa, *Biochem. Biophys. Res. Commun.*, 1979, **86**, 27.
[124] T. E. Cawston and J. A. Tyler, *Biochem. J.*, 1979, **183**, 647.
[125] C.-C. Huang, C.-H. Wu, and M. Abramson, *Biochim. Biophys. Acta*, 1979, **570**, 149.

molecular weight of this collagenase was estimated to be 71 000, and the enzyme was inhibited by EDTA.

The mode of action of bacterial (clostridial) collagenase on the synthetic substrate (Pro-Pro-Gly)$_5$ has been investigated;[126] Pro-Pro from the *N*-terminus was found to be released most rapidly, yielding the tridecapeptide. The enzyme then hydrolyses the inner linkages of this peptide (*i.e.* acts as an endopeptidase) preferentially at the *N*-terminal or *C*-terminal positions (occupied by Gly).

A specific inhibitor of collagenase has been purified from the serum-free medium of human skin fibroblasts.[127] The inhibitor has an apparent molecular weight of 31 000 by electrophoresis on Na$^+$ dodecylsulphate gel. Tight binding of the inhibitor to collagenase was only observed in the presence of collagen, and the inhibitor was found to bind to this substrate. This inhibitor was found to be effective against a wide variety of vertebrate collagenases, but non-collageno-lytic proteases and invertebrate collagenases were not inhibited. In contrast, the collagenase inhibitor that had previously been isolated from rabbit bone explants has also been shown to inhibit other neutral metalloproteinases that are pro-duced by this tissue.[128]

Three new assays for collagenases have been proposed. One of these utilizes Cbz-Gly-Pro-Gly-Gly-Pro-Ala as substrate, and the liberation of the tripeptide is monitored.[129] Another involves the use of ^{3}H-labelled collagen as the sub-strate,[130] whereas the third utilizes ^{14}C-labelled collagen as the substrate.[131]

Alcohol Dehydrogenase.—Equine liver alcohol dehydrogenase (LADH) contains two Zn^{II} ions per subunit (the enzyme is dimeric). These two metal ions have distinct functions: one, referred to as the catalytic Zn^{II}, is located in the active site of the enzyme and is essential for catalysis, whereas the other, referred to as the non-catalytic Zn^{II}, is distinct from the active site and is required for the stabilization of the tertiary structure of the protein.[132] Recently, conditions for the selective replacement of the catalytic or non-catalytic Zn^{II} by ^{65}Zn^{II} (ref. 133) or Co^{II} (ref. 134) have been developed. These studies have now been extended to the replacement of the Zn^{II} by ^{109}Cd^{II} (ref. 135) to yield [(LADH)^{109}Cd$_2$Zn$_2$], *i.e.* a derivative in which the non-catalytic Zn^{II} is specifically replaced by the ^{109}Cd^{II}, and of [(LADH)^{109}Cd$_2$^{109}Cd$_2$]. The u.v. difference spectra of the Cd^{II} enzymes *vs.* the native Zn^{II} enzyme reveal maxima at 240 nm which are strikingly similar to the charge-transfer absorbance that is seen in metallothionein. The $\Delta\varepsilon_{240}$ values are 1.6×10^4 l mol^{-1} cm^{-1} for the non-catalytic Cd^{II} and 0.9×10^4 l mol^{-1} cm^{-1} for the catalytic Cd^{II}, consistent with the co-ordination of the metals by four and two thiolate ligands respectively. The carboxymethylation of Cys-46, which is one of the ligands to the catalytic metal ion, lowers the

[126] G. Oshima, H. Shimabukuro, and K. Nagasawa, *Biochim. Biophys. Acta*, 1979, **567**, 392.
[127] H. G. Welgus, G. P. Stricklin, A. Z. Einen, and E. A. Bauer, *J. Biol. Chem.*, 1979, **254**, 1938.
[128] A. Sellers, G. Murphy, M. C. Meikle, and J. J. Reynolds, *Biochem. Biophys. Res. Commun.*, 1979, **87**, 581.
[129] H. U. Siebeneick, *Arzneim.-Forsch.*, 1979, **29**, 172.
[130] M. F. Lefevre, G. A. Slegers, and A. E. Claeys, *Clin. Chim. Acta*, 1979, **92**, 167.
[131] T. E. Cawston and A. J. Barrett, *Anal. Biochem.*, 1979, **99**, 340.
[132] D. E. Drum and B. L. Vallee, *Biochemistry*, 1970, **9**, 4078.
[133] A. J. Sytkowski and B. L. Vallee, *Proc. Natl. Acad. Sci. USA*, 1976, **73**, 344.
[134] A. J. Sytkowski and B. L. Vallee, *Biochemistry*, 1978, **17**, 2850.
[135] A. J. Sytkowski and B. L. Vallee, *Biochemistry*, 1979, **18**, 4095.

overall stability constant at this site and results in the loss of catalytic Cd^{II} or Co^{II}, but not of catalytic Zn^{II}, from the enzyme.

The specific replacement of the Zn^{II} by Co^{II} for the enzyme in solution and in the crystalline state and for the agarose-immobilized enzyme has been investigated.[136] In solution, an anaerobic column chromatography procedure permits the replacement of the Zn^{II} by Co^{II} in a much shorter time than previously possible. Treatment of crystal suspensions of the enzyme with dipicolinic acid results in the selective removal of the catalytic Zn^{II}. This protein could be reconstituted with Zn^{II} or Co^{II}. In contrast, the catalytic Zn^{II} cannot be selectively removed from the agarose-immobilized enzyme. Copper(II) has also been substituted for the Zn^{II} in equine liver alcohol dehydrogenase.[137] The e.p.r. spectra of the resulting Cu^{II} enzyme are similar to those of type I copper proteins, and are indicative of a tetrahedral co-ordination geometry. The presence of coenzyme induces a conformational change in the protein, but the metal remains in a tetrahedral geometry, whereas the binding of the inhibitor pyrazole results in e.p.r. spectra which are indicative of square-planar co-ordination geometry.

The binding of the chromophoric substrate *trans*-4-(NN-dimethylamino)-cinnamaldehyde and of NADH or 1,4,5,6-tetrahydronicotinamide-adenine dinucleotide to alcohol dehydrogenase, to give a ternary complex, results in a red shift of the visible spectrum of the substrate.[138] This spectral shift is 14 nm larger for the enzyme with Co^{II} at the catalytic sites than that observed with the Zn^{II} enzyme, and the d–d band of the catalytic Co^{II} ion increases greatly when a substrate is bound. The presence of the catalytic metal ion is an absolute requirement for the binding of the substrate. Taken together, these results suggest that the carbonyl group of the substrate directly co-ordinates to the catalytic metal ion. The binding of the dinucleotides to the Co^{II}/Zn^{II} enzyme (with Co^{II} in the catalytic sites) results in a red shift of the d–d band of the Co^{II} that is at 650 nm. This suggests that the binding of the coenzyme triggers a conformational change in the protein that involves the catalytic metal ion.

Temperature-jump studies on several metal derivatives of alcohol dehydrogenase indicate the presence of a fast conformational change during the catalytic mechanism of the enzyme.[139] The rate-constant for this conformational transition is dependent on the nature of the metal ion occupying the catalytic sites on the enzyme.

The inactivation or inhibition of equine liver and of yeast alcohol dehydrogenases with halogeno-(5-imidazolyl) derivatives has been investigated.[140] These reagents are thought to act as metal-directed affinity labels; according to this scheme, inhibition is due to the binding of the imidazole to the metal ion

[136] W. Maret, I. Andersson, H. Dietrich, H. Schneider-Bernloehr, R. Einarsson, and M. Zeppezauer, *Eur. J. Biochem.*, 1979, **98**, 501.

[137] W. Maret, H. Dietrich, H. H. Ruf, and M. Zeppezauer, *Metalloproteins, Autumn Meet. Ger. Biochem. Soc.*, 1979, 254.

[138] H. Dietrich, W. Maret, L. Wallen, and M. Zeppezauer, *Eur. J. Biochem.*, 1979, **100**, 267.

[139] I. Giannini, G. Baroncelli, and P. Renzi, *J. Mol Catal.*, 1979, **6**, 123.

[140] K. H. Dahl, J. S. McKinley-McKee, H. C. Beyerman, and A. Noordam, *FEBS Lett.*, 1979, **99**, 308.

at the active site, and inactivation is due to the subsequent alkylation of an amino-acid residue by the metal-bound reagent. The most potent inhibitors for the liver and yeast enzymes were 5-(2-bromoethyl)imidazole and methyl 2-chloro-3-(5-imidazolyl)propionate (ClPMe), respectively. The differences in the efficacy of these reagents in inactivating the liver and yeast enzymes indicate that the active sites of these two alcohol dehydrogenases differ. In addition, the inactivation of equine liver dehydrogenase with ClPMe and 2-chloro-3-(5-imidazolyl)propionic acid was found to be highly stereoselective, the (*R*)-enantiomers being totally ineffective;[141] this suggests that the active site of the enzyme is asymmetrical.

The NAD⁺ analogue 4-(3-bromoacetylpyridinio)butyldiphosphoadenosine inactivates equine liver and yeast alcohol dehydrogenases by the modification of amino-acid side-chains at the active sites of the enzymes.[142] The inactivation followed pseudo-first-order kinetics and the stoicheiometry was one analogue incorporated per subunit. The liver enzyme was inactivated by keto-alkylation of Cys-46, which is one of the ligands to the catalytic metal ion.

Equine liver and yeast alcohol dehydrogenases are also rapidly inactivated by but-3-yn-1-ol.[143] This inactivation is only observed in the presence of NAD⁺, and is due to an electrophilic product of the oxidation of but-3-yn-1-ol by the enzyme, possibly the 1,2-butadiene aldehyde (2) (see Scheme 4); nucleophiles

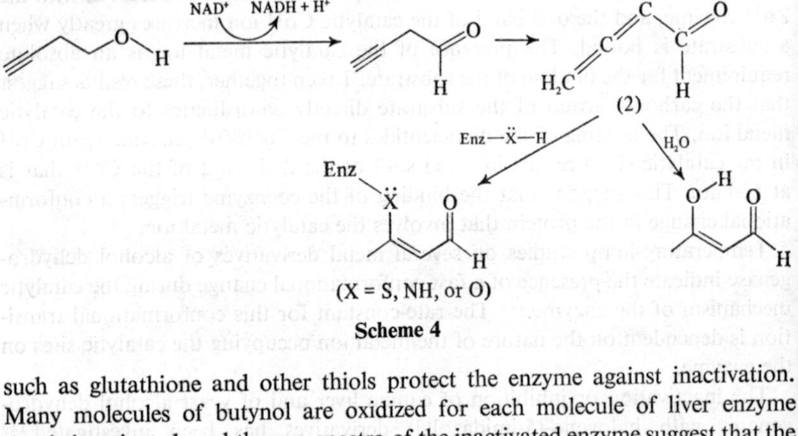

(X = S, NH, or O)

Scheme 4

such as glutathione and other thiols protect the enzyme against inactivation. Many molecules of butynol are oxidized for each molecule of liver enzyme that is inactivated, and the u.v. spectra of the inactivated enzyme suggest that the protein is alkylated at many sites. The inhibition of mouse liver alcohol dehydrogenase by dimethylformamide has been investigated.[144] Both inhibitors act in a non-competitive manner.

The three-dimensional structure of a ternary complex of equine liver alcohol dehydrogenase with NADH and the inhibitor dimethyl sulphoxide has been

[141] K. H. Dahl, J. S. McKinley-McKee, H. C. Beyerman, and A. Noordam, *FEBS Lett.*, 1979, **99**, 313.
[142] C. Woenckhaus, R. Jeck, and H. Joernvall, *Eur. J. Biochem.*, 1979, **93**, 65.
[143] T. A. Alston, L. Mela, and H. J. Bright, *Arch. Biochem. Biophys.*, 1979, **197**, 516.
[144] M. Sharkawi, *Toxicol. Lett.*, 1979, **4**, 493.

determined to a resolution of 4.5 Å.[145] Unlike previous difference X-ray structural analyses on the binding of nucleotides and of inhibitors, this study does not rely on the X-ray structure of the apoenzyme; it is, however, in agreement with the previous studies. Both subunits of the dimeric enzyme were found to bind coenzyme and inhibitor to the same extent. The two coenzyme-binding domains form the centre of the dimeric molecule, and their position is unchanged by the binding of coenzyme. On the other hand, the catalytic domains in the ternary complex are rotated such that the front sides of the domains move toward the central core. In the ternary complex the active site becomes shielded from the solution by a combination of this rotation, local movements of a loop from residues 53 to 57, and binding of coenzyme and of inhibitor. The nicotinamide ring of the coenzyme is positioned close to the zinc ion in the active site, and the inhibitor is bound to this zinc ion. These studies therefore indicate that the role of the catalytic metal in alcohol dehydrogenase is to polarize the C—O bond of the susceptible group. The binding of the dye Cibacron Blue F3GA to alcohol dehydrogenase has also been investigated by X-ray diffraction.[146] The dye was found to bind similarly to the coenzyme NAD$^+$, except for ring A of the dye, which binds differently from the corresponding ribose ring of the coenzyme.

Absorption microspectrophotometric measurements on single crystals of equine liver alcohol dehydrogenase that contain bound coenzymes have been reported.[147] The spectra of crystals containing NADH were identical to those obtained for the enzyme in solution. Importantly, the crystalline enzyme–NADH complex was shown to be catalytically active; the addition of any one of several aldehyde substrates results in the oxidation of the coenzyme to NAD$^+$, and in the concomitant decrease of the absorption band of NADH. The subsequent addition of ethanol reverses the spectral changes.

N^6-[N-(6-aminohexyl)carbamoylmethyl]-NAD has been covalently coupled to equine liver alcohol dehydrogenase, and the interaction of the coenzyme–enzyme complex with lactate dehydrogenase has been investigated.[148] The addition of ethyl alcohol causes the reduction of the bound NAD$^+$ to NADH, and this is accompanied by an increase in the fluorescence of the coenzyme complex. The addition of lactate dehydrogenase and oxalate results in a further increase in fluorescence, indicating the formation of an oxalate–lactate dehydrogenase–NADH–alcohol dehydrogenase complex. The addition of pyruvate decreases the fluorescence to its initial value, owing to the reduction of the NADH to NAD$^+$, mediated by lactate dehydrogenase. It has been suggested that the coenzyme that is covalently bound to the alcohol dehydrogenase can swing out of the active site of this enzyme and interact with the lactate dehydrogenase in solution. The mean cycling rate of the coenzyme was estimated to be 5 min^{-1}. A series of cyclic compounds, comprising berberine alkaloids, tricyclic psychopharmaceuticals, and acridine derivatives, have been used as probes of the substrate-binding pocket of alcohol dehydrogenase.[149] The results

[145] H. Eklund and C.-I. Branden, *J. Biol. Chem.*, 1979, **254**, 3458.
[146] J. F. Biellman, J. P. Samama, C. I. Branden, and H. Eklund, *Eur. J. Biochem.*, 1979, **102**, 107.
[147] E. Bignetti, G. L. Rossi, and E. Zeppezauer, *FEBS Lett.*, 1979, **100**, 17.
[148] M. O. Mansson, P. O. Larsson, and K. Mosbach, *FEBS Lett.*, 1979, **98**, 309.
[149] J. Kovor, E. Duerrova, and L. Skursky, *Eur. J. Biochem.*, 1979, **101**, 5507.

suggest that these compounds, which all inhibit the equine EE isoenzyme, bind at three distinct sites on the enzyme, with the alkaloids binding at the most hydrophobic site and the acridines at the most polar site.

Chlorine-35 n.m.r. quadrupole relaxation experiments indicate that at least two anion-binding sites exist within the coenzyme-binding domains of equine liver alcohol dehydrogenase.[150] It is suggested that the anion-binding sites are positively charged amino-acid residues, and, in conjunction with previous X-ray crystallographic data, it is postulated that these sites are Arg-47 and Arg-271; no evidence for any binding of anions to the Zn^{II} ions was obtained. The two anion-binding sites are not independent, but show co-operativity. The fact that halides do not bind at the Zn^{II} in alcohol dehydrogenase has also been suggested in a recent study, which follows the fate of the iodide that is liberated during the carboxymethylation of the enzyme with iodoacetate.[151] It was found that, subsequent to alkylation, no iodide was bound to the protein, and hence previous X-ray diffraction studies, which had indicated that I^- is bound to the Zn^{II} in the active site following carboxymethylation, were re-evaluated.

The transition-state kinetics of ligand-displacement reactions in liver alcohol dehydrogenase have been investigated.[152] In agreement with previous studies,[153,154] it was demonstrated that the rate of dissociation of NADH from the enzyme is independent of pH, whereas the rate of dissociation of NAD^+ is dependent on a pK_a of 7.6. The association rates for both NAD^+ and NADH are dependent on a pK_a of 9.2. It has been known for some time that this pK_a is reduced to 7.6 by the binding of NAD^+. The group that is responsible for these pK_a values has been proposed to be the metal-bound water molecule[155] or an amino-acid residue in the active site, possibly a tyrosyl residue.[156] A recent calorimetric study attempts to distinguish between these two possibilities.[157] The enthalpy of binding for NADH is 0 ± 0.5 kcal mol^{-1}, and is essentially independent of pH. On the other hand, the enthalpy of binding for NAD^+ varies with pH in a sigmoidal fashion, with a pK_{app} of 7.6. The enthalpy of ionization of a proton of the group that is responsible for this pK_a was calculated to be about 9.3 kcal mol^{-1}. The enthalpy for the ionization of a metal-bound water molecule in model compounds is 6.0—8.3 kcal mol^{-1},[158] and the corresponding value for a tyrosyl residue is 6.3 kcal mol^{-1}.[159] On the basis of these data it has been concluded that the enzymatic group in question is a zinc-bound water molecule; this implies that the role of the catalytic metal ion in alcohol dehydrogenase is to lower the pK_a of a bound water molecule substantially. This interpretation is therefore in conflict with the X-ray structural

150 I. Andersson, M. Zeppezauer, T. Bull, R. Einarsson, J.-E. Norne, and B. Lindman, *Biochemistry*, 1979, **18**, 3407.
151 J. F. Bielmann and P. R. Goulas, *Eur. J. Biochem.*, 1979, **100**, 461.
152 J. Kvassman and G. Pettersson, *Eur. J. Biochem.*, 1979, **100**, 115.
153 K. Dalziel, *J. Biol. Chem.*, 1963, **238**, 2850.
154 M. C. DeTraglia, J. Schmidt, M. F. Dunn, and J. T. McFarland, *J. Biol. Chem.*, 1977, **252**, 3493.
155 S. Taniguchi, H. Theorell, and A. Akeson, *Acta Chem. Scand.*, 1967, **21**, 1903.
156 W. R. Laws and J. D. Shore, *J. Biol. Chem.*, 1978, **253**, 8593.
157 S. Subramanian and P. D. Ross, *J. Biol. Chem.*, 1979, **254**, 7827.
158 P. Woolley, *Nature (London)*, 1975, **258**, 677.
159 D. D. F. Shiao and J. M. Sturtevant, *Biopolymers*, 1976, **15**, 1201.

studies quoted above. Other authors reconcile these two opposing views by proposing that the role of the catalytic Zn^{II} in alcohol dehydrogenase is both to polarize the susceptible C—O bond, as proposed by those who have used X-ray crystallography to study the enzyme, and to lower the pK_a of a water molecule at the active site.[160] According to this scheme, the substrate binds to the Zn^{II} *without* displacing the metal-bound water molecule, thus giving rise to a five-co-ordinate intermediate (Scheme 5).

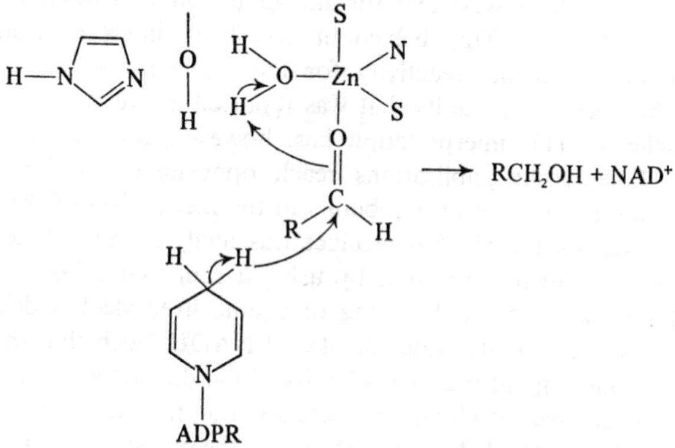

$$\longrightarrow RCH_2OH + NAD^+$$

A proposed mechanism for reduction of an aldehyde RCHO via a five-co-ordinate metal intermediate

Scheme 5

A number of kinetic studies on liver alcohol dehydrogenases have been published; these include an evaluation of non-linear regression and jacknife techniques for the analysis of kinetic data[161] and an investigation of the activation of the enzyme by deoxycholate.[162] The latter work involves the interconversion of ethanol and acetaldehyde, as catalysed by the enzyme from rat liver, and indicates that the activation by deoxycholate is a result of an overall change of the reaction mechanism for this interconversion. The effect of deoxycholate on alcohol dehydrogenase is species-dependent.[163] It has been suggested that for some species, including Man, the bile acids could control both the oxidation of ethanol and their own formation, since liver alcohol dehydrogenase is involved in both metabolic pathways. High concentrations of ethanol and NAD^+ have also been reported to activate equine liver alcohol dehydrogenase.[164] This activation was observed over the pH range 6.0—7.9, and has been attributed to negative co-operativity between the two subunits of the enzyme. Temperature-jump experiments on equine liver alcohol dehydrogenase indicate that, over a wide range of pH, the reaction can be described in terms of two catalysis-linked

[160] R. T. Dworschack and B. V. Plapp, *Biochemistry*, 1977, **16**, 2716.
[161] A. Cornish-Bowden and J. T.-F. Wong, *Biochem. J.*, 1979, **175**, 969.
[162] G. M. Hanozet, M. Simonetta, D. Barisio, and A. Guerritore, *Arch. Biochem. Biophys.*, 1979, **196**, 46.
[163] M. Simonetta, A. Ansaloni, and G. M. Hanozet, *Comp. Biochem. Physiol.*, B, 1979, **64**, 363.
[164] B. M. Kershengol'ts and V. V. Rogozhin, *Biokhimiya* (*Moscow*), 1979, **44**, 661.

protons that are not associated with the transfer of electrons.[165] The rates of the protonation steps in buffered solutions were extremely fast, and could not be measured in the temperature-jump experiments. The scheme most consistent with the kinetic data involves proton release subsequent to the binding of alcohol to give an alcoholate anion, which then undergoes hydride transfer to form the aldehyde.

It has been reported that, under conditions of a single catalytic turnover, two kinetic processes are observed[166] for the reduction of certain aldehydes by alcohol dehydrogenase. This behaviour has been interpreted in terms of coenzyme-induced half-site reactivity for the enzyme, and the negative co-operativity between the subunits that was reported above would be consistent with this scheme. This interpretation has, however, been disputed by other authors.[167] Two recent publications reach opposite conclusions about the presence of interactions between subunits in the mechanism of the enzyme. In one of these studies the alcohol product was analysed immediately after the initiation of the reduction reaction by using a rapid sampling device.[168] The experiment involved the rapid mixing of equine liver alcohol dehydrogenase containing a very low molar ratio of $[4\alpha\text{-}^3H]NADH$ (such that the coenzyme was bound to only one of the two subunits of the dimeric enzyme) with excess 4-(2'-imidazolylazo)benzaldehyde substrate and non-radioactive NADH. About 90% of the available ^3H label was transferred to the product during this reaction, indicating full-site reactivity for the enzyme. On the contrary, in a detailed kinetic re-investigation of the reduction of aromatic aldehydes by the enzyme, the original authors re-assert the presence of subunit–subunit interactions.[169] In this study the time courses of reactions were monitored at 330 nm, *via* stopped-flow rapid-mixing spectrophotometry, and the reaction was limited to a single turnover by the addition of pyrazole, which forms a strong dead-end complex with the enzyme-bound NAD^+ that is produced during the reaction. In all the cases investigated, the reaction was markedly biphasic; the formation of a rapid transient preceded a slower step, the latter corresponding to turnover. It was concluded that the results could only be interpreted in terms of subunit–subunit interactions. The deuterium solvent isotope effect for the two transients that were observed during the reduction of aromatic aldehydes has been investigated in another publication.[170] The isotope effect (k_{H_2O}/k_{D_2O}) for the rapid transient is 1.0 ± 0.1, whereas the slow transient shows an inverse isotope effect of 0.5—0.3. The transient kinetic rate of oxidation of alcohol and the rate of binding of phenanthroline to the Zn^{II} at the active site also show inverse solvent isotope effects. Furthermore, the pK_a of the base-catalytic group of the enzyme is perturbed by 0.3—0.5 unit in D_2O, and there is no solvent isotope effect at equivalent pH (pD) values. These results have been interpreted according to a

[165] G. H. Czerlinski, J. O. Erickson, and H. Theorell, *Physiol. Chem. Phys.*, 1979, **11**, 537.
[166] S. A. Bernhard, M. F. Dunn, P. L. Luisi, and P. Schack, *Biochemistry*, 1970, **9**, 185.
[167] J. Kvassman and G. Pettersson, *Eur. J. Biochem.*, 1976, **69**, 279.
[168] R. J. Kordaland and S. M. Parsons, *Arch. Biochem. Biophys.*, 1979, **194**, 439.
[169] M. F. Dunn, S. A. Bernhard, D. Anderson, A. Copeland, R. G. Morris, and J.-P. Roque, *Biochemistry*, 1979, **18**, 2346.
[170] J. Schmidt, J. Chen, M. DeTraglia, D. Minkel, and J. T. McFarland, *J. Am. Chem. Soc.*, 1979, **101**, 3634.

mechanism whereby the substrate binds to the metal ion in the active site *without* displacing the solvent molecule, the latter acting as an acid–base catalyst (see above, Scheme 5). The reverse isotope effects that are observed during this experiment suggest that transfer of protons between the bound solvent molecule and the bound alcoholate anion is not concerted with transfer of hydride, but occurs as a distinct step.

A recent stopped-flow study indicates that the effective substrate in the reduction of acetaldehyde by equine liver alcohol dehydrogenase is the free carbonyl form, and that the hydrate form is not a substrate.[171] Trifluoroacetaldehyde hydrate was a competitive inhibitor of the enzyme, whereas acetaldehyde hydrate was not inhibitory. Equine liver dehydrogenase has been utilized to effect the efficient and stereospecific reduction of S-heterocyclic ketone analogues to the corresponding alcohols.[172] Subsequent treatment with Raney nickel yields the corresponding acyclic alcohols in good yields; acyclic secondary alcohols cannot be obtained directly through enzymatic reduction, as the corresponding ketones are poor substrates for the enzyme. In a further paper, the synthetic potential of the acetimidylated and hydroxybutyrimidylated enzymes was evaluated.[173] The modified enzymes were found to have some limited synthetic advantages over the native enzyme. In a recent publication it has been reported that, in addition to its oxidoreductase activity, liver alcohol dehydrogenase is capable of hydrolysing octanoate esters.[174] Alcohol dehydrogenase has also been found to dismutate aldehydes into the corresponding acids and alcohols, in a Cannizzaro-type reaction.[175] Thus, heptanal is converted into heptanoic acid and heptanol, with the formyl proton being transferred from one half of the sample to the other.

The decrease in protein fluorescence of liver alcohol dehydrogenase at alkaline pH or in the presence of NAD$^+$ and substrate has been attributed to energy transfer from Trp-314 to ionized Tyr-286. Ultraviolet difference spectra of the enzyme confirm the presence of ionized tyrosine residues in these situations, and also in the ternary complex with NADH and isobutyramide.[176] It has been postulated that the formation of a ternary complex, with either oxidized or reduced coenzyme, causes a conformational change which results in a partial ionization of tyrosine residues in regions of the enzyme that are far from the active site. The phosphorescence decay of the tryptophan residue that is located in the nucleotide-binding domain of the liver enzyme has been measured between 1 and 40 °C.[177] The results indicate that solvent–protein interactions are important in determining the conformational flexibility of the enzyme.

A number of recent studies concern the immobilization of the enzyme on solid supports. Thus the EE and SS isoenzymes of equine alcohol dehydrogenase were separately immobilized on Sepharose 4B which was weakly activated by

[171] M. A. Abdallah, J. F. Biellmann, and P. Lagrange, *Biochemistry*, 1979, **18**, 836.
[172] J. Davies and J. B. Jones, *J. Am. Chem. Soc.*, 1979, **101**, 5405.
[173] J. B. Jones and D. R. Dodds, *Can. J. Biochem.*, 1979, **57**, 2533.
[174] C. S. Tsai, *Biochem. Biophys. Res. Commun.*, 1979, **86**, 808.
[175] A. R. Battersby, D. G. Buckley, and J. Staunton, *J. Chem. Soc., Perkin. Trans. 1*, 1979, 2559.
[176] W. R. Laws and J. D. Shore, *J. Biol. Chem.*, 1979, **254**, 2582.
[177] S. Kishner, E. Trepman, and W. C. Galley, *Can. J. Biochem.*, 1979, **57**, 1299.

cyanogen bromide,[178] and the resulting immobilized dimeric enzymes were inactivated by 6M-urea. The enzymatic activity could be restored by treatment of the inactivated, immobilized enzyme with solutions of either the native enzyme or of the separated subunits. The results indicate that treatment with urea yields immobilized single subunits, and that these subunits are inactive: hence it appears that the formation of dimers is a prerequisite for enzymatic activity. In the presence of isobutyramide, the equine liver enzyme is strongly bound as a ternary complex to the affinity resin N^6-[N-(6-aminohexyl)carbamoyl-methyl]-NADH-Sepharose 6B.[179] A low degree of substitution of coenzyme on to the gel was used so that the enzyme–gel interaction was restricted to a single subunit. The bound enzyme was modified by iodoacetate and then eluted from the gel. The resulting enzyme lost 50% of its NADH-binding capacity and was only half as active as the native enzyme. Hence it appears that the above procedure restricts modification to only one of the subunits of the enzyme and the results further suggest that the two subunits of the native enzyme are kinetically equivalent. Alcohol dehydrogenase has also been immobilized on silica gel and on silica gel that was coated with serum albumin, cholesterol, or lecithin.[180,181] The specific activity of the enzyme when it was immobilized on the carriers decreased with increasing degree of substitution. The K_M values for NAD$^+$ and ethanol were increased several-fold for the enzyme that was immobilized on SiO_2, compared to the values observed in solution, and inhibition of the formation of products by ethanol, which is observed in solution, was absent in the immobilized enzyme.

A method for the preparation of large amounts of very pure EE isoenzyme of alcohol dehydrogenase from horse liver has been described.[182] It has been reported that the mouse enzyme polymerized upon lyophilization; therefore any mechanistic studies done on lyophilized samples must be suspect.[183] Three variants of alcohol dehydrogenase have been partially purified from mouse liver and characterized.[184]

In contrast to the equine enzyme, human alcohol dehydrogenase occurs in multiple molecular forms; at least ten distinct forms have been identified to date. The number and amounts of these forms in any individual are partly genetically dictated, and partly dictated by the state of health of the donor.[185] A recent survey of post-mortem liver samples, conducted at Indianapolis, revealed that 16% of the specimens that were studied contained hitherto unknown forms of the enzyme.[186] These forms exhibited unusual electrophoretic mobilities,

178 L. Andersson and K. Mosbach, *Eur. J. Biochem.*, 1979, **94**, 557.
179 L. Andersson and K. Mosbach, *Eur. J. Biochem.*, 1979, **94**, 565.
180 Z. Mikelsone, A. N. Mitrofanova, O. M. Poltorak, and A. Arens, *Vestn. Mosk. Univ.*, Ser. 2: Khim., 1979, **20**, 109.
181 A. N. Mitrofanova, Z. Mikelsone, O. M. Poltorak, and A. Arens, *Vestn. Mosk. Univ.*, Ser. 2: Khim., 1979, **20**, 114.
182 D. C. Anderson and F. W. Dahlquist, *Anal. Biochem.*, 1979, **99**, 392.
183 J. B. A. Ross, E. L. Chang, and D. C. Teller, *Biophys. Chem.*, 1979, **10**, 217.
184 T. Haseba, K. Hirakawa, M. Nihira, M. Hayashida, M. Kurosu, Y. Tomita, Y. Ide, and T. Watanabe, *Arukoru Kenkyu*, 1979, **14**, 324.
185 T.-K. Li, *Adv. Enzymol.*, 1977, **45**, 427.
186 W. F. Bosron, T.-K. Li, and B. L. Vallee, *Biochem. Biophys. Res. Commun.*, 1979, **91**, 1549.

possessing cathodic bands not seen in the normal samples, and had two pH optima for oxidation of ethanol, *i.e.* the usual optimum at 10.0 and an additional optimum at 7.0. Furthermore, these studies revealed that, amongst the normal samples, the specific activities of livers with the ADH_3 2 phenotype were significantly higher than of those with the ADH_3 1 or ADH_3 2-1 phenotypes.

The kinetic and molecular properties of human liver π-alcohol dehydrogenase (π-ADH), a form of the enzyme that has been recognized only recently, have been reported.[187] Its general characteristics are similar to those of other mammalian alcohol dehydrogenases. Thus it has a molecular weight of 78 000, contains 4 gram ions of Zn^{II} per mole, and is dimeric. The kinetics of π-ADH follow an ordered 'bibi' mechanism in which the cofactor binds first (to form a binary enzyme complex) followed by the substrate (to give the ternary complex). In contrast to the other molecular forms, π-ADH has a higher K_M for ethanol and acetaldehyde and is much less sensitive to inhibition by pyrazole and 4-methylpyrazole. Hence the differentiation of ADH-independent pathways for oxidation of ethanol in Man cannot be based solely upon the lack of inhibition by these reagents, as has customarily been done in the past. However, π-ADH is strongly inhibited by other pyrazole derivatives, including 4-bromo-, 4-nitro-, and 4-pentyl-pyrazole.

Human alcohol dehydrogenase shows a much broader substrate specificity than that observed with the enzyme from most other species. The latest substrates that have been discovered for the enzyme are the sterols digitoxigenin, digoxigenin, and gitoxigenin,[188] which are oxidized to the corresponding 3-keto-derivatives. This oxidation is competitive with that of ethanol. Since these sterols are the active constituents of the cardiac glycosides, their oxidation by alcohol dehydrogenase is important in the pharmacology of these agents.

The characterization of two alcohol dehydrogenase isoenzymes that had been isolated from human stomach has been reported.[189] The reduction of biogenic aldehydes by aldehyde reductase and by alcohol dehydrogenase from human liver has been compared.[190] The former enzyme preferred aldehydes derived from β-hydroxylated amines, whereas the latter enzyme preferred substrates with no hydroxyl groups. The amount of alcohol dehydrogenase in human liver is approximately 40 times as high as that of aldehyde reductase.

Two methods for the detection of alcohol dehydrogenase in human serum have been reported. One of these utilizes liquid chromatography with electrochemical detection[191] and the other involves the use of the chromophoric substrate p-nitrosodimethylaniline in a recycling reaction.[192]

Comparatively little is known as yet about plant alcohol dehydrogenases. The enzyme from rape (*Brassica napus*) has been isolated and found to contain two coenzyme-binding sites per molecule, indicating that the enzyme is dimeric.[193] This enzyme is inhibited by heavy-metal ions, the most effective being Cu^{II} and

[187] W. F. Bosron, T.-K. Li, W. P. Dafeldecker, and B. L. Vallee, *Biochemistry*, 1979, **18**, 1101.
[188] W. A. Frey, and B. L. Vallee, *Biochem. Biophys. Res. Commun.*, 1979, **91**, 1543.
[189] J. D. Hempel and R. Pietruszko, *Alcohol.: Clin. Exp. Res.*, 1979, **3**, 95.
[190] B. Wermuth and J. D. B. Munch, *Biochem. Pharmacol.*, 1979, **28**, 1431.
[191] G. C. Davies, K. L. Holland, and P. T. Kissinger, *J. Liq. Chromatogr.*, 1979, **2**, 663.
[192] L. Skursky, J. Kovar, and M. Stachova, *Anal. Biochem.*, 1979, **99**, 65.
[193] M. Stiborova, R. Lapka, and S. Leblova, *FEBS Lett.*, 1979, **104**, 309.

Ag^{I}.[194] The kinetics for the oxidation of ethanol and the reduction of acetaldehyde by this enzyme followed a 'bibi' ordered mechanism, and the pH optima for these substrates are 8.5 and 7.0 respectively.[195] Rape alcohol dehydrogenase is inhibited by carboxylic acids and their amides, the former being competitive with ethanol and the latter competitive with acetaldehyde.[196] Dimethyl sulphoxide is also an inhibitor of the reduction of acetaldehyde by the enzyme. An alcohol dehydrogenase that has been isolated from pea (*Pisum arvense*) seeds has identical pH optima for oxidation of ethanol and acetaldehyde.[197] Phenanthroline and ATP are inhibitors of pea alcohol dehydrogenase, being competitive with NAD^{+}.[198] Broad bean alcohol dehydrogenase exhibits a 'bibi' ordered mechanism, and it is inhibited by adenosine and by AMP.[199] Isoenzyme 2 of cinnamyl alcohol dehydrogenase from soybean suspension cultures has been purified to apparent homogeneity.[200] The molecular weight of the enzyme on SDS disc gel electrophoresis was estimated to be 40 000. It is strongly inhibited by reagents that contain thiol groups and by metal-chelating agents. Atomic absorption spectroscopy has revealed the presence of Zn^{II} in the enzyme, and in steady-state kinetics a 'bibi' pathway is followed, with the coenzyme NADP(H) being the first substrate to bind. Hence this enzyme is very similar to the mammalian alcohol dehydrogenases. An analysis of maize alcohol dehydrogenase by native SDS two-dimensional electrophoresis and autoradiography has revealed that the subunits of the isoenzyme ADH-2 have a higher molecular weight than those of ADH-1.[201] An alcohol dehydrogenase has also been isolated from wheat (*Triticum monococcum*).[202] The native enzyme has a molecular weight of 116 000 and is dimeric.

The structures of the two major isoenzymes of yeast (*Saccharomyces cerevisiae*) alcohol dehydrogenase have been compared.[203] Analysis of 82% of the primary structure revealed fifteen amino-acid exchanges (out of 284 compared) between the two isoenzymes; these appear to result from single base changes in the gene. Four of the differences of an amino-acid residue occur in positions close to the interface between the subunits; no residues with known catalytic functions were exchanged. It has been postulated that these two isoenzymes arose by gene duplication. Yeast alcohol dehydrogenase is inactivated by *p*-azidophenacyl iodoacetate.[204] It has been suggested that this inactivation is due to a co-operative process whereby the reagent modifies both of the active-site residues Cys-43 and Cys-153. The modification of carboxymethylated yeast alcohol dehydrogenase (which is modified at Cys-43) by the above reagent results

[194] M. Stiborova and S. Leblova, *Biochem. Physiol. Pflanz.*, 1979, **174**, 39.
[195] M. Stiborova and S. Leblova, *Phytochemistry*, 1979, **18**, 23.
[196] M. Stiborova and S. Leblova, *Biochem. Physiol. Pflanz.*, 1979, **174**, 446.
[197] S. Leblova, R. Lapka, and N. Novakova, *Collect. Czech. Chem. Commun.*, 1979, **44**, 986.
[198] M. Stiborova, R. Lapka, N. Novakova, and S. Leblova, *Collect. Czech. Chem. Commun.*, 1979, **44**, 986.
[199] S. Leblova, P. Nemec, P. Kaspar, and M. Stiborova, *Biochem. Physiol. Pflanz.*, 1979, **174**, 418.
[200] D. Wyrambik and H. Grisebach, *Eur. J. Biochem.*, 1979, **97**, 503.
[201] R. J. Ferl, S. R. Dlouhy, and D. Schwartz, *Mol. Gen. Genet.*, 1979, **169**, 7.
[202] P. J. Langston, G. E. Hart, and C. N. Pace, *Arch. Biochem. Biophys.*, 1979, **196**, 611.
[203] C. Wills and H. Joernvall, *Eur. J. Biochem.*, 1979, **99**, 323.
[204] S. H. Hixson, S. F. Burroughs, T. M. Caputo, B. B. Crapster, M. V. Daly, A. W. Lowrie, and M. L. Wasko, *Arch. Biochem. Biophys.*, 1979, **192**, 296.

in the modification of Cys-153. The affinity determination of yeast alcohol dehydrogenase with 4-isothiocyanatobenzaldehyde has been described.[205] This aldehyde rapidly inactivates the enzyme by modifying thiol groups.

The primary sequence of alcohol dehydrogenase from the yeast mutants *S-AA-5* and *C-40* has been investigated.[206] These mutants showed substitutions at positions 44 (histidine→arginine) and 316 (proline→arginine) respectively. The *S-AA-5* enzyme had altered NAD^+-binding capacity and the *C-40* enzyme showed weakened substrate-binding capacity. The effect of shear stresses on yeast alcohol dehydrogenase has been reported.[207] The enzyme was extraordinarily stable to shearing at 683 s^{-1} and 3440 s^{-1}, with only a small decrease in specific activity being observed after 15 h at 5 °C. The alcohol dehydrogenase of *Bacillus stearothermophilus* is inactivated by the NAD^+ analogue 4-(3-bromo-acetylpyridinio)butyldiphosphoadenosine;[208] this inactivation is accompanied by the covalent attachment of the coenzyme analogue to the enzyme. The modification of the modified bacterial enzyme by iodoacetate results in the carboxymethylation of Cys-38, which is homologous to Cys-43 in the yeast enzyme and Cys-46 in the equine enzyme; unlike these two cysteine residues, however, Cys-38 is not modified by iodoacetate in the native enzyme. A novel alcohol dehydrogenase, specific for secondary alcohols, has been isolated from cell-free extracts of *Pseudomonas* sp. ATCC 21439.[209] The enzyme is dimeric, with a molecular weight of 48 000 for the subunit, and it is inhibited by metal-chelating agents. An alcohol dehydrogenase which is specific for secondary alcohols and is inhibited by metal-chelating agents has also been isolated from various species of yeasts.[210] It is possible that these enzymes are zinc-containing metalloenzymes, like their primary alcohol dehydrogenase counterparts.

Alkaline Phosphatase.—The metal-binding properties of alkaline phosphatase from *Escherichia coli* have in the past generated a lot of controversy. The reported metal content of this enzyme ranges from 2 to 6 gram atoms of metal per mole of dimer. However, it is becoming increasingly clear that the correct metal stoicheiometry is that reported by Vallee and co-workers; namely 4 gram atoms of Zn^{II} plus 2 gram atoms of Mg^{II} per mole.[211] These authors have suggested that the enzyme contains three distinct metal-binding sites. The first pair of Zn^{II} ions bind at equivalent sites, and they generate the catalytic activity of the enzyme; these two sites are therefore referred to as the catalytic sites. The second pair of Zn^{II} ions bind less tightly, and their presence modulates the activity of the enzyme; sites in this second set are referred to as the structural sites. The binding of two Mg^{II} ions at a third pair of sites further modulates the

[205] A. Breier and P. Sulo, *Zb. Stud. Ved. Odb. Pr.* (*Slov. Vys. Sk. Tech. Bratislave, Chemicko-technol. Fak.*), 1979, 145.
[206] C. Wills and H. Joernvall, *Nature (London)*, 1979, **279**, 734.
[207] C. R. Thomas, A. W. Nienow, and P. Dunnill, *Biotechnol. Bioeng.*, 1979, **21**, 2263.
[208] R. Jeck, C. Woenckhaus, J. I. Harris, and M. J. Runswick, *Eur. J. Biochem.*, 1979, **93**, 57.
[209] C. T. Hou, R. N. Patel, A. I. Laskin, N. Barnabe, and I. Marczak, *FEBS Lett.*, 1979, **101**, 179.
[210] R. N. Patel, C. T. Hou, A. I. Lasken, P. Derelanko, and A. Felix, *Eur. J. Biochem.*, 1979, **101**, 401.
[211] W. F. Bosron, F. S. Kennedy, and B. L. Vallee, *Biochemistry*, 1975, **14**, 2275.

activity of the enzyme.[212] A number of papers published during 1979 confirm the above conclusions. A study (using scanning calorimetry) of the reconstitution of apo-alkaline phosphatase indicates that the binding of Zn^{II} at the catalytic sites is highly co-operative, with no mono-Zn^{II} species (containing 1 mole of Zn^{II} per mole of dimer) being evident.[213] The binding of the catalytic Zn^{II} induces structural changes in the protein which result in the stabilization of the enzyme. Zinc(II) is much more efficient than Mn^{II}, Co^{II}, and Cd^{II} in this role; in the absence of Zn^{II}, the addition of Mg^{II} has no effect on the stability of the protein. The addition of 2 moles of Zn^{II} or of Mg^{II} to each mole of the di-Zn^{II} enzyme further increases the stability of the enzyme, indicating that the occupancy of the structural sites is required for maximal stability. On the other hand, the binding of the third pair of metal ions appears to have negligible structural consequences, though the occupation of these sites is required for maximal catalytic activity. The binding of inorganic phosphate to the native and to the Cd^{II}-substituted enzymes produces a marked stabilization of the protein. Most of this stabilization is attained on the binding of one phosphate ion per dimer, which indicates that the binding of phosphate proceeds in a negatively co-operative fashion. The dimeric apoenzyme can be reversibly dissociated into stable monomers in the presence of formamide. The thermally induced unfolding of the monomers indicates that these are metastable relative to the dimeric apoenzyme.

Phosphorus-31 n.m.r. spectroscopy has also been used to monitor the binding of metals and of ligands to alkaline phosphatase.[214] When the apoenzyme is reconstituted with 2 moles of Zn^{II} or Cd^{II} per mole of dimer, such that only the catalytic sites are occupied, the ^{31}P n.m.r. spectra indicate that only 1 mole of phosphate per mole of dimer binds to the enzyme; hence, under these conditions the enzyme shows absolute negative co-operativity for the binding of phosphate. On the other hand, when, in addition to the catalytic sites, the structural sites are also occupied, two phosphate ions bind per dimer. Thus, the presence of the structural metal ions induces the tight binding of phosphate at the second phosphate-binding site in the dimeric enzyme. This may be the basis for the modulation of the enzymatic activity by these metal ions. The correspondence between the phosphate-binding stoicheiometry and the metal content of the enzyme probably explains the large variations for the former parameter that have been reported in the literature, because many of these studies were done on samples with ill-defined metal contents.

The binding of Mn^{II} to apo-alkaline phosphatase has been followed by e.p.r. spectroscopy.[215] This study reveals that the first 2 moles of Mn^{II} that are bound to a mole of the enzyme occupy identical, low-symmetry sites, presumed to be the catalytic sites. Phosphorus-31 n.m.r. spectroscopy indicates that the di-Mn^{II} enzyme binds 1 mole of inorganic phosphate per mole of enzyme. The binding of phosphate induces dramatic changes in the e.p.r. spectra of the

[212] W. F. Bosron, R. A. Anderson, M. C. Falk, F. S. Kennedy, and B. L. Vallee, *Biochemistry*, 1977, **16**, 610.

[213] J. F. Chlebowski and S. Mabrey, *J. Biol. Chem.*, 1979, **254**, 5745.

[214] J. D. Otvos, I. M. Armitage J. F. Chlebowski, and J. E. Coleman, *J. Biol. Chem.*, 1979, **254**, 4707.

[215] R. E. Weiner, J. F. Chlebowski, P. H. Haffer, and J. E. Coleman, *J. Biol. Chem.*, 1979 **254**, 9739.

enzyme. These changes suggest that the symmetry around the catalytic metal ion is altered by the binding of phosphate. The addition of Zn^{II} to the di-Mn^{II} enzyme, or of Mn^{II} to the di-Zn^{II} enzyme, results in the binding of the added metal ions to the structural sites on the enzyme. In both cases, this induces the binding of phosphate at its second site on the enzyme; this is in agreement with the results on the native and Cd^{II}-substituted enzymes that were quoted above.[214] This additional phosphate causes little further change in the e.p.r. spectra of the Mn^{II} enzyme. It has been suggested that the structural metal ions may act as allosteric activators of alkaline phosphatase.

The binding of phosphate to alkaline phosphatase is known to result in the formation of two complexes, *i.e.* a non-covalent complex and a covalent complex, the relative amounts of which depend on the pH and on the nature of the metal ion that is present at the catalytic sites.[216] In a recent ^{31}P n.m.r. investigation, the rate-constants for the formation and breakdown of these complexes were estimated by saturation-transfer experiments.[217] In addition, it was shown that, when $^{113}Cd^{II}$ (which has a spin of $\frac{1}{2}$) is substituted for the native Zn^{II}, the signal corresponding to the non-covalent complex is split into a doublet, owing to $^{113}Cd-^{31}P$ spin coupling ($J = 30$ Hz) (Figure 2). This result provides the first unequivocal evidence for a direct metal–phosphate interaction in the non-covalent phosphate-containing intermediate of alkaline phosphatase.

The binding of Zn^{II} and of Mg^{II} to the alkaline phosphatase of *Escherichia coli* has also been investigated by ^{35}Cl n.m.r.[218] This study confirmed that four Zn^{II} ions bind to each dimer of the enzyme. Two Mg^{II} ions bind at a pair of sites that are distinct from the Zn^{II}-binding sites, but ^{35}Cl relaxation studies indicate that the binding of Zn^{II} and that of Mg^{II} are interdependent. Since these authors had previously reported a similar study purporting to show that only two Zn^{II} ions per dimer bind to the enzyme,[219] it has now been suggested[218] that the discrepancy may be due to the existence of two forms of the enzyme, which differ in their metal-binding properties. However, this suggestion constitutes an *ad hoc* hypothesis, and there is no other evidence or reason to believe that the alkaline phosphatase of *E. coli* is heterologous in its metal-binding properties. Over the temperature range 4—30 °C, the lifetime of the bound Cl^- was found to be very short (*i.e.* fast exchange conditions prevailed over this temperature range); the activation energy for the binding of Cl^- was estimated to be 23 kJ mol^{-1}. It was inferred that the Cl^- is bound at an amino-acid residue rather than at a Zn^{II} ion, because the lifetime of Cl^- that is bound to the metal is expected to be significantly longer than that observed.

A study of the inactivation of porcine kidney alkaline phosphatase by arginine-specific reagents has indicated that an essential arginine residue is present in the active site of this enzyme.[220] During the modification of the enzyme by these reagents, the K_m for the substrate *p*-nitrophenyl phosphate is uneffected, but

[216] M. L. Applebury, B. P. Johnson, and J. E. Coleman, *J. Biol. Chem.*, 1970, **245**, 4968.
[217] J. D. Otvos, J. R. Alger, J. E. Coleman, and I. M. Armitage, *J. Biol. Chem.*, 1979, **254**, 1778.
[218] J.-E. Norne, H. Szajn, H. Csopak, P. Reimarsson, and B. Lindman, *Arch. Biochem. Biophys.*, 1979, **196**, 553.
[219] H. Csopak, K. E. Falk, and H. Szajn, *Biochim. Biophys. Acta*, 1972, **258**, 466.
[220] M. N. Woodroofe and P. J. Butterworth, *Biochem. J.*, 1979, **181**, 137.

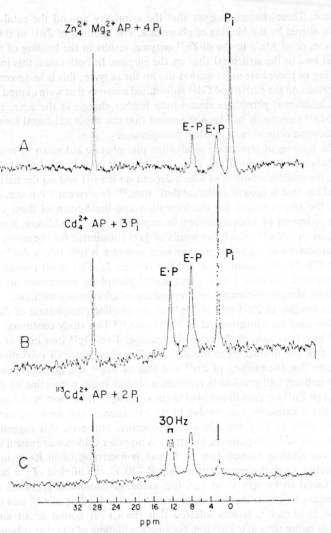

Figure 2 *Phosphorus-31 n.m.r. spectra of Zn^{II} and Cd^{II} alkaline phosphatases in the presence of added phosphate: (A) Zn^{II} enzyme at pH 5.5; (B) Cd^{II} enzyme at pH 8.9 (natural-abundance Cd^{II}); (C) $^{113}Cd^{II}$ enzyme at pH 9.1 (enriched $^{113}Cd^{II}$). The labels E-P, E·P, and P_i refer to the covalent complex, the non-covalent complex, and inorganic phosphate, respectively. The resonance at 29.3 p.p.m. is due to methylphosphonate, which was used as an external standard* (Reproduced by permission from *J. Biol. Chem.*, 1979, **254**, 1778)

V_{max} falls progressively. This indicates that the modified enzyme is completely inactive. Inorganic phosphate, which is a competitive inhibitor of the enzyme, and the substrate AMP protect the enzyme against inactivation. It has been proposed that the residue that is responsible for the inactivation constitutes the binding site for the phosphate portion of the substrates. The total number of arginine residues that are modified by these reagents was not estimated, but the electrophoretic mobility of the modified enzyme indicates that it is large. The partially modified enzyme has a reduced sensitivity towards inhibition by the non-competitive inhibitor NADH, which suggests that NADH binds at an arginine residue that is not located in the active site. However, since NADH also protects (in a partially competitive manner) against the modification of the essential arginine residue, it appears that the binding site for NADH is close to the active site.

The effects of anions on calf intestinal alkaline phosphatase, utilized as a model enzyme, have been thoroughly investigated.[221] The influence of the halides on the activation volume and the free energy of activation was in the order $F^- > Cl^- > Br^- > I^-$, reflecting the Hofmeister (lyotropic) series. The wavelength maxima for the fluorescence emission spectra of the enzyme also follow the same order. When different halides were used in combination, their effects were counteractive. The observed effects are due to alterations in the structure of the enzyme that are induced by the anions.

A series of phosphonitrile–polyol copolymers have been tested as substrates for alkaline phosphatase.[222] Tetrachloro-spiro-compounds were found to be efficient substrates, and their hydrolysis followed Michaelis–Menten kinetics. Polymers from acyclic polyols were inert as substrates but were effective inhibitors of the enzyme.

Two new inhibitors of human alkaline phosphatase have been reported; these are sulphonamide and vanadate. The enzyme is inhibited by various sulphonamides that are of therapeutic use.[223] It has been suggested that the mechanism of inhibition may involve binding to the Zn^{II} ion in the active site of the enzyme. It has also been postulated that some of the pharmacological actions of these drugs may be due to inhibition of alkaline phosphatase, rather than of carbonic anhydrase. The inhibition by vanadate is competitive ($K_i < 1$ μmol l^{-1}) and is more potent than that observed with phosphate.[224] It has been postulated that vanadate resembles a transition-state analogue of phosphate. Similar results have previously been reported for the inhibition of the alkaline phosphatase of *E. coli* by vanadate.[225]

The kinetic parameters for the hydrolysis of a number of physiologically important phospho-esters by human liver alkaline phosphatase have been determined.[226] Phosphodiesters and phosphonates were not hydrolysed by the enzyme, but the latter were inhibitors. Calcium(II) and Mg^{II} inhibited the hydrolysis of pyrophosphate and of ATP, and the metal complexes of these substrates are not hydrolysed by the enzyme.

A new chromatographic technique for the preparation of human liver alkaline phosphatase has been reported.[227] This method involves the use of a phospho-

[221] G. S. Greaney and G. N. Somero, *Biochemistry*, 1979, **18**, 5322.
[222] I. Tabushi, K. Fujita and S. Matsuo, *J. Polym. Sci., Polym. Lett. Ed.*, 1979, **17**, 357.
[223] G. H. Price, *Clin. Chim. Acta*, 1979, **94**, 211.
[224] L. E. Seargeant and R. A. Stinson, *Biochem. J.*, 1979, **181**, 247.
[225] V. Lopez, T. Stevens, and R. N. Lindquist, *Arch. Biochem. Biophys.*, 1976, **175**, 31.
[226] L. E. Seargeant and R. A. Stinson, *Can. J. Biochem.*, 1979, **57**, 1000.
[227] L. E. Seargeant and R. A. Stinson, *J. Chromatogr.*, 1979, **173**, 101.

nic acid–Sepharose 4B affinity ligand. The enzyme binds to the column at pH 6.0, and can be selectively eluted with substrate. Another method of preparation of alkaline phosphatase (from pig kidney), by affinity chromatography, utilizes the general ligand Cibacron Blue, coupled to cross-linked agarose beads.[228]

It has been reported that alkaline phosphatase solutions that have been prepared from lyophilized protein show great variations in enzymatic activity, and that, if reproducible results are to be obtained, the lyophilized material should be mixed for 30 minutes, at ambient temperature, after reconstitution and then stored for more than 4 hours at 4 °C before being assayed.[229]

A number of reports detailing the isolation of alkaline phosphatases from new sources have been published. The sources include rat uterine deciduomata,[230] intestinal epithelium,[231] the mucosae of rat small intestine,[232] mouse uterus,[233] chicken tissues,[234] and bovine foetal epiphyseal cartilage.[235] A novel alkaline phosphatase, termed alkaline phosphatase S, has been isolated from canine cardiac muscle.[236] In contrast to other well-characterized alkaline phosphatases, phosphatase S is of cytosolic origin, and it requires the presence of both Mg^{II} and sulphydryl compounds for activity. The phosphatase is closely associated with a phosphoprotein phosphatase, and it has been suggested that the alkaline phosphatase may play a role in the regulation of phosphorylation–dephosphorylation reactions of proteins. The two enzymes can be partially separated by hydrophobic interaction chromatography. Another unusual alkaline phosphatase has been isolated from *Halobacterium cutirubrum*.[237] This enzyme is highly stimulated by Mn^{II} and is reported to have a molecular weight of only 15 500.

It has long been known that human alkaline phosphatases occur in multiple molecular forms, these being characteristic of the tissue of origin;[238, 239] the biological role of the different forms is not known. Since the human enzymes are glycoproteins, it is likely that many of these forms are the products of secondary modification, but it has been suggested that at least three separate genes code for human alkaline phosphatase.[240] Recent structural studies on the enzymes isolated from various human tissues confirm this hypothesis. Thus, tryptic digests of the placental and renal enzymes that had been labelled with radioactive phosphate at their active centres yielded different peptide maps, as

[228] P. J. Butterworth and D. T. Plummer, *Proc. Anal. Div. Chem. Soc.*, 1979, **16**, 182.
[229] D. G. Bullock, F. M. McSweeney, H. Saidi, and T. P. Whitehead, *Ann. Clin. Biochem.*, 1979, **16**, 271.
[230] U. Tarachand and P. J. Heald, *Biol. Reprod.*, 1979, **20**, 617.
[231] S. D. Hanna, A. K. Mircheff, and E. M. Wright, *J. Supramol. Struct.*, 1979, **11**, 451.
[232] H. Nakasaki, T. Matsushima, S. Sato, and T. Kawachi, *J. Biochem. (Tokyo)*, 1979, **86**, 1283.
[233] R. N. Murdoch, D. J. Kay, and W. J. Capper, *Aust. J. Biol. Sci.*, 1979, **32**, 153.
[234] I. Debruyne and J. Stockx, *Int. J. Biochem.*, 1979, **10**, 981.
[235] R. Fortuna, H. C. Anderson, R. Carty, and S. W. Sajdera, *Biochim. Biophys. Acta*, 1979, **570**, 291.
[236] H.-C. Li, K.-J. Hsiao, and S. Sampathkumar, *J. Biol. Chem.*, 1979, **254**, 3368.
[237] P. S. Fitt and P. Baddoo, *Biochem. J.*, 1979, **181**, 347.
[238] H. N. Fernley, in 'The Enzymes', ed. P. D. Boyer, Academic Press, New York, 1971, Vol. 4, p. 417.
[239] W. H. Fishman, *Am. J. Med.*, 1974, **56**, 617.
[240] H. H. Sussman, P. A. Small, and E. Cotlove, *J. Biol. Chem.*, 1968, **234**, 160.

detected by autoradiography.[241,242] Since the labelled peptides originate from the active-site regions of the enzymes, this result indicates that the proteins from these sources are structurally distinct. On the basis of this and other evidence, including the immunochemical characterization of the native enzymes by monospecific antisera, it has been suggested that human alkaline phosphatases can be divided into three classes.[242] Those from the liver, bone, and kidney form one class, the placental enzymes constitute the second class, and the intestinal enzymes form the third class. This classification has been confirmed by an independent study, in which it was shown that the enzymes obtained from the liver, kidney, and serum of a patient with Paget's disease are the products of the same structural gene, and are distinct from the placental and intestinal enzymes.[243] In this study the alkaline phosphatases were radiolabelled with [125]I (which is mainly incorporated into tyrosine residues) and then digested with trypsin. The tryptic digests were resolved into peptide maps by thin-layer electrophoresis followed by thin-layer chromatography in a second direction. The authors suggest that this technique may be generally useful for the structural characterization of alkaline phosphatases. In another study the alkaline phosphatase from human milk was shown to be similar (but not identical) to the liver enzyme, and hence the enzyme from milk belongs to the first class of the above classification.[244] Several reports deal with the characterization of alkaline phosphatases from human placenta. Alkaline phosphatase from human first-trimester placenta has been shown to exist in three forms, *i.e.* I, IIa, and IIb.[245] Form IIb appears to be a tetramer, consisting of two dimers of phosphatase II or IIa. Form I is indistinguishable from the liver enzyme in several of its physical and chemical properties and seems to be specific for the first-trimester placenta, whereas form II is indistinguishable from the full-term placental enzyme. An atypical variant of the human placental enzyme, with a reduced electrophilic mobility, has been partially purified.[246] A comparison of the kinetic properties of human placental and foetal brain alkaline phosphatases has been published.[247] A method for analysing for differences amongst placental alkaline phosphatases by complement fixation has been reported.[248]

The different isoenzymes of alkaline phosphatase can be distinguished by inhibition studies, using certain amino-acids. Thus the liver, intestinal, and placental types of alkaline phosphatases can be measured separately by using high concentrations of bromotetramisole, 1-phenylalaninamide, and L-phenylalanine. In one study, this procedure was used to measure intestinal alkaline phosphatase in maternal serum.[249] No evidence for the presence of meconial alkaline phosphatase was found in the serum of the mother after meconial passage

[241] K. B. Whitaker and D. W. Moss, *Biochem. J.*, 1979, **183**, 189.
[242] M. J. McKenna, T. A. Hamilton, and H. H. Sussman, *Biochem. J.*, 1979, **181**, 67.
[243] L. E. Seargeant and R. A. Stinson, *Nature (London)*, 1979, **281**, 152.
[244] T. A. Hamilton, S. Z. Gornichi, and H. H. Sussman, *Biochem. J.*, 1979, **177**, 197.
[245] T. Sakiyama, J. C. Robinson, and J. Y. Chou, *J. Biol. Chem.*, 1979, **254**, 935.
[246] F. Guilleux, M. Hayer, N. Thomas, and M. B. DeBornier, *Pathol. Biol.*, 1979, **27**, 79.
[247] T. K. Chatterjee, P. Banerjee, S. K. Banerjee, and J. J. Ghosh, *Indian J. Biochem. Biophys.*, 1979, **16**, 268.
[248] A. H. Rule, A. DiNapoli, S. Green, W. H. Fishman, and G. J. Doellgast, *J. Immunol. Methods*, 1979, **29**, 35.
[249] G. J. Doellgast and P. J. Meis, *Clin. Chem.*, 1979, **25**, 1230.

in utero. The isoenzymes of alkaline phosphatase can also be differentiated by studying the influence of Mg^{II} on the enzymatic activity,[250] the liver and bone isoenzymes being stimulated to a greater extent than the intestinal isoenzyme. Through a combination of inhibition, thermostability, and electrophoretic studies, it has been shown that the alkaline phosphatase from canine placenta closely resembles the enzymes from canine and human liver, bone, and kidney, but is markedly different from human placental alkaline phosphatase.[251] This is a surprising result, for it indicates that the gene locus for alkaline phosphatase in canine placenta is not homologous to that expressed in human placenta; rather, it appears to be homologous to the liver (or class 1) enzyme. Other recent studies indicate that this is also true for placental enzymes from other mammalian species. Thus the enzymes from the placenta of rodents, carnivores, and ungulates, and from many primates (baboon, african green monkey, spider monkey, rhesus monkey, and gorilla) were found to be unlike the human placental enzyme.[252, 253] In fact, the only species that were found to possess placental alkaline phosphatases that are similar to the human enzyme are the chimpanzee and the orang utang.[253] Since these two species are biologically very close to Man, the authors suggest that the human placental enzyme is of late evolutionary origin. Three different methods of measuring the activity of placental alkaline phosphatase in human serum have been compared.[254] These methods are: (i) inactivation by heating; (ii) differential inactivation with bromotetramisole; (iii) immunological precipitation. A good correlation was found between all these methods. Another report details the analysis of alkaline phosphatase isoenzymes by electrophoresis on polyacrylamide gel.[255] The inhibition of human alkaline phosphatases by carbohydrates has been investigated.[256] With all the carbohydrates tested, the inhibition of the placental isoenzyme is proportional to the carbohydrate concentration, whereas the hepatic and intestinal isoenzymes are activated at low concentrations of carbohydrate and inhibited at high concentrations. Equine alkaline phosphatase isoenzymes, like the human enzymes, can be differentiated by their inhibition and by their electrophoretic properties.[257] It is known that the three main isoenzymes of the alkaline phosphatase that is found in *Escherichia coli* are structurally related, and that isoenzyme 1 can be converted into isoenzymes 2 and 3 by the removal of arginine residues from the N-terminus of the protein.[258] Unexpectedly, it has been found that the conversion of the isoenzymes is not effected by inhibitors of aminopeptidases, but instead is suppressed by inhibitors of endopeptidases.[259]

Some time ago it was reported that an inhibitor of alkaline phosphatase is

250 K. Jung and M. Pergande, *Enzyme*, 1979, **24**, 322.
251 G. Moak and H. Harris, *Proc. Natl. Acad. Sci. USA*, 1979, **76**, 1948.
252 D. J. Goldstein and H. Harris, *Nature (London)*, 1979, **280**, 602.
253 G. J. Doellgast and K. Benirschke, *Nature (London)*, 1979, **280**, 601.
254 P. M. Bayer, F. Gabl, and E. Knoth, *J. Clin. Chem. Clin. Biochem.*, 1979, **17**, 605.
255 O. Ideta and Y. Ito, *Eisei Kensa*, 1979, **28**, 508.
256 S. Iino and L. Fishman, *Clin. Chim. Acta*, 1979, **92**, 197.
257 B. G. Froscher and L. A. Nagode, *Am. J. Vet. Res.*, 1979, **40**, 1514.
258 M. J. Schlesinger, W. Block, and P. M. Kelly, in 'Isoenzymes', ed. C. L. Market, Academic Press, New York, 1975, Vol. 1, p. 333.
259 A. Nakata, H. Shinagawa, and J. Kawamata, *FEBS Lett.*, 1979, **105**, 147.

present in commercial preparations of the buffer 2-amino-2-methylpropan-1-ol; this buffer is commonly used in the assay of alkaline phosphatase. In more recent publications it has been reported that this inhibitor exhibits a differential effect on the isoenzymes of alkaline phosphatase, the intestinal isoenzyme being inhibited to a greater extent than the other isoenzymes.[260] Similar conclusions were reached by two other groups, who also reported that glycine too has a differential inhibitory effect on the isoenzymes; no inhibition was found with Tris, sodium carbonate, diethanolamine, or 2-amino-2-methylpropane-1,3-diol.[261, 262] The inhibitor that is present in the 2-amino-2-methylpropan-1-ol buffer has been suggested to be an ethylenediamine derivative, which inhibits the enzyme by chelating Zn^{II}, and indeed it has been found that the addition of Zn^{II} to the buffer results in increased activity for the enzyme.[263] The amount of Zn^{II} that must be added to the buffer for re-activation is critical, because excess Zn^{II} is inhibitory. The effects of wavelength error and spectral bandwidth on the measurement of alkaline phosphatase activity have been re-examined.[264] This study negates earlier reports which had suggested that these two sources contributed significantly to the errors of measurement of the enzymatic activity. Two recent publications report the preferred methods for the assay of alkaline phosphatase activity in human tissues.[265, 266] The recommended method involves the use of *p*-nitrophenyl phosphate as the substrate in 2-amino-2-methylpropan-1-ol buffer at pH 10.5, in the presence of Mg^{II}, at 30 °C. Several reports deal with new methods for the estimation of alkaline phosphatase activity. These include a modified SMAC method,[267] a high-performance liquid chromatography assay,[268] an assay utilizing continuous monitoring microfluorimetry with 4-methylumbelliferyl phosphate as the substrate,[269] colorimetric methods utilizing glycerophosphate,[270] ammonium thymolphthalein phosphate,[271] or phenolphthalein monophosphate[272] as substrates, and a method that is dependent on the electrophoretic separation of the isoenzymes on cellulose acetate, coupled with the detection of alkaline phosphatase activity with 4-methylumbelliferyl phosphate.[273] A method for the estimation of alkaline phosphatase activity in freshwater sediments has also been reported.[274]

Nucleotidyl Polymerases and Nucleases.—The presence of Zn^{II} in nucleotidyl polymerases is now well established. For the RNA polymerase of *E. coli*, which is one of the best characterized enzymes in this group, indirect evidence suggests

[260] J. M. Rattenbury, K. C. Miller, and I. J. Hogg, *Clin. Chim. Acta*, 1979, **93**, 295.
[261] K. Jung, M. Pergande, and E. Egger, *Enzyme*, 1979, **24**, 18.
[262] A. Zubek and R. Lemanczyk, *Dtsch. Gesundheitswes.*, 1979, **34**, 1259.
[263] J. M. Pekelharing, P. J. Noordeloos, and B. Leijnse, *Clin. Chim. Acta*, 1979, **98**, 61.
[264] R. S. Schifreen and R. W. Burnett, *Clin. Chem. (Winston-Salem, N. C.)*, 1979, **25**, 429.
[265] N. Gochman, *Clin. Enzymol.*, 1979, 81.
[266] R. B. Johnson, *Clin. Enzymol.*, 1979, 93.
[267] S. I. Hansen, E. West-Nielsen, and J. Lyngbye, *Scand. J. Clin. Lab. Invest.*, 1979, **39**, 279.
[268] A. M. Krstulovic, R. A. Hartwick, and P. R. Brown, *J. Chromatogr.*, 1979, **163**, 19.
[269] K. Matsumoto, M. Tanaka, S. Kano, and S. Kanno, *Takashi Rinsho Kagaku*, 1979, **7**, 278.
[270] Q.-H. Wu, *Chung-Hua I Hsueh Chien Yen Tsa Chih*, 1979, **2**, 134.
[271] R.-F. Feng, *Chung-Hua I Hsueh Chien Yen Tsa Chih*, 1979, **2**, 137.
[272] R. J. Georges and E. Persona, *S. Afr. J. Med. Lab. Technol.*, 1979, **25**, 63.
[273] F. Watanabe, M. Takano, F. Tanaka, N. Amino, C. Hayashi, and K. Miyai, *Clin. Chim Acta*, 1979, **91**, 273.
[274] G. S. Sayler, M. Puziss, and M. Silver, *Appl. Environ. Microbiol.*, 1979, **38**, 922.

11

that the Zn^{II} is located in the β and β' subunits of the pentameric enzyme (subunit structure is $\alpha_2\beta\beta'\sigma$).[275] This has been confirmed in a recent study, in which the Zn^{II} content of the separated subunits of the enzyme was directly measured by flameless atomic absorption spectroscopy.[276] This study also suggests, in contradiction to earlier studies, that the Zn^{II} can be removed from the RNA polymerase of *E. coli* by prolonged dialysis against 1,10-phenanthroline, and that the resultant apoenzyme can be reconstituted, with complete restoration of activity, by adding Zn^{II}.

In addition to the polymerase activity, nucleotidyl polymerases possess a $3' \to 5'$ nuclease activity.[277] This nuclease activity is thought to be important in minimizing incorporation errors by proof-reading the newly synthesized nucleotide chain. Reverse transcriptases (RNA-directed DNA polymerases) also possess another nuclease activity, RNase H, which is specific for the RNA template strand.[278] Two recent investigations have demonstrated that, for both DNA polymerase I and reverse transcriptase, 1,10-phenanthroline (which is a potent inhibitor of the polymerase reaction) does not inhibit the accompanying nuclease activity.[279, 280] These observations reinforce earlier suggestions that the synthetic and hydrolytic reactions of nucleotidyl polymerases proceed *via* different mechanisms, and indicate that Zn^{II} does not play any role in the hydrolysis of nucleic acids by these enzymes. Furthermore, one of these studies[279] confirms previous reports that the inhibition of the DNA polymerase I of *E. coli* by 1,10-phenanthroline is potentiated by Cu^I, and that the Cu^I–1,10-phenanthroline complex is a powerful inhibitor of nucleotidyl polymerases.[281] This latest study indicates that the inhibition by the metal–chelate complex is non-competitive with respect to the deoxynucleotide triphosphate substrate, but competitive with the primer/template substrate, and it is suggested that the complex competes with the primer terminus for the active site of the polymerase activity (but not with that of the exonuclease activity) of DNA polymerase I. The inhibition of the DNA polymerase I of *E. coli* by the 1,10-phenanthroline–Cu^I complex has been investigated in greater detail.[282] It has been postulated that the inhibition is due to the cleavage of the template by the metal complex in an oxygen-dependent reaction, yielding products which are effective inhibitors of the enzyme. The role of oxygen in this process is revealed by the reversal of the inhibition in the presence of catalase. It is suggested that the cleavage of the DNA template by the copper–chelate complex resembles that observed with

275 C.-W. Wu. F. Y.-H. Wu, and D. C. Speckhard, *Biochemistry*, 1977, **16**, 5449.
276 J. A. Miller, G. F. Serio, R. A. Howard, J. L. Bear, J. E. Evans, and A. P. Kimball, *Biochim. Biophys. Acta*, 1979, **579**, 291.
277 T. Kornberg and A. Kornberg, in 'The Enzymes', ed. P. D. Boyer, Academic Press, New York, 1974, Vol. 10, p. 119.
278 K. Molling, D. P. Bolognesi, W. Bauer, W. Busen, H. W. Plassmann, and P. Hansen, *Nature (London)*, 1971, **234**, 240.
279 B. G. Que, K. M. Downey, and A. G. So, *Biochemistry*, 1979, **18**, 2064.
280 J. M. Modak and A. Srivastava, *J. Biol. Chem.*, 1979, **254**, 4756.
281 V. D'Aurora, A. M. Stern, and D. S. Sigman, *Biochem. Biophys. Res. Commun.*, 1977, **78**, 170.
282 D. S. Sigman, D. R. Graham, V. D'Aurora, and A. M. Stern, *J. Biol. Chem.*, 1979, **254**, 12 269.

the anti-tumour drugs bleomycin and neocarzinostatin. (However, copper inhibits the former drug and is not required for the action of the latter.)

RNA polymerase I that had been isolated from mouse ascites sarcoma SR-C3H/He cells has been shown to be inhibited by 1,10-phenanthroline, whereas RNA polymerase II from the same source was not.[283] The latter represents the only polymerase that (to date) has been reported to be unaffected by this chelating agent. Contrary to the work reported above,[279] the inhibition of RNA polymerase I was found to be competitive with regard to nucleoside triphosphates. The non-chelating analogue 4,7-dimethyl-1,10-phenanthroline had no effect on either polymerase. Two new nucleases, isolated from *Penicillium citrinum*[284] and *Aspergillus oryzae*,[285] have been shown to be inhibited by EDTA, the subsequent restoration of activity being achieved in each case by the addition of Zn^{II}. This suggests that these nucleases are zinc-containing metalloenzymes.

Although Zn^{II} is essential for the activity of nucleotidyl polymerases, excess Zn^{II} has been found to inhibit the RNA polymerase I of *E. coli*.[286] This inhibition is specific for the initiation of RNA synthesis, excess Zn^{II} having no effect on the elongation reaction. Copper(II), Pb^{II}, and Cd^{II} also have been found to inhibit human DNA polymerase β to a significant extent.[287] This inhibition is not due to the displacement of the Zn^{II} at the active site by the inhibitory metal ions, and it is non-competitive with respect to the template/primer and deoxynucleoside triphosphate, which indicates that it is not due to the formation of a complex between the metal ion and the substrates.

Besides the Zn^{II} at the active site, nucleotidyl polymerases require the presence of an activating metal ion, generally Mg^{II} (although Mn^{II} and Co^{II} are also effective). A series of publications concerning the effects of different activation cations on the fidelity of replication of DNA by the DNA polymerase I of *E. coli* has appeared.[288-290] It was found that, in the presence of Mg^{II}, the error rate for the incorporation of dGMP and dCTP into poly[d(A-T)] was 1 in 80 000 and 1 in 8000, respectively; the dGMP was invariably substituted for dAMP, and its level of incorporation was affected by the concentration of dATP in the medium. The frequency of incorporation of dGMP was invariant with respect to the concentration of Mg^{II}. When the Mg^{II} is replaced by Co^{II} as the activating cation, the error rate for incorporation of dGMP increases two- to three-fold, and again is independent of the concentration of the metal ion, whereas when Mn^{II} is substituted for Mg^{II} the error frequency is two- to fifteen-fold higher, and is dependent on the concentration of Mn^{II}. Interestingly, the nuclease activity of the enzyme is not dependent on which activating cation is used. The above results, obtained for the synthesis of DNA by the polymerase I of *E. coli* on the artificial template poly[d(A-T)], were extended to other DNA poly-

[283] H. Misumi, *Okayama Igakkai Zasshi*, 1979, **91**, 491.
[284] G. U. Ko, *Choson Minjujuui Ihmin Konghwaguk Kwahagwon Tongbo*, 1979, **27**, 85
[285] R. E. Abramov and Kh. O. Bezirdzhyan, *Biokhimiya*, 1979, **44**, 990.
[286] Y. Nagamine, D. Mizuno, and S. Natori, *FEBS Lett.*, 1979, **99**, 29.
[287] E. A. Popenoe and M. A. Schmaeler, *Arch. Biochem. Biophys.*, 1979, **196**, 109.
[288] S. S. Agarwal, D. K. Dube, and L. A. Loeb, *J. Biol. Chem.*, 1979, **254**, 101.
[289] M. A. Sirover, D. K. Dube, and L. A. Loeb, *J. Biol. Chem.*, 1979, **254**, 107.
[290] T. A. Kundel and L. A. Loeb, *J. Biol. Chem.*, 1979, **254**, 5718.

merases, including human placental DNA polymerases α and β,[291] and to the natural template $\phi\chi174$.[290, 292] In all cases it was found that the replacement of Mn^{II} by Mn^{II} and Co^{II} increases the frequency of misincorporations.

A theoretical treatment of the kinetics of the reaction of DNA polymerase has been published.[293] The kinetic analysis of this system presents formidable mathematical difficulties, because the product of the reaction is also a substrate for the next catalytic cycle, and the enzyme possess both polymerase and exonuclease activities; hence the chain which is being synthesized may be either elongated or trimmed, so that its growth has the character of a random walk. A strategy for the estimation of the kinetic parameters of the polymerase reaction from the measurement of the amount of nucleotide which is incorporated or hydrolysed at any particular time has been proposed, and several models have been analysed. The most elementary model, which serves as a limiting case for the other models, involves just two rate constants, one for incorporation and another for excision.

The steady-state kinetics of DNA polymerase from mouse myeloma have been investigated.[294] The results suggest an ordered 'bibi' mechanism for the polymerization reaction, with the binding and release of DNA constituting the first and last steps, respectively, of the reaction. Activation by Mg^{II}, although essential for the reaction, does not alter the K_M values of the substrates (DNA and nucleotide triphosphate).

Diastereoisomeric phosphorothioate analogues of dATP have been used[295] to determine the stereochemistry of action of the DNA polymerase I of *E. coli*. In the presence of Mg^{II}, the (S_P)-diastereoisomers of 2'-deoxyadenosine 5'-O-[α-thio]triphosphate (dATP[αS]) and 2'-deoxyadenosine 5'-O-[β-thio]triphosphate (dATP[βS]) are substrates for the enzyme, whereas the (R_P)-diastereoisomers are not. However, when Co^{II}, Mn^{II}, or Zn^{II} is used as the activating cation, the stereoselectivity at the β-phosphate is lost, so that (R_P)-dATP[βS] is incorporated into the DNA chain. Since these three metal ions can bind to the deoxynucleotide through both sulphur and oxygen, whereas Mg^{II} binds only through oxygen,[296] it has been concluded that Mg^{II}dATP is bound to the enzyme as a β,γ-bidentate complex, in the D configuration. Furthermore, this study shows that an inversion of absolute configuration occurs at the α-phosphorus during synthesis; this is consistent with an in-line mechanism, where both the attacking 3'-OH of the primer terminus and the leaving pyrophosphate group occupy the apical positions of a trigonal bipyramid (Figure 3).

2',3'-Dideoxythymidine 5'-triphosphate (ddTTP) has been found to inhibit DNA polymerase from mouse myeloma strongly when Mn^{II} is used as an activator, but not when Mg^{II} is the activating cation.[297] The inhibition is competitive with the substrate dTTP, and is characterized by a K_i of 35 nmol l^{-1}. In the

291 G. Seal, C. W. Shearman, and L. A. Loeb, *J. Biol. Chem.*, 1979, **254**, 5229.
292 L. A. Weymouth and L. A. Loeb, *Proc. Natl. Acad. Sci. USA*, 1978, **75**, 1924.
293 F. Bernardi, M. Saghi, M. Dorizzi, and J. Ninio, *J. Mol. Biol.*, 1979, **129**, 93.
294 K. Tanabe, E. W. Bohn, and S. H. Wilson, *Biochemistry*, 1979, **18**, 3401.
295 P. M. J. Burgers and F. Eckstein, *J. Biol. Chem.*, 1979, **254**, 6889.
296 E. K. Jaffe and M. Cohn, *J. Biol. Chem.*, 1978, **253**, 4823.
297 K. Ono, M. Ogasawara, and A. Matsukage, *Biochem. Biophys. Res. Commun.*, 1979, **88**, 1255.

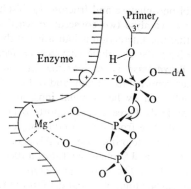

Figure 3 *A model of the absolute configuration of the Mg^{II}–dATP complex at the active site of DNA polymerase I*
(Reproduced by permission from *J. Biol. Chem.*, 1979, **254**, 6889)

presence of Mn^{II}, 2′-deoxy-2′-fluorouridine triphosphate has been found to be a substrate for the RNA polymerase of *E. coli*;[298] it replaces UTP, the level of transcription for the fluoro-derivative being 55% of that observed with UTP.

Descriptions of new purification procedures for the RNA polymerase of *Bacillus subtilis* and for the DNA polymerase I of *E. coli* have been published.[299,300] The former procedure utilizes heparin-agarose to absorb the polymerase rapidly and quantitatively whilst the latter uses polymin P to separate the DNA polymerase from RNA polymerase.

Other Zinc Metalloenzymes.—The activity of **fructose 1,6-bisphosphatase** from rabbit liver has been shown to be modulated by Zn^{II}, and recently this observation has been extended to the enzyme from rabbit muscle.[301] In the latter enzyme, only two Zn^{II}-binding sites per subunit (as compared to three for the liver enzyme) are present. Binding of Zn^{II} at the four high-affinity sites (one per subunit; the enzyme is tetrameric) results in the inhibition of the enzyme. This inhibition is relieved when Zn^{II} binds at the second set of sites, which appear to be identical with the activator-binding site. Zinc(II) is a much more efficient activator of the enzyme than either Mg^{II} or Mn^{II}. AMP is an allosteric inhibitor of fructose 1,6-biphosphatase, and the enzyme has one AMP-binding site per subunit. The presence of Zn^{II} enhances the inhibition by AMP ($K_i = 2$ μmol l^{-1} in the presence of Zn^{II} and 15 μmol l^{-1} in its absence).[302] Two AMP derivatives, *i.e.* 8-azidoadenosine 5′-monophosphate and periodate-oxidized AMP, have been shown to act as affinity labels for the regulatory sites of the enzyme.[303,304] Another report suggests that there is a specific interaction

[298] D. Pinto, M. T. Sarocchi-Landousy, and W. Guschlbauer, *Nucleic Acids Res.*, 1979, **6**, 1041.

[299] B. L. Davison, T. Leighton, and J. C. Rabinowitz, *J. Biol. Chem.*, 1979, **254**, 9220.

[300] G. Rhodes, K. D. Jentsch, and T. M. Jovin, *J. Biol. Chem.*, 1979, **254**, 7465.

[301] S. Pontremoli, B. Sparatore, F. Salamino, E. Melloni, and B. L. Horecker, *Arch. Biochem. Biophys.*, 1979, **194**, 481.

[302] S. Pontremoli, E. Melloni, F. Salamino, B. Sparatore, M. Michetti, and B. L. Horecker, *Biochem. Biophys. Res. Commun.*, 1979, **88**, 656.

[303] F. Marcus and B. E. Haley, *J. Biol. Chem.*, 1979, **254**, 259.

[304] R. B. Maccioni, E. Hubert, and J. C. Slebe, *FEBS Lett.*, 1979, **102**, 29.

between fructose 1,6-bisphosphatase and fructose 1,6-bisphosphate aldolase.[305] Since these two enzymes catalyse two consecutive reactions of the glycolytic cycle, it is postulated that this interaction is indicative of the presence of a multi-enzyme complex, involving the glycolytic enzymes, *in vivo*.

δ-Aminolaevulinate dehydratase is an enzyme in the haem biosynthetic pathway which has been reported to contain, and to be activated by, Zn^{II}. This enzyme is markedly inhibited by heavy-metal ions, and has been used as a marker of exposure to lead. Three recent reports detail the purification of δ-aminolaevulinate dehydratase from human erythrocytes.[306-308] One of these procedures[306] utilized chromatography on DEAE-Sephacel followed by gel filtration on Sephadex G-100 and Sepharose CL-6B; a 1000-fold purification was achieved, with a yield of 85%. The second procedure[307] involves precipitation by a salt, ion-exchange chromatography, and gel filtration; it gives a 9000-fold purification. The third[308] uses chromatography on DEAE-cellulose, hydrophobic chromatography on octyl- and phenyl-Sepharose, and gel filtration on Sephadex G-200 to achieve a 38 000-fold purification, with a 69% yield. The molecular weight of the native enzyme was established as 252 000 by gel filtration, and, under denaturing conditions, the enzyme showed a single subunit in polyacrylamide gel electrophoresis, with a molecular weight of 31 000; this indicates that the enzyme is composed of eight identical subunits. It was confirmed that Pb^{II} is a potent inhibitor ($K_1 = 1.7$ μmol l^{-1}) of the enzyme. A new, direct, spectrophotometric assay, which is suitable for the highly purified enzyme, has also been described.[309] Another publication reports the isolation of δ-aminolaevulinic acid dehydratase from bovine liver.[309] The enzyme is reported to contain one Zn^{II} ion per subunit, and two cysteine residues and two histidine residues were found to be present in its active site. The thiol groups are reported to be necessary for enzymatic activity, but it is claimed that the Zn^{II} *is not* essential for the activity, and that its function is to prevent the oxidation of the thiol groups, perhaps through co-ordination to them. It is suggested that the histidine residues in the active site may also act as Zn^{II}-binding ligands. Two reports describe investigations of the inhibition of δ-aminolaevulinic acid dehydratase by heavy-metal ions.[310, 311] Lead(II), Hg^{II}, Cd^{II}, Cu^{II}, and Ag^I were found to inhibit the enzyme, and Zn^{II} reverses the inhibition that is caused by Pb^{II}, but not that which is observed with the other metal ions.

A description of a new purification procedure for phospholipase C from *Bacillus cereus* has been published.[312] This procedure involves hydrophobic chromatography on palmitoyl-cellulose; the enzyme is strongly absorbed on this support, and can be eluted with a detergent. Phospholipase C is extremely

305 S. Pontremoli, E. Melloni, F. Salamino, B. Sparatore, M. Michetti, V. N. Singh, and B. L. Horecker, *Arch. Biochem. Biophys.*, 1979, **197**, 356.
306 D. Huckel and D. Beyersmann, *Anal. Biochem.*, 1979, **97**, 277.
307 N. Despaux, E. Comoy, C. Bohuon, and C. Boudene, *Biochimie*, 1979, **61**, 1021.
308 P. M. Anderson and R. J. Desnick, *J. Biol. Chem.*, 1979, **254**, 6924.
309 I. Tsukamoto, T. Yoshinaga, and S. Sano, *Biochim. Biophys. Acta*, 1979, **570**, 167.
310 K. Tomokuni, *J. Toxicol. Sci.*, 1979, **4**, 11.
311 T. Koreeda, K. Kashiwabara, S. Shirouzu, and M. Yuguchi, *Kyorin Igakkai Zasshi*, 1979, **10**, 15.
312 S. Imamura and Y. Horiuchi, *J. Lipid Res.*, 1979, **20**, 519.

table towards denaturation by 8M-urea; indeed the enzyme is catalytically active in this medium.[313] However, a recent study shows that the enzyme is readily denatured by 2M-guanidinium chloride.[314] This denaturation is accompanied by the release of the Zn^{II} from the protein, and by the exposure (to chemical modification) of all the histidine residues in the molecule; this latter observation suggests that the enzyme is fully unfolded by the guanidinium chloride. The addition of Zn^{II} protects the enzyme against inactivation and induces the re-folding of the denatured enzyme. A new, simplified, assay for phospholipase C activity has been described.[315] This assay used alkaline phosphatase to convert phosphorylcholine, which is a product of the hydrolysis of phospholipids by phospholipase C, into inorganic phosphate, which is then determined by standard procedures.

Yeast **inorganic pyrophosphatase** is known to require a bivalent metal ion for activity; Mg^{II}, Zn^{II}, Co^{II}, and Mn^{II} can fulfill this requirement, whereas Ca^{II} Cd^{II}, Cu^{II}, and Ni^{II} are inhibitory. The effects of Zn^{II} on the activity of the enzyme have recently been investigated in detail.[316] The metal has a dual role; t binds to the protein, thereby activating it, and it forms a metal–pyrophosphate complex, which then acts as a substrate for the enzyme. Yeast inorganic pyrophosphatase has been found to catalyse the rapid exchange of oxygen atoms between inorganic phosphate and solvent water.[317] The order of effectiveness amongst the activating metal ions is $Mg^{II} > Zn^{II} > Co^{II} > Mn^{II}$. Kinetic studies indicate that the exchange of oxygen can be explained by the rapid formation and breakdown of enzyme-bound pyrophosphate. The pyrophosphatase (exonuclease) from *Crotalus adamanteus* venom has also been shown to be activated by metal ions, and to utilize a metal–phosphate complex as a substrate.[318] Unlike yeast inorganic pyrophosphatase, the role of the metals in exonuclease is differentiated, with Zn^{II} being a specific activator and Mg^{II} being specific for substrate binding.

Glyoxalase I from various mammalian sources and from yeast has recently been shown to be a zinc-containing metalloenzyme, containing one atom of zinc per subunit (the enzyme is dimeric).[319] This has been confirmed in two recent publications, one concerning the enzyme from rat liver[320] and the other the enzyme from yeast.[321] These publications describe the purification of these two enzymes to homogeneity by affinity chromatography on *S*-hexylglutathione-Sepharose. For both enzymes, the Zn^{II} is essential for catalytic activity. The subunit of the rat liver enzyme has a molecular weight of 27 000 whereas the yeast enzyme, which is monomeric, has a molecular weight of 32 000. The isoelectric points for the yeast and mammalian enzymes are at pH 7.0 and 4.8,

13 C. Little, *Biochem. J.*, 1978, **175**, 977.
14 C. Little and S. Johansen, *Biochem. J.*, 1979, 509.
15 E. L. Krug, N. J. Truesdale, and C. Kent, *Anal. Biochem.*, 1979, **97**, 43.
16 O. A. Moe, S. Pham, T. Dang, and L. Styer, *Arch. Biochem. Biophys.*, 1979, **197**, 73.
17 C. A. Janson, C. Degani, and P. D. Boyer, *J. Biol. Chem.*, 1979, **254**, 3743.
18 R. Vasileva and L. Dolapchiev, *Dokl. Bolg. Akad. Nauk.*, 1979, **32**, 1109.
19 A. C. Aronsson, E. Marmstal, and B. Mannervik, *Biochem. Biophys. Res. Commun.*, 1978, **81**, 1235.
20 E. Marmstal and B. Mannervik, *Biochim. Biophys. Acta*, 1979, **566**, 362.
21 E. Marmstal, A.-C. Aronson, and B. Mannervik, *Biochem. J.*, 1979, **183**, 23.

respectively. Peptide maps of tryptic digests indicate that the enzymes from these two sources are dissimilar in structure, and it has been suggested that the enzymes have separate evolutionary origins. The similarities demonstrated in the kinetic and mechanistic properties of the enzymes are proposed to reflect convergent evolution.

Erythrocyte **superoxide dismutase** is generally considered to be a copper metalloenzyme; however, in addition to one atom of copper per subunit, this enzyme contains one atom of zinc per subunit, the function of which is obscure. The binding of Zn^{II} to apo-superoxide dismutase from bovine erythrocytes has been investigated by ^1H n.m.r. spectroscopy.[322] This study indicates that the binding of Zn^{II} at the zinc-binding sites pre-forms the copper-binding sites, so that one of the functions of the Zn^{II} appears to be the structural stabilization of the active site of the enzyme. This is in accord with the X-ray diffraction studies on the enzyme, which have shown that the two metal-binding sites in each subunit of superoxide dismutase are very close to one another, and that the Cu^{II} and Zn^{II} share one ligand (a histidine residue) which bridges the two metal ions as an imidazolate anion.[323] The enzyme undergoes reversible spectral changes below pH 4.5, and these have been attributed to a disruption of the imidazolate bridge by protonation; this disruption has been suggested to be of mechanistic significance. However, a recent report shows that the Zn^{II} is reversibly lost from the enzyme at low pH, and that the observed spectral changes simply reflect this loss of metal.[324] It is suggested that the loss of metal is due to a protein conformational change. A further report by the same group indicates that, at high pH, the Cu^{II} in zinc-free bovine erythrocyte superoxide dismutase can migrate to the vacant zinc-binding site.[325] This observation puts in doubt some earlier kinetic studies on the zinc-free enzyme, which had indicated that this species has only about 50% of the activity of the native enzyme; a re-examination of the activity of the zinc-free enzyme at pH 6.0, where the loss of Cu^{II} from the active site is minimal, indicates that the copper-only enzyme retains at least 80% of the activity of the native enzyme. This suggests that the presence of Zn^{II} in the enzyme is of little functional significance, though it could still, as suggested above, stabilize the structure of the protein.

Unlike all other pencillinases that have been characterized to date, β-lactamase II from *Bacillus cereus* is a zinc-containing metalloenzyme that contains one gram atom of zinc per mole of protein.[326] Recent ^1H n.m.r. studies have suggested that the zinc is bound in part by three histidine residues, and that the rate of exchange of protons at C-2 of these histidine residues is markedly reduced when Zn^{II} is bound to the protein.[327] The latter conclusion has now been

[322] A. E. G. Cass, H. A. O. Hill, J. V. Bannister, and W. H. Bannister, *Biochem. J.*, 1979, **177**, 477.

[323] J. S. Richardson, K. A. Thomas, and D. C. Richardson, *Biochem. Biophys. Res. Commun.* 1975, **63**, 986.

[324] M. W. Pantoliano, P. J. McDonnell, and J. S. Valentine, *J. Am. Chem. Soc.*, 1979, **101**, 6454.

[325] J. S. Valentine, M. W. Pantoliano, P. J. McDonnell, A. R. Burger, and S. J. Lippard, *Proc. Natl. Acad. Sci. USA*, 1979, **76**, 4245.

[326] R. B. Davies and E. P. Abraham, *Biochem. J.*, 1974, **143**, 129.

[327] G. S. Baldwin, A. Galdes, H. A. O. Hill, B. E. Smith, S. G. Waley, and E. P. Abraham, *Biochem. J.*, 1978, **175**, 441.

confirmed by ^{1}H–^{3}H exchange studies, which also indicate that, by digesting the ^{3}H-exchanged protein with trypsin, the peptides that contain the histidine ligands can be identified.[328] This will therefore enable the histidine residues that act as Zn^{II}-binding ligands in β-lactamase II to be recognized when the primary structure of this enzyme is determined.

The **pyruvate carboxylase** of a thermophilic species of *Bacillus* has been characterized.[329] The enzyme is a tetramer, with a molecular weight of 558 000, and it contains four molecules of D-biotin. Like the yeast enzyme, it contains one atom of Zn^{II} per subunit.

Methionyl-tRNA synthetase from *Escherichia coli* has been shown to be a zinc metalloenzyme.[330] The enzyme is a dimer, with a molecular weight of 170 000, and it contains one Zn^{II} ion per subunit. The enzymatic activity is reversibly inhibited by 1,10-phenanthroline but not by its non-chelating analogues. The enzyme dissociates into monomers, without loss of Zn^{II}, when treated with trypsin. Another publication describes the modification of the methionyl-RNA synthetase of *E. coli* by Co^{II} complexes of ATP.[331] This inactivation is accompanied by the incorporation of one cobalt atom and one ATP molecule per active site. Two other enzymes that, during 1979, have been shown to be zinc metalloenzymes are **dihydropyrimidine amidohydrolase**[332] and **creatinine amidohydrolase**.[333] Both enzymes are reported to contain one atom of zinc per subunit (the former enzyme is tetrameric, the latter octameric) and to be reversibly inactivated by certain metal chelators. **Aminoacylase** from *Aspergillus oryzae* also appears to be a zinc metalloenzyme,[334] although the metal content was not firmly established in this study.

[28] G. S. Baldwin, S. G. Waley, and E. P. Abraham, *Biochem. J.*, 1979, **179**, 459.

[29] S. Libor, T. K. Sundaram, R. Warwick, J. A. Chapman, and S. M. W. Grundy, *Biochemistry*, 1979, **18**, 3647.

[30] L. H. Posorske, M. Cohn, N. Yanagisawa, and D. S. Auld, *Biochim. Biophys. Acta*, 1979, **576**, 128.

[31] T. Kalogerakos, S. Blanquet, and J. P. Waller, *Eur. J. Biochem.*, 1979, **93**, 339.

[32] K. P. Brooks, B. D. Kim, and E. G. Sander, *Biochim. Biophys. Acta*, 1979, **570**, 213.

[33] K. Rikitake, I. Oka, M. Ando, T. Yoshimoto, and D. Tsuru, *J. Biochem. (Tokyo)*, 1979, **85**, 1109.

[34] I. Gentzen, H. G. Loeffler, and F. Schneider, *Metalloproteins, Autumn Meet. Ger. Biochem. Soc.*, 1979, 270.

7
Manganese Metalloproteins and Manganese-Activated Enzymes

A. R. McEUEN

Introduction.—This Report considers those proteins which contain tightly bound manganese and those proteins which have a looser association with this element, usually manifested by the stimulation of enzyme activity by the addition of manganese(II) ions to the assay medium. In some instances, this stimulation of activity is specific for Mn^{2+}; in other instances, Mg^{2+} can stimulate as well, and often to a greater degree than Mn^{2+}. For the purposes of this Report, therefore, manganese metalloproteins and manganese-activated enzymes are classified into two groups: manganese-specific proteins and (Mn^{2+}/Mg^{2+})-activated proteins.

Of general interest is a review article with 58 references on the physiology and biochemistry of manganese in mammals.[1] It gives a brief summary of the absorption, excretion, transport, and function of manganese, with emphasis on the correlation of gross physiological symptoms of Mn deficiency with molecular processes. Another review article deals specifically with the manganese nutrition of ruminants.[2]

In view of the importance of e.p.r. in the study of manganese(II), it is noteworthy that yet another set of authors have extended the analysis of Mn^{II}-protein e.p.r. spectra to include third-order perturbation terms.[3] The results are similar, but not identical, to those of Meirovitch and Poupko,[4] published in 1978. Four experimentally obtained spectra are shown, along with computer simulations based on their equations. The examples used were MnADP + formyltetrahydrofolate synthetase, Mn + oxalate + pyruvate kinase, MnGDP + adenylosuccinate synthetase, and MnADP + creatine kinase.

1 Manganese-Specific Proteins

Concanavalin A and Related Lectins.—Concanavalin A (Con A), from the jack bean, is the best known and most thoroughly studied of the lectins, or saccharide binding proteins. It exists in both dimeric (below pH 6.0) and tetrameric (above pH 7.0) forms. Each subunit ($M_r = 27\,000$) contains a saccharide-binding site, a transition-metal-binding site, S_1 (which in Nature binds Mn^{2+}), and a calcium-binding site, S_2. Attempts to label the S_1 site with kinetically inert Co^{II}

1 R. M. Leach and M. S. Lilburn, *World Rev. Nutr. Diet.*, 1978, **32**, 123.
2 M. Hidiroglou, *Can. J. Anim. Sci.*, 1979, **59**, 217.
3 G. D. Markham, B. D. N. Rao, and G. H. Reed, *J. Magn. Reson.*, 1979, **33**, 595.
4 E. Meirovitch and R. Poupko, *J. Phys. Chem.*, 1978, **82**, 1920.

proved unsuccessful.[5] A stable Co^{III}–protein complex was formed, but this complex was still capable of binding an additional transition-metal ion (Co^{2+}, Ni^{2+}, or Mn^{2+}) and Ca^{2+}, to form a biologically active lectin. Also, when $Co^{II}CaConA$ was treated with H_2O_2, no oxidation of Co^{II} occurred unless it was in molar excess over the protein subunits. The authors concluded that Co^{III} bound to another site on the protein subunit, possibly the lanthanide-binding site, S_3. The failure of Co^{III} to bind to S_1 may be because the ligands of S_1 do not meet the requirements for co-ordination of Co^{III} or possibly because the site is inaccessible to H_2O_2.

As isolated, Con A is a mixture of intact and nicked subunits. Most papers reviewed in this Report describe studies using preparations of Con A in which the nicked subunits had been removed, but one paper[6] describes the binding of metals to nicked Con A. When Con A was eluted on a column of CM-cellulose (20mM-HEPES buffer, at pH 7.1) with a NaCl gradient, three distinct fractions were obtained. The first (F_1) consisted almost entirely of fragments, the second (F_2) was a mixture of nicked and intact subunits, and the third (F_3) contained only intact subunits. Each sample was desalted on Bio-Gel P-10 and analysed for metal content: F_1 contained 0.2 Ca^{2+} and 0.1 Mn^{2+} ions per tetramer and had almost no lectin activity, F_2 contained 2.0 Ca^{2+} and 0.8 Mn^{2+} ions per tetramer and had moderate lectin activity, and F_3 contained 4.0 Ca^{2+} and 1.8 Mn^{2+} ions per tetramer and had twice the lectin activity of F_2. If the initial sample of Con A was incubated with excess Ca^{2+} and Mn^{2+} prior to chromatography on CM-cellulose, a similar elution profile was obtained, and metal analysis after desalting revealed that, in each fraction, the Ca^{2+} content was unchanged but the content of Mn^{2+} had been doubled by this treatment. Surprisingly, doubling the Mn^{2+} content increased the lectin activity of each fraction by only 10%. Thus, calcium content appeared to be more important than manganese content for determining biological activity. The lectin activity appeared to be due entirely to the intact subunits present in each fraction. However, if Mn^{2+} and Ca^{2+} were added back after desalting, the nicked subunits also displayed lectin activity, suggesting that the basis for their lower lectin activity was a greatly reduced affinity for the two bivalent cations.

In 1977, Brown, *et al.*[7] proposed that the binding of cations induces a conformational change from the 'unlocked' to the 'locked' conformation and that only 'locked' Con A has any saccharide-binding activity. This proposal continued to excite interest in the 1979 literature. Evidence in favour has come from c.d.[8] and kinetic[9] studies, and evidence against has come from other kinetic studies[10] and from direct studies of the binding of metals.[11]

Different forms of Con A (apo-Con A, CoCon A, EDTA-sensitive CaCoCon A, and EDTA-resistant CaCoCon A) had similarly shaped c.d. spectra in the

[5] M. S. Urdea, D. J. Christie, G. R. Munske, J. A. Magnuson, and J. I. Legg, *Biochem. Biophys. Res. Commun.*, 1979, **91**, 1045.
[6] F. Obata, R. Sakai, and H. Shiokawa, *J. Biochem. (Tokyo)*, 1979, **85**, 1037.
[7] R. D. Brown, III, C. F. Brewer, and S. H. Koenig, *Biochemistry*, 1977, **16**, 3883.
[8] A. D. Cardin, W. D. Behnke, and F. Mandel, *J. Biol. Chem.*, 1979, **254**, 8877.
[9] C. A. Stark and A. D. Sherry, *Biochem. Biophys. Res. Commun.*, 1979, **87**, 598.
[10] G. M. Alter and J. A. Magnuson, *Biochemistry*, 1979, **18**, 29.
[11] D. J. Christie, G. R. Munske, and J. A. Magnuson, *Biochemistry*, 1979, **18**, 4638.

near-u.v., but different and characteristic molar extinction coefficients at 275 nm. Difference spectra revealed a broad symmetrical transition at 277 nm. Addition of Mn^{2+} or Ni^{2+} to apo-Con A produced results identical to Co^{2+}. Addition of EDTA to the EDTA-sensitive CaCoCon A resulted in the spectrum that is characteristic of apo-Con A, but the addition of EDTA to EDTA-resistant CaCoCon A produced no change. The EDTA-sensitive CaCoCon A slowly evolved into the EDTA-resistant form with a rate-constant ($k_1 = 1.3 \times 10^{-2}$ s^{-1}) and energy of activation (96 kJ mol^{-1}) that are in very good agreement with the results of Brown *et al.*[7] and of Sherry *et al.*[12] The kinetics of this process are consistent with the rate-limiting step being a conformational change in the protein. Examination of the X-ray structure led the authors to propose Tyr-12 as the residue that is responsible for the observed changes in c.d. spectra. There was no change in the far-u.v. c.d. spectrum upon any treatment.

The transformation of apo-Con A into a saccharide-binding form after the addition of metal ions was monitored by the fluorescence quenching of 4-methyl-umbelliferyl α-D-mannopyranoside (MUM).[9] The rates were independent of the concentration of bivalent cation or of MUM when in five- to ten-fold excess of protein, but the rates were always first-order with respect to protein. The results are consistent with a rapid equilibrium between bound and unbound metal ion, followed by a slow irreversible change in conformation of the protein. Several metal ions were tested, alone and in combination, at three different pH values. Most rapid conversion was achieved by the combinations $Mn^{2+} + Ca^{2+}$, $Zn^{2+} + Ca^{2+}$, and $Co^{2+} + Ca^{2+}$ ($k_1 = 1.5 \times 10^{-2}$ s^{-1} at pH 5.6), and the final products had the same affinity for MUM as did the native protein. Lower rates of conversion were observed when each of the cations was added singly. At pH 5.6, the mono-metal ConA's had less affinity for MUM than native Con A, but at pH values of 6.4 and 7.2, the affinity was approximately the same. Addition of excess EDTA to MnCon A and CoCon A resulted in rapid removal of the metal ($t_{\frac{1}{2}} \approx 20$ s), but in the presence of 1M-NaCl the Con A retained full MUM-binding activity, which decayed with a half-life of 2.5 h at pH 6.4. When apo-Con A was similarly prepared from CaCon A, it also bound MUM, but with a lower affinity than apo-Con A that had been prepared from MnCon A.

The conclusion by Brown *et al.*,[7] based on proton relaxation rate (p.r.r.) data, that Mn^{2+} ion can bind to the S_2 site as well as to the S_1 was challenged by results from equilibrium dialysis binding studies.[11] Christie *et al.*[11] found only a single Mn^{2+}-binding site, even with a ten-fold molar excess of Mn^{2+}. They repeated the p.r.r. work of Brown *et al.*[7] and confirmed that, at 23 °C and at pH 6.4, the addition of Mn^{2+} in excess of one gram atom per mole of subunit decreased the relaxation rate of protons, reaching a minimum around a stoicheio-metry of 2:1 (Figure 1). However, determination of the concentration of free Mn^{2+} in these samples by e.p.r. showed that only 1.0 Mn^{2+} ion was bound per subunit. Furthermore, at 5 °C, at the same pH, such a decrease in proton relaxa-tion rate was not observed. A temperature-dependent conformational change was invoked to explain these differences. The kinetic schemes of Brown *et al.*[7]

[12] A. D. Sherry, A. E. Buck, and C. A. Peterson, *Biochemistry*, 1978, **17**, 2169.

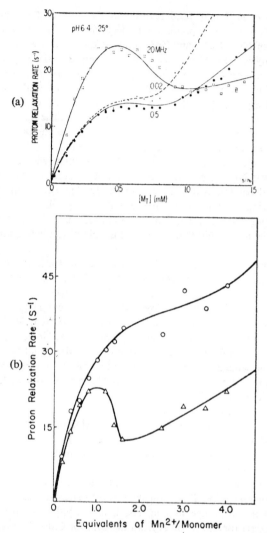

Figure 1 *Relaxation rates of protons of the solvent in solutions of apo-Con A in the presence of varying concentrations of added Mn^{2+} ion and no Ca^{2+} ions, at pH 6.4. (a) The apo-Con A was 0.40 mmol l^{-1} of monomer in 0.1 M-KOAc and 0.9 M-KCl. A separate aliquot of enzyme was used for each value of $[M_T]$, and the values of the magnetic fields used correspond to Larmor frequencies of protons of 0.02, 0.5, and 20 MHz. (b) Measurements at 20.5 MHz and 23 °C for samples of 0.37 mmol l^{-1} of monomer of apo-Con A in 0.9 M-KCl and 0.05 M-MOPS buffer, prepared at 5 °C and then allowed to warm to 23 °C ($\bigcirc$) and in 0.9 M-KCl and 0.1 M-KOAc, prepared at 23 °C ($\triangle$). The samples that were prepared at 5 °C were incubated at that temperature for 72 hours and then warmed to 23 °C just before spectra were obtained. They were then incubated at 23 °C for 7 days and spectra were remeasured. In both cases, the lines through the points indicate the general trend of the data*

(Reproduced by permission from *Biochemistry*, 1979, **18**, 3883 and 4638)

and of Koenig et al.[13] may need revision in the light of these findings, but Christie et al.[11] are probably premature in declaring them totally invalid.

The binding of Ni^{2+}, Co^{2+}, and Zn^{2+} to apo-Con A has also been studied by equilibrium dialysis.[11] Evidence was found for only one site per subunit, and binding constants agreed with previously published results. The same stoicheiometry for binding of Co^{2+} was also determined by near-u.v. c.d.[8] In agreement with Stark and Sherry,[9] Christie et al.[11] found that Mn^{2+} or Ca^{2+} alone was able to convert apo-Con A into an MUM-binding form. The number of MUM-binding sites per subunit was found to be equal to the number of Mn^{2+} ions bound per subunit, up to a maximum of 1.0 each. In contrast to Stark and Sherry,[9] Christie et al.[11] did not obtain an MUM-binding form with either Co^{2+} or Zn^{2+}; neither was Ni^{2+} able to induce apo-ConA to bind MUM, even though the experimental conditions described were nearly identical for the two sets of authors. Surprisingly, addition of Mn^{2+} to NiCon A produced an active lectin without displacing the Ni^{2+}. The authors proposed that Ni^{2+} was not using all the ligands in the S_1 site and that Mn^{2+} was utilizing ligands from both S_1 and S_2 sites.

After monitoring the kinetics of binding of Mn^{2+} to CaCon A by e.p.r. and p.r.r. measurements, Alter and Magnuson[10] concluded that an 'unlocked' to 'locked' conformational change was not the best interpretation of their data. Their estimates of activation enthalpy and entropy were in good agreement with the results of Brown et al.,[7] as was their observation that Mn^{2+} was more accessible to solvent in MnCon A than in CaMnCon A. However, they found that the reaction rates for the disappearance of free Mn^{2+} that were measured by e.p.r. were the same as those measured for the formation of the final CaMn-Con A by p.r.r. In order for their data to be compatible with the model of Brown et al.,[7] the affinity of Mn^{2+} for the second S_2 site of the dimer must be low until conformational change occurs, otherwise the rates would be markedly different. They propose an alternative scheme whereby the binding of Mn to the second S_1 site of the dimer is the rate-limiting step. Thus, the controversy over the 'unlocked' to 'locked' model of concanavalin A continues, and probably will continue for a few more years yet.

The binding of saccharide and glycopeptides to Con A was monitored by nuclear magnetic relaxation dispersion (NMRD).[14] The paramagnetic contribution to $1/T_1$ at a wide range of magnetic field strengths was determined by subtracting the data for CaZnCon A from the data for CaMnCon A. The binding of saccharides reduced the paramagnetic contribution by 25% but did not affect the diamagnetic contribution to $1/T_1$. A computer analysis of the data indicated an increase of forty per cent in the residence time of exchanging water molecules on the Mn^{2+}. Results from other studies led the authors to conclude that the increase in residence time is not due to direct binding of the saccharide to the Mn^{2+}, but to conformational changes that are induced by binding the saccharide. At saturating concentrations, all oligosaccharides tested (one to four residues, and unbranched) produced identical NMRD

[13] S. H. Koenig, C. F. Brewer, and R. D. Brown, III, *Biochemistry*, 1978, **17**, 4251.
[14] C. F. Brewer and R. D. Brown, III, *Biochemistry*, 1979, **18**, 2555.

profiles, indicating a single saccharide-binding site, and suggesting the absence of an extended binding site such as is found in lysozyme. In contrast, the branched glycopeptide from ovalbumin[15] was only 40% as effective as simple saccharides in reducing $1/T_1$ at saturating concentrations, although a qualitatively similar NMRD profile was produced. Since the glycopeptide is bivalent, a ratio of 2:1 for subunit:glycopeptide had the same NMRD profile as a 1:1 ratio. In a separate study,[16] a wide variety of glycopeptides was examined and a correlation was sought between structure and affinity for Con A.

The amino-acid sequences of lectins from other species of Leguminosae (Fabaceae) have been reported, and they were found to be homologous to Con A.[17-20] The α subunits from pea lectin (*Pisum sativum*)[17] and from favin (*Vicia faba*)[18] have been sequenced in their entirety and found to be homologous to residues 72—120 of Con A, with 25 and 24 identities, respectively. However, both were more closely related to lentil lectin, with 40 and 37 identities, respectively. This region of the polypeptide chain does not contain any ligands for the S_1 and S_2 sites, but does contain the ligands for the lanthanide-binding site, S_3; these were conserved in both lectins. Two types of lectins were found in seeds of *Vicia cracca*;[19] a glucose-specific lectin (VC-Glc) which has an $\alpha + \beta$ subunit structure similar to lentil lectin and an N-acetylgalactosamine-specific lectin (VC-GalNAC) which has subunits composed of a single polypeptide chain. The first 25 residues from the N-terminus were sequenced for each: the α subunit of VC-Glc was nearly identical to the α subunits of pea lectin and lentil lectin, the β subunit of VC-Glc was identical to the N-terminus of the β-subunit of pea lectin, and VC-GalNAc was homologous to the β-chains of VC-Glc, pea, and lentil lectins, which in turn are homologous to Con A beginning at around residue 122.[17-19]

By far the most significant paper on the sequencing of lectins is the first report of the complete amino-acid sequence of a β-chain, *i.e.* the β-chain of favin.[20] Its homology begins at residue 120 of Con A, extends to the C-terminus, and continues without interruption through the N-terminal 69 residues of Con A. The α-chain is homologous to residues 70–119, so that, together, favin $\alpha + \beta$ is a circular permutation of Con A. (The slight shift in assignment of the α-chain with Con A arises from a few insertions and deletions near each terminus.) In view of the close homology of favin with numerous other lectins, it is probable that they, too, are circular permutations of Con A. All the direct metal ligands of Con A are retained in favin, and a preliminary determination of metals present indicates that favin binds stoicheiometric amounts of Mn^{2+} and Ca^{2+}. Overall, about 40% of the residues are identical, with a slightly higher degree of conservation in the β-sheet regions of Con A, which suggests a similar tertiary

[15] C. F. Brewer, *Biochem. Biophys. Res. Commun.*, 1979, **90**, 117.
[16] J. U. Baenziger and D. Fiete, *J. Biol. Chem.*, 1979, **254**, 2400.
[17] C. Richardson, W. D. Behnke, J. M. Freisheim, and K. M. Blumenthal, *Biochim. Biophys. Acta*, 1978, **537**, 310.
[18] J. J. Hemperly, T. P. Hopp, J. W. Becker, and B. A. Cunningham, *J. Biol. Chem.*, 1979, **254**, 6803.
[19] C. Baumann, H. Rüdiger, and A. D. Strosberg, *FEBS Lett.*, 1979, **102**, 216.
[20] B. A. Cunningham, J. J. Hemperly, T. P. Hopp, and G. M. Edelman, *Proc. Natl. Acad. Sci. USA*, 1979, **76**, 3218.

structure. Furthermore, the site of the proposed *cis–trans* isomerization under-lying the 'locked' to 'unlocked' conformational change[7, 21] has also been re-tained in favin, so a comparative study of the kinetics of binding of metals should prove to be very interesting.

Phosphoglycerate Phosphomutase (EC 5.4.2.1).—This enzyme has been purified to homogeneity from *Bacillus subtilis*[22] and *B. megaterium*[23] and found to be absolutely dependent on Mn^{2+} for activity: Mg^{2+}, Ca^{2+}, and Fe^{3+} had no effect on the enzyme of *B. subtilis* and Mg^{2+}, Ca^{2+}, Fe^{2+}, Co^{2+}, and Zn^{2+} had no effect on the enzyme of *B. megaterium* on their own, but in the presence of sub-optimal concentrations of Mn^{2+}, $1mM$-Mg^{2+} stimulated the enzyme to full activity.[23] The apparent K_m values for Mn^{2+} were 4.5 μmol l^{-1} (*B. subtilis*) and 40 μmol l^{-1} (*B. megaterium*). Since neither report mentions correction for chelation of Mn^{2+} by other components of the assay medium, this difference in K_m values may be attributable to different assay conditions. The enzyme that was isolated from spores appeared to be identical to the enzyme that was isolated from vegetative cells.[23] Phosphoglycerate phosphomutase from these two species resembles the enzyme from wheat germ and differs from the enzymes from *Escherichia coli* and mammals in that it has no requirement for 2,3-diphospho-glycerate, nor was any found to be tightly bound to the protein. The require-ment for manganese appears to be unique to the genus *Bacillus*; it has not been reported for the purified enzyme from any other source, and an examination of unfractionated cell-free extracts revealed a manganese dependence for phospho-glycerate phosphomutase from *B. subtilis*, *B. megaterium*, and *B. cereus* but not for that from wheat germ or from rabbit liver.[24]

The importance of Mn^{2+} in the activity of this enzyme is underscored by the growth of *B. subtilis* on Mn-deficient media.[24] With glucose as the sole carbon source, growth was twice as fast in Mn-supplemented medium than in Mn-deficient medium; with malate as the carbon source, growth rates were the same with and without added $MnCl_2$. In further investigations of growth on glucose or glycerol in the absence of added Mn^{2+}, it was found that cessation of growth was accompanied by the accumulation of 3-phosphoglycerate (3-PGA). Upon addition of malate to the growth medium, growth resumed and the concentration of 3-PGA declined. (Catabolism of glucose and of glycerol proceeds through 3-PGA as an intermediate, but catabolism of malate does not.) Also, sporulation in the presence of glucose was found to require Mn^{2+}, but spore formation could proceed without it if malate was the major carbon source.

Developing spores accumulate 3-PGA and catabolize it rapidly upon germina-tion; therefore phosphoglycerate phosphomutase must be inactive in the develop-ing and dormant spore, but active in the germinating spore. The possibility that it is inhibited by sulpholactate, which also accumulates in spores, was ruled out by studies with the purified enzyme from *B. subtilis*.[22] Further evidence has been presented that a low intracellular concentration of free Mn^{2+} is re-

21 G. N. Reeke, Jr., J. W. Becker, and G. M. Edelman, *Proc. Natl. Acad. Sci. USA*, 1978, **75**, 2286.
22 K. Watabe and K. Freese, *J. Bacteriol.*, 1979, **137**, 773.
23 R. P. Singh and P. Setlow, *J. Bacteriol.*, 1979, **137**, 1024.
24 N. Vasantha and E. Freese, *J. Gen. Microbiol.*, 1979, **112**, 329.

sponsible for switching off this enzyme.[25] Extracts of cells at different stages of development were assayed with and without added Mn^{2+}. The enzyme activity in vegetative cells and in germinating spores was stimulated about three-fold by the addition of Mn^{2+}, but the activity in dormant spores and isolated forespores was stimulated greater than 25-fold. Treatment of forespores with Mn^{2+} and the ionophore X-537A had previously been shown to result in the depletion of accumulated 3-PGA.[26] This effect was studied in greater detail.[25] At a constant extracellular concentration of Mn^{2+}, increasing the amount of X-537A increased the rates of depletion of 3-PGA and of accumulation of Mn^{2+}. The depletion of 3-PGA was not due to leakage into the medium. Neither the ionophore alone nor ionophore in combination with Mg^{2+} or Ca^{2+} had any effect. Surprisingly, the forespores accumulated some manganese in the absence of ionophore, but manganese in forespores was very slow to exchange, even in the presence of X-537A or toluene. In contrast, the manganese in vegetative cells and in germinating spores exchanged quite rapidly, so it appears that manganese is tightly bound by some component of the spore. Since uptake of Mn^{2+} precedes accumulation of dipicolinate, some other component is likely to be involved.

Manganese-containing Superoxide Dismutase (MnSOD; EC 1.15.1.1).—(The iron-containing and copper,zinc-containing superoxide dismutases are described elsewhere in this volume.)

The MnSODs from *Escherichia coli* and *Bacillus stearothermophilus* were studied by electrochemical titration[27] with methylviologen as mediator titrant, under anaerobic conditions. The MnSOD of *E. coli* had $E^{\circ\prime} = +0.31$ V, with one electro-active Mn atom per dimer and a complicated dependence on pH above pH 7.0. The MnSOD of *B. stearothermophilus* had $E^{\circ\prime} = +0.26$ V, also with one electro-active Mn atom per dimer, but with a simple pH dependence ($\Delta E/\Delta pH = 50$ mV), indicating the coupling of one H^+ to reduction of the enzyme.

The apoprotein has been prepared from the MnSOD of *E. coli* and reversibly reconstituted with Mn^{2+}.[28] This reconstitution was inhibited by the presence of Zn^{2+}, Fe^{2+}, Ca^{2+}, Co^{2+}, and Ni^{2+}, slightly inhibited by La^{3+} and Pb^{2+}, and not at all by Cr^{3+}, Li^+, Ca^{2+}, Mg^{2+}, and Tl^+. Further studies showed that Fe^{2+}, Zn^{2+}, Ni^{2+}, and Co^{2+} did indeed bind to the protein; iron had a stoicheiometry of four gram atoms per dimer, while the other three metals and manganese had a stoicheiometry of 1.0 gram atoms per dimer. The visible spectrum for the cobalt-substituted enzyme was also reported.

The reliability of the azide[29] and hydrogen peroxide[30] tests for distinguishing between FeSOD and MnSOD has been strengthened by studies with purified enzymes from additional sources.[31] Of the two methods, the selective inactivation of FeSOD by H_2O_2 has proved to be the more popular because of its ability

[25] R. P. Singh and P. Setlow, *J. Bacteriol.*, 1979, **139**, 889.
[26] R. P. Singh and P. Setlow, *Biochem. Biophys. Res. Commun.*, 1978, **82**, 1.
[27] G. D. Lawrence and D. T. Sawyer, *Biochemistry*, 1979, **18**, 3045.
[28] D. E. Ose and I. Fridovich, *Arch. Biochem. Biophys.*, 1979, **194**, 360.
[29] H. P. Misra and I. Fridovich, *Arch. Biochem. Biophys.*, 1978, **189**, 317.
[30] L. Britton, D. P. Malinowski, and I. Fridovich, *J. Bacteriol.*, 1978, **134**, 229.
[31] S. Kanematsu and K. Asada, *Arch. Biochem. Biophys.*, 1979, **195**, 535.

to characterize the individual isozymes that have been revealed by gel electrophoresis. By this method, the presence of MnSOD has been indicated in *Neisseria ovis*,[32] *Klebsiella pneumoniae*,[33] *Enterobacter cloacae*,[33] *Bacillus circulans*,[33] *B. polymyxa*,[33] *B. macerans*,[33] *Euglena gracilis*,[31] *Plectonema boryanum*,[34] *Anabaena variabilis*,[34] and *Anacystis nidulans*.[34] In the last four species, which are algae, the MnSOD was shown to be localized in the thylakoids.[31, 34] This membrane-bound SOD was *not* released by treatments which remove the manganese that is associated with photosystem II, but was released by detergents.[34] This membrane-bound SOD accounted for 12% of the total SOD activity of the cell and probably for only 2% of the manganese in chloroplasts. In contrast, the membrane-bound SOD in tomato leaf chloroplasts was removed by treatments which remove the manganese that is associated with photosystem II.[35]

In a study of the effect of composition of growth media on levels of SOD in *E. coli*,[33] it was found that statically grown cells had only one third the amount of SOD that is in cells that have been grown in vigorously shaken media. Supplementation with Mn^{2+} had only a small effect on enzyme levels in static cultures, but a dramatic effect in shaken cultures. Full induction of the MnSOD and hybrid SOD was achieved with $0.3\mu M$-Mn^{2+}, whereas $0.03\mu M$-Mn^{2+} and up to $300\mu M$-Fe^{2+} had no effect. The induction of MnSOD in *E. coli* by compounds which produce superoxide *in vivo* has also been examined in depth.[36, 37]

In the fungus *Dactylium dendroides*, MnSOD was induced by low levels of copper in the growth medium.[38] Neither growth nor total SOD content was affected by copper concentrations ranging from 10^{-8} to 10^{-3} mol l^{-1}, but, below 10^{-5} mol l^{-1}, the proportion of MnSOD increased as copper supply decreased, until at 10^{-8} mol l^{-1} it represented 87% of the total SOD activity in cell-free extracts. At copper concentrations of 10^{-5} mol l^{-1} or greater, MnSOD represented only 13% of the total SOD, and was apparently confined to mitochondria; as the copper supply decreased, it appeared in the cytoplasm as well. This increased biosynthesis was reflected in a 3.4-fold higher uptake of manganese at 10^{-8} mol l^{-1} than at copper concentrations of 10^{-5} mol l^{-1}, although the uptake of Fe, Mg, Ca, and Zn was invariant. The low levels of Cu,ZnSOD were shown to be due to decreased biosynthesis, as no inactive (apo-)Cu,ZnSOD was detected by immunoprecipitation with anti-Cu,ZnSOD.

Cyanide-resistant SOD (probably MnSOD) has been localized in the mitochondria of spinach (*Spinacia oleracea*),[39] maize (*Zea mays*),[40] and pike (*Esox lucius*).[41] In maize,[40] the mitochondrial SOD was the predominant isozyme in the scutellum of developing kernels, but not in germinating seeds nor in

[32] P. Norrod and S. A. Morse, *Biochem. Biophys. Res. Commun.*, 1979, **90**, 1287.
[33] K. Yano and H. Nishie, *J. Gen. Appl. Microbiol.*, 1978, **24**, 333.
[34] S. Okada, S. Kanematsu, and K. Asada, *FEBS Lett.*, 1979, **103**, 106.
[35] Z. Kaniuga, J. Zabek, and W P. Michalski, *Planta*, 1979, **145**, 145.
[36] H. M. Hassan and I. Fridovich, *Arch. Biochem. Biophys.*, 1979, **196**, 385.
[37] H. M. Hassan and I. Fridovich, *J. Biol. Chem.*, 1979, **254**, 10 846.
[38] A. R. Shatzman and D. J. Kosman, *J. Bacteriol.*, 1979, **137**, 313.
[39] C. Jackson, J. Dench, A. L. Moore, B. Halliwell, C. H. Foyer, and D. O. Hall, *Eur. J. Biochem.*, 1978, **91**, 339.
[40] J. A. Baum and J. G. Scandalios, *Differentiation*, 1979, **13**, 133.
[41] J. A. Healy and M. F. Mulcahy, *Comp. Biochem. Physiol., B*, 1979, **62**, 563.

different parts of the growing or mature plant. In pike,[41] genetic studies revealed two different alleles for the gene that codes for (Mn)SOD, which in heterozygotes resulted in hybridization between the two types of monomers to form five isozymes. Such heterozygosity for tetrameric proteins is rare, and this is the first reported instance of it in SOD. The frogs *Xenopus laevis* and *X. borealis* appeared to have the same isozyme of CN^--resistant SOD in all tissues,[42] but rat and mouse appeared to have different isozymes in brain and liver[43] (CN^--resistant SOD from rat liver has been isolated and shown to contain Mn).[44] Other comparative studies on CN^--resistant SOD from eukaryotes have been presented.[45-47]

Photosystem II.—Photosystem II in chloroplasts has 4—8 gram-atoms of Mn per reaction centre. A chlorophyll-containing protein has been isolated from spinach chloroplasts and analysed for manganese content.[48] Two values are given; one is in the text and one in the legend of a Figure, and they do not agree, possibly due to a typographical error. However, both values give far less than stoicheiometric amounts. The manganese in intact chloroplasts from spinach has been studied by e.p.r. spectroscopy.[49] The membrane-bound manganese was virtually e.p.r.-silent, but rapid temperature shock released Mn^{2+}. The pattern of release of Mn^{2+} after various light regimes indicated a change in oxidation state during photosynthesis. Addition of H_2O_2 released more Mn^{2+}, and the dependence of this reaction on pH suggested that Mn^{IV} is possibly present. After a series of brief flashes, the pattern of release of Mn^{2+} oscillated over a period of four flashes, which is very similar to the pattern of the relationship between flashes of light and the yield of O_2. There was successive oxidation in the first two flashes, but partial reduction in the third. Slightly different results were obtained with chloroplast fragments from lettuce leaves:[50] maximal oxidation occurred after three flashes, not two. Both sets of authors concluded that manganese is involved in the S-states of evolution of oxygen.

Nitrogenase-activating System.—The purple non-sulphur bacteria *Rhodospirillum rubrum* and *Rhodopseudomonas capsulata* were found to have two forms of nitrogenase:[51, 52] these are the A form, which is fully active in cell-free extracts, and the R form, which requires an activating system composed of Mn^{2+}, ATP, and a component which is bound to the chromatophore membrane. The A form is found in nitrogen-starved cells and the R form in cells where fixed nitrogen is still available. Studies with *R. rubrum* showed that it was the Fe protein, and not the MoFe protein, of nitrogenase which was modified.[51] The physio-

[42] E. Schonne, C. R. Jacquemyns, and J. J. Picard, *Arch. Int. Physiol. Biochem.*, 1978, **86**, 958.

[43] B. Lönnerdal, C. L. Keen, and L. S. Hurley, *FEBS Lett.*, 1979, **108**, 51.

[44] M. L. Salin, E. D. Day, Jr., and J. D. Crapo, *Arch. Biochem. Biophys.*, 1978, **187**, 223.

[45] J. N. A. Van Balgooy and E. Roberts, *Comp. Biochem. Physiol., B*, 1979, **62**, 263.

[46] F. Mazeaud, J. Maral, and A. M. Michelson, *Biochem. Biophys. Res. Commun.*, 1979, **86**, 1161.

[47] Y. Kobayashi, S. Okahata, K. Sakano, K. Tanabe, and T. Usui, *FEBS Lett.*, 1979, **98**, 391.

[48] C. H. Foyer and D. O. Hall, *FEBS Lett.*, 1979, **101**, 324.

[49] T. Wydrzynski and K. Sauer, *Biochim. Biophys. Acta*, 1979, **589**, 56.

[50] Y. Siderer and S. Malkin, *FEBS Lett.*, 1979, **104**, 335.

[51] R. P. Carithers, D. C. Yoch, and D. I. Arnon, *J. Bacteriol.*, 1979, **137**, 779.

[52] D. C. Yoch, *J. Bacteriol.*, 1979, **140**, 987.

logical importance of Mn^{2+} was demonstrated [52] by the inability of either species to grow in the absence of Mn^{2+} when N_2 was the sole nitrogen source, but, when either ammonia or glutamate was present, growth was the same whether or not Mn^{2+} was present (Figure 2). Manganese deficiency did not affect the

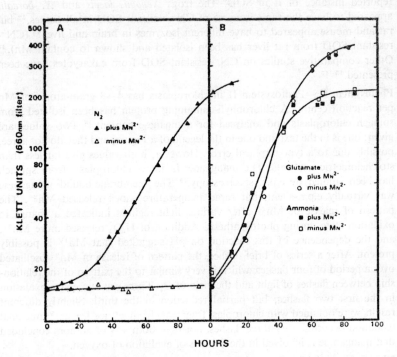

Figure 2 *The effect of* Mn^{2+} *on the rate of growth of* Rhodospirillum rubrum *in static culture with (A) gaseous nitrogen (1 atm) or (B) glutamate or ammonium ion (10 mmol l⁻¹) as the nitrogen source*
(Reproduced by permission from *J. Bacteriol.*, 1979, **140**, 987)

biosynthesis of nitrogenase or of glutamine synthetase, nor the turnover of glutamine synthesis. Studies *in vitro* showed that Fe^{2+} and Co^{2+} could substitute for Mn^{2+} to some degree: the apparent K_m values for Mn^{2+}, Fe^{2+}, and Co^{2+} were 0.19, 0.50, and 1.62 mmol l⁻¹, respectively, and Mg^{2+} alone had absolutely no effect. However, *in vivo*, neither Fe^{2+} nor Co^{2+} could support growth on N_2 in the absence of Mn^{2+}. Chromatophore membranes which contained the activating factor had 1.5 μg of manganese per mg of bacteriochlorophyll, but membranes which had been treated to remove the activating factor contained only one-tenth as much manganese. The membrane-bound Mn^{2+} was e.p.r.-detectable and gave a six-line spectrum similar to that of concanavalin A. This nitrogenase-activating system appears to be unique to the purple non-sulphur bacteria, as it has not been reported elsewhere, and is even lacking in the purple sulphur bacterium *Chromatium vinosum*.[52]

UDP-Galactose : Glycoprotein Galactosyltransferase (EC 2.4.1.38).—This protein has been purified to homogeneity (22 000-fold) from bovine cornea, and compared to bovine milk A-protein.[53] They were found to have very similar molecular weights (42 000—43 000), very similar specificities for metal ions (Mg^{2+} gave only 1 % of the activity of Mn^{2+}, both at 4 mmol l^{-1}), very similar K_m and V_{max} values for substrates and cofactors, and very similar substrate specificities (utilizing any terminal non-reducing N-acetylglucosamine residue as a galactose acceptor). In the presence of α-lactalbumin, the specificity of both enzymes was shifted in favour of glucose as acceptor; *i.e.*, they both became effective lactose synthases (EC 2.4.1.22). The authors concluded that the galacto-syltransferase from cornea and milk A-protein are quite possibly identical gene products, but they assume different roles due to the presence or absence of α-lactalbumin.

Two isozymes, GT-I (which is the normal enzyme) and GT-II (which is the cancer-associated enzyme), have been purified to homogeneity from pooled effusions from human malignancies.[54] Both isozymes had an absolute dependence on Mn^{2+}: after treatment with EDTA, neither Mg^{2+}, Zn^{2+}, Ca^{2+}, nor K^+ produced any stimulation, but activity in both was partially restored ($\sim 25\%$) by Cd^{2+} or Co^{2+}. The optimum concentration of Mn^{2+} was 10 mmol l^{-1} for both, but no K_m value was reported. The two isozymes had identical pH optima and identical K_m values for UDP-galactose and ovalbumin, but GT-II had a ten-fold greater K_m value for asialo-agalacto-fetuin. Isozyme GT-I was readily converted into lactose synthase by α-lactalbumin, but GT-II was only partially converted, producing nearly equal amounts of lactose and N-acetyl-lactosamine. Other differences noted between GT-I and GT-II were molecular weight, sac-charide content, amino-acid composition, and peptide maps; GT-I and GT-II were not interconverted by effusions from either malignant or benign cells.

A membrane-bound galactosyltransferase in plasmalemma-enriched fractions from the transformed mouse fibroblast cell line Balb/c 3T12 has been partially characterized.[55] The apparent K_m value for Mn^{2+} was 3 mmol l^{-1}, but this rather high value may be partially attributable to the presence of 3 mM-AMP in the assay medium to prevent the hydrolysis of UDP-galactose by pyrophos-phatases that are present in the extract. No other bivalent cation was examined. Both N-acetylglucosamine and asialo-agalacto-fetuin could function as galactose acceptors, but, in the presence of α-lactalbumin, synthesis of N-acetyl-lactos-amine was inhibited by over 85 % while synthesis of lactose was stimulated more than 60-fold. Dialdehyde-UDP was found to inactivate this membrane-bound galactosyltransferase and also the galactosyltransferase in foetal calf serum. The possibility of using this compound to probe or label any glycosyl-UDP-utilizing enzyme was proposed.

UDP-Glucose : Galactosylhydroxylysyl-collagen Glucosyltransferase (formerly EC 2.4.1.66).—This enzyme has been highly purified from chick embryos and its metal-binding properties have been studied by enzyme kinetics.[56] Evidence was

[53] J. E. Christner, J. J. Distler, and G. W. Jourdian, *Arch. Biochem. Biophys.*, 1979, **192**, 548.
[54] D. K. Podolsky and M. M. Weiser, *J. Biol. Chem.*, 1979, **254**, 3983.
[55] R. D. Cummings, J. A. Cebula, and S. Roth, *J. Biol. Chem.*, 1979, **254**, 1233.
[56] R. Myllylä, H. Anttinen, and K. I. Kivirikko, *Eur. J. Biochem.*, 1979, **101**, 261.

found for two Mn^{2+}-binding sites (both activating), one with a dissociation constant (K_D) of 3.5 μmol l^{-1} and one with a K_D of 50—70 μmol l^{-1}. Binding of the second manganese decreased the K_m value for UDP-glucose, but did not affect the value for V_{max}. The kinetic data were consistent with an ordered equilibrium binding of Mn^{2+}, UDP-glucose, and collagen at low concentrations of Mn^{2+} and of Mn^{2+}, Mn^{2+}, UDP-glucose, and collagen at high Mn^{2+} concentrations. The Mn^{2+}-activated enzyme was inhibited by Zn^{2+}, Cu^{2+}, Ni^{2+}, and Ca^{2+}. Neither Mg^{2+} nor Ca^{2+} could stimulate the EDTA-treated enzyme, alone or in the presence of Mn^{2+}. Low concentrations of Fe^{2+} $(K_D=5—7$ μmol $l^{-1})$ and Co^{2+} $(K_D=30$ μmol $l^{-1})$ activated the enzyme, but they inhibited it at higher concentrations. The authors concluded that Mn^{2+} and perhaps Fe^{2+} may be the physiological activators of this enzyme.

Phosphoenolpyruvate Carboxykinase (EC 4.1.1.32).—The cytosolic isozyme from bullfrog (*Rana catesbiana*) liver has been purified to homogeneity.[57] When assayed in the direction of phosphoenolpyruvate carboxylation, the enzyme was found to have an absolute requirement for Mn^{2+}. The apparent K_m value for Mn^{2+} was 2 mmol l^{-1}, with maximum activity at 10—30 mmol l^{-1}. No activity at all was found when Mn^{2+} was replaced by 5—30 mM-Mg^{2+}. When the enzyme was assayed in the direction of oxaloacetate decarboxylation, the optimum concentration for Mn^{2+} was reduced to 2 mmol l^{-1} and the apparent K_m value for Mn^{2+} was reduced to 0.4 mmol l^{-1}, even in the presence of 3 mM-ATP: Mg^{2+} also supported this reaction but had a higher optimum concentration (25 mmol l^{-1}). The Mn^{2+}-supported and Mg^{2+}-supported reactions had different pH optima; at their respective optima of concentration and pH, each cation was as effective as the other. No mention was made of ferroactivator.

Arginase (EC 3.5.3.1).—While working out an improved technique for purifying this manganese metalloenzyme from chick kidney,[58] it was found that the addition of Mn^{2+} to a final concentration of 50 mmol l^{-1} stabilized the enzyme during heat treatment, and that the presence of 10 mM-Mn^{2+} stabilized the partially purified enzyme during storage at $-30\,°C$.

2 (Mn^{2+}/Mg^{2+})-Activated Proteins

Glutamine Synthetase (EC 6.3.1.2).—A review article[59] with 29 references summarizes what has been learned about glutamine synthetase (GS) from *E. coli* through physical techniques, with particular emphasis on the n_1 and n_2 metal-binding sites.

In Gram-negative bacteria, this key enzyme in nitrogen metabolism is regulated by covalent modification with adenylyl groups, which convert the fully active Mg^{2+}-specific enzyme into the less active Mn^{2+}-specific form. Two papers have put forward methods for determining the average degree of adenylylation: one describes six enzymatic methods[60] and the other describes a method that

[57] Y. Goto, J. Shimizu, T. Okazaki, and R. Shukuya, *J. Biochem.* (*Tokyo*), 1979, **86**, 71.
[58] H. Kadowaki and M. C. Nesheim, *Int. J. Biochem.*, 1979, **10**, 303.
[59] J. J. Villafranca and M. S. Balakrishnan, *Int. J. Biochem.*, 1979, **10**, 565.
[60] E. R. Stadtman, P. Z. Smyrniotis, J. N. Davis, and M. E. Wittenberger, *Anal. Biochem.*, 1979, **95**, 275.

utilizes discontinuous electrophoresis on sodium dodecylsulphate gel.[61] The latter method was shown to work with GS from *E. coli* and from *Klebsiella pneumoniae*, but the former method was limited to *E. coli*.

The stoicheiometric incorporation of Co^{III} and of Cr^{III} into the n_1 metal-binding sites in un-adenylylated GS from *E. coli* has been reported.[62] Both derivatives were inactive. The Co^{III} derivative was able to bind Mn^{2+} to the n_2 sites, but with a greatly increased dissociation constant ($K_D = 0.67 \pm 0.11$ mmol l^{-1} *vs* a value of $K_D = 0.045$ mmol l^{-1} for the native enzyme). The Cr^{III} derivative was also able to bind Mn^{2+} at the n_2 sites, but no dissociation constant was given. The perturbation of the e.p.r. signal of Mn^{II} by substrates and substrate analogues was investigated and it was found that the transition-state analogue L-methionine-(SR)-sulphoximine, which co-ordinates to the metal in the n_1 site, also perturbs the environment of Mn^{II} in the n_2 site. Comparison of the quaternary enzyme–Cr^{III}–Mn^{II}–ADP complex, which has spin–spin interaction between the two metal centres, with the enzyme–Co^{III}–Mn^{II}–ADP complex, which does not, permitted the calculation of the distance between n_1 and n_2 as 7 ± 2 Å.

Glutamine synthetase from *E. coli* has been subjected to limited proteolysis by subtilisin.[63] Although the nicked protein was catalytically inactive, it appeared to retain both metal-binding sites. Another set of authors also investigated the effects of limited proteolysis on the GS of *E. coli*,[64] but did not consider the effects on the metal-binding sites. Neither set of authors examined the effects of Mg^{2+} or Mn^{2+} on susceptibility to proteolysis.

Glutamine synthetase has been purified to homogeneity from *Azotobacter vinelandii*.[65] Electron microscopy revealed a quaternary structure similar to that of the enzyme from *E. coli*, *i.e.* twelve subunits in two superimposed hexagonal rings. Other similarities to the enzyme of *E. coli* include regulation by adenylylation, the magnitude of the sedimentation coefficient, and the requirements for metal cofactors if the biosynthetic reaction (1) is to be effected. It differs from the enzyme of *E. coli* in the molecular weight of the subunit (56 500 *vs* 50 000) and in the kinetics for the transferase reaction (2). The Mn^{2+}-supported reaction in un-adenylylated enzyme is very low, in marked contrast to *E. coli*.

$$\text{glutamate} + NH_4^+ + ATP \xrightarrow{M^{2+}} \text{glutamine} + ADP + P_i \qquad (1)$$

$$\text{glutamine} + NH_2OH \xrightarrow[P_i \text{ or arsenate}]{M^{2+}, ADP} \gamma\text{-glutamylhydroxamate} + NH_3 \qquad (2)$$

Although blue-green algae have Gram-negative cell walls, no evidence for regulation by adenylylation was found in crude extracts of *Anacystis nidulans*[66] or in the purified enzyme from *Anabaena* sp. CA.[67] In the transferase reaction, GS from both species favoured Mn^{2+}. Under optimal conditions for each cation, the Mg^{2+}-supported reaction was only 49% and 10% of the Mn^{2+}-supported reaction for enzymes of these *Anacystis* and *Anabaena* species, respectively.

[61] R. A. Bender and S. L. Streicher, *J. Bacteriol.*, 1979, **137**, 1000.
[62] M. S. Balakrishnan and J. J. Villafranca, *Biochemistry*, 1979, **18**, 1546.
[63] A. Dautry-Varsat, G. N. Cohen, and E. R. Stadtman, *J. Biol. Chem.*, 1979, **254**, 3124.
[64] M. Lei, U. Aebi, E. G. Heidner, and D. Eisenberg, *J. Biol. Chem.*, 1979, **254**, 3129
[65] J. Siedel and E. Shelton, *Arch. Biochem. Biophys.*, 1979, **192**, 214.
[66] D. Emond, N. Rondeau, and R. J. Cedergren, *Can. J. Biochem.*, 1979, **57**, 843.

However, in the biosynthetic reaction, Mn^{2+} was far more effective than Mn^{2+}.[67] Maximal biosynthetic activity with Mn^{2+} occurred when it was in equimolar amounts with ATP, but maximum activity with Mg^{2+} occurred when it was in molar excess over ATP.

Two separate isozymes have been isolated from etiolated hypocotyls of *Glycine max* (soybean) seedlings.[68] They were similar in many physical and kinetic parameters, but different in thermal stability and K_m values for glutamine in the transferase reaction. Both showed a marked dependence on Mn^{2+} for the transferase reaction; Mg^{2+} was a poor substitute. The relative effectiveness of the two cations in the biosynthetic reaction was not examined. Two separate isozymes have also been isolated from barley (*Hordeum vulgare*);[69] one is localized in the chloroplast and confined to green tissues, and the other is localized in the cytoplasm and found in all tissues. The effects of Mg^{2+} and Mn^{2+} were not examined (all assay media contained Mn^{2+}).

In crude extracts of rat liver, Mn^{2+} was found to be a potent inhibitor of glutamine synthetase.[70] In the presence of 10 mM-Mg^{2+}, Mn^{2+} had an apparent K_i value of 20 μmol l^{-1}. Addition of phosphate, bicarbonate, or carbamoyl phosphate increased the inhibitory effects of Mn^{2+}, but had no effect on enzyme activity in its absence. Since the rate of synthesis of glutamine in perfused rat liver is only one-fortieth of the potential catalytic activity present, the authors have proposed that Mn^{2+} and phosphate may be physiological inhibitors of this enzyme.

Ribulose Bisphosphate Carboxylase/Oxygenase (EC 4.1.1.39).—A comparative study of the effects of Mn^{2+} and Mg^{2+} on the carboxylase and oxygenase activities of the enzymes from *Glycine max* has been published[71] (see Table 1). These results agree almost exactly with previously published results[72] that, at atmospheric oxygen tension and in the presence of 25 mM-HCO_3^-, substitution of Mn^{2+} for Mg^{2+} produced a ten-fold inhibition of carboxylase activity and a 1.5-fold stimulation of oxygenase activity. Coexistence of Mn- and Mg-forms of

Table 1 *Kinetic parameters of ribulose bisphosphate carboxylase/oxygenase from* Glycine max *as a function of activating cation*[a, b]

	Mn^{2+}	Mg^{2+}
K_m (HCO_3^-)/mmol l^{-1}	0.85	2.48
K_m (O_2)[c]	1.7%	37%
V_{max} (carboxylation reaction)/μmol min^{-1}	0.29	1.7
V_{max} (oxygenation reaction)/μmol min^{-1}	0.29	0.61

(a) From *Biochem. J.*, 1979, **183**, 747; (b) oxygenase and carboxylase assays were carried out simultaneously; (c) values refer to the concentration in the gas phase.

[67] G. Stacey, C. Van Baalen, and F. R. Tabita, *Arch. Biochem. Biophys.*, 1979, **194**, 457.

[68] S. Stasiewicz and V. L. Dunham, *Biochem. Biophys. Res. Commun.*, 1979, **87**, 627.

[69] A. F. Mann, P. A. Fentem, and G. R. Stewart, *Biochem. Biophys. Res. Commun.*, 1979, **88**, 515.

[70] S. K. Joseph, N. M. Bradford, and J. D. McGivan, *Biochem. J.*, 1979, **184**, 477.

[71] J. T. Christeller and W. A. Laing, *Biochem. J.*, 1979, **183**, 747.

[72] G. F. Wildner and J. Henkel, *FEBS Lett.*, 1978, **91**, 99.

the enzyme would explain the biphasic response to O_2 tension of the synthesis of glycollate by isolated chloroplasts.[73]

The complementary kinetic study was performed with the enzyme from *Rhodospirillum rubrum*,[74] *i.e.* the K_m and V_{max} values for Mg^{2+} and Mn^{2+} were determined at constant concentrations of HCO_3^- and O_2 (Table 2). The carboxylase reaction showed a biphasic response to Mg^{2+} and Mn^{2+}, hence the two values for K_m and V_{max}. Although Co^{2+} did not support any carboxylase activity, it was a potent inhibitor of the Mg^{2+}-supported reaction, with a K_i value of 0.01 mmol l⁻¹. Neither Zn^{2+}, Ni^{2+}, Cu^{2+}, nor Ca^{2+} supported either enzymic activity. The results are comparable to those obtained with the enzyme from spinach,[72] except that the K_m value for Mg^{2+} for the carboxylase reaction is much lower for the bacterial enzyme.

Table 2 *Kinetic constants for the carboxylation and oxygenation reactions of ribulose bisphosphate carboxylase/oxygenase from* Rhodospirillum rubrum *when different activating cations are used*[a, b]

Cation	Carboxylation		Oxygenation	
	$K_m{}^c$/mmol l⁻¹	V_{max}/μmol min⁻¹	$K_m{}^d$/mmol l⁻¹	V_{max}/μmol min⁻¹
Mg^{2+} [e]	0.15	1.22	4.0	0.54
	0.52	1.55		
Mn^{2+} [e]	0.05	0.22	0.22	0.61
	0.8	0.61		
Co^{2+}	no activity		0.77	0.25

(a) From *Biochemistry*, 1979, **18**, 4453; (b) assays for carboxylase and oxygenase were not performed simultaneously, nor under identical conditions; (c) values refer to total concentration of cations in the presence of 0.8 mM-ribulose bisphosphate, 20 mM-NaHCO₃, and 40—50 mM-MOPS–KOH buffer, at pH 7.8; (d) as for (c), but without NaHCO₃; (e) biphasic response in the carboxylation reaction.

It was found with the *R. rubrum* enzyme that activity was dependent on pre-incubation with bicarbonate and the cation in question.[74] Maximal activity was usually obtained after 5 minutes. Highest carboxylase activity was obtained with 10 mM-dithiothreitol, 4 mM-EDTA, and 10 mM-MgCl₂ (67% higher than with 10 mM-MgCl₂ alone).

β-**Galactosidase (EC 3.2.1.23).**—The binding of a number of bivalent cations to β-galactosidase from *E. coli* was studied by equilibrium dialysis.[75] Neither Zn^{2+}, Ni^{2+}, Fe^{2+}, nor Co^{2+} bound to the enzyme, and the apparent activation of the enzyme by these metals was shown to be due to Mg^{2+} impurities. The dissociation constants for Mn^{2+}, Mg^{2+}, and Ca^{2+} were 0.011, 0.280, and 15 μmol l⁻¹, respectively. Calcium did not activate the enzyme, but did inhibit Mn^{2+}- and Mg^{2+}-supported activities. Competition studies indicated that all three cations bound to the same site. The binding of Mn^{2+} was co-operative (Hill coefficient = 3.4) whereas that of Mg^{2+} was non-co-operative (Hill coefficient =

[73] J. D. Eisenbusch and E. Beck, *FEBS Lett.*, 1973, **31**, 225.

[74] P. D. Robinson, M. N. Martin, and F. R. Tabita, *Biochemistry*, 1979, **18**, 4453.

[75] R. E. Huber, H. Parfett, H. Woulfe-Flanagan, and D. J. Thompson, *Biochemistry*, 1979, **18**, 4090.

1.0). Both forms of the enzyme had similar activities towards allolactose and synthetic substrates, but the Mn-form had a lower reactivity towards lactose for both the hydrolysis and transgalactolysis reactions. The K_m value for lactose was unaffected, but the V_{max} value was halved.

The relationship between the metal ion and the active site in the enzyme from *E. coli* was probed by ^{1}H n.m.r. spectroscopy.[76] Broadening of the proton resonance of the thiomethyl group of the substrate analogue methyl 1-thio-β-D-galactopyranoside occurred on binding to the Mn-form of the enzyme but not to the Mg-form. The average distance between the manganese atom and these protons was calculated to be 9 Å. The resonances of protons at C-6 and C-5 had the same linewidth in both the Mg- and the Mn-form of the enzyme, indicating that they are further away from the metal. Similar results were obtained with another substrate analogue, namely β-D-galactopyranosyltrimethylammonium ion. The authors concluded that Mn^{2+} is too far away to be directly involved in catalysis, but near enough for it to modulate, for example, a conformational change. Further studies indicated that there is a single ionizable group near the active site and that the binding of the Mn^{2+} deprotonates one or more ligands.

β-Galactosidase has been purified to near homogeneity from the yeast *Kluyveromyces lactis*.[77] It was found to be very similar to the enzyme from *E. coli* in molecular weight (135 000) and in the types and relative amounts of reaction products, although there was no immunological cross-reaction between the two proteins. They differed, however, in their requirements for metal: the enzyme from *K. lactis* required univalent cations (40 mM-K^+ or 40—100 mM-Na^+) for full activity, and the Mn-supported activity was three times higher than the Mg-supported activity. Addition of 2-mercaptoethanol to the Mg-form of the enzyme stimulated its activity three-fold but, when it was added in the presence of Mn^{2+}, the Mn^{2+} precipitated.

Adenylate Cyclase (EC 4.6.1.1).—The effects of limited proteolysis[78] and of solubilization by detergents[79] on the plasma-membrane-bound adenylate cyclase from rat liver have been reported. Limited proteolysis by papain, α-chymotrypsin, subtilisin, or thermolysin greatly stimulated basal and activated activities when Mg^{2+} was the metal cofactor, but had little or no effect on activities when Mn^{2+} was the cofactor. The effects were similar whether the cations were added before the protease or after proteolysis had been stopped by the addition of soybean trypsin inhibitor. If Mg^{2+} was added in various amounts in excess over MnATP, there was still no stimulation by proteolysis. The results are compatible with the primary effect of proteolysis being on the regulatory components of the enzyme. Evidence has been presented that adenylate cyclase which has been solubilized by Lubrol PX or deoxycholate from rat liver membranes has been stripped of regulatory components:[79] there was little stimulation by guanylyl-imidodiphosphate [Gpp(NH)p] and little inhibition by adenosine. Interestingly, the enzyme was also changed from a form which utilized either MgATP or MnATP as substrate to a form with a marked preference for MnATP.

[76] R. S. T. Loeffler, M. L. Sinnott, B. D. Sykes, and S. G. Withers, *Biochem. J.*, 1979, **177**, 145.
[77] R. C. Dickson, L. R. Dickson, and J. S. Markin, *J. Bacteriol.*, 1979, **137**, 51.
[78] D. Stengel, M. L. Lacombe, M. C. Billon, and J. Hanoune, *FEBS Lett.*, 1979, **107**, 105.
[79] C. Londos, P. M. Lad, T. B. Nielson, and M. Rodbell, *J. Supramol. Struct.*, 1979, **10**, 31.

Also, submillimolar quantities of Mn^{2+} no longer stimulated activity with MgATP, implicating the loss of the regulatory site for bivalent cations. In all these properties, the solubilized enzyme strongly resembled the soluble adenylate cyclase of rat testis. If the enzyme was pretreated with Gpp(NH)p prior to solubilization, sensitivity to Gpp(NH)p and adenosine was retained, as well as the original requirement for cations. In another study of solubilization of the adenylate cyclase of rat liver plasma membrane,[80] Arrhenius plots of enzyme activity were found to be markedly dependent on the detergent that was used. The authors concluded that the detergents bind tightly to the enzyme and can modulate its activity.

A Mn^{2+}-specific activator site was retained when the membrane-bound adenylate cyclase from bovine brain was solubilized by Triton X-100.[81] Addition of Mn^{2+} greatly stimulated Mg^{2+}-supported activity, while the addition of 250 μM-ethylene glycol bis(aminoethyl ether)-NN'-tetra-acetic acid (EGTA) in the presence of 6—12 mM-Mg^{2+} inhibited activity by 78%. In contrast, 1 mM-EDTA had no effect. After incubation with EGTA for 2 minutes, only Mn^{2+} could appreciably restore activity; either Ca^{2+} or Ca^{2+} plus calmodulin was much less effective. EGTA had no effect on the Stokes radius, as measured by gel filtration. The Mn^{2+}- and Gpp(NH)p-activation sites appeared to be independent in that Gpp(NH)p could still stimulate the EGTA-treated enzyme, but not to the levels that were achieved in the absence of EGTA.

Adenylate cyclase from sarcolemma of the skeletal muscle of the guinea-pig was found to differ in its specificities for metal ions from the enzyme in cardiac sarcolemma from the same organism, and these differences have been studied at length.[82, 83] The optimal Mn^{2+}-supported basal activity for skeletal muscle adenylate cyclase was much less than the optimal Mg^{2+}-supported basal activity (Figure 3).[82] In contrast, adenylate cyclase from cardiac sarcolemma had equal activity at the respective optimum concentrations of Mn^{2+} and Mg^{2+}. Preactivation with Gpp(NH)p and isoproterenol eliminated this discrepancy (Figure 3). A number of possibilities for the lower activity of Mn^{2+} in adenylate cyclase of skeletal muscle were examined and eliminated: Mn^{2+} did not potentiate inhibition by adenosine, did not inhibit the ATP-regenerating system used in the assay, nor did it destabilize the enzyme. There was no evidence for an endogenous inhibitor and no evidence for the presence of multiple isozymes. The apparent K_m value for Mn^{2+} was the same for the enzymes of heart and skeletal muscle, and the enzyme of skeletal muscle had similar K_m values for MnATP and MgATP, but the V_{max} value for MnATP was only one quarter of that for MgATP.[82, 83] Further studies showed that the Mn^{2+}-supported activity had a lower pH optimum than the Mg^{2+}-supported activity, as well as different responses to β-adrenergic agents and Gpp(NH)p.[83] The authors concluded that differences at the allosteric metal-binding site are the basis for the unusual properties of the adenylate cyclase of skeletal muscle.[82]

Adenylate cyclase from the chytridiomycete *Blastocladiella emersonii* was

[80] I. D. Dipple and M. D. Houslay, *Biochem. Biophys. Res. Commun.*, 1979, **90**, 663.
[81] E. J. Neer, *J. Biol. Chem.*, 1979, **254**, 2089.
[82] N. Narayanan, J. W. Wei, and P. V. Sulakhe, *Arch. Biochem. Biophys.*, 1979, **197**, 18.
[83] J. W. Wei, N. Narayanan, and P. V. Sulakhe, *Int. J. Biochem.*, 1979, **10**, 109.

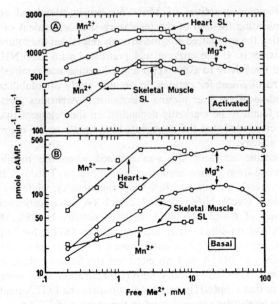

Figure 3 *A comparison of the effects on the activity of adenylate cyclase in sarcolemmal*
membranes (SL) from heart and skeletal muscle of varying the concentrations of
free Mg^{2+} and Mn^{2+}. (A) Enzyme activities for preparations that had been pre-
activated with GMP-P(NH)P and isoproterenol. (B) Basal activities of non-
activated enzyme preparations
(Reproduced by permission from *Arch. Biochem. Biophys.*, 1979, **197**, 18)

found to resemble the enzyme from other lower eukaryotic sources in its high
specificity for Mn^{2+} ion:[84] at concentrations that were 50—100 times greater than
those required for activation by Mn^{2+}, Mg^{2+} supported only very low amounts
of enzyme activity. At a constant concentration of free Mn^{2+}, the utilization of
MnATP was co-operative, with a Hill coefficient of 1.7. An excess of Mn^{2+} at
low concentrations of MnATP activated the enzyme by decreasing the K_m
value for MnATP, and altered the kinetics by decreasing the degree of co-
operativity. Free ATP, *i.e.* not co-ordinated to a bivalent cation, did not appear
to inhibit the enzyme, as is the case in higher organisms.

Incubation of frog erythrocyte membranes with 5—20 mM-Mn^{2+} resulted in a
progressive loss of hormone- and Gpp(NH)p-stimulated activities, while basal
and NaF-stimulated activities were relatively preserved.[85] The Mn^{2+} ion was
shown not to interfere with the binding of hormones to the β-adrenergic receptor,
so it must uncouple activity in another manner.

The requirements of the particulate enzyme from *Mycobacterium smegmatis*
for metal have been investigated.[86] In the absence of bivalent cations, there was
a residual activity which was abolished by EDTA. The residual activity was

[84] S. L. Gomes and J. C. da Costa Maia, *Biochim. Biophys. Acta*, 1979, **567**, 257.
[85] L. E. Limbird, A. R. Hickey, and R. J. Lefkowitz, *J. Biol. Chem.*, 1979, **254**, 2677.
[86] C. H. Lee, *J. Gen. Microbiol.*, 1978, **108**, 325.

stimulated by a number of cations, Mn^{2+} (61-fold), Co^{2+} (43-fold), and Fe^{2+} (26-fold) being the most effective. Maximum activation with Mn^{2+} occurred at concentrations of 8.5—26.6 mmol l^{-1}. In contrast, Mg^{2+} stimulated the activity only 2.8-fold, with an optimal concentration range of 3.0—13.5 mmol l^{-1}. Virtually all other cations that were tested gave more activity than Mg^{2+}, the order of activity being $Mn^{2+} > Co^{2+} > Fe^{2+} > Ca^{2+} > Cd^{2+} > Zn^{2+} > Ni^{2+} > Ba^{2+} > Mg^{2+} > Cu^{2+} > Fe^{3+}$. The effects of fluoride, hormones, and metabolic intermediates on enzyme activity were also reported.

In a study of adenylate cyclase from the brain of the moth *Mamestra configurata* it was noted that 5 mM-$MnCl_2$ greatly increased basal activity as compared to 15 mM-$MgCl_2$, but effectively abolished stimulation by octopamine.[87]

The manganese-dependent adenylate cyclase from yeast (*Saccharomyces cerevisiae*) was inactivated by the arginine-modifying agents butane-2,3-dione, cyclohexane-1,2-dione, and phenylglyoxal.[88] Adenylylimidodiphosphate [App-(NH)p] slowed the rate of inactivation by 34% in the presence of EDTA. In the presence of 4.5 mM-$MgCl_2$, App(NH)p slowed inactivation by 67%; in the presence of 4.5 mM-$MnCl_2$, it retarded it by 96%.

Guanylate Cyclase (EC 4.6.1.2).—A review article has been published on guanylate cyclase and 3',5'-cyclic guanosine monophosphate (cGMP), with an examination of their possible roles *in vivo*.[89]

The soluble (cytoplasmic) form of the enzyme has been purified to homogeneity from rat liver[90] and rat lung.[91] Both isolates had a molecular weight of about 150 000, and both had much higher activity with Mn^{2+} as cation than with Mg^{2+}. The specific activity of the liver enzyme was 276 nmol of cGMP formed per minute per mg of protein in the presence of 4.8 mM-Mn^{2+}, but only 23.8 nmol min^{-1} mg^{-1} in the presence of 4.8 mM-Mg^{2+}. The difference was even greater for the lung enzyme (700 nmol min^{-1} mg^{-1} *vs* 21 nmol min^{-1} mg^{-1}). The lung enzyme was also activated by Ca^{2+} (40 nmol min^{-1} mg^{-1}) and, in the presence of Ca^{2+}, 9.4 μM-Mn^{2+} stimulated to a much greater degree than would be expected.[91] When the concentration of MnGTP was held constant, velocity *versus* [free Mn^{2+}] exhibited negative co-operativity, but the Hill coefficient was not calculated.

The effect of activators on purified soluble guanylate cyclase from rat liver has been investigated.[90, 92] The Mg-form of the enzyme was activated by arachidonate, linoleate, oleate, sodium nitroprusside, and (to a small, but reproducible, amount) by superoxide dismutase, but the resultant activities were less than that of the Mn-form of the enzyme.[90] None of these agents had any effect on the Mn^{2+}-supported activity of the enzyme. Exposure to 165 μl of nitric oxide stimulated the Mg^{2+}-supported activity so that it equalled the levels of Mn^{2+}-supported activity, but had no effect on the Mn-form of the enzyme.[90, 92]

[87] R. P. Bodnaryk, *Can. J. Biochem.*, 1979, **57**, 226.
[88] K. Varimo and J. Londesborough, *FEBS Lett.*, 1979, **106**, 153.
[89] F. Murad, W. P. Arnold, C. K. Mittal, and J. M. Braughler, in 'Advances in Cyclic Nucleotide Research, Vol. 11', ed. P. Greengard and G. A. Robison, Raven Press, New York, 1979, p. 175.
[90] J. M. Braughler, C. K. Mittal, and F. Murad, *Proc. Natl. Acad. Sci. USA*, 1979, **76**, 219.
[91] D. L. Garbers, *J. Biol. Chem.*, 1979, **254**, 240.
[92] J. M. Braughler, C. K. Mittal, and F. Murad, *J. Biol. Chem.*, 1979, **254**, 12 450.

However, smaller amounts of NO (4—8 μl) did stimulate the Mn-form of the enzyme 2.5-fold.[92] During purification, guanylate cyclase lost its responsiveness to NO, but this was restored by the addition of dithiothreitol, methaemoglobin, albumin, or sucrose. These compounds broadened the dose–response curve for the Mn-form, but merely heightened the dose–response curve for the Mg-form of the enzyme.

The membrane-bound guanylate cyclase from rat liver was stimulated six- to eight-fold by limited proteolysis.[93] This treatment did not alter the K_m value for MnGTP nor the degree of co-operativity (Hill coefficient = 1.6), but did increase the value of V_{max}. Limited proteolysis had no effect on the soluble guanylate cyclase from rat liver.

The guanylate cyclase from bovine retinal rod axoneme was stimulated by Mg^{2+} when its concentration exceeded that of GTP, but Mn^{2+} had little effect under the same conditions.[94] Detailed kinetic analysis revealed that the small amount of free Mn^{2+} that is present when Mn^{2+} is in less than a 1:1 ratio with GTP is sufficient to activate the enzyme. The kinetics are consistent with an ordered equilibrium mechanism in which the cation binds first. The dissociation constants that were calculated were 0.11 mmol l^{-1} for Mn^{2+} and 0.95 mmol l^{-1} for Mg^{2+}. The theoretical values of V_{max} were the same for both cations, but in practice they were not achieved, because both Mn^{2+} and Mg^{2+} inhibit the reaction at higher concentrations. The pH dependence of the enzyme suggested the presence of a single ionized group, with a pK_a of around 7.0, which must be ionized for the enzyme to be active. The kinetics of inhibition by Ca^{2+} were also investigated.

In contrast to the usual pattern, the guanylate cyclase from particulate fractions of the protozoan *Tetrahymena pyriformis*[95] was more active in the presence of Mg^{2+} than Mn^{2+}. In the presence of 1mM-GTP, the activity of the enzyme at the optimal concentration of Mg^{2+} (3 mmol l^{-1}) was twice that at the optimal concentration of Mn^{2+} (0.5 mmol l^{-1}). However, at a constant concentration of free bivalent cation, the K_m value for MnGTP (20 μmol l^{-1}) was lower than the K_m value for MgGTP (50 μmol l^{-1}). Evidence was presented for the activation of this particular guanylate cyclase by calmodulin.

Cytidylate Cyclase.—The report of a Mn^{2+}-stimulated cytidylate cyclase activity in mouse liver homogenates[96] has been challenged. Gaion and Krishna[97] have repeated the work and found that the product that was previously identified as cyclic CMP was mainly, if not entirely, 5′-CMP and CDP. They found no evidence for the presence of cCMP or for cytidylate cyclase.

Phosphoprotein Phosphatase (EC 3.1.3.16).—Quite a few different isozymes have been identified in extracts of different tissues, some of these being independent of a metal ion, some (Mn^{2+}/Mg^{2+})-activated, and some Mn^{2+}-specific. Of the two major isozymes in canine heart, the activity of phosphorylase *a* phosphatase

[93] M.-L. Lacombe and J. Hanoune, *J. Biol. Chem.*, 1979, **254**, 3697.
[94] D. Fleischman and M. Denisevich, *Biochemistry*, 1979, **18**, 5060.
[95] K. Nakazawa, H. Shimonaka, S. Nagao, S. Kudo, and Y. Nozawa, *J. Biochem. (Tokyo)*, 1979, **86**, 321.
[96] S. Y. Cech and L. J. Ignarro, *Biochem. Biophys. Res. Commun.*, 1978 **80**, 119.
[97] R. M. Gaion and G. Krishna, *Biochem. Biophys. Res. Commun.*, 1979, **86**, 105.

is independent of any cation, but the glycogen synthase D phosphatase is activated equally by Mn^{2+} and by Mn^{2+} (no details given).[98] A partially purified phosphoprotein phosphatase from the glycogen particles in porcine polymorphonuclear leucocytes [99] is a rather unusual (Mn^{2+}/Mg^{2+})-enzyme in that Mg^{2+} is more effective at lower concentrations than Mn^{2+}. Another phosphoprotein phosphatase from rat liver nuclear membranes has no absolute dependence on bivalent cations but is equally stimulated by 1 mM-Mn^{2+} and 1 mM-Mg^{2+}.[100] Finally, two isozymes from rabbit reticulocytes, which are quite proficient at dephosphorylating some of the factors involved in protein synthesis, have been reported to be absolutely dependent on Mn^{2+}, although no details are given.[101]

A rather interesting account has been given of the co-purification of alkaline phosphatase activity with the major phosphorylase *a* phosphatase isozyme from bovine adrenal cortex.[102] The protein was apparently homogeneous. The phosphorylase phosphatase activity was inhibited by concentrations of 1 mmol l^{-1} of Mn^{2+}, Mg^{2+}, Co^{2+}, and Ca^{2+}, but the alkaline phosphatase activity had an absolute dependence on bivalent cations, with K_m values of 5.5 μmol l^{-1} for Mn^{2+} and 25 mmol l^{-1} for Mg^{2+}, although the value of V_{max} for Mg^{2+} was twice that for Mn^{2+}, The alkaline phosphatase reaction also required dithiothreitol. Other cations had little or no effect. The phosphorylase phosphatase, but not the alkaline phosphatase, was inactivated by incubation with pyrophosphoryl compounds. After desalting on Sephadex G-25, only Co^{2+} and Mn^{2+} restored the activity, the Co^{2+} ion being three times more effective at one-tenth the concentration of Mn^{2+}. The inactivation reaction was inhibited by Mg^{2+}, Mn^{2+}, and Ca^{2+}.

Carbamoyl Phosphate Synthetase (EC 2.7.2.9).—This enzyme catalyses the reaction:

$$2\ MgATP + HCO_3^- + \text{L-glutamine} \xrightarrow{M^{2+}} \begin{array}{c} 2\ MgADP + P_i + \text{L-glutamate} \\ + \\ \text{carbamoyl phosphate} \end{array}$$

Kinetic studies with the enzyme from *E. coli*[103] indicated an equilibrium ordered mechanism, with Mg^{2+} binding before the MgATP, demonstrating that the bivalent cation is an absolute requirement for enzyme activity and not just an allosteric effector. The Mn^{2+} ion also supported enzymatic activity, and, at the optimal concentration of Mn^{2+} (0.2 mmol l^{-1}, in the presence of 0.1 mM-ATP), enzyme activity was 50% greater than it was when the enzyme was activated at the optimal concentration of Mg^{2+} (6—10 mmol l^{-1}). The dissociation constant of a single high-affinity binding site for Mn^{2+} was determined to be 35

---

[98] J. F. Binstock and H.-C. Li, *Biochem. Biophys. Res. Commun.*, 1979, **87**, 1226.

[99] F. Pegueroles and R. Cusso, *Int. J. Biochem.*, 1979, **10**, 989.

[100] R. C. Steer, M. J. Wilson, and K. Ahmed, *Biochem. Biophys. Res. Commun.*, 1979, **89**, 1082.

[101] M. Mumby and J. A. Traugh, *Biochemistry*, 1979, **18**, 4548.

[102] H. C. Li, *Eur. J. Biochem*, 1979, **102**, 363.

[103] F. M. Raushel, C. J. Rawding, P. M. Anderson, and J. J. Villafranca, *Biochemistry*, 1979, **18**, 5562.

μmol l^{-1} by e.p.r. spectroscopy. Competition for Mn^{2+} gave a K_D of 5.2 mmol l^{-1} for Mg^{2+}, in reasonable agreement with the kinetically determined value for K_D of 4.2 mmol l^{-1}. Other metals also bound to the high-affinity site; in decreasing order of K_D values, they were $Ca^{2+} > Mg^{2+} > Ni^{2+} > Co^{2+} > Zn^{2+} > Cd^{2+} > Gd^{3+} > Mn^{2+}$.

Three sulphydryl groups in the enzyme of *E. coli* were spin-labelled,[103] and all three nitroxide groups were more than 20 Å away from the Mn^{2+}- and MnATP-binding sites. Subunit–subunit interaction was demonstrated in that the mobility of the spin-label on the small subunit was modified by ATP, which binds to the large subunit. Further kinetic studies have also been reported.[104]

Malic Enzyme (NAD$^+$-Linked) (EC 1.1.1.38).—Evidence has been presented that Mn^{2+} and Mg^{2+} stabilize two different forms of the enzyme from *E. coli*,[105] which differ from each other in their response to substrates and allosteric effectors. The Mn-form of the enzyme was only 1.3 times more active than the Mg-form, but the kinetics of utilization of substrate were markedly different. In the presence of Mg^{2+}, the velocity *versus* [malate] curve displayed both negative and positive co-operativity, the velocity *versus* [free M^{2+}] curve exhibited positive co-operativity, and the velocity *versus* [NAD$^+$] curve was a simple hyperbola. However, in the presence of Mn^{2+}, all three curves were simple hyperbolas. The Mg-form of the enzyme was more sensitive to allosteric inhibition by coenzyme A and to activation by aspartate, while the Mn-form was more sensitive to inhibition by ATP. These results are similar to results obtained with NAD$^+$-linked isocitrate dehydrogenase,[106] thus giving more weight to the hypothesis that the $Mg^{2+}:Mn^{2+}$ ratio may be yet another level of control of cellular metabolism.

In studies on the inactivation of malic enzymes of *E. coli* by *N*-ethylmaleimide, it was found that Mn^{2+} offered partial protection to the NAD$^+$-linked enzyme,[107] but accelerated the rate of inactivation of the NADP-linked enzyme (EC 1.1.1.40).[108]

The Anthranilate Synthase–Phosphoribosyltransferase Enzyme Complex (EC 4.1.3.27 · EC 2.4.2.18).—This enzyme complex from *Salmonella typhimurium* catalyses the first two steps in the biosynthesis of tryptophan. It is a tetramer, with two subunits of each enzyme. The anthranilate synthase moiety catalyses the reaction:

$$\text{chorismate} + NH_3 \xrightarrow{M^{2+}} \text{anthranilate} + \text{pyruvate} + H^+$$

The phosphoribosyltransferase moiety catalyses two reactions: the *N*-terminal portion cleaves glutamine to glutamate (to provide ammonia for anthranilate synthase) and the *C*-terminal portion catalyses the reaction:

$$\begin{array}{c} \text{anthranilate} \\ + \\ \text{5-phosphoribosyl-1-pyrophosphate} \end{array} \xrightarrow{M^{2+}} N\text{-(5$'$-phosphoribosyl)anthranilate} + PP_i$$

[104] F. M. Raushel and J. J. Villafranca, *Biochemistry*, 1979, **18**, 3424.
[105] J. A. Milne and R. A. Cook, *Biochemistry*, 1979, **18**, 3604.
[106] D. G. Barratt and R. A. Cook, *Biochemistry*, 1978, **17**, 1561.
[107] M. Yamaguchi, *J. Biochem. (Tokyo)*, 1979, **86**, 325.
[108] M. Iwakura, M. Tokushige, and H. Katsuki, *J. Biochem. (Tokyo)*, 1979, **86**, 1239.

The differences in K_m values for metal ions between the complexed and uncomplexed enzymes have been investigated,[109] and the results are summarized in Table 3. Anthranilate synthase displayed biphasic kinetics when in isolation, but simple hyperbolic kinetics in the complex. Phosphoribosyltransferase had the same kinetics and K_m values whether it was complexed or dissociated, but was far more susceptible to feedback inhibition by tryptophan when complexed.

Table 3 *Values of K_m for metal ions for activities of the uncomplexed subunits and the native anthranilate synthase–phosphoribosyltransferase enzyme complex from* Salmonella typhimurium[a, b]

	Apparent K_m values[b]/μmol l^{-1}		
	Mg^{2+}	Mn^{2+}	Co^{2+}
Enzyme subunit			
Uncomplexed anthranilate synthase[c]	480	1.9	14
	2070	36	—[d]
Uncomplexed phosphoribosyltransferase	126	16	14
Enzyme complex			
Ammonia-linked anthranilate synthase activity	100	1.5	9.1
Phosphoribosyltransferase activity	133	16	15

(a) From *Arch. Biochem. Biophys.*, 1979, **193**, 242; (b) all values have been calculated on the basis of the total concentration of cations; (c) the kinetics are biphasic; (d) the value could not be determined, owing to inhibition.

The binding of Mn^{2+} to the enzyme complex has been studied by e.p.r. spectroscopy.[110] The results indicate the presence of 1 to 2 sites with a K_D of 3—5 μmol l^{-1} and 5 to 6 sites with a K_D of 40—70 μmol l^{-1}. The apparent K_m value for Mn^{2+} in the glutamine-linked anthranilate synthase reaction decreased as the concentration of chorismate increased. Extrapolation to zero concentration of chorismate gave an activator constant for Mn^{2+} of 9 ± 3 μmol l^{-1}. A similar dependence of K_m (for Mn^{2+}) on the concentration of anthranilate gave an activator constant of 4 ± 2 μmol l^{-1} for Mn^{2+} in the phosphoribosyltransferase reaction. Both activator constants are in the range for the high-affinity sites. The uncomplexed anthranilate synthase had a much lower affinity for Mn^{2+} ($K_D = 300 \pm 100$ μmol l^{-1}), but a comparable activator constant (4 ± 1 μmol l^{-1}), which suggests that the binding of substrate is a prerequisite for binding of Mn^{2+} in the dissociated enzyme.

The effect of substrates and tryptophan on the proton relaxation rate of the Mn^{2+}–enzyme complex was also investigated.[110] Chorismate dramatically decreased the enhancement of p.r.r. that is attributable to Mn^{2+} in the high-affinity site, but had no effect on the Mn^{2+} in the low-affinity sites. Tryptophan markedly alleviated the effects of chorismate, but had only a little effect on its own. E.p.r. spectroscopy showed that Mn^{2+} was not released by either treatment. Glutamine, pyruvate, and anthranilate had no effect on enhancement of the

[109] P. D. Robison and H. R. Levy, *Arch. Biochem. Biophys.*, 1979, **193**, 242.
[110] P. D. Robison, T. Nowak, and H. R. Levy, *Arch. Biochem. Biophys.*, 1979, **193**, 252.

12

p.r.r., but 5-phosphoribosyl-1-pyrophosphate also decreased the enhancement of p.r.r. Possible interpretations of the data are discussed by the authors.

Pyridoxal 5′-phosphate inhibited the phosphoribosyltransferase reaction, but not the glutamine-dependent anthranilate synthase reaction.[111] Neither Mg^{2+} nor 5-phosphoribosyl-1-pyrophosphate alone had any protective effect, but the two in combination did.

Orotate Phosphoribosyltransferase (EC 2.4.2.10).—The binding of Mn^{2+} to the purified enzyme from yeast was studied by e.p.r. and p.r.r. spectroscopy.[112] This enzyme, which has an absolute requirement for bivalent cations, was found to bind four Mn^{2+} ions in a co-operative fashion, with an effective K_m value of 50 μmol l^{-1}. The kinetics of the enzyme reactions were complex (biphasic with respect to Mg^{2+} and Mn^{2+}), but they fit a scheme where both metal-free and metal-containing enzyme can catalyse the reaction, but at different rates.

Isocitrate Dehydrogenase (NADP-Dependent) (EC 1.1.1.42).—The production of citrate by mycelia of *Aspergillus niger* that had grown on sucrose was found to depend on the concentration of Mn^{2+} in the growth medium.[113] The lowest concentration of Mn^{2+} that was tested (1 p.p.m.) produced the greatest amount of citrate, but growth was optimal at 10 p.p.m. The concentration of manganese in the harvested mycelia was comparable to the concentration of manganese in the growth medium at all concentrations tested. In an effort to understand the biochemical basis of this phenomenon, the effects of Mn^{2+} and Mg^{2+} on mitochondrial isocitrate dehydrogenase were compared. The double-reciprocal plot of velocity *versus* the total Mn^{2+} concentration was linear, giving an apparent K_m value of 3.3 mmol l^{-1} in the presence of 0.7 mM-isocitrate, but the double-reciprocal plot of velocity *versus* total Mg^{2+} was not linear. The K_m value for Mg^{2+} was not estimable, but was probably greater than that for Mn^{2+}. At isocitrate concentrations less than 0.5 mmol l^{-1}, citrate totally inhibited the Mn^{2+}-supported activity. The authors concluded that isocitrate dehydrogenase is the locus of the effect of Mn^{2+} on accumulation of citrate in the growth medium.

$(Na^+ + K^+)$-ATPase (EC 3.6.1.3).—The binding of Mn^{2+} to the purified enzyme from sheep kidney was studied by e.p.r. spectroscopy.[114] A single high-affinity site was found ($K_D = 0.21$ μmol l^{-1}) and 24 ± 3 low-affinity sites ($K_D = 185$ μmol l^{-1}). The native enzyme contains 230 ± 20 moles of phospholipid per mole of enzyme, and the removal of 50% of these lipids resulted in the loss of the low-affinity sites. The removal of these lipids also altered the spectrum of the Mn^{2+} that is in the high-affinity site and the effects of nucleotides on that spectrum. The authors concluded that delipidation, which inactivates the enzyme, produces a conformational change in the vicinity of the active site. The effects of nucleotides on the e.p.r. spectrum indicated that ATP, and not

111 T. H. Grove and H. R. Levy, *Biochem. Biophys. Res. Commun.*, 1979, **86**, 387.
112 J. Victor, A. Leo-Mensah, and D. L. Sloan, *Biochemistry*, 1979, **18**, 3597.
113 I. Bowes and M. Mattey, *FEMS Microbiol. Lett.*, 1979, **6**, 219.
114 S. E. O'Connor and C. M. Grisham, *Biochemistry*, 1979, **18**, 2315.

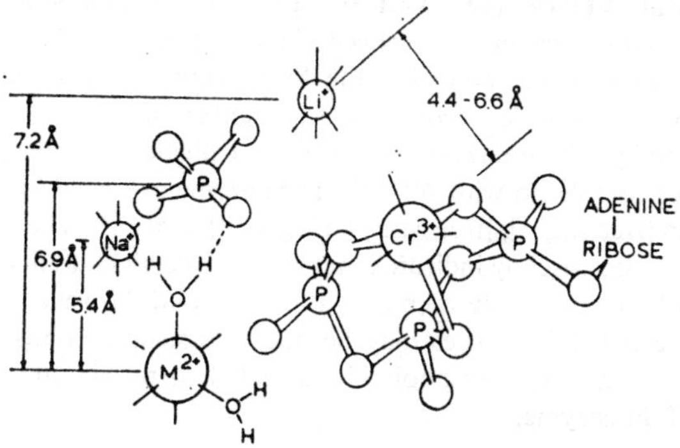

Figure 4 *The structure of the active site of* $(Na^+ + K^+)$-*ATPase, as determined by* ^1H, $^{205}Tl^+$, ^{31}P, *and* $^7Li^+$ *n.m.r.,* Mn^{2+} *e.p.r., and kinetic studies* Reproduced by permission from *Biochemistry*, 1979, **18**, 2315)

MnATP, is the substrate. The results of this study have been combined with the results of previous studies to propose a structure of the active site (Figure 4).

Citrate Lyase (EC 4.1.3.6).—This enzyme has been purified to homogeneity from *Klebsiella aerogenes*.[115] In the presence of Mg^{2+} or Mn^{2+} it catalyses the hydrolysis of citrate, but in the presence of EDTA it catalyses its synthesis. The Mn-, Mg-, and apo-forms of this protein all had the same sedimentation coefficient. The binding of Mn^{2+} was studied by equilibrium dialysis. There were eighteen binding sites per molecule, and binding was co-operative (Hill coefficient = 2.27). The microscopic dissociation constant for the last three sites was 45 μmol l^{-1}. The authors extrapolated these results back to the situation with Mg^{2+} as cofactor, apparently without being aware that binding of Mn^{2+} can have markedly different kinetics from binding of Mg^{2+}.[75, 113]

Protein Kinase (cAMP-Dependent) (EC 2.7.1.37).—The binding of Mn^{2+} to the purified catalytic subunit from bovine heart was studied by e.p.r. and p.r.r. spectroscopy.[116] In the absence of nucleotides, the enzyme had a rather low affinity for Mn^{2+} ($K_D \geqslant 1$ mmol l^{-1}). However, in the presence of nucleotide or of nucleotide plus heptapeptide substrate, two binding sites appeared: the high-affinity site ($K_D = 6$—10 μmol l^{-1}) was provided by the nucleotide, whereas that of lower affinity ($K_D = 50$—60 μmol l^{-1}) was provided by the protein. Kinetic studies revealed that low concentrations of Mn^{2+} activated the enzyme with a K_m value of 3 μmol l^{-1}, but higher concentrations inhibited with a K_i value of 29 μmol l^{-1}. Similar results were obtained with Mg^{2+}, but the K_m and K_i values were in the millimolar range. When Mn^{2+} was bound to the inhibitory site, the binding affinity of the enzyme for nucleotides was increased by two orders of magnitude, but the reaction rate was drastically reduced.

[115] H. Sivaraman and C. Sivaraman, *FEBS Lett.*, 1979, **105**, 267

[116] R. N. Armstrong, H. Kondo, J. Granot, E. T. Kaiser, and A. S. Mildvan, *Biochemistry*, 1979, **18**, 1230.

Mg²⁺-Dependent ATPase (EC 3.6.1.3).—This membrane-bound enzyme has been purified to homogeneity from sheep kidney medulla.[117] At a concentration of bivalent cations of 0.5 mmol l⁻¹, the Mn^{2+}-supported enzyme activity was 77% of the Mg^{2+}-supported activity, but a number of other cations also supported the activity of the enzyme; it required a bivalent cation in addition to that which was complexed with ATP.[118] The enzyme exhibited biphasic kinetics with respect to [free Mg^{2+}], [free Mn^{2+}], [MgATP], and [MnATP]. Results from e.p.r. and p.r.r. spectroscopy indicated that there is one high-affinity site for Mn^{2+} ($K_D = 2$ μmol l⁻¹) per moiety of mol. wt 469 000 and 34 weak sites ($K_D = 550$ μmol l⁻¹). The low-affinity sites may be due to associated phospholipids, but in no other system have phospholipid-associated metal-binding sites affected the kinetics of an enzyme.

RNA and DNA Polymerases (EC 2.7.7.6 and EC 2.7.7.7, respectively).—A review article (with 123 references) has been published on the role of metal ions (Zn^{2+}, Mg^{2+}, and Mn^{2+}) in RNA and DNA polymerases.[119]

RNA polymerase II has been purified to apparent homogeneity from the edible mushroom *Agaricus bisporus*.[120] In contrast to most nucleic acid polymerases, Mg^{2+} did not support any enzyme activity at all, and, in fact, it appeared to be antagonistic to Mn^{2+}-supported activity. The enzyme was absolutely dependent on Mn^{2+}, with optimal activity at a concentration of 2.5 mmol l⁻¹. Another unique property of the enzyme is its 650-fold greater resistance to α-amanitin than mammalian RNA polymerases II.

RNA polymerase A from *Saccharomyces cerevisiae*[121] and RNA polymerase I from wheat germ[122] were found to be absolutely dependent on Mn^{2+} for the initiation step of synthesis of RNA, but Mn^{2+} and Mg^{2+} were equally effective in the elongation step.

RNA polymerases I, II, and III from wheat germ were found to have an absolute requirement for bivalent cations.[122] Optimal activity was obtained at concentrations of Mn^{2+} of 1.0—1.5 mmol l⁻¹ and of Mg^{2+} of 5—10 mmol l⁻¹. The relative activities for the two cations were the same for polymerases II and III, regardless of the DNA template used, but the relative activities for polymerase I depended on the template: with DNA of cauliflower mosaic virus, the Mn^{2+}-supported activity was 30 times greater than the Mg^{2+}-supported activity, but with the DNA of calf thymus it was only 2.2 times greater.

The rate of error in replication of $\phi\chi174$ DNA by DNA polymerase I from *E. coli* was affected by the choice and concentration of the bivalent cation cofactor.[123] The lowest rate of error (1 in 22 400) was achieved at 1 mM-Mg^{2+}. As the concentration increased, so did the rate of DNA synthesis and the rate of incorporation of errors. The error rate continued to increase past the optimum concentration of Mg^{2+} (7.5 mmol l⁻¹) and reached its highest value at 15 mM-

[117] M. L. Gantzer and C. M. Grisham, *Arch. Biochem. Biophys.*, 1979, **198**, 263.
[118] M. L. Gantzer and C. M Grisham, *Arch. Biochem Biophys.*, 1979, **198**, 268.
[119] A. S. Mildvan and L. A. Loeb, *CRC Crit. Rev. Biochem.*, 1979, **6**, 219.
[120] A. C. Vaisius and P. A. Horgen, *Biochemistry*, 1979, **18**, 795.
[121] C. S. Cooper and R. V. Quincey, *Biochem. J.*, 1979, **181**, 301.
[122] M. Teissere, R. Durand, J. Ricard, R. Cooke, and P. Penon, *Biochem. Biophys. Res. Commun.*, 1979, **89**, 526.
[123] T. A. Kunkel and L. A. Loeb, *J. Biol. Chem.*, 1979, **254**, 5718.

Mg^{2+} (1 in 2270). In contrast, there was little variation in error rate as the concentration of Mn^{2+} was varied from 0.25 mmol l^{-1} (1 in 2060), through its optimum concentration (1.0 mmol l^{-1}), to 1.75 mmol l^{-1} (1 in 2680).

DNA polymerases A and C from wheat germ were inhibited by Mn^{2+} when natural DNA served as the template, but were activated by it when synthetic DNA was the template.[124] DNA polymerase B was activated by Mn^{2+} regardless of template, but Mn^{2+} did alter its relative activity with different templates as compared to Mg^{2+}.

An endonuclease that is specific for double-stranded DNA was found to co-purify with the reverse transcriptase of avian myeloblastosis virus.[125] It had an absolute requirement for Mn^{2+} and had optimal activity at 0.5 mmol l^{-1} of the cation. The authors concluded that this endonuclease activity was an intrinsic property of the reverse transcriptase.

The relative reactivities with Mn^{2+} or Mg^{2+} of a number of polymerases from other sources[126-128] are given in Table 4.

Miscellaneous.—Quite a number of papers that were published in 1979 compared the Mg^{2+}- and Mg^{2+}-supported activities of (Mn^{2+}/Mg^{2+})-activated enzymes.[129-144] The results are summarized in Table 4.

Gluconolactonase (EC 3.1.1.17) has been purified to homogeneity from bovine liver.[145] Activity was lost during purification, but was fully restored by the addition of Mn^{2+}, and to a lesser extent by Mg^{2+}. No numerical data were given other than the optimal concentration for Mn^{2+}, which was 1.3 mmol l^{-1}.

Unlike the enzyme from *Pseudomonas* sp. ATCC 14676 (see Vol. 2, p. 260), **guanidinoacetate amidinohydrolase** (EC 3.5.3.2) from *Flavobacterium* sp. GE-1 contains zinc as the prosthetic metal ion instead of manganese.[146]

[124] M. Castroviejo, D. Tharaud, L. Tarrago-Litvak, and S. Litvak, *Biochem. J.*, 1979, **181**, 183.
[125] K. P. Samuel, T. S. Papas, and J. G. Chirikjian, *Proc. Natl. Acad. Sci. USA*, 1979, **76**, 2659.
[126] B. Christian and J. Begueret, *J. Biol. Chem.*, 1979, **254**, 11 566.
[127] K. Tanabe, E. W. Bohn, and S. H. Wilson, *Biochemistry*, 1979, **18**, 3401.
[128] G. R. Banks, J. A. Boezi, and I. R. Lehman, *J. Biol. Chem.*, 1979, **254**, 9886.
[129] Y. Shibano, T. Narita, and T. Komano, *Agric. Biol. Chem.*, 1979, **43**, 1055.
[130] Y. Shibano and T. Komano, *Agric. Biol. Chem.*, 1979, **43**, 1117.
[131] K. Matsumoto, T. Ando, H. Saito, and Y. Ikeda, *J. Biochem. (Tokyo)*, 1979, **86**, 627.
[132] Y. Sawai, S. Uchida, J. Saito, N. Sugano, and K. Tsukada, *J. Biochem. (Tokyo)*, 1979, **85**, 1301.
[133] C. Niedergang, H. Okazaki, and P. Mandel, *Eur. J. Biochem.*, 1979, **102**, 43.
[134] C. L. Villemez and P. L. Carlo, *J. Biol. Chem.*, 1979, **254**, 4814.
[135] P. L. Carlo and C. L. Villemez, *Arch. Biochem. Biophys.*, 1979, **198**, 117.
[136] F. Yoshizaki and K. Imahori, *Agric. Biol. Chem.*, 1979, **43**, 527.
[137] P. S. Agutter, J. B. Cockrill, J. E. Lavine, B. McCaldin, and R. B. Sim, *Biochem. J.*, 1979, **181**, 647.
[138] K. Iwai, K. Shibata, H. Taguchi, and T. Itakura, *Agric. Biol. Chem.*, 1979, **43**, 345.
[139] K. Iwai, K. Shibata, and H. Taguchi, *Agric. Biol. Chem.*, 1979, **43**, 351.
[140] A. G. Tomaselli and L. H. Noda, *Eur. J. Biochem.*, 1979, **93**, 263.
[141] S. Omura, H. Ikeda, and C. Kitao, *J. Biochem. (Tokyo)*, 1979, **86**, 1753.
[142] A. Guranowski, *Arch. Biochem. Biophys.*, 1979, **196**, 220.
[143] F. J. Green and R. A. Lewis, *Biochem. J.*, 1979, **183**, 547.
[144] R. Croteau and F. Karp, *Arch. Biochem. Biophys.*, 1979, **198**, 512.
[145] G. D. Bailey, B. D. Roberts, C. M. Buess, and W. R. Carper, *Arch. Biochem. Biophys.*, 1979, **192**, 482.
[146] T. Yorifuji, N. Komaki, K. Oketani, and E. Entani, *Agric. Biol. Chem.*, 1979, **43**, 55.

Table 4 Comparison of Mg^{2+}- and Mn^{2+}-supported activities for a number of enzymes

Enzyme	Source	$[Mn^{2+}]$/mmol l^{-1}	$[Mg^{2+}]$/mmol l^{-1}	Activity ratio[a]	Ref.
RNA Polymerase I	Podospora anserina	3*	5—10*	600	126
RNA Polymerase II	Podospora anserina	4*	5—10*	300	126
RNA Polymerase III	Podospora anserina	4*	5—10*	250	126
DNA Polymerase β	Mouse myeloma	0.5*	2*	<100	127
DNA Polymerase	Drosophila melanogaster embryos	0.1*	12*	14	128
Deoxyribonuclease A	Bacillus subtilis	0.5—3.0*	2—5*	100	129
Deoxyribonuclease B	Bacillus subtilis	4*	5*	118	130
Deoxyribonuclease MII	Bacillus subtilis	5—20*	5—20*	130	131
Ribonuclease H	Carrot cell culture	0.2*	10—15*	230	132
Poly(ADP-ribose) polymerase	Calf thymus	2*	8*	110	133
UDP-glucose:dolichyl phosphate glucosyltransferase (EC 2.4.1.-)	Acanthamoeba castellani	1	8*	86	134
GDP-mannose:dolichyl phosphate mannosyltransferase (EC 2.4.1.-)	Acanthamoeba castellani	5—8*	1	180	135
Pyruvate kinase (EC 2.7.1.40)	Thermus thermophilus	1	10*	160 ⎫	136
		5*	1	700 ⎭	
		1	10*	80 ⎫	
		5*	1	2500 ⎭	
ATPase (EC 3.6.1.3)	Nuclear envelope, rat liver	1	1	70	137
ATPase (EC 3.6.1.3)	Nuclear envelope, pig liver	1	1	74	137
ATPase (EC 3.6.1.3)	Nuclear envelope, SV40-transformed mouse embryo 3T3 cells	1	1	60	137
Quinolinate phosphoribosyl transferase (EC 2.4.2.19)	Alcaligenes eutrophus ssp. quinolinicus	2*	1.5*	80	138
Quinolinate phosphoribosyltransferase (EC 2.4.2.19)	Pig liver	0.3*	2*	100	139
GTP-AMP Phosphotransferase	Beef heart mitochondria	5	5	80[b], 36[c]	140
Spiramycin I 3-hydroxyl acylase	Streptomyces ambofaciens	1	1	92	141
Adenosine kinase (EC 2.7.1.20)	Lupinus luteus	0.5*	1.0*	115	142
Deoxyguanosine kinase	Skin of newborn pig	10	10	92	143
Bornyl pyrophosphate synthetase	Salvia officinalis	1*	10*	10	144

* Optimum concentration for enzyme activity.

(a) Activity ratio is defined as (Mn^{2+}-supported activity/Mg^{2+}-supported activity) × 100; (b) for ADP synthesis; (c) for AMP synthesis.

Equilibrium dialysis studies of the binding of Mn^{2+} to **phosphofructokinase** (EC 2.7.1.11) from yeast revealed 3.0 independent and equivalent binding sites per subunit, with a dissociation constant of 2.26 mmol l^{-1}.[147] No mention was made of ionic strength or buffer concentration.

As a warning to the pitfalls in drawing conclusions from activation studies, purified **creatinine amidohydrolase** (EC 3.5.2.–) from *Pseudomonas putida* contains one zinc atom per subunit.[148] However, Mn^{2+} was six times more effective in reconstituting the apoenzyme than was Zn^{2+}, and Mg^{2+} was also more effective than Zn^{2+}. The discussion reviews other zinc-containing proteins which show similar behaviour.

[147] W. H. Peters, K. Nissler, W. Schellenberger, and E. Hofmann, *Biochem. Biophys. Res. Commun.*, 1979, **90**, 561.
[148] K. Rikitake, I. Oka, M. Ando, T. Yoshimoto, and D. Tsuru, *J. Biochem. (Tokyo)*, 1979, **86**, 1109.

8

Trace Elements in Animal Nutrition

J. R. ARTHUR, I. BREMNER & J. K. CHESTERS

Introduction.—This review surveys some of the essential roles of trace elements in the animal. Work with toxic concentrations has only been included when normal functions of other elements have been shown to be affected.

As this chapter only considers work published in 1979, the reader is referred to the reviews cited for more detailed background information on the wide range of subjects discussed.

1 Chromium

A case of Cr deficiency in Man following prolonged total parenteral nutrition has been reported. The symptoms included glucose intolerance and weight loss, and were reversible by supplementing with chromic chloride.[1] The Cr concentrations of the hair of insulin-treated human diabetics were lower in women than in men.[2] Retention of ^{51}Cr that had been administered intravenously in tracer quantities was reduced in male diabetic rats compared with non-diabetic male rats, but was increased to nearer normal values following injection of insulin.[3]

2 Cobalt

Young[4] has written a monograph discussing all aspects of the biochemistry and biology of cobalt, including a section on the current knowledge of the role of the element in animal nutrition.

The use of Co bullets, oral Co drenches, and injections of vitamin B_{12} in the effective prevention of Co deficiency in sheep continues to be investigated.[5, 6]

3 Copper

Incidence and Symptoms of Copper Deficiency States.—Copper deficiency continues to be one of the most important of trace-metal disorders in animal nutrition. However, difficulty is still experienced in diagnosing marginal Cu deficiency and in assessing when Cu therapy is likely to be beneficial. Thus oral

[1] H. Freund, S. Atamian, and J. E. Fischer, *J. Am. Med. Assoc.*, 1979, **241**, 496.
[2] J. W. Rosson, K. J. Foster, R. J. Walton, P. P. Munro, T. G. Taylor, and K. G. M. M. Alberti, *Clin. Chim. Acta*, 1979, **93**, 299.
[3] J. L. Kraszeski, S. Wallach, and R. L. Verch, *Endocrinol.*, 1979, **104**, 881.
[4] R. S. Young, 'Cobalt in Biology and Biochemistry', Academic Press, London, 1979.
[5] A. Whitelaw and A. J. F. Russel, *Vet. Rec.*, 1979, **104**, 8.
[6] D. I. Givens, P. J. Cross, W. B. Shaw, and P. E. Knight, *Vet. Rec.*, 1979, **104**, 508

or parenteral administration of Cu to pregnant cattle, or to calves with low Cu concentrations in the plasma or liver, was found to have no beneficial effect on growth in some experiments, although it did restore Cu concentrations in plasma and liver to normal.[7–9] In contrast, the growth rate of other hypocupraemic calves and of lambs was increased, and other symptoms of Cu deficiency were abolished, after Cu therapy.[9, 10] The Cu status of calves, as indicated by Cu concentrations in their liver and plasma, was effectively improved over a period of several months by oral administration of a single dose of CuO needles.[7]

Despite the claims that intakes of Cu in humans are frequently inadequate,[11, 12] the incidence of clinical disease that is directly attributable to Cu deficiency is very low. Hypocupraemia has been reported in patients undergoing total parenteral nutrition[13] and in an infant with a low birthweight who also exhibited extensive bone changes and neutropenia.[14]

The biochemical, pathological, and teratogenic effects of Cu deficiency in animals have been reviewed.[15, 16] The development of bone lesions in Cu-deficient children[14] or animals[17] has been attributed to possible disturbances in metabolism of collagen. The changes in the mechanical properties in the tibiae of Cu-deficient chicks, with increases in stress relaxation and decreases in tolerance to deformation, have been attributed to an increase in the solubility of bone collagen and a decrease in the degree of cross-linking in the protein.[18]

Reductions in the degree of cross-linking have frequently been demonstrated in elastin from Cu-deficient animals,[16] and these are thought to result from the reduced activity of lysyl oxidase.[19] When the role of Cu in the induction of this enzyme was studied *in vitro*, the activation of the enzyme only occurred in metabolically active tissue, and appeared to involve the binding of Cu to newly synthesized protein.[19] Cycloheximide, but not actinomycin D, inhibited the incorporation of [^{3}H]lysine and of ^{64}Cu into the enzyme.

The binding of copper to another amine oxidase, pig plasma benzylamine oxidase, has been investigated by n.m.r. and kinetic methods.[20] The enzyme appears to consist of two subunits, each containing a single Cu^{2+} ion. It was suggested that copper participates in the re-oxidation of the reduced enzyme by molecular oxygen.

[7] M. P. B. Deland, P. Cunningham, M. L. Milne, and D. W. Dewey, *Aust. Vet. J.*, 1979, **55**, 493.

[8] L. R. Gurnett and N. G. Lawrence, *Exp. Husb.*, 1978, **37**, 97.

[9] A. MacPherson, R. C. Voss, and J. Dixon, *Anim. Prod.*, 1979, **29**, 91.

[10] A. Whitelaw, R. H. Armstrong, C. C. Evans, and A. R. Fawcett, *Vet. Rec.*, 1979, **104**, 455.

[11] N. W. Solomons, *Am. J. Clin. Nutr.*, 1979, **32**, 856.

[12] K. E. Mason, *J. Nutr.*, 1979, **109**, 1979.

[13] S. F. Lowry, J. T. Goodgame, J. C. Smith, M. M. Maher, R. W. Makuch, R. I. Henkin, and M. F. Brennan, *Ann. Surg.*, 1979, **189**, 120.

[14] P. Yuen, H. J. Lin, and J. H. Hutchinson, *Arch. Dis. Child.*, 1979, **54**, 553.

[15] L. S. Hurley and C. L. Keen, in 'Copper in the Environment', Part II, ed. J. O. Nriagu, John Wiley and Sons, New York, 1979, p. 33.

[16] C. F. Gallagher, in ref. 15, p, 57.

[17] A. C. Okonkwo, P. K. Ku, E. R. Miller, K. K. Keahey, and D. E. Ullrey, *J. Nutr.*, 1979, **109**, 939.

[18] R. S. Riggins, A. G. Cartwright, and R. B. Rucker, *J. Biomech.*, 1979, **12**, 197.

[19] J. K. Rayton and E. D. Harris, *J. Biol. Chem.*, 1979, **254**, 621.

[20] R. Barker, N. Boden, G. Cayley, S. C. Charlton, R. Henson, M. C. Holmes, I. D. Kelly, and P. F. Knowles, *Biochem. J.*, 1979, **177**, 289.

The activity of the Cu,Zn-superoxide dismutase in chicks[21] and in rats[22] decreased rapidly when the animals were maintained on Cu-deficient diets. Liver superoxide dismutase activity decreased by about 50% in rats within one week of receiving the low-Cu diet, but smaller decreases occurred in the erythrocytes and in the heart.[22] In the chick, erythrocyte superoxide dismutase activity was found to give a reasonable indication of the Cu status of tissues.[21] It has been suggested that the enhanced peroxidation of lipids in the mitochondria in the liver of Cu-deficient rats could be related to differences in fatty-acid composition of membranes and to decreased activities of superoxide dismutase and of catalase.[23, 24] However, the rats that were used in these studies could also have been deficient in nutrients other than Cu, since they were reared on an unsupplemented diet of powdered milk.

The effects of Cu deficiency on the metabolism of cholesterol in rats have been confirmed.[25] Thus cholesterol levels in plasma were increased, as was the incorporation of [³H]mevalonate into the total lipids and cholesterol esters of the liver. There was a transient decrease in cholesterol in the plasma in swine when the dietary Cu content was increased from 8 to 125 mg per kg,[26] but variations over a ten-fold range in dietary Cu and Zn contents had no effect on the cholesterol level in the plasma of rats during a period of three weeks.[27]

In an investigation of the effects of Cu on vascular reactivity, physiological levels of Cu were found to inhibit pressor response to noradrenaline, but not to potassium, in a rat mesenteric preparation.[28] The effects were similar to those induced by prostaglandin I₂. There was a dose-related inhibition of mitochondrial monoamine oxidases in the brain and liver by Cu *in vitro*.[29] Copper also inhibited the microsomal oxidative demethylation of *NN*-dimethylaniline *in vitro*, suggesting some involvement of the metal in mixed-function oxidases.[30]

Metabolism of Copper.—General aspects of the absorption, transport, and storage of Cu have been reviewed.[31, 32] The homeostatic adaptation of absorption and excretion of Cu in response to changes in the dietary supply of Cu was demonstrated in a study where both Cu-deficient and Cu-adequate rats received a single oral or intravenous dose of ⁶⁴Cu.[33] The absorption of the oral ⁶⁴Cu was greater and the excretion of the injected ⁶⁴Cu was smaller in the Cu-deficient animals. It has been suggested that the regulation of the absorption of Cu

21 W. J. Bettger, J. E. Savage, and B. L. O'Dell, *Nutr. Rep. Int.*, 1979, **19**, 893.
22 D. I. Paynter, R. J. Moir, and E. J. Underwood, *J. Nutr.*, 1979, **109**, 1570.
23 E. Russanov and E. Ivancheva, *Acta Physiol. Pharmacol. Bulg.*, 1979, **5**, 67.
24 E. Russanov, P. Balevska, and S. Leutchev, *Acta Physiol. Pharmacol. Bulg.*, 1979, **5**, 73.
25 K. G. D. Allen and L. M. Klevay, *Atherosclerosis*, 1978, **31**, 259.
26 J. H. Eisemann, W. G. Pond, and M. L. Thonney, *J. Anim. Sci.*, 1979, **48**, 1123.
27 W. O. Caster and J. M. Doster, *Nutr. Rep. Int.*, 1979, **19**, 773.
28 S. C. Cunnane, H. Zinner, D. F. Horrobin, M. S. Manku, R. O. Morgan, R. A. Zarmali, A. I. Ally, M. Karmazyn, W. E. Barnette, and K. C. Nicolaou, *Can. J. Physiol. Pharmacol.*, 1979, **57**, 35.
29 S. Magour, O. Cumpelik, and M. Paulus, *J. Clin. Chem. Clin. Biochem.*, 1979, **17**, 777.
30 V. P. Kurchenko, S. A. Usanov, and D. I. Metelitsa, *React. Kinet. Catal. Lett.*, 1979, **12**, 31.
31 G. W. Evans, in ref. 15, p. 163.
32 N. Marceau, in ref. 15, p. 177.
33 F. J. Schwarz and M. Kirchgessner, *Z. Tierphysiol. Tierernaehr. Futtermittelkd.*, 1979, **41**, 335.

involves the induction of synthesis of copper-metallothionein in the intestinal mucosa.[34]

Copper Requirements of Animals.—Further attempts have been made to define the requirements of Cu of animals, using an empirical approach in which purified diets containing different amounts of Cu were fed to baby pigs,[17] to pregnant and lactating rats,[35] and to growing rats.[36] The criteria used to assess adequacy of Cu intake determined the estimate of requirement of Cu; for example, less Cu is required to support growth than to maintain normal tissue levels of Cu or to prevent bone lesions and anaemia.[17, 36] Minimum dietary requirements of Cu to satisfy all criteria were estimated to be about 6 mg per kg of diet for baby pigs[5] and rats during pregnancy[6] but 9 mg per kg of diet for rats during both pregnancy and lactation.[6] An alternative approach for the determination of Cu requirements has been based on factorial analysis. This requires information on the amount of Cu that is deposited in the carcass during growth. Carcass analysis of lambs and calves has indicated that the Cu requirement for growth in these animals is < 1 mg per kg of body weight.[37] It was also estimated that the availability of Cu from milk is about 23%, which is ten times greater than that in weaned animals.[37]

Factors Governing the Availability of Copper.—It is well established that the development of Cu deficiency cannot be related solely to a low Cu intake, and that other dietary factors influence the availability of dietary Cu.[38] For example, hypocupraemia developed in grazing cattle that were receiving one type of herbage with 6.6 mg of Cu per kg but not in animals receiving another herbage with only 4.6 mg of Cu per kg.[39] It has been suggested that the fibre content of diets may influence the availability of Cu to Man,[40] but increasing the cellulose content of rat diets had no major effect on Cu status.[41] The hepatic concentration of Cu in rats was increased when the dietary protein content was increased from 7.5 to 15 or 30%.[42] The addition of methionine or cysteine to chick diets prevented or restricted the increase in plasma, spleen, liver, and biliary concentrations of Cu which occurred as the dietary Cu content was increased to 500 mg per kg.[43] Supplementation with these sulphur-containing amino-acids also counteracted the toxic effects of the added Cu, especially when the basal diet was deficient or only marginally adequate in sulphur.

The hepatic accumulation of Cu by pigs receiving diets with 250 mg of Cu per kg was also reduced by supplementing the diet with up to 500 mg of sulphide per kg (as ferrous sulphide).[44] Greater amounts of sulphide further decreased concentrations of Cu in the liver but abolished the beneficial effect of Cu on

[34] G. W. Evans, in 'Metallothionein', ed. J. H. R. Kagi and M. Nordberg, Birkhauser Verlag, Basel, 1979, p. 321.
[35] F. L. Cerklewski, *J. Nutr.*, 1979, **109**, 1529.
[36] J. L. McNaughton and E. J. Day, *J. Nutr.*, 1979, **109**, 559.
[37] N. F. Suttle, *Br. J. Nutr.*, 1979, **42**, 89.
[38] M. Kirchgessner, F. J. Schwarz, E. Grassmann, and H. Steinhart, in ref. 15, p. 433.
[39] M. J. Stoszek, J. E. Oldfield, G. E. Carter, and P. H. Weswig, *J. Anim. Sci.*, 1979, **48**, 893.
[40] C. Reilly, *Biochem. Soc. Trans.*, 1979, 7, 202.
[41] R. C. Y. Tsai and K. Y. Lei, *J. Nutr.*, 1979, **109**, 1117.
[42] A. C. Magee and F. P. Grainger, *Nutr. Rep. Int.*, 1979, **20**, 771.
[43] L. S. Jensen and D. V. Maurice, *J. Nutr.*, 1979, **109**, 91.
[44] T. J. Prince, V. W. Hays, and G. L. Cromwell, *J. Anim. Sci.*, 1979, **49**, 507.

growth and food-conversion efficiency. The effects of the ferrous sulphide probably arose from the inhibition by sulphide of the absorption of Cu and were independent of the increase in intake of Fe. This latter point was confirmed in other studies, where the addition of 150 mg of Fe per kg of diet had no effect on the concentration of Cu in the liver of pigs.[45]

Changes in Cu metabolism have been reported in Fe-deficient animals.[38] It appeared from studies involving the use of isolated intestinal loops in rats that the uptake of ^{64}Cu by the mucosa was increased and the transfer of ^{64}Cu to the body was decreased in Fe deficiency.[46] However, the amounts of metal added to the gut loops were excessively large. The level of erythropoietic activity in rats also influenced the metabolism of Cu.[47] Thus differences in Cu content of plasma and caeruloplasmin content and in the clearance of ^{64}Cu from plasma were observed in anaemic and polycythemic rats. The turnover of ^{64}Cu in plasma and the uptake of ^{64}Cu by the bone marrow were greatest in the anaemic animals, perhaps because of their greater requirement for Cu for the production of red cells.

Increased dietary intakes of Cd and Zn also decreased the availability of Cu.[38, 48, 49] For example, when pregnant ewes received a diet with 750 mg of Zn per kg, a severe Cu deficiency developed, with decreased plasma and liver concentrations of Cu and decreased transfer of Cu to the developing foetuses.[50] Similarly, increased Zn supply to Japanese quail receiving a diet that was only marginally adequate in Cu resulted in poor growth, feather depigmentation, mild perosis, and decreased concentrations of Cu in the liver.[51] These effects were abolished by supplementation with Cu. Supplementation with Zn has caused reduced liver Cu in rats[42] and pigs[45] and has caused hypocupraemia in humans.[52, 53] In one instance, the antagonistic effect of Zn on the metabolism of Cu was utilized to decrease the concentration of Cu in the liver of a patient with Wilson's disease.[54] Alterations in Cu metabolism have also been reported in Zn-deficient animals, with increases in Cu in the liver of rats[42] and in Cu in the milk of cows.[55]

Investigation of the mechanism of the interaction between Zn and Cu has confirmed that Zn inhibits the intestinal absorption of ^{64}Cu.[56] This effect was associated with increased mucosal uptake of Cu, although decreased mucosal Cu content was recorded in another study.[57] However, in both papers[56, 57] it

45 E. B. Greer, C. E. Lewis, and M. G. Croft, *Aust. J. Exp. Agric. Anim. Husb.*, 1979, **19**, 312.
46 F. A. El-Shobaki and W. Rummel, *Res. Exp. Med.*, 1979, **174**, 187.
47 S. Iwanska and D. Strusinska, *Acta Physiol. Pol.*, 1978, **29**, 465.
48 I. Bremner, *Proc. Nutr. Soc.*, 1979, **38**, 235.
49 I. Bremner and C. F. Mills, in 'Management and Control of Heavy Metals in the Environment', CEP Consultants, Edinburgh, 1979, p. 139.
50 J. K. Campbell and C. F. Mills, *Environ. Res.*, 1979, **20**, 1.
51 R. P. Hamilton, M. R. S. Fox, B. E. Fry, A. O. L. Jones, and R. M. Jacobs, *J. Food Sci.*, 1979, **44**, 738.
52 M. Abdulla, *Lancet*, 1979, **1**, 616.
53 M. Abdulla and S. Svensson, *J. Lab. Clin. Invest.*, 1979, **39**, 31.
54 T. U. Hoogenraad, C. J. A. Van den Hamer, R. Koevoet, and E. G. W. M. de Ruyter Korver, *Lancet*, 1978, **2**, 1262.
55 M. Kirchgessner, F. J. Schwarz, H. P. Roth, and W. A. Schwarz, *Arch. Tierernaehr.*, 1978, **28**, 723.
56 A. C. Hall, B. W. Young, and I. Bremner, *J. Inorg. Biochem.*, 1979, **11**, 57.
57 T. Ogiso, N. Ogawa, and T. Miura, *Chem. Pharm. Bull.*, 1979, **27**, 515.

was established that Zn caused increased incorporation of ^{64}Cu into intestinal metallothionein, the synthesis of which was probably induced by the Zn. It seems likely that Cu which was bound to this protein in the intestinal mucosa was not immediately available for transport into the blood.

The most important antagonists of the metabolism of Cu in ruminant animals are undoubtedly Mo and S.[38, 48, 49] High Mo intakes in sheep (45—50 mg per kg of Cu-supplemented diet) caused a decrease in liver Cu content but increased kidney and serum concentrations of Cu.[58] It has been suggested that these changes result from the formation, in the rumen, of thiomolybdates which may inhibit the absorption of Cu or induce changes in the systemic metabolism of Cu.[48, 49] Inclusion of tetrathiotungstate in rat diets also inhibited the absorption of ^{64}Cu and modified the systemic metabolism of Cu in a similar manner to tetrathiomolybdate.[59] Whether or not the inhibition of absorption of ^{64}Cu resulted from direct interaction between Cu and tetrathiotungstate has still to be established, but a copper–thiotungstate complex has been synthesized *in vitro* and characterized.[60]

Although increased intakes of Cd induced Cu deficiency in animals, with decreased liver and plasma concentration of Cu and decreased placental transfer of Cu to the developing foetus,[48, 49, 61, 62] Cd also caused, in some species, an increase in concentrations of Cu in the kidney.[63, 64] The additional renal Cu accumulated with Cd on metallothionein, but the origin of the Cu was not established.[65, 66]

Copper Proteins.—The isolation, synthesis, physico-chemical properties, and possible roles of copper-metallothionein have been reviewed.[67–69] The gel-chromatographic and anion-exchange chromatographic behaviour of copper-metallothionein was compared with that of the (cadmium, zinc)-containing proteins.[70] There were differences between species in the proportion of hepatic Cu that was bound to metallothionein,[67] with less Cu present in this form in goat[71] than in pig[72] liver. The Zn content of the liver also influenced the binding of hepatic Cu to metallothionein.[67, 71, 72] The biological half-life of hepatic and renal cadmium-metallothionein in rats decreased as the Cu content of the protein increased.[73]

The physiological role of metallothionein has yet to be established. Although

[58] J. B. J. Van Ryssen, *S. Afr. J. Anim. Sci.*, 1979, **9**, 21.
[59] I. Bremner, B. W. Young, and C. F. Mills, *Biochem. Soc. Trans.*, 1979, **7**, 677.
[60] A. Mueller, T. K. Hwang, and H. Boegge, *Angew. Chem.*, 1979, **91**, 656.
[61] H. G. Petering, H. Choudhury, and K. L. Stemmer, *Environ. Health Perspect.*, 1979, **28**, 97.
[62] W. S. Webster, *J. Nutr.*, 1979, **109**, 1646.
[63] J. A. Szymanska and A. J. Zelazowski, *Environ. Res.*, 1979, **19**, 121.
[64] K. T. Suzuki, *Arch. Environ. Contam. Toxicol.*, 1979, **8**, 255.
[65] K. T. Suzuki, T. Maitani, and S. Takenaka, *Chem. Pharm. Bull.*, 1979, **27**, 647.
[66] K. T. Suzuki and S. Takenaka, *Chem. Pharm. Bull.*, 1979, **27**, 1753.
[67] I. Bremner, in ref. 34, p. 273.
[68] H. Rupp, R. Cammack, H.-J. Hartmann, and U. Weser, *Biochim. Biophys. Acta*, 1979, **578**, 462.
[69] H. Rupp and U. Weser, in ref. 34, p. 231.
[70] K. T. Suzuki and M. Yamamura, *Arch. Environ. Contam. Toxicol.*, 1979, **8**, 471.
[71] M. Mjor-Grimsrud, N. E. Soli, and T. Sivertsen, *Acta Pharmacol. Toxicol.*, 1979, **44**, 319
[72] M. Webb, S. R. Plastow, and L. Magos, *Life Sci.*, 1979, **24**, 1901.
[73] K. Cain and D. E. Holt, *Chem.-Biol. Interact.*, 1979, **28**, 91.

it has been suggested that homeostatic control of absorption of Cu involves induction of the synthesis of mucosal metallothionein by Cu,[34] no correlation between oral Cu intake and mucosal copper-metallothionein content was observed in two studies.[56, 57] Nevertheless, the increase in absorption of ^{64}Cu in tumour-bearing rats and the decrease in absorption of ^{64}Cu in oestrogen-treated rats could be inversely related to the binding of the metal to intestinal metallothionein, suggesting that, in some circumstances, this protein may indeed regulate the absorption of Cu.[74] It was not established how metallo-thionein levels were controlled in these animals, but it was shown that they could not be related to plasma concentrations of caeruloplasmin.

The physico-chemical properties and the possible physiological roles of caeruloplasmin, as a transport protein, a ferroxidase, and an amine oxidase, have been reviewed.[75] In studies of the regulation of caeruloplasmin in Cu-deficient rats, it was established that the availability of Cu controls the rate of synthesis, the activation, and the plasma concentrations of the protein.[76] It was suggested that caeruloplasmin may play a protective role against inflamma-tion, since experimental inflammation was more readily induced in Cu-deficient than in Cu-adequate rats and since the injection of caeruloplasmin alleviated the condition.[77] The increase in plasma concentrations of Cu in endotoxin-treated chicks and the decrease in hydrocortisone-treated birds were almost certainly accounted for by changes in caeruloplasmin concentrations.[78] Two chromatographic and electrophoretic techniques have been described for the separation and determination of Cu complexes in plasma.[79, 80]

The purification of a Cu protein in brain, neurocuprein, has been described.[81] It was suggested that this protein plays an essential role in neuronal processes, especially in the detoxification of adrenochrome-like compounds and in the neutralization of excess catecholamines. [81]

Genetic Influence on the Metabolism of Copper.—The influence of genetic factors on the incidence of disorders in Cu metabolism in sheep has been reviewed.[82] Breed differences in the rate of increase in plasma Cu levels following the adminis-tration of Cu to hypocupraemic lambs were attributed to differences in the efficiency of absorption of Cu.[83]

Menkes' Kinky Hair Disease (MKHD) in children is, in essence, an example of a genetically induced Cu deficiency.[15] On treatment of a child that had this disease with Cu and histidine, serum Cu was increased but caeruloplasmin levels were unchanged.[84] Concentrations of Cu in fibroblasts from MKHD patients

[74] D. I. Cohen, B. Illowsky, and M. C. Linder, *Am. J. Physiol.*, 1979, **236**, E309.
[75] E. Frieden, in ref. 15, p. 241.
[76] M. C. Linder, P. A. Houle, E. Isaacs, J. R. Moor, and L. E. Scott, *Enzyme*, 1979, **24**, 23.
[77] C. W. Denko, *Agents Actions*, 1979, **9**, 333.
[78] B. Sas and I. Bremner, *J. Inorg. Biochem.*, 1979, **11**, 67.
[79] J. Teape, H. Kamel, D. H. Brown, J. M. Ottaway, and W. E. Smith, *Clin. Chim. Acta*, 1979, **94**, 1.
[80] N. Kahn and J. C. Van Loon, *J. Liq. Chromatogr.*, 1979, **2**, 23.
[81] V. S. Gasparov, R. M. Nalbandyan, and H. Ch. Buniatian, *FEBS Lett.*, 1979, **79**, 37.
[82] G. Wiener, *Livestock Prod. Sci.*, 1979, **6**, 223.
[83] G. Wiener, N. F. Suttle, A. C. Field, J. G. Herbert, and J. A. Wooliams, *J. Agric. Sci.*, 1978, **91**, 433.

were considerably higher than normal.[84] Although it was reported that extracellular Cu did not induce the synthesis of metallothionein in the fibroblasts from MKHD or control subjects, no attempt was made to measure the degree of incorporation of [^{35}S]cysteine into the purified Cu-binding proteins.[85]

It has been suggested that the primary metabolic defect in the metabolism of Cu in brindled mice (which are an animal model for MKHD) may lie in some abnormal structure or rate of synthesis of metallothionein.[86] Even though brindled mice are clinically Cu-deficient, the concentrations of Cu in tissues such as the kidney, intestinal mucosa, and pancreas were found to be greater than normal.[87, 88] The Cu accumulated in these tissues principally as a Cu-protein with similar properties to metallothionein. Although it was reported that the Cu proteins had markedly different amino-acid compositions from metallothionein,[89, 90] no attempt was made to isolate them under anaerobic conditions, as is required for the successful purification of copper-metallothionein.[67] Concentrations of Cu were much greater in the ileum than in the duodenum or stomach of the brindled mice, and were greater in affected animals than in heterozygotes.[87] Absorption of Cu was severely inhibited in adult brindled mice, but this effect was less evident in suckling animals, perhaps because of their ability to take up Cu by pinocytosis.[87, 91] Subcutaneous injection of copper into neonatal brindled mice transiently increased tissue concentrations of copper, restored caeruloplasmin and lysyl oxidase activities, and eliminated some clinical signs of copper deficiency, especially if Cu^+ was given with sebacic acid.[92]

Defects in Cu metabolism were also reported in crinkled mice, with decreases in liver Cu in the neonatal animals.[93] Liver superoxide dismutase activities were also decreased, and remained so even when normal levels of Cu in the liver had been restored at 60 days of age.[94]

Tyrosinase activities in hair bulbs from mottled mice were also decreased, which perhaps explains the reduced melanogenesis in these animals.[95] Addition of Cu to the supernatant fraction from hair bulbs of the mottled mice (and also from Cu-deficient animals) restored the activity of the T_1 variety of tyrosinase, indicating that the apoenzyme was still present.[95]

Wilson's Disease, which is a form of Cu toxicity in Man, is also genetically induced and is characterized by greatly increased levels of Cu in the liver and

[84] I. T. Lott, R. Dipaolo, S. S. Raghaven, P. Clopath, A. Milunsky, W. C. Robertson, and J. N. Kanfer, *Pediatr. Res.*, 1979, **13**, 845.
[85] W.-Y. Chan, A. D. Garnica, and O. M. Rennert, *Pediatr. Res.*, 1979, **13**, 197.
[86] C. J. A. Van den Hamer, H.-W. Prins, and J. L. Nooijen, in 'Models for the Study of Inborn Errors of Metabolism', ed. F. A. Hommes, Elsevier–North Holland, Amsterdam, p. 95.
[87] J. Camakaris, J. R. Mann, and D. M. Danks, *Biochem. J.*, 1979, **180**, 597.
[88] H.-W. Prins and C. J. A. Van den Hamer, *J. Inorg. Biochem.*, 1979, **10**, 19.
[89] D. M. Hunt and A. E. Port, *Life Sci.*, 1979, **24**, 1453.
[90] A. E. Port and D. M. Hunt, *Biochem. J.*, 1979, **183**, 721.
[91] J. R. Mann, J. Camakaris, and D. M. Danks, *Biochem. J.*, 1979, **180**, 613.
[92] J. R. Mann, J. Camakaris, D. M. Danks, and E. G. Walliczek, *Biochem. J.*, 1979, **180**, 605.
[93] C. L. Keen and L. S. Hurley, *J. Inorg. Biochem.*, 1979, **11**, 269.
[94] C. L. Keen and L. S. Hurley, *Proc. Soc. Exp. Biol. Med.*, 1979, **162**, 152.
[95] T. J. Holstein, R. Q. Fung, W. C. Quevedo, and T. C. Bienieki, *Proc. Soc. Exp. Biol. Med.* 1979, **162**, 264.

decreased levels of caeruloplasmin in serum.[96] Differentiation between Wilson's disease and other liver diseases which are characterized by high concentrations of Cu in the liver can be difficult.[97, 98] Certain breeds of dogs have been found to be genetically prone to chronic Cu toxicosis.[99]

Toxicity of Copper.—General aspects of Cu toxicosis in domestic and laboratory animals have been reviewed.[100] Sheep are particularly prone to this disease, and the development of pathological lesions in the liver of these animals has been described.[101] As in the children with Indian Childhood Cirrhosis,[102, 103] in patients with cholestasis,[104] and in Bedlington terriers,[99] the hepatic Cu accumulated in lysosomes. Positive orcein staining indicated the presence of sulphydryl-rich proteins,[99, 102-104] and this was confirmed in the sheep liver, where both Cu and S were detected by electron probe microanalysis in the residual bodies that were derived from lysosomes.[101] Subcellular fractionation of the liver from Cu-poisoned sheep showed that most of the Cu sedimented with the nuclear fraction, although the metal was probably present in dense lysosomes rather than in nuclei.[105]

The uptake of Cu by sheep liver seemed to be stimulated when liver damage was induced in lupinosis.[106] Although an increased intake of ascorbic acid had a general protective effect against certain metal toxicities in the chick, it had no beneficial effect against Cu toxicity.[107] Male rats were found to be more susceptible to Cu toxicity than females, probably because concentrations of Cu in the liver were greater.[108]

The toxicity of Cu in sheep and other species generally results in haemolysis of red blood cells.[100] In investigations of the haemolytic action of Cu, it appeared that the primary cytotoxic effect of Cu occurred through interaction with thiol groups in the erythrocyte membrane.[109] Haemolysis could be prevented, however, by lipid antioxidants. Copper was also found to cause a reduction in erythrocyte deformability, which was associated with an increase in the permeability of the membrane and with osmotic fragility, without any apparent oxidative damage to the cell.[110]

[96] I. H. Schienberg, in ref. 15, p. 17.
[97] J. Evans, S. Newman, and S. Sherlock, *Gastroenterology*, 1978, **75**, 875.
[98] J. A. Perman, S. L. Werlin, R. J. Grand, and J. B. Watkins, *J. Pediatr.*, 1979, **94**, 564.
[99] D. C. Twedt, I. Sternlieb, and S. R. Gilbertson, *J. Am. Med. Vet. Assoc.*, 1979, **175**, 269.
[100] I. Bremner, in ref. 15, p. 286.
[101] T. P. King and I. Bremner, *J. Comp. Pathol.*, 1979, **89**, 515.
[102] A. Popper, S. Goldfischer, I. Sternlieb, N. C. Nayak, and T. V. Madhaven, *Lancet*, 1979, **1**, 1205.
[103] M. S. Tanner, B. Portmann, A. P. Mowat, R. Williams, A. N. Pandit, C. F. Mills, and I. Bremner, *Lancet*, 1979, **1**, 1203.
[104] Y. Nakanuma, T. Karino, and G. Ohta, *Virchows Arch. A: Pathol. Anat. Histol.*, 1979, **382**, 21.
[105] S. R. Gooneratne, J. McC. Howell, and J. Gawthorne, *Res. Vet. Sci.*, 1979, **27**, 30.
[106] K. P. Croker, J. G. Allen, D. S. Petterson, H. G. Masters, and R. F. Frayne, *Aust. J. Agric. Res.*, 1979, **30**, 551.
[107] C. H. Hill, *J. Nutr.*, 1979, **109**, 84.
[108] S. Haywood, *J. Comp. Pathol.*, 1979, **89**, 481.
[109] P. Hochstein, K. S. Kumar, and S. J. Forman, in 'The Red Cell', ed. G. J. Brewer, Alan R. Liss Inc., New York, 1979, p. 669.
[110] K. F. Adams, G. Johnson, K. E. Hornowski, and T. H. Lineberger, *Biochim. Biophys. Acta*, 1979, **550**, 279.

4 Iodine

Iodide ion within the thyroid gland is derived from two sources, *i.e.* from iodide in plasma and from the de-iodination of iodotyrosine residues within the gland. The extent to which the iodide from the two sources forms a common pool or is retained in separate compartments within the thyroid has been a matter of controversy. A recent investigation appears to have shown conclusively that, at least in one situation, metabolically separate pools do exist.[111] When given radioiodide in their drinking water, normal rats always had lower specific activities of iodine in thyroglobulin than in the total iodide pool of the thyroid gland. However, this relationship only applied for the initial period after introduction of radioiodide in hypophysectomized rats. Thereafter the specific activity of iodine in thyroglobulin exceeded that of the total iodide pool. In these circumstances, therefore, the iodine that was incorporated into thyroglobulin must have been derived from a distinct pool within the gland that was relatively minor compared to the total iodide pool, but which was in ready equilibrium with the iodide with a high specific activity in the plasma.

The difference in the sequence of labelling between control and hypophysectomized rats that is described above suggests that, when the thyroid gland was subject to stimulation by thyroid-stimulating hormone (TSH), a greater proportion of the iodide that was used for iodination of thyroglobulin was derived from the de-iodination of iodotyrosine residues within the gland. A similar conclusion was reached by a separate group, who showed that the rise in thyroid iodide following the administration of TSH was prevented by administering an inhibitor of the dehalogenation of iodotyrosine.[112]

Further investigations have suggested that iodine is preferentially incorporated into newly synthesized thyroglobulin and that iodination occurs in apical vesicles before the protein is secreted into the colloid lumen.[113] Within the follicles, the proportion of large thyroglobulin molecules was reduced by dietary iodine deficiency, but then recovered rapidly when iodide was given in the diet.[114] The appearance of large thyroglobulin molecules within the follicle occurred even in the presence of cycloheximide, and was thought to involve an iodide-mediated aggregation of thyroblobulin molecules rather than their synthesis *de novo*.

Iodine deficiency also caused an increase in the ratio of mono- to di-iodotyrosine in the thyroid, but did not change the ratio of T3/T4 in the plasma.[115] A system has been described for studying thyroid function *in vitro* in which the relative proportions of the products of iodination were maintained at the values that are present *in vivo*.[116] Extracts of thyroid tissue have been shown to contain glycopeptides of low molecular weight that contain iodine.[117] The similarity in

[111] C. Simon and P. Bastiani, *Endocr. Res. Commun.*, 1979, **6**, 149.

[112] J. D. Hildebrandt, J. R. Scranton, and N. S. Halmi, *Endocrinology*, 1979, **105**, 618.

[113] S. Matsukawa and T. Hosoya, *J. Biochem. (Tokyo)*, 1979, **85**, 1009.

[114] S. Smeds, R. Ekholm, and H. Studer, *J. Endocrinol.*, 1979, **82**, 199.

[115] V. Boonnamsiri, J. C. Kermode, and B. D. Thompson, *J. Endocrinol.*, 1979, **82**, 227.

[116] K. Okamura, K. Inoue, T. Nakashima, A. Shiroozu, and M. Yoshinari, *Acta Endocrinol. (Copenhagen)*, 1979, **92**, 286.

[117] A. Haeberli, H. Engler, C. von Grunigen, H. Kohler, and H. Studer, *Acta Endocrinol. (Copenhagen)*, 1979, **92**, 105.

chemical composition between these peptides and thyroglobulin suggests that they are the products of hydrolysis of the latter.

Although the response of adenyl cyclase in thyroid tissue to TSH is reduced by exogenous iodide,[118] the latter did not alter the binding of TSH to the tissue.[119] The authors suggest that iodide probably acts on the coupling mechanism between the binding of TSH and the activation of adenyl cyclase. Administration of iodide also inhibited the synthesis of prostaglandins in canine thyroid.[120]

Evidence is available that mammary tissue requires iodine for normal function, and a recent study suggests that iodination of protein occurs within the mammary gland.[121] A high ambient temperature altered the balance of iodide metabolism in goats by decreasing the proportion taken up by the thyroid and, possibly as a consequence of this, increasing the proportion that is secreted in the milk.[122]

Glucosinolates appeared to be responsible for the goitrogenic properties of certain varieties of rape-seed meal, but there was no evidence of their transfer to the milk of cows that were fed on these diets.[123] The goitrogens of soya beans were located in the material that was precipitated by calcium ions from extracts that were soluble in hot water. They were not susceptible to destruction by proteolytic digestion.[124]

5 Manganese

Improvements in the direct estimation of manganese in blood or serum by flameless atomic absorption have been found when the Zeeman effect was used to correct for non-specific absorption.[125] The role of Mn in ruminant nutrition has been reviewed,[126] and studies with rats have shown differences in absorption of Mn with diet.[127] At a biochemical level, Mn appears to be necessary for the activation of brain adenylate cyclase[128] and for galactosylhydroxylysyl glucosyltransferase, which is an enzyme that is involved in the biosynthesis of collagen.[129] Manganese deficiency in rats also resulted in impaired immune response.[130]

6 Molybdenum

Most interest in the metabolism of Mo in animals has centred on its antagonistic effects on Cu metabolism, which have already been discussed (see Section 3). The metabolism of Mo in sheep was shown to be greatly influenced by the dietary

118 J. R. Arthur, I. Bremner, and J. K. Chesters, in 'Inorganic Biochemistry' (Specialist Periodical Reports), ed. H. A. O. Hill, The Royal Society of Chemistry, London, 1981, Vol. 2, p. 292.
119 H. Uchimura, S. M. Amir, and S. H. Ingbar, *Endocrinology*, 1979, **104**, 1207.
120 J. M. Boeynaems, N. Galand, and J. E. Dumont, *Endocrinology*, 1979, **105**, 996.
121 B. A. Eskin, C. E. Sparks, and B. I. La Mont, *Biol. Trace Elements Res.*, 1979, **1**, 101.
122 F. W. Lengemann, *J. Dairy Sci.*, 1979, **62**, 412.
123 A. Papas, J. R. Ingalls, and L. D. Campbell, *J. Nutr.*, 1979, **109**, 1129.
124 J. Suwa, T. Koyanagi, and S. Kimura, *J. Nutr. Sci. Vitaminol.*, 1979, **25**, 309.
125 P. A. Pleban and K. H. Pearson, *Clin. Chem.*, 1979, **25**, 1915.
126 M. Hidiroglou, *Can. J. Anim. Sci.*, 1979, **59**, 217.
127 B. D. King, J. W. Lassiter, M. W. Neathery, W. J. Miller, and R. P. Gentry, *J. Anim. Sci.*, 1979, **49**, 1235.
128 E. J. Neer, *J. Biol. Chem.*, 1979, **254**, 2089.
129 R. Myllyla, H. Anttinen, and K. I. Kivirikko, *Eur. J. Biochem.*, 1979, **101**, 261.
130 J. H. McCoy, M. A. Kenney, and B. Gillham, *Nutr. Rep. Int.*, 1979, **19**, 165.

sulphur content.[131] Thus increasing the daily S intake from 1.0 to 3.2 g decreased the absorption, decreased the urinary and faecal endogenous excretion, and increased the overall retention of Mo. There were also changes in the binding of Mo in the plasma, and it was suggested that the manifold effects of S on the metabolism of Mo were related to a common interaction in the rumen, leading to the formation of Mo complexes (possibly thiomolybdates) which were poorly absorbed but even more poorly excreted.

The formation of thiomolybdates in aqueous media which resembled rumen fluid in pH and mineral composition has been studied.[132] Although di- and tri-thiomolybdates were formed readily at S^{2-}:Mo ratios that were similar to those found in the rumen, tetrathiomolybdates were not. It was reported that potassium tetrathiomolybdate hydrolysed rapidly in aqueous solution to give a mixture of oxythiomolybdates, HS^-, and H_2S.[133] When rats received diets containing small amounts of ammonium tetrathiomolybdate (6 mg of Mo per kg), severe skeletal lesions rapidly developed at growth plates of long bones, at muscle insertions, and beneath the periosteum.[134] In addition, there was cartilage dysplasia and slowing of osteogenesis.

When rats were given diets with 1 g of molybdenum (as molybdate) per kg, the activities of alkaline phosphatase, acid phosphatase, and 5-nucleotidase in the kidney increased.[135] Supplementation with copper caused inhibition of these enzymes, indicating some antagonism between the two metals. The biliary excretion of molybdenum in rats was found to depend on the oxidation state of the metal when it was administered by intravenous injection.[136] When high doses of molybdenum were injected, the excretion of Mo^V was less than that of Mo^{VI}.

A probable genetic defect in Mo metabolism in Man has been reported.[137, 138] This was characterized by extremely low levels of xanthine oxidase and sulphite oxidase in the liver and intestine of a newborn girl. Levels of Mo in the serum were normal, however, and it was suggested that the metabolic defect occurred in the incorporation of Mo into flavoproteins rather than in its absorption.

7 Nickel

Schnegg and Kirchgessner have comprehensively reviewed the subject of Ni in animal nutrition.[139] Diets that were low in Ni (0.03 mg per kg) caused decreased weight gains in lambs that were fed liquid diets based on skimmed-milk powder.[140] When steers and lambs were fed solid diets that had a low Ni content, weight gains tended to be depressed in comparison to animals that were fed a supple-

[131] N. D. Grace and N. F. Suttle, *Br. J. Nutr.*, 1979, **41**, 125.
[132] N. J. Clarke and S. H. Laurie, *J. Inorg. Biochem.*, 1979, **12**, 37.
[133] K. M. Weber, D. D. Leaver, and A. G. Wedd, *Br. J. Nutr.*, 1979, **41**, 403.
[134] B. F. Fell, D. Dinsdale, and T. T. El-Gallad, *J. Comp. Pathol.*, 1979, **89**, 495.
[135] S. V. S. Rana and A. Kumar, *Ind. Health*, 1979, **17**, 11.
[136] J. Lener and B. Bibr, *Toxicol. Appl. Pharmacol.*, 1979, **51**, 259.
[137] C. Van der Heiden, F. A. Beemer, W. Brink, S. K. Wadman, and M. Duran, *Clin. Biochem.*, 1979, **12**, 206.
[138] M. Duran, F. A. Beemer, C. Van der Heiden, J. Korteland, P. K. De Bree, M. Brink, S. K. Wadman, and I. Lombeck, *J. Inher. Metab. Dis.*, 1978, **1**, 175.
[139] A. Schnegg and M. Kirchgessner, *Ubers. Tierernaehr.*, 1979, **7**, 179.
[140] J. W. Spears, E. E. Hatfield, and G. C. Fahey, Jr., *Nutr. Rep. Int.*, 1978, **18**, 621.

ment of 5 mg of Ni per kg. This effect was especially apparent when dietary concentrations of crude protein were reduced from between 12 and 13% to approximately 7.5%.[141] Another effect of the supplementation with Ni was to raise ruminal urease activity in the lambs that were fed concentrations of 13.1% and 7.5% of crude protein and the steers that were fed 7.4% of crude protein. The role of Ni in ruminal urease has been the subject of investigation.[142, 143] Incubation of metal ions with rumen fluid from sheep that had been fed low-Ni diets (0.06 mg of Ni per kg of dry matter) showed that urease activity was stimulated by Ba, Ni, and Mn but inhibited by Cu, Zn, and Cd. When lambs were fed a diet with 5 mg of Ni per kg (9.5% crude protein), ruminal urease activity was increased in comparison to that found in animals receiving 0.32 mg of Ni per kg of diet. Ruminal NH_3 was higher three hours post feeding in the Ni-supplemented animals, which also tended to have better growth rates.[142] In rats, Ni deficiency increased the specific activities of pancreatic proteinase and leucine aminopeptidase but decreased the activity of β-amylase. Total tissue protein concentrations were also increased, which indicated that the observed enzyme changes were not the result of a general lack of protein in Ni deficiency.[144]

Investigations have continued into the interaction between Fe and Ni in the rat.[145-147] The Fe status of rats has been shown to have large effects on the severity and symptoms of Ni deficiency. Second-generation rats receiving 60 mg of Fe per kg in the diet exhibited worse effects of Ni deficiency than rats with 30 mg of Fe per kg in the diet.[145] In another series of experiments, manipulation of Fe and Ni contents of rat diets allowed the effects of Ni to be demonstrated in rats from dams that had a normal Ni status. Nickel deficiency reduced growth rate and decreased haematocrit, the concentration of haemoglobin in the blood, and levels of alkaline phosphatase and total lipids in the plasma. The amounts of total lipids and of Mn, Cu, and Ni in the liver were also affected.[146] In cases of severe Fe deficiency, supplementation with Ni partially alleviated the effects on haemoglobin and haematocrit and Cu concentration; however, the amount of Fe in the liver was not increased and growth was decreased.[146] Concentrations of Cu, Fe, Mn, and Zn in rat liver were found to be affected by deprivation of Ni, the direction and extent of changes being dependent on the Fe status.[147]

The transport of ^{63}Ni across membranes has been investigated in rat liver slices. It was not affected by Ca^{II}, by adrenaline, or by the metabolic inhibitors iodoacetate and N-ethylmaleimide. However, the uncoupling agent 2,4-dinitrophenol stimulated flux in and out of the liver. It was concluded that nickel was

[141] J. W. Spears, E. E. Hatfield, and R. M. Forbes, *J. Anim. Sci.*, 1979, **48**, 649.
[142] J. W. Spears and E. E. Hatfield, *J. Anim. Sci.*, 1978, **47**, 1345.
[143] A. Hennig, G. Jahreis, M. Anke, M. Partschefeld, and M. Grün, *Arch. Tierernaehr.*, 1978, **28**, 267.
[144] M. Kirchgessner and A. Schnegg, *Nutr. Metab.*, 1979, **23**, 62.
[145] F. H. Nielsen, T. J. Zimmerman, M. E. Collings, and D. R. Myron, *J. Nutr.*, 1979, **109**, 1623.
[146] F. H. Nielsen, T. R. Schuler, T. J. Zimmerman, M. E. Collings, and E. O. Uthus, *Biol. Trace Element Res.*, 1979, **1**, 325.
[147] F. H. Nielsen and T. R. Schuler, *Biol. Trace Element Res.*, 1979, **1**, 337.

primarily excluded from cells by a membrane transport process th at is closely
linked to oxidative phosphorylation.[148]

8 Selenium

The major interest in Se continues to be in the role of the element in the enzyme
glutathione peroxidase (GSHpx) and in how this may be related to the prevention of Se-responsive disorders.

**Peroxide Metabolism, Glutathione Peroxidase, and the Metabolism of Selenium
and Factors that Affect it.**—The biochemistry of selenium has been
reviewed,[149,150] the article by Stadtman[150] emphasizing its importance in the
bacterial enzymes formate dehydrogenase and glycine reductase as well as in
the GSHpx of animal cells. The role of GSHpx in the metabolism of peroxides
has been discussed in a section of an extensive review on metabolism of hydroperoxides by Chance, Sies, and Boveris.[151] During the destruction of peroxides
by GSHpx in rat liver mitochondria, previously accumulated Ca^{2+} ions were
expelled from the matrix of the organelle, with a concurrent oxidation of
NADPH. Mitochondria from livers of selenium-deficient rats would not expel
the Ca^{2+} ions on the addition of peroxide, probably due to low GSHpx activity.
Effects of the deficiency of selenium were also apparent in Arrhenius plots of
the uptake of Ca^{2+}. Changes in gradient of the plots occurred at 24 °C or at
11 °C with the mitochondria from normal rats and at 24 °C in the mitochondria
from selenium-deficient animals.[152] The capacity of rat liver cytosolic GSHpx
to metabolize peroxides has been calculated to be great enough to metabolize
the amounts of peroxide that are thought to be produced in the organ.[153] Metabolism of linoleic acid hydroperoxide by guinea pig liver cytosol does not appear
to be dependent on selenium, and is probably due to non-Se GSHpx activity of
gluthathione-S-transferases.[118,154] In the rabbit, non-Se GSHpx activity has
been found in liver and kidney but not in spleen, heart, lung, and erythrocytes.[155]
Supplementation of the diet with vitamin E had no effect on the decline of
GSHpx activity in the cardiac tissue of mice that were offered low-selenium
diets.[156] Non-Se GSHpx activity was not detectable in the cardiac tissue of
these animals. GSHpx activity in the blood of mice was, however, increased after
the feeding of 300 mg of vitamin E per kg in the diet for 18 months. The GSHpx
activity was not affected in any other of the body tissues.[157]

The structure of bovine erythrocyte GSHpx has been determined at 2.8 Å
resolution by *X*-ray crystallography. The enzyme molecule was thought to

[148] A. B. Chausmer, C. H. Rogers, and A. V. Colucci, *Nutr. Rep. Int.*, 1978, **18**, 249.
[149] C. P. Downes, C. A. McAuliffe, and M. R. C. Winter, *Inorg. Persp. Biol. Med.*, 1979,
2, 241.
[150] T. C. Stadtman, *Adv. Enzymol.*, 1979, **48**, 1.
[151] B. Chance, H. Sies, and A. Boveris, *Physiol. Rev.*, 1979, **59**, 527.
[152] H. R. Lotscher, K. H. Winterhalter, E. Carafoli, and C. Richter, *Proc. Natl. Acad. Sci.
USA*, 1979, **76**, 4340.
[153] J. W. Forstrom, F. H. Stults, and A. L. Tappel, *Arch. Biochem. Biophys.*, 1979, **193**, 51.
[154] M. R. Shreve, P. G. Morrissey, and P. J. O'Brien, *Biochem. J.*, 1979, **177**, 761.
[155] Y. H. Lee, D. K. Layman, and R. R. Bell, *Nutr. Rep. Int.*, 1979, **20**, 573.
[156] G. Y. Locker, J. H. Doroshow, J. C. Baldinger, and C. E. Myers, *Nutr. Rep. Int.*, 1979,
19, 671.
[157] L. C. Su, K. L. Ayaz, and A. S. Csallany, *Nutr. Rep. Int.*, 1979, **20**, 461.

consist of four identical subunits, each of 178 amino-acid residues.[158] The α-helices and β-pleated sheets in the structure were found to be similar in some respects to those found in flavodoxin, thioredoxin, rhodanese, and some dehydrogenases. The structure supported previous evidence that Se is in selenocysteine at the active centre of the enzyme.[118] It also appeared that each active site of GSHpx consisted of portions from two of the subunits.[158] Kinetic data also suggested a co-operative interaction between subunits of GSHpx during metabolism of peroxide.[159] The Se in GSHpx was postulated to cycle between the selenic and seleninic acid forms during enzymic activity.[158, 159] Suggestions have also been repeated that the Se can cycle between the selenol and selenic acid form during enzyme activity.[118, 160] Experiments *in vitro* have shown that cadmium(II) and zinc(II) ions inhibit bovine erythrocyte GSHpx activity.[161]

Vitamin B_6 deficiency in rats was shown to affect the incorporation of Se into GSHpx when selenomethionine was used as the dietary source of Se. When Se in the diet was supplied as selenite, there was no effect of the vitamin deficiency on GSHpx synthesis.[162]

In experiments with mouse fibroblasts grown in tissue culture, it was also shown that Se from selenite appeared to be more available for the synthesis of GSHpx than Se from selenomethionine. Selenomethionine was taken up by the cells but then became incorporated into several proteins, whereas Se as selenite appeared predominantly in a protein fraction that contained GSHpx activity.[163] Riboflavin deficiency in the baby pig caused a reduction in the hepatic concentration of Se and in the GSHpx activity. The GSHpx activity in muscle was also reduced, but without a significant change in the concentration of Se.[164] High dietary concentrations of Pb, Cd, Hg, and Ag had no effect on non-Se GSHpx activity in rat liver, and only Hg and Ag reduced the activity of Se-containing GSHpx.[165] The effects of S on the metabolism of Se[118] continue to be investigated. Increasing the dietary S content from 0.05 to 0.24% had no significant effect on the faecal excretion of an intraruminal dose of ^{75}Se in sheep, but increased the urinary excretion of the isotope over a period of 10 days from 12% to 22% of the dose.[166] In rats, increasing the dietary concentration of S (from 0.03 to 0.5%) decreased the GSHpx activity in small intestinal tissue but increased that in colonic tissue.[167]

The adverse effects of adding 40 mg of Cd per litre to the drinking water of mice throughout pregnancy were alleviated by 0.05, 2, 20, or 200 mg of Se per kg in the diet.[62]

[158] R. Ladenstein, O. Epp, K. Bartels, A. Jones, R. Huber, and A. Wendel, *J. Mol. Biol.*, 1979, **134**, 119.
[159] A. G. Splittgerber and A. L. Tappel, *J. Biol. Chem.*, 1979, **254**, 9807.
[160] J. W. Forstrom and A. L. Tappel, *J. Biol. Chem.*, 1979, **254**, 2888.
[161] A. G. Splittgerber and A. L. Tappel, *Arch. Biochem. Biophys.*, 1979, **197**, 534.
[162] K. Yasumoto, K. Iwama, and M. Yoshida, *J. Nutr.*, 1979, **109**, 760.
[163] C. L. White and W. G. Hoekstra, *Biol. Trace Element Res.*, 1979, **1**, 243.
[164] P. S. Brady, L. J. Brady, M. J. Parsons, D. E. Ullrey, and E. R. Miller, *J. Nutr.*, 1979, **109**, 1615.
[165] R. S. Black, P. D. Whanger, and M. L. Tripp, *Biol. Trace Element Res.*, 1979, **1**, 313.
[166] A. L. Pope, R. J. Moir, M. Somers, E. J. Underwood, and C. L. White, *J. Nutr.*, 1979, **109**, 1448.
[167] H. W. Lane, K. L. Shirley, and J. J. Cerda, *J. Nutr.*, 1979, **109**, 444.

The metabolism of Se has been studied, using systems that operate *in vivo* and *in vitro*. On administration of tracer and toxic doses of ^{75}Se to mice, there was an initial rapid dose-dependent loss of ^{75}Se from the body. Subsequent loss of radioactivity occurred at a similar rate in all animals, with a half-life of 14.4 days.[168] Other studies, using [^{75}Se]selenomethionine that was administered orally to mice, revealed that diurnal oscillations occurred in the extent of up-take of radioactivity by liver, pancreas, blood, kidney, skeletal muscle, bones, and brain. Uptake by liver, pancreas, and kidney was affected not only by the time of day at which a dose was administered but also by the nutritional state of the animals.[169] Selenite that was added to sheep blood *in vitro* became attached to a plasma protein by a process that was dependent on the initial uptake of the Se by erythrocytes and on the possible formation of a —Se—S— derivative.[170] Selenite was taken up by human lymphocytes after previous binding to a plasma protein. This binding of Se to the protein required erythrocytes and —SH groups but no source of energy.[171] Selenite has also been shown to inhibit the uptake of amino-acids by isolated hepatocytes.[172] Accumulation of Se in the liver of Merino sheep has been shown to be promoted by toxins from lupins.[173]

Selenium Deficiency.—Dietary Se deficiency can cause several clinical and sub-clinical changes in animals, many of which have been the subject of investigation.

The Se contents of livers from pigs with hepatosis dietetica (liver necrosis) and/or nutritional myopathy were found to be lower than in animals with mulberry heart disease. It was suggested that mulberry heart disease is associated with vitamin E deficiency whereas hepatosis dietetica and nutritional myopathy are manifestations of a combined deficiency of Se and of vitamin E.[174] Further evidence that low concentrations of Se in blood are associated with low concentrations of Se in feedstuff and the occurrence of myopathy in cattle has been presented.[175, 176]

Unlike the very low concentrations of Se in muscle in animals with nutritional muscular dystrophy, Se concentrations and GSHpx activities in muscle tissue from mice with genetic muscular dystrophy were greater than in muscle from control animals that were unaffected by the disease. When mice were transferred from Se-sufficient to Se-deficient diets, the muscle from the genetically dystrophic animals retained Se more readily than muscle from control animals.[177]

The influence of Se deficiency on peroxidation and other biochemical changes in animals has also been investigated. Selenium status did not alter the accumulation of thiobarbituric-acid-reactive metabolites (thought to be products of peroxidation of lipids) in the muscle of vitamin-E-deficient rats that were sub-

[168] J. C. Hansen and P. Kristensen, *J. Nutr.*, 1979, **109**, 1223.
[169] M. N. Eakins, *J. Nutr.*, 1979, **109**, 1865.
[170] C. H. McMurray and W. B. Davidson, *Biochim. Biophys. Acta*, 1979, **583**, 332.
[171] E. K. Porter, J. A. Karle, and A. Shrift, *J. Nutr.*, 1979, **109**, 1901.
[172] J. Hogberg and A. Kristoferson, *FEBS Lett.*, 1979, **107**, 77.
[173] J. G. Allen, *Vet. Rec.*, 1979, **105**, 434.
[174] D. C. Moir and H. G. Masters, *Aust. Vet. J.*, 1979, **55**, 360.
[175] U. Dotta, O. Abate, R. Gulielmino, and G. Mondo, *Schweiz. Arch. Tierheilkd.*, 1979, **121**, 395.
[176] G. A. Chalmers, M. DeCaire, C. J. Zachar, and M. W. Barrett, *Can. Vet. J.*, 1979, **20**, 105.
[177] N. W. Revis, C. Y. Horton, and S. Curtis, *Proc. Soc. Exp. Biol. Med.*, 1979, **160**, 139.

jected to swimming stress. Normal intakes of vitamin E did, however, prevent the accumulation of the metabolites.[178] The production of ethane has been used as an index of the peroxidation of lipids that is induced by the administration of drugs and CCl_4 to Se-deficient and Se-supplemented rats. Examination with an optical microscope revealed that the liver necrosis that is caused by CCl_4 in both groups of rats was not made more severe by the deficiency of Se.[179]

The effects of Se deficiency on the oxidative metabolism of drugs in the liver have been reviewed.[180] In livers from Se-deficient male and female rats, the metabolism of drugs by mixed-function oxidases was reduced in comparison to that in Se-supplemented control animals. This supports previous evidence that Se may be involved in the oxidation of foreign compounds by the liver.[181]

Peroxidation caused the appearance of fluorescent lipid compounds in the pigment of retinal material from Se-deficient rats; changes that were accompanied by increased thresholds in electroretinograms, which also showed decreased amplitudes.[182] The concentration of polyunsaturated fatty acids in retinal epithelium was also decreased by Se deficiency in rats, presumably as a result of peroxidation.[183] Selenium deficiency, as evidenced by low concentrations of Se in serum, has been shown to occur concurrently with raised γ-glutamyl transferase activity in the serum of thoroughbred horses. The raised enzyme activity was postulated to be due to a sub-clinical Se-responsive hepatic insufficiency.[184]

Neutrophils from Se-deficient cattle were less able to kill ingested yeast cells *in vitro* than were neutrophils from Se-supplemented animals.[185] Low GSHpx activity in whole blood, and presumably therefore low Se status, have been associated with an increased incidence of gut diseases in pigs.[186] Selenium has also been shown to stimulate immune-response mechanisms in rats even when administered in excess of the usually accepted requirements. Selenium deficiency in rats and mice depressed immune systems, especially cell-mediated processes.[187]

The Role of Selenium in Reproduction.—After being maintained on Se-deficient diets for 11 to 12 months, 80% of male rats showed defects in sperm.[188] Radioactive Se (as $^{75}SeO_3{}^{2-}$), when administered intraperitoneally to rats, has been shown to accumulate in the testis over a period of seven days. Sixty per cent of the radioactivity was found in the cytosol on a protein of mol. wt 59 000. Chromatography of cytosol that had been treated with sodium dodecylsulphate (SDS) revealed the binding of Se to proteins with molecular weights of 57 000, 45 000, and 15 000. There was strong binding of the ^{75}Se to the protein of mol.

[178] P. S. Brady, L. J. Brady, and D. E. Ullrey, *J. Nutr.*, 1979, **109**, 1103.
[179] R. F. Burk and J. M. Lane, *Toxicol. Appl. Pharmacol.*, 1979, **50**, 467.
[180] A. T. Diplock, *Adv. Pharmacol. Ther., Proc. Int. Congr. Pharmacol. 7th (1978)*, 1979, **8**, 25.
[181] L. R. Shull, G. W. Buckmaster, and P. R. Cheeke, *J. Environ. Pathol. Toxicol.*, 1979, **2**, 1127.
[182] W. L. Stone, M. L. Katz, M. Lurie, M. F. Marma, and E. A. Dratz, *Photochem. Photobiol.*, 1979, **29**, 725.
[183] C. C. Farnsworth, W. L. Stone, and E. A. Dratz, *Biochim. Biophys. Acta*, 1979, **522**, 281.
[184] D. J. Blackmore, K. Willett, and D. Agness, *Res. Vet. Sci.*, 1979, **26**, 76.
[185] R. Boyne and J. R. Arthur, *J. Comp. Pathol.*, 1979, **89**, 151.
[186] P. F. Jorgensen and I. Wegger, *Acta Vet. Scand.*, 1979, **20**, 610.
[187] B. E. Sheffy and R. D. Schultz, *Fed. Proc.*, 1979, **39**, 2139.
[188] A. S. H. Wu, J. E. Oldfield, L. R. Shull, and P. R. Cheeke, *Biol. Reprod.*, 1979, **20**, 793.

wt 15 000, the radioactivity not being removed by dialysis against water, 2M-NaCl, β-mercaptoethanol, 8M-urea, or selenite. This strong binding of ^{75}Se to a specific protein suggested that it has a physiological role.[189] Selenium has also been found in a structural protein, of mol. wt 20 000, from bull sperm mitochondria. This protein was a component of a complex of proteins which were insoluble in SDS. The complex was postulated to stabilize the outer membrane of the mitochondrion.[190] Selenium that was administered to bulls as ^{75}SeO$_3$$^{2-}$ was found in seminal plasma, along with GSHpx activity. The ^{75}Se appeared to be retained in the reproductive tract in preference to all body tissues apart from the kidneys.[191] In contrast to previous reports, administration of Se had no effect on the incidence of retained placentae in dairy cattle. However, it was not stated whether the cattle were of low Se status before administration of extra Se.[192] Supplementation with Se by oral drenching had inconsistent effects on levels of progesterone in the plasma of ewes; in some cases, the concentration of the steroid was increased.[193]

Measurements of the Selenium Status of Animals.—As there are many Se-responsive disorders in livestock, it is important to be able to assess the Se status of animals. It may then be possible to predict whether there is any risk of a Se-deficiency syndrome arising.

The Se status of an animal can conveniently be measured by the activity of the selenoenzyme GSHpx in tissues. The enzymic assay is easier to perform than the determination of total Se concentration, and has the advantage of only measuring the Se that is available for biosynthesis of GSHpx. The GSHpx activity of whole blood has been used to assess the Se status of sheep in Britain[194] and of cattle in N.E. Scotland,[195] Devon,[196] Sweden,[197] and the Netherlands.[198] These results were obtained by using conventional spectrophotometric methods for the determination of GSHpx. Recently, tests relying on the defluorescence of reaction mixtures in the presence of GSHpx from blood have been developed for field use.[199, 200] These tests, however, can only be used to assign animals to very broad categories of Se concentration and GSHpx activity in the blood. The peroxidation of lipids in erythrocytes and GSHpx activity in young pigs were found to be influenced by dietary concentrations of Se and vitamin E, and thus were thought to provide useful indices of the Se/vitamin E status of the

[189] K. P. McConnell, R. M. Burton, T. Kute, and P. J. Higgins, *Biochim. Biophys. Acta*, 1979, **588**, 113.
[190] V. Pallini and E. Bacci, *J. Submicrosc. Cytol.*, 1979, **11**, 165.
[191] D. G. Smith, P. L. Senger, J. F. McCutchan, and C. A. Landa, *Biol. Reprod.*, 1979, **20**, 377.
[192] F. C. Gwazdauskas, T. L. Bibb, M. L. Gilliard, and J. A. Lineweaver, *J. Dairy Sci.*, 1979, **62**, 978.
[193] S. K. Walker, J. M. Obst, D. H. Smith, G. P. Hall, and P. F. Flavel, *Aust. J. Biol. Sci.*, 1979, **32**, 221.
[194] P. H. Anderson, S. Berrett, and D. S. P. Patterson, *Vet. Rec.*, 1979, **104**, 235.
[195] J. R. Arthur, J. Price, and C. F. Mills, *Vet. Rec.*, 1979, **104**, 340.
[196] P. A. Bloxham, G. W. Davis, and R. L. Stephenson, *Vet. Rec.*, 1979, **105**, 201.
[197] G. Carlstrom, G. Jonsson, and B. Pehrson, *Swed. J. Agric. Res.*, 1979, **9**, 43.
[198] W. T. Binnerts, *Neth. Milk Dairy J.*, 1979, **33**, 24.
[199] S. Berrett and C. N. Hebert, *Vet. Rec.*, 1979, **105**, 145.
[200] K. A. Backall and R. W. Scholz, *Am. J. Vet. Res.*, 1979, **40**, 733.

animals.[201] The peroxidation of haemoglobin that was initiated by hydrogen peroxide and GSHpx activity were significantly correlated in whole blood from sheep. The possibility of using this as an indicator of the Se status of the sheep has been discussed.[202] Concentrations of Se in plasma have been used to predict its concentrations in the tissues of pigs. The activity of the enzymes aspartate aminotransferase and alanine aminotransferase in plasma, however, proved to be unsuitable for this purpose.[203] These enzymes are released into the plasma when liver cells are damaged during Se deficiency. Concentrations of Se in tissues were related to the GSHpx activity in the whole blood of rabbits that had been fed both Se-supplemented and low-Se diets.[204] The distribution of GSHpx and Se in the blood of dairy cows has been examined in detail; 98% of GSHpx activity in whole blood was found in the red blood cells.[205] Measurement of Se concentrations and GSHpx activities in blood and other body tissues of various animals has been reviewed.[206] Changing the concentration of Se in the diet, although it changed the GSHpx activity, had no influence on some catalase and superoxide dismutase activities of rat erythrocytes.[207]

Supplementation with Selenium.—Selenium at a concentration of 0.05 mg per kg, in the presence of adequate vitamin E, in the diet of laying and breeding hens is thought to be needed for optimal reproductive performance and growth.[208]

Selenium depletion and repletion by dietary means decreased and then increased the GSHpx activity in plasma and the concentration of Se in the blood and body tissues of growing pigs.[209-211] Increasing the amounts of Se in the diets of pigs to <0.1 mg per kg of dry matter increased the gains in weight.[212] Oral drenching of sheep[213] with Se and injection of calves[214] with Se also increased weight gains. The growth of mouse fibroblasts that were cultured in a medium based on foetal calf serum was stimulated by Se (<0.1 nmol l^{-1}).[215] Although Se supplementation was shown in some cases to have no effect on the growth and reproduction of pigs, these animals had a normal Se status at the start of the experiment.[216]

The effects of Se supplementation on the prevention of several forms of cancer have been reviewed in detail.[217] Various anti-oxidants have been com-

201 P. Thode-Jensen, V. Danielsen, and H. E. Nielsen, *Acta Vet. Scand.*, 1979, **20**, 92.
202 J. F. Wilkins, *Aust. J. Biol. Sci.*, 1979, **32**, 451.
203 M. G. Simesen, H. E. Nielsen, V. Danielsen, G. Gissel-Nielsen, W. Hjarde, T. Leth, and A. Basse, *Acta Vet. Scand.*, 1979, **20**, 289.
204 E. Wiesner, F. Berschneider, and S. Willer, *Arch. Exp. Veterinaermed.*, 1979, **33**, 299.
205 R. W. Scholz, and L. J. Hutchinson, *Am. J. Vet. Res.*, 1979, **40**, 245.
206 J. Baart and H. Deelstra, *Farm. Tijdschr. Belg.*, 1979, **56**, 219.
207 C. K. Chow, *Int. J. Vit. Nutr. Res.*, 1979, **49**, 182.
208 G. F. Combs and M. L. Scott, *Poultry Sci.*, 1979, **58**, 871.
209 E. R. Chavez, *Can. J. Anim. Sci.*, 1979, **59**, 67.
210 E. R. Chavez, *Can. J. Anim. Sci.*, 1979, **59**, 761.
211 A. J. Froseth, *Acta Agric. Scand., Suppl.*, 1979, **21**, 219.
212 H. E. Nielsen, V. Danielsen, M. G. Simesen, G. Glissel-Nielsen, W. Hjarde, T. Leth, and A. Basse, *Acta Vet. Scand.*, 1979, **20**, 276.
213 D. I. Paynter, J. W. Anderson, and J. W. McDonald, *Aust. J. Agric. Res.*, 1979, **30**, 703.
214 W. H. Johnson, B. B. Norman, and J. R. Dunbar, *Calif. Agric.*, 1979, **33**, 14.
215 A. S. M. Giasuddin and A. T. Diplock, *Arch. Biochem. Biophys.*, 1979, **196**, 270.
216 T. L. Piatkowski, D. C. Mahan, A. H. Cantor, A. L. Moxon, J. H. Cline, and A. P. Grifo, *J. Anim. Sci.*, 1979, **48**, 1357.
217 A. C. Griffin, *Adv. Cancer Res.*, 1979, **29**, 419.

pared with Se as to their effectiveness in the prevention of chemically induced cancers of the colon, liver, and lung of rats.[218] Selenium inhibited the formation of the tumours in all three tissues whereas the anti-oxidants were most effective against the hepatic tumours.

9 Silicon

Even though chicks that were given silicon-deficient diets were able to grow normally, the development of their skulls was abnormal, the major defect appearing to result from reduced synthesis of collagen.[219] In other studies with chicks, concentrations of Si in plasma were inversely related to the dietary intake of molybdenum.[220]

The problems of unambiguous identification of Si as a natural constituent of biological samples have been investigated, using *X*-ray microanalysis, and the authors concluded that the presence of Si in blood vessels was not an artefact of the method of preparation.[221] This view is supported by research which showed that there is a suppression of experimental atheromas in rabbits by administration of Si as lysyl silicate.[222]

In Man, diets that are high in fibre from fruit and vegetable sources increased the dietary intakes of Si but reduced its retention.[223] The proportion of dietary Si that was excreted in the urine decreased with the high-fibre diet, but the absolute amount in urine increased. Urinary excretion of Si increased following the oral administration of a range of Si compounds to rats.[224]

10 Zinc

Analysis and Assessment of Zinc Status.—Although it has been common practice to precipitate proteins from plasma or serum before Zn is estimated by atomic absorption,[118] it has now been proposed that the standard method should involve the direct measurement of Zn in samples that have been diluted 1 to 5 with water.[225] Already this proposal has met with criticism,[226] and its adoption would appear to be uncertain. Reports that concentrations of Zn in serum are higher than values in plasma have been refuted, any differences being considered to be caused by haemolysis and liberation of Zn from erythrocytes during clotting.[227]

The value of the concentration of Zn in plasma as an indicator of Zn status has been questioned for a variety of reasons in the past,[118] and again recently on another basis. Occlusion of the blood vessel prior to sampling by venepuncture increased both the protein and Zn contents of the plasma, but the relation-

[218] M. M. Jacobs and A. C. Griffin, *Biol. Trace Element Res.*, 1979, **1**, 1.
[219] E. M. Carlisle, *J. Nutr.*, 1980, **110**, 352.
[220] E. M. Carlisle, *Fed. Proc.*, 1979, **38**, 553.
[221] C. H. Becker and A. G. S. Janossy, *Micron*, 1979, **10**, 267.
[222] J. Loeper, J. Goy-Loeper, L. Rozensztajn, and M. Fragny, *Atherosclerosis*, 1979, **33**, 397.
[223] J. L. Kelsey, K. M. Behall, and E. S. Prather, *Am. J. Clin. Nutr.*, 1979, **32**, 1876.
[224] G. M. Benke and T. W. Osborn, *Food Cosmet. Toxicol.*, 1979, **17**, 123.
[225] J. C. Smith, G. P. Butrimovitz, W. C. Purdy, R. L. Boeckx, R. Chu, M. E. McIntosh, K-D. Lee, J. K. Lynn, E. C. Dinovo, A. S. Prasad, and H. Spencer, *Clin. Chem.*, 1979, **25**, 1487.
[226] J. R. Kelson, *Clin. Chem.*, 1980, **26**, 349.
[227] D. J. Kosman and R. I. Henkin, *Lancet*, 1979, **1**, 1410.

ships between protein and Zn were too variable to permit ready correction for this effect.[228] The problems of diagnosis of Zn deficiency have been reviewed[11] and attempts have been made to develop more sensitive indicators of this deficiency. The uptake of ^{65}Zn by erythrocytes *in vitro* proved less useful as a diagnostic aid than the change in serum alkaline phosphatase after injection of Zn.[229-231] Activities of alkaline phosphatase in serum decreased only in animals whose intakes of Zn were marginal to deficient, but the effects of other factors that are known to influence alkaline phosphatase activity were not investigated. Other workers have found differences between patients in the response of their concentrations of Zn in plasma to a standard large oral dose of Zn, but the relevance of these responses to the patients' requirements for Zn remains uncertain.[232] Several groups have investigated the measurement of the concentration of Zn in saliva as a non-invasive method of determining Zn status. Studies with rats gave little hope that the method would prove useful,[233] but other investigations have shown marked differences between the soluble and particulate fractions of saliva.[234-236] The former contained a much lower concentration of Zn than the particulate material, but appeared to be more responsive to Zn intake. Changes in taste acuity and taste preferences may also be helpful in confirming a tentative diagnosis of Zn deficiency.[237]

Future advances in the understanding of the relationships between the concentration of Zn in plasma and the Zn status of animals may depend on an appreciation of the factors that control the binding of Zn to the plasma proteins. Studies, *in vitro*, of the relative affinities for Zn of human albumin and transferrin suggested that the latter would not compete effectively for Zn *in vivo*, and that albumin would be the major binding agent for labile Zn in plasma.[238] Comparison of the binding of Zn by synthetic peptides of known structure with that by albumin suggested that the protein's binding site for Zn involves histidyl and carboxyl residues.[239] In addition to albumin, α_2-macroglobulin also binds Zn in plasma, and a third Zn-binding component has been observed in the rabbit.[240] In a number of human diseases, the amount of Zn on α_2-macroglobulin remained constant despite wide changes in the total concentration of Zn in the plasma, and in each instance the molar ratio of Zn to α_2-macroglobulin was close to unity.[241]

A new technique for the isolation of Zn proteins involves affinity chromato-

[228] B. E. Walker, I. Bone, B. H. Mascie-Taylor, and J. Kelleher, *Int. J. Vitamin Nutr. Res.*, 1979, **49**, 411.
[229] H-P. Roth and M. Kirchgessner, *Z. Tierphysiol. Tierernaehr. Futtermittelkd.*, 1979, **42**, 95.
[230] H-P. Roth and M. Kirchgessner, *Res. Exp. Med. (Berlin)*, 1979, **174**, 283.
[231] H-P. Roth and M. Kirchgessner, *Zentralbl. Veterinaermed., Reihe A*, 1979, **26**, 835.
[232] J. F. Sullivan, M. M. Jetton, and R. E. Burch, *J. Lab. Clin. Med.*, 1979, **93**, 485.
[233] G. A. Everett and J. Apgar, *J. Nutr.*, 1979, **109**, 406.
[234] A. Baratieri, A. Picarelli, and D. Piselli, *J. Dental Res.*, 1979, **58**, 440.
[235] J. L. Greger and V. S. Sickles, *Am. J. Clin. Nutr.*, 1979, **32**, 1859.
[236] F. J. Langmyhr and B. Eyde, *Anal. Chim. Acta*, 1979, **107**, 211.
[237] F. A. Catalanotto, *J. Nutr.*, 1979, **109**, 1079.
[238] P. A. Charlwood, *Biochim. Biophys. Acta*, 1979, **581**, 260.
[239] H. Lokustra and B. Sarkar, *J. Inorg. Biochem.*, 1979, **11**, 303.
[240] K. Hayakawa, T. Imanari, S. Asazuma, R. Kizu, and M. Miyazaki, *Chem. Pharm. Bull.*, 1979, **27**, 2849.
[241] M. K. Song and N. F. Adham, *Clin. Chim. Acta*, 1979, **99**, 13.

graphy of the proteins on the Zn chelate of iminodiacetic acid that is covalently linked to Sepharose.[242] The inherent Zn-binding capacity of the matrix during gel chromatography tends to cause undesirable dissociation of loosely bound complexes, and a recent report recommends equilibration of Sephadex with Zn-containing buffers prior to separation of the labile Zn complexes.[243] The usefulness of this approach appears limited to identification of the full range of Zn-binding agents that are present rather than those that are specifically present *in vivo*. The extent to which compounds that are used as pH buffers may themselves bind Zn has also been catalogued.[244]

Although the above methods may clarify our understanding of factors influencing the quantities and distribution of Zn in plasma, two recent studies provide a wider insight into the metabolism of Zn within the whole animal.[245, 246] In these investigations, patients were given ^{69m}Zn either orally or intravenously, and the kinetics of absorption, redistribution, and excretion of Zn were studied. Absorption of the oral dose had virtually ceased within four hours, two thirds of the absorbed Zn being taken up initially by the liver. The subsequent patterns of redistribution and loss of Zn were similar, regardless of the route of dosing, with as many as five terms being required in the exponential equations that are used to describe the loss of isotope.

The Effect of Zinc on Growth and Cell Division.—Previous knowledge of the role of Zn in biochemical and physiological functions has been reviewed[247-249] and possibly new insights have been gained into the function of Zn in controlling cell replication.[250-254]

Zinc deficiency in cultures of *Candida albicans* reduced cell numbers in stationary-phase cultures without causing a concomitant reduction in the total cell mass, suggesting that the deficiency permitted the continued accumulation of cell substance but restricted cell division.[253] Mice that had been inoculated with Ehrlich ascites cells have been shown to provide another useful model for the study of Zn deficiency. The concentrations of Zn in ascites fluid were found to depend on those in the diet, although the intracellular concentrations of Zn in the ascites cells were independent of Zn supply.[251] Zinc deficiency that was imposed prior to or at the time of inoculation restricted the growth of the tumour

[242] T. Kurecki, L. F. Kress, and M. Laskowski, *Anal. Biochem.*, 1979, **99**, 415.
[243] G. W. Evans, P. E. Johnson, J. G. Brushmiller, and R. W. Ames, *Anal. Chem.*, 1979, **51**, 839.
[244] H. B. Collier, *Clin. Chem.*, 1979, **25**, 495.
[245] R. L. Aamodt, W. F. Rumble, G. S. Johnston, D. Foster, and R. I. Henkin, *Am. J. Clin. Nutr.*, 1979, **32**, 559.
[246] D. M. Foster, R. L. Aamodt, R. I. Henkin, and M. Berman, *Am. J. Physiol.*, 1979, **237**, R340.
[247] D. S. Auld, *Adv. Chem. Ser.*, 1979, **172**, 112.
[248] A. S. Prasad, *Annu. Rev. Pharmacol. Toxicol.*, 1979, **20**, 393.
[249] P. J. Aggett and J. T. Harries, *Arch. Dis. Child.*, 1979, **54**, 909.
[250] D. H. Petering and L. A. Saryan, *Biol. Trace Element Res.*, 1979, **1**, 87.
[251] D. T. Minkel, P. J. Dolhun, B. L. Calhoun, L. A. Saryan, and D. H. Petering, *Cancer Res.*, 1979, **39**, 2951.
[252] L. A. Saryan, D. T. Minkel, P. J. Dolhun, B. L. Calhoun, S. Wielgus, M. Schaller, and D. H. Petering, *Cancer Res.*, 1979, **39**, 2457.
[253] G. W. Bedell and D. R. Soll, *Infect. Immun.*, 1979, **26**, 348.
[254] G. J. Brewer, J. C. Aster, C. A. Knutsen, and W. C. Kruckeberg, *Am. J. Hematol.*, 1979, **7**, 53.

cells, but, when imposed after the establishment of the tumour, the latter appeared to be able to compete successfully with the animal for the available Zn. When the cells were harvested and their ability to utilize labelled thymidine and uridine was investigated, Zn deficiency did not appear to have much influence on RNA synthesis, but it restricted both the phosphorylation and the incorporation of thymidine into DNA.[250, 252] The effects were not consistent with a reduction in the rate of utilization of thymidine by the population as a whole, but suggested that Zn deficiency reduced the proportion of cells that are capable of synthesizing DNA. However, those cells which did synthesize DNA appeared to do so normally, but the overall respiration rates of the cultures were markedly increased by Zn deficiency.

It has been suggested that regulation of cell metabolism by Zn may involve the antagonism of those effects of Ca which are mediated by calmodulin, but, in general, the concentrations of Zn that are required to produce these effects were substantially greater than those that are likely to be present *in vivo*.[254] Zinc deficiency has been shown to delay the onset of oral cancer in rats that had been treated with 4-nitroquinoline *N*-oxide[255] and to restrict the activity of thymidine kinase in connective tissue. On the other hand, the major Zn proteins of lymphocytes did not appear to be influenced by leukaemic or lectin-induced transformation,[256] and the inhibition of growth of the muscles of weanling rats that were given Zn-deficient diets was mainly associated with the reduction in food intake which accompanied the deficiency.[257]

The Effect of Zinc on Reproduction and Neonatal Development.—The concentrations of Zn in the accessory genital glands of male sheep were lower in castrates than in intact rams, but the differences failed to disappear following a single dose of testosterone.[258] Zinc in sperm is mainly located in the tails, recent evidence suggesting that it is associated with the sulphydryl-rich polypeptides of tail keratin.[259]

The importance of Zn to the foetus and neonate has been reviewed[260] and the effects of low Zn intake on pregnancy in the ewe have been studied.[261] Although the ewes showed signs of Zn deficiency, and four out of ten of the deficient ewes died, the birth-weights of the lambs were not different in the control and deficient groups, and there were no congenital malformations in the lambs. In contrast, the foetuses of rats that were fed Zn-deficient diets were lighter in weight, and had malformations of the skeleton.[262] In a study of women at high risk of giving birth to babies with Down's syndrome or neural tube defects, the concentrations of Zn in amniotic fluid did not correlate with those in plasma, and neither differed significantly from control values.[263] However, the concentra-

[255] K. Wallenius, A. Mathur, and M. Abdulla, *Int. J. Oral Surg.*, 1979, **8**, 56.
[256] T. Guenther, R. Averdunk, and H. Ruehl, *Z. Immunitaetsforch.*, 1979, **155**, 269.
[257] M. J. O'Leary, C. J. McClain, and P. V. J. Hegarty, *Br. J. Nutr.*, 1979, **42**, 487.
[258] M. Hidiroglou, C. J. Williams, and L. Tryphonas, *Am. J. Vet. Res.*, 1979, **40**, 103.
[259] H. I. Calvin, *Biol. Reprod.*, 1979, **21**, 873.
[260] J. C. L. Shaw, *Am. J. Dis. Child.*, 1979, **133**, 1260.
[261] J. Apgar and H. F. Travis, *J. Anim. Sci.*, 1979, **48**, 1234.
[262] W. Hickory, R. Nanda, and F. A. Catalanotto, *J. Nutr.*, 1979, **109**, 883.
[263] T. R. Shearer, E. W. Lis, K. S. Johnson, J. R. Johnson, and G. H. Prescott, *Nutr. Rep. Int.*, 1979, **19**, 209.

tion of Zn in the hair of parents of achondroplastic children tended to be lower than control values, and that in the hair of the children themselves was even lower still.[264]

Neonatal rats had high concentrations of Zn metallothionein in their livers, but these declined to negligible amounts after 4—5 weeks.[265, 266] Concentrations of trace elements in the brain and spinal cord of rats appeared to be low at birth, and Zn accumulated to adult values over the initial 2—3 weeks.[267] The deposits of Zn in the mossy fibres of the hippocampus appeared to be related to neuronal function, both synaptic response and Zn content being reduced by dietary Zn deficiency.[268] However, the effects of the deficiency on the fibres were reversed within 48 hours of re-feeding Zn despite a previous period of five weeks on the low-Zn diet. The nature of the Zn deposits remains unknown, but extracts of bovine cerebral white matter have yielded a dimeric protein that contains one atom of Zn and eleven sulphydryl groups per dimer.[269] Despite high levels of Zn in the hippocampus of normal rats, a prolonged dietary deficiency of Zn caused changes in the behaviour of rats which were characteristic of excessive glucocorticoid levels rather than those associated with lesions in the hippocampus.[270] These rats had been subjected to low Zn intake since they were 32 days old. Others that had been reared by dams which were fed Zn-deficient diets and then weaned onto normal diets showed retarded development of long-term memory.[271]

Absorption of Zinc.—Measurements of absorption of Zn are often confounded by the secretion of endogenous Zn into the gut, reducing the apparent availability of Zn from the diet. A method for correcting for endogenous secretion has been published previously,[118] and a recent report, utilizing the intramuscular injection of ^{65}Zn to label the endogenous Zn, has confirmed the importance of this phenomenon.[272] Zinc homeostasis appears to depend to a considerable extent on the control of endogenous secretion rather than on the regulation of the initial absorption of Zn.

As animals age, the proportion of their energy intake that is used for growth decreases; since their requirement for Zn is largely controlled by their growth rate, the minimum dietary ratio of Zn to the energy that is required for normal growth also decreases. Studies with rats that were offered diets of constant Zn content have shown that the apparent reduction of availability of Zn from the diet as the rats aged was a consequence of excretion, *via* the faeces, of dietary Zn that was in excess of requirement, and did not indicate that there is a change in the true absorption of Zn from the diet.[273]

[264] P. J. Collipp, S. Y. Chen, V. T. Maddaiah, S. Amin, and M. Castro-Magana, *J. Pediatr.*, 1979, **94**, 609.
[265] J. U. Bell, *Toxicol. Appl. Pharmacol.*, 1979, **50**, 101.
[266] S. H. Oh and P. D. Wanger, *Am. J. Physiol.*, 1979, **237**, E18.
[267] M. Kozma and A. Ferke, *Acta Histochem.*, 1979, **65**, 219.
[268] G. W. Hesse, *Science*, 1979, **205**, 1005.
[269] A. A. Shaldzhyan, S. G. Sharoyan, R. M. Nalbandyan, and G. K. Bunyatyan, *Dokl. Akad. Nauk SSSR*, 1979, **249**, 1480.
[270] G. W. Hesse, K. A. F. Hesse, and F. A. Catalanotto, *Physiol. Behav.*, 1979, **22**, 211.
[271] E. S. Halas, M. D. Heinrich, and H. H. Sandstead, *Physiol. Behav.*, 1979, **22**, 991.
[272] G. W. Evans, E. C. Johnson, and P. E. Johnson, *J. Nutr.*, 1979, **109**, 1258.
[273] E. Weigand and M. Kirchgessner, *Biol. Trace Element Res.*, 1979, **1**, 347.

A number of factors do, however, influence the true availability of Zn from diets, one of the most important being phytic acid. Zinc from zinc phytate appeared to be fully available to rats,[274] but the molar ratio of phytate to Zn in these diets would have been less than one. However, as this ratio was increased to values of 15 and greater, the availability of Zn from the diets decreased.[275] Ratios of phytate to Zn of 25 and higher have been found in food products that are based on soya protein,[276, 277] and these are probably the prime cause of the low availability of Zn from these products.[277, 278] Another postulated cause of the low availability of Zn from diets is the presence of fibre components. The inclusion of cellulose at up to 16% of a diet that was given to rats produced little effect on the availability of Zn,[41] but a similar study with boys resulted in increased faecal excretion of Zn.[279] High-fibre diets that are based on fruit and vegetables also reduced the Zn balance of human subjects,[280] as did the inclusion of hemicelluloses.[281]

Absorption of Zn within the intestine has often been considered to be maximal in the duodenum, but perfusion experiments *in vivo* appeared to indicate a higher absorptive efficiency for Zn in the ileum of the rat.[282] Furthermore, studies of the transfer of Zn across everted sacs of rat gut failed to reveal differences in uptake by segments from duodenum and jejunum.[283]

The processes that regulate the absorption of Zn from the gut lumen and its transfer to the plasma have been reviewed.[284] In the past there has been considerable speculation regarding ligands of low molecular weight in the gut which have been postulated to facilitate the absorption of Zn. Recent evidence has suggested that some of these may be artefacts, caused by the breakdown of metallothionein during isolation procedures.[285] On the other hand, at least three Zn-binding components have been recognized following gel exclusion chromatography of duodenal secretions from Man.[286] Studies with patients suffering from acrodermatitis enteropathica (A.E.) showed a normal complement of Zn-binding ligands in the duodenal secretions, but the efficiency with which one of these, of mol. wt 5—7000, bound added Zn seemed to be reduced in the patients.[286] Mucosal biopsies of patients with A.E. showed reduced rates of accumulation of Zn *in vitro*,[287] and studies of the drug diiodoquin, which had previously been used to treat A.E., suggest that it might have acted as an iono-

[274] E. Weigand and M. Kirchgessner, *Z. Tierphysiol. Tierernaehr. Futtermittelkd.*, 1979, **42**, 137.
[275] N. T. Davies and S. E. Olpin, *Br. J. Nutr.*, 1979, **41**, 590.
[276] Anon., *Nutr. Rev.*, 1979, **37**, 365.
[277] N. T. Davies and H. Reid, *Br. J. Nutr.*, 1979, **41**, 579.
[278] R. M. Forbes, K. E. Weingartner, H. M. Parker, R. R. Bell, and J. W. Erdman, *J. Nutr.*, 1979, **109**, 1652.
[279] L. M. Drews, C. Kies, and H. M. Fox, *Am. J. Clin. Nutr.*, 1979, **32**, 1893.
[280] J. L. Kelsay, R. A. Jacob, and E. S. Prather, *Am. J. Clin. Nutr.*, 1979, **32**, 2307.
[281] C. Kies, H. M. Fox, and D. Beshgetoor, *Cereal Chem.*, 1979, **56**, 133.
[282] D. L. Antonson, A. J. Barak, and J. A. Vanderhoof, *J. Nutr.*, 1979, **109**, 142.
[283] N. Gruden and B. Momcilovic, *Nutr. Rep. Int.*, 1979, **19**, 483.
[284] R. J. Cousins, *Nutr. Rev.*, 1979, **37**, 97.
[285] R. J. Cousins, *Am. J. Clin. Nutr.*, 1979, **32**, 339.
[286] C. E. Casey, K. M. Hambidge, and P. A. Walravens, *J. Pediatr.*, 1979, **95**, 1008.
[287] D. J. Atherton, P. R. Muller, P. J. Aggett, and J. T. Harries, *Clin. Sci.*, 1979, **56**, 505.

phore which facilitated the uptake of Zn.[288] Frequently, young babies suffering from A.E. develop symptoms of Zn deficiency only after transfer from human to bovine milk, and the concentrations of Zn in the plasma of normal babies who were fed breast milk were higher than in those given bovine milk.[289] A recent report suggests that the difference may be associated with the binding of Zn mainly to relatively large Ca/casein micelles in bovine milk, but to a ligand of low molecular weight in human milk.[290] The latter has recently been identified as citric acid.[291] The addition of Fe reduced the absorption of Zn by rats from Zn-fortified bovine milk but not from unfortified milk.[283, 292]

Both *in vivo* and *in vitro*, the transport of Zn across the mucosa was influenced by prostaglandins.[293] Prostaglandin E_2 stimulated the transport of Zn from the mucosal to the serosal surface of the gut, whereas prostaglandin F_2 activated the reverse process. Prostaglandin E_2 also overcame the inhibitory effects of indomethacin on the uptake of Zn. Zinc deficiency, superimposed on low intakes of essential fatty acids, accentuated the development of skin lesions,[294] and the effects of Zn deficiency were apparently overcome by administration of essential fatty acids, especially γ-linolenic acid.[295] In studies of the pressor effects of noradrenalin, angiotensin, and K^+ on rat mesenteric preparations, the addition of 0.8 μg of Zn per ml to the perfusion medium produced the same response as the addition of prostaglandin E_1.[296] However, since the perfusate did not contain any protein, the concentrations of ionic Zn in these preparations would have been far in excess of physiological levels.

Metallothionein.—The Zn-binding protein metallothionein is normally present in relatively low concentrations in the liver of adult rats, but is synthesized in response to the injection of Zn [297] and to partial hepatectomy.[298] Stresses which caused a reduction in the concentration of Zn in plasma with a parallel increase in the content of Zn in the liver also induced the synthesis of metallothionein in the liver.[78, 299, 300] The latter was inhibited by actinomycin D,[300] and the cell-free translation of liver mRNA suggested that the increased rates of synthesis of metallothionein in rats that were injected with Zn were associated with increased quantities of metallothionein mRNA in the liver.[301] A similar rise in

[288] P. J. Aggett, H. T. Delves, and J. T. Harries, *Biochem. Biophys. Res. Commun.*, 1979, **87**, 513.
[289] K. M. Hambidge, P. A. Walravens, C. E. Casey, R. M. Brown, and C. Bender, *J. Pediatr.*, 1979, **94**, 607.
[290] J. E. Piletz and R. E. Ganschow, *Am. J. Clin. Nutr.*, 1979, **32**, 275.
[291] L. S. Hurley, B. Lonnerdal, and A. G. Stanislowski, *Lancet*, 1979, **1**, 677.
[292] B. Momcilovic and D. Kello, *Nutr. Rep. Int.*, 1979, **20**, 429.
[293] M. K. Song and N. F. Adham, *J. Nutr.*, 1979, **109**, 2152.
[294] W. J. Bettger, P. G. Reeves, E. A. Moscatelli, G. Reynolds, and B. L. O'Dell, *J. Nutr.*, 1979, **109**, 480.
[295] S. C. Cunnane, D. F. Horrobin, K. B. Ruf, and G. Sella, *J. Physiol.*, 1979, **296**, 83P.
[296] M. S. Manku, D. F. Horrobin, M. Karmazyn, and S. C. Cunnane, *Endocrinology*, 1979, **104**, 774.
[297] K. T. Suzuki and M. Yamamura, *Biochem. Pharmacol.*, 1979, **28**, 2852.
[298] H. Ohtake and M. Koga, *Biochem. J.*, 1979, **183**, 683.
[299] P. Z. Sobocinski, G. L. Knutsen, W. J. Canterbury, and E. C. Hauer, *Toxicol. Appl. Pharmacol.*, 1979, **50**, 557.
[300] P. Z. Sobocinski, W. J. Canterbury, E. C. Hauer, and F. A. Beall, *Proc. Soc. Exp. Biol. Med.*, 1979, **160**, 175.
[301] C. L. Hew and P. E. Penner, *Can. J. Biochem.*, 1979, **57**, 1030.

concentration of metallothionein mRNA in liver accompanied the increase in Zn metallothionein following the administration of dexamethasone to adrenalecto-mized rats.[302] The latter was not associated with an increase in the concentration of Zn in liver. In contrast, other investigations have shown an increase in both cytosolic Zn and Zn metallothionein in the liver after adrenalectomy, but subsequent administration of corticosterone did not alter the concentration of Zn metallothionein.[303] Although the elevated concentration of Zn in liver cytosol after adrenalectomy was not altered by dosing with aldosterone, this did reduce the levels of Zn metallothionein to those in sham operated controls.[303]

The already low concentrations of the metallothionein in normal rat liver were reduced still further by dietary Zn deficiency.[304] The synthesis of metallo-thionein is not limited to cells of tissues which normally accumulate trace metals, since HeLa cells in culture also produced the protein when challenged with Zn.[305] Similarly, lymphocytes that were cultured in the presence of Zn synthesized metallothionein.[306] The ability to synthesize this protein is obviously wide-spread, and an extensive review of the subject has recently been published.[307]

Hormonal and Immunological Effects of Zinc.—In mice, Zn deficiency has been shown to increase the concentration of corticosterone in plasma relative to that of controls which were fed *ad lib*.[308] However, a restriction of food intake due to Zn deficiency may have been at least partially responsible for the change. Furthermore, about half of the reduction in the lymphocyte T-cell-helper function which was observed in these animals occurred before the concentrations of corticosterone in plasma increased. Reduction in T lymphocyte function in Zn-deficient animals has been demonstrated[309-313] by several groups of inves-tigators, but the production of a circulating antibody is relatively unaffected.[311] The reduction of T-cell function was associated with increased numbers of autologous rosette-forming cells (immature T cells), and this change was reversed by giving Zn but not by adrenalectomy.[314] This again suggests that the inhibition of T-cell function does not result from high concentrations of corticosteroid in plasma. Zinc deficiency also reduced the levels of thymic hormone in mice.[313] Other studies with mice suggested that their Zn requirement was lower than

[302] K. R. Etzel, S. G. Shapiro, and R. J. Cousins, *Biochem. Biophys. Res. Commun.*, 1979, **89**, 1120.
[303] F. O. Brady and P. C. Bunger, *Biochem. Biophys. Res. Commun.*, 1979, **91**, 911.
[304] M. Panemangalore and F. O. Brady, *J. Nutr.*, 1979, **109**, 1825.
[305] C. J. Rudd and H. R. Herschman, *Toxicol. Appl. Pharmacol.*, 1979, **47**, 273.
[306] J. L. Phillips, *Biol. Trace Element Res.*, 1979, **1**, 359.
[307] 'Metallothionein', ed. J. H. R. Kagi and M. Nordberg, Birkhauser Verlag, Basel, 1979.
[308] P. De Pasquale-Jardieu and P. J. Fraker, *J. Nutr.*, 1979, **109**, 1847.
[309] J. M. Oleske, M. L. Westphal, S. Shore, D. Gorden, J. D. Bogden, and A. Nahmias, *Am. J. Dis. Child.*, 1979, **133**, 915.
[310] R. L. Gross, N. Osdin, L. Fong, and P. M. Newberne, *Am. J. Clin. Nutr.*, 1979, **32**, 1260.
[311] G. Fernandes, M. Nadir, K. Gnoe, T. Tanaka, R. Floyd, and R. A. Good, *Proc. Natl. Acad. Sci. USA*, 1979, **76**, 457.
[312] R. S. Pekarek, H. H. Sandstead, R. A. Jacob, and D. F. Barcome, *Am. J. Clin. Nutr.*, 1979, **32**, 1466.
[313] T. Owata, G. S. Incefy, T. Tanaka, G. Fernandes, C. J. Menendez-Botet, K. Pih, and R. A. Good, *Cell. Immunol.*, 1979, **47**, 100.
[314] L. Nash, T. Iwata, G. Fernandes, R. A. Good, and G. S. Incefy, *Cell. Immunol.*, 1979, **48**, 238.

that of rats for both growth and immune function.[315] The involvement of Zn in the immune response has also been reviewed recently.[316]

Zinc deficiency in rats resulted in a complex spectrum of changes in pituitary hormone concentrations, both in the gland and in the plasma.[317] Most consistent was a reduction in growth hormone and an increase in luteinizing hormone. There were no significant effects on thyroid-stimulating hormone or prolactin.

The actions of zinc on the pressor responses of rat mesenteric preparations to noradrenalin and angiotension mimicked those of prolactin,[296] but zinc has been shown to inhibit the action of prolactin in stimulating the synthesis of RNA and of casein in mammary gland explants.[318] In both instances, the concentrations of Zn^{2+} that were used were well above those that are likely to be present *in vivo*.

The administration of prednisolone to rats restricted their growth and increased the Zn content of the femurs of the rats which had been offered Zn-deficient diets. Although the hormone also restricted the growth of rats that were given Zn-adequate diets, it did not alter the concentration of Zn in their femurs.[319]

Recent reports have tended to refute previous suggestions that dietary Zn/Cu ratios influence the concentrations of cholesterol in plasma[26, 27] and have also questioned the importance of Zn in stabilizing membranes against oxidative damage *in vivo*.[320]

[315] R. W. Luecke and P. J. Fraker, *J. Nutr.*, 1979, **109**, 1373.
[316] L. H. Schloen, G. Fernandes, J. A. Garofalo, and R. A. Good, *Clin. Bull.*, 1979, **9**, 63.
[317] A. W. Root, G. Duckett, M. Sweetland, and E. O. Reiter, *J. Nutr.*, 1979, **109**, 958.
[318] J. A. Rillema, *Proc. Soc. Exp. Biol. Med.*, 1979, **162**, 464.
[319] J. Turnlund and S. Margen, *J. Nutr.*, 1979, **109**, 467.
[320] H. K. Kang, and R. A. Harnish, *Bull. Environ. Contam. Toxicol.*, 1979, **21**, 206.

9

Inorganic Elements in Biology and Medicine

N. J. BIRCH & P. J. SADLER

Introduction.—The objective of our present chapter, like that of its predecessors in Volumes 1 and 2 of these Reports,[1,2] is to review the biological and medical aspects of inorganic biochemistry. We have omitted consideration of inorganic nutrition since this is dealt with in detail in this and the previous volume of these Reports.[3,4] The scope of the present chapter is therefore that of the physiological and biochemical aspects of inorganic elements and the emerging areas of inorganic toxicology and inorganic pharmacology. We have attempted to extend the coverage of our previous Reports yet update certain aspects of them; because of this, our literature survey is not limited to 1979.

The biological and medical uses and the significance of inorganic elements constitute an area of interest that is blossoming, and it is important that, in the near future, we are able to provide a better description of the objects of our study. The apparent tautology and conflict is clearly represented in the title of this chapter, which was the subject of much editorial discussion in the first volume. Other major areas of descriptive conflict come in the indiscriminate use of the terms 'trace metals' and 'heavy metals'. Many of the elements involved are present in appreciable quantities, have relatively low atomic weights, and may not even be metals. The distribution of a range of elements has been discussed.[1]

Another point is of concern to us, that of the implied assumption that each element has characteristic properties in living systems. While this may be true of the lighter elements that readily form ions, it is important to appreciate that, for the bulk of the elements in Nature, the biological properties depend largely on the molecular environment of the element and hence its co-ordination chemistry. The two extremes of behaviour may be seen in the elements which are most widely used as inorganic drugs. Lithium and magnesium are absorbed as simple ions, and the form in which they are administered has very little effect on the final pharmacological action; the ions or their complexes with naturally occurring biological ligands are the active species. By contrast, gold and platinum

[1] N. J. Birch and P. J. Sadler, in 'Inorganic Biochemistry' (Specialist Periodical Reports) ed. H. A. O. Hill, The Chemical Society, London, 1979, Vol. 1, Ch. 9.
[2] N. J. Birch and P. J. Sadler, in 'Inorganic Biochemistry' (Specialist Periodical Reports), ed. H. A. O. Hill, The Royal Society of Chemistry, London, 1981, Vol. 2, Ch. 8.
[3] J. R. Arthur, I. Bremner, and G. K. Chesters, in 'Inorganic Biochemistry' (Specialist Periodical Reports) ed. H. A. O. Hill, The Royal Society of Chemistry, London, 1981, Vol. 2, Ch. 7.
[4] J. R. Arthur, I. Bremner, and G. K. Chesters, in this volume, Ch. 8.

compounds have a variety of actions, and these depend on the configuration of the particular compound; the whole compound, and not just the metal, is absorbed, and may be the active species.

Toxicology.—A comprehensive review of those biological effects of elements that are relevant to the understanding and evaluation of their toxicity has been published.[5] The first part of the book deals with general aspects of their toxicology, and the second part deals with 28 elements in turn: Al, Sb, As, Ba, Be, Bi, Cd, Cr, Co, Cu, Ge, In, Fe, Pb, Mn, Hg, Mo, Ni, Se, Ag, Te, Tl, Sn, Ti, W, U, V, and Zn. The emphasis is placed on a description of the biological effects rather than explanations in molecular terms. Under Ni, for example, we find information about analysis, production and uses, environmental exposure, metabolism, levels in tissues and fluids, effects and dose–response relationships (dermatitis due to nickel and its compounds is common in industry and in the general population), and carcinogenicity.

Inorganic Elements in the Biological Environment.—A number of very valuable reports have been published by the Canadian National Research Council of which European inorganic biochemists may not be aware. These reports, listed in Table 1, provide good bibliographic introductions to the areas in question and summarize the Canadian studies.

1 Lithium

Aspects of the use of lithium were considered in Volume 1 of these Reports,[1] but a number of major developments and controversies have occurred in the interval, and the present Report seeks to update the position to the end of 1980.

Lithium is used in the preventative treatment of bipolar periodic affective disorder, in which both manic and depressed phases have been seen in the patient's history, when a dose of from one to two grams per day is given as a prophylactic measure. It is also effective in unipolar depressive affective disorder.[6] Other psychiatric indications have been reviewed by Schou.[7] It has been suggested for use in recurrent schizo-affective disorder, pathological impulsive aggression, premenstrual tension syndrome, alcoholism, and drug addiction. Non-psychiatric uses have also been reviewed,[8] and the following have been reported:

  (i) Movement disorders (Huntington's Chorea, tardive dyskinesia)
  (ii) Thyroid diseases (hyperthyroidism, thyroid cancer)
  (iii) Inappropriate secretion of antidiuretic hormone
  (iv) Granulocytopenia.

The use of lithium in the treatment of granulocytopenia has developed rapidly, and a recent symposium[9] brings together much of the basic information.

[5] 'Handbook on the Toxicology of Metals', ed. L. Freberg, G. F. Nordberg, and V. B. Vouk, Elsevier/North Holland Biomedical Press, Amsterdam, 1979, p. 1.
[6] P. C. Baastrup, in 'Handbook of Lithium Therapy', ed. F. N. Johnson, M.T.P. Press, Lancaster, U.K., 1980, p. 26.
[7] M. Schou, in ref. 6, p. 68.
[8] R. F. Prien and M. Schou, in 'Lithium: Controversies and Unresolved Issues', ed. T. B. Cooper, S. Gershon, N. S. Kline, and M. Schou, Excerpta Medica, Amsterdam, 1979, p. 157.
[9] 'Lithium Effects on Granulopoiesis and Immune Function', ed. A. H. Rossof and W. A. Robinson, Plenum Press, New York, 1980 (Advances in Experimental Biology No. 127).

Table 1 *Special Reports of the Environmental Secretariat of the National Research Council of Canada which are of relevance to inorganic biochemistry*[a]

NRCC No.	Subject	Date	Pages
12 226	Environmental Fluoride	1971	39
13 501	The Effects of Pulp and Paper Wastes on Aquatic Life	1973	30
13 682	Lead in the Canadian Environment	1973	116
13 566	Radioactivity in the Environment	1974	53
13 690	Phosphorus Loading Concept in Eutrophication Research	1974	42
14 100	Waterborne Dissolved Oxygen Requirements	1975	111
15 015	Sulphur and Its Inorganic Derivatives in the Canadian Environment	1977	426
15 017	Effects of Chromium in the Canadian Environment	1976	168
15 019	The Effects of Alkali Halides in the Canadian Environment	1977	171
15 023	NTA (Nitrilotriacetic Acid) – An Ecological Appraisal	1976	46
15 391	Effects of Arsenic in the Canadian Environment	1978	349
16 081	Environmental Fluoride (1977)	1978	152
16 452	Effects of Asbestos in the Canadian Environment	1979	185
16 454	Copper in the Aquatic Environment	1979	227
16 736	Effects of Lead in the Environment – 1978 Quantitative Aspects	1979	779
16 739	Effects of Mercury in the Canadian Environment	1979	290
16 743	Effects of Cadmium in the Canadian Environment	1979	148
16 745	Effects of Lead in the Canadian Environment (1978)	1979	24
17 581	Water Hardness, Human Health, and the Importance of Magnesium	1979	119
18 132	Effects of Vanadium in the Canadian Environment	1980	94

(a) All publications are also available in French, and may be purchased from Publications, NRCC, Ottawa, Ontario, Canada K1A 0R6.

Regular monitoring and recording of concentrations of lithium in plasma are essential in long-term use, and the monitoring of thyroid and renal function is strongly advised. The objective is to maintain the concentration within the range 0.6—1.2 mmol l^{-1}. Toxic symptoms may be seen above 2 mmol l^{-1}. Elderly patients, whose renal function is gradually reduced, may require doses of as little as 400 mg of lithium carbonate per day to maintain the therapeutic level, while younger patients may require much higher doses.[10, 11]

Lithium Intoxication.—Minor side-effects have been discussed previously.[1] The development of lithium intoxication has been studied by Thomsen.[12, 13] The final stages of lithium intoxication affect mainly the nervous system, but are the result of the rapid rise in the concentration of lithium in plasma arising from the effects of lithium on renal function. Lithium is excreted through the kidney, and in a normal state the daily dose is totally excreted after the first few

[10] D. S. Hewick, P. Newbury, S. Hopwood, G. Naylor, and J. Moody, *Br. J. Clin. Pharmacol.*, 1977, **4**, 201.
[11] N. J. Birch, A. A. Greenfield, and R. P. Hullin, *Int. Pharmacopsychiatry*, 1980, **15**, 91.
[12] K. Thomsen, *J. Pharmacol Exp. Ther.*, 1976, **199**, 483.
[13] K. Thomsen and O. V. Olesen, *Gen. Pharmacol.*, 1978, **9**, 85.

days of treatment. Lithium is filtered at the glomerulus, and this filtrate passes to the proximal tubule, where about 80% of lithium is re-absorbed together with water and sodium. The remaining 20% of filtered lithium passes to the distal renal tubule, where very little is absorbed and most of it is excreted in the urine. This mechanism differs from that of the re-absorption of sodium, where 80% is absorbed in the proximal tubule and the fine control of excretion of sodium is achieved in the distal tubule, where up to 95% of the remaining sodium may be re-absorbed. If there is a fall in dietary sodium, the body conserves sodium by re-absorbing more in the distal tubule. However, should this mechanism not be adequate, proximal tubular re-absorption of sodium is increased. In lithium-treated patients, increased proximal re-absorption of sodium causes increased re-absorption of lithium. Thus lithium toxicity occurs, since lithium inhibits the re-absorption of sodium in the distal tubule. The renal response to this is to increase the proximal re-absorption of sodium and hence the proximal absorption of lithium. The increased absorption of lithium results in increased concentrations of lithium in plasma, which in turn exacerbates the loss of sodium.

Normally, the concentration of lithium in plasma rises after each successive dose of lithium, and the sodium-conservation mechanism returns towards normal as the concentration of lithium in plasma then falls as it is excreted. Concomitant re-absorption of lithium in the proximal tubule is therefore also reduced. However, if the dose of lithium is increased, or when an additional stress (such as depletion of water or sodium) is added, the effective clearance of lithium by the kidney is progressively decreased, since the sodium-conservation mechanism is operative for a larger proportion of the between-dose interval. The level of lithium in plasma thus rises and further stimulates the retention of sodium in the proximal tubule, leading to rapidly increasing concentrations of lithium in plasma; lithium toxicity becomes established. Haemodialysis to remove lithium from plasma is the treatment of choice.

The Distribution of Lithium in Tissues.—During prophylactic treatment with lithium, the concentration in most tissues is similar to that of plasma, and is related to the rate at which lithium enters and leaves particular tissues.[14]

Lithium accumulates in bone[15] and also in the pituitary gland,[16] and this is of particular importance when one recognizes that this endocrine organ is the major interface between the brain, *via* the hypothalamus, and the rest of the endocrine regulatory system. It is also concentrated in the thyroid and adrenal glands.[17]

Patt *et al.*[18] have investigated the possibility that lithium may be an essential element in the rat, and have maintained animals for three generations on lithium-depleted diets. After one generation, the pituitary gland contained ten times more lithium than any other tissue with the exception of the adrenal gland, which had approximately one-half of the content of the pituitary. Both the

[14] M. Schou, *Pharmacol. Rev.*, 1957, **9**, 17.
[15] N. J. Birch, *Clin. Sci. Mol. Med.*, 1974, **46**, 409.
[16] M. S. Ebadi, V. J. Simmonds, M. J. Hendrickson, and P. S. Lacy, *Eur. J. Pharmacol.*, 1974, **27**, 324.
[17] S. Stern, A. Frazer, J. Mendels, and C. Frustaci, *Life Sci.*, 1977, **20**, 1669.
[18] E. L. Patt, E. E. Piulett, and B. L. O'Dell, *Bioinorg. Chem.*, 1978, **9**, 299.

pituitary and the adrenals contained a similar quantity of lithium in the depleted group and in the group that was treated with commercial diets. In the control group, bone had the highest content of lithium, and might be considered to be an indicator of lithium exposure. This latter is in general agreement with other studies on bone.[15] In second-generation lithium-depleted animals the pituitary had the same concentration of lithium as both the first-generation lithium-depleted animals and both sets of controls. The second-generation animals had a somewhat reduced concentration in the adrenal glands, but it was still higher than in the remainder of the tissues. These findings suggest that the pituitary and adrenal glands accumulate lithium, and may indeed protect themselves from lithium depletion. The second- and third-generation females had reduced fertility compared with the first generation and the normal control group, although there was no effect on the growth rate.

Subcellular Actions of Lithium.—In a study of the exchange of ATP and ADP in mitochondria, lithium was compared with sodium, potassium, magnesium, and Tris buffer; no difference was seen between the effects of the different cations.[19] At low concentrations of lithium in mitochondria of rat brain, ADP-deficient respiration was inhibited at < 10 mmol l^{-1} but was stimulated above this concentration.[20] Other workers have reported that, in rat heart mitochondria, lithium caused a stimulation of calcium efflux, though this was not as effective as the stimulation that is caused by sodium.[21]

At concentrations below 1 mmol l^{-1}, lithium promotes the polymerization of tubulin in the presence of magnesium, and this may have a role in the stabilization of microtubules in neurosecretory systems.[22]

Glycolysis has a particular importance in metabolism in the brain, since, under normal circumstances, glucose is the major energy source of the brain. A number of enzymes in the cycle have been investigated. Lithium inhibits pyruvate kinase, at pharmacological concentrations, both in the enzyme that is obtained from rabbit muscle and in that from a purified rat brain preparation.[23] Inhibition characteristics of these two different sources were identical, and lithium was found to be competitive with respect to ADP and non-competitive with respect to other substrates, *e.g.* phosphoenol pyruvate, potassium, and magnesium. Pyruvate kinase is also inhibited by calcium, and the characteristics of the inhibition are identical except that inhibition by calcium is competitive with respect to magnesium. Lithium might interfere in the magnesium–ADP complex which is the substrate for the reaction.[24] Lithium did not inhibit manganese-activated pyruvate kinase, while calcium did.[25] This finding suggests that lithium and calcium inhibit at different sites.

A number of studies of other magnesium-dependent glycolytic enzymes have been reported. Phosphofructokinase (PFK) is inhibited competitively with respect

[19] H. Meissner, *Biochemistry*, 1971, **10**, 3485.
[20] A. R. Krall, *Life Sci.*, 1967, **6**, 1339.
[21] M. Crompton, M. Capano, and E. Carafoli, *Eur. J. Biochem.*, 1976, **69**, 453.
[22] B. Bhattacharya and J. Wolff, *Biochem. Biophys. Res. Commun.*, 1976, **73**, 383.
[23] P. K. Kajda, N. J. Birch, M. J. O'Brien, and R. P. Hullin, *J. Inorg. Biochem.*, 1979, **11**, 361.
[24] N. J. Birch and I. Goulding, *Anal. Biochem.*, 1975, **66**, 293.
[25] N. J. Birch, G. Foster, and P. K. Kajda, *Br. J. Pharmacol.*, 1981, **72**, 506P.

to ADP and magnesium but non-competitively with respect to potassium and fructose 6-diphosphate.[26] The action of PFK constitutes the major regulatory step in glycolysis. The inhibition by lithium of the reverse reaction to that catalysed by PFK, *i.e.* that mediated by fructose 1,6-bisphosphatase, was 47% at 2 mmol 1^{-1} of lithium, and this may suggest that lithium disturbs the relationship between glycolysis and gluconeogenesis or perhaps prevents the operation of futile cycles at this point.[27] Lithium inhibits glycolysis in human erythrocytes.[28]

The studies on lithium and glycolysis are put into perspective by previous reports of changes in carbohydrate metabolism in patients who were receiving lithium[29-31] and by studies on utilization of glucose and synthesis of glycogen in rats.[32] These studies suggest that there is an increase in the utilization of glucose and in the synthesis of glycogen, though there is disagreement with regard to the mechanisms involved. The importance of ions in the regulation of metabolism has been extensively reviewed by Rasmussen and his colleagues.[33]

Adenyl cyclase and cyclic AMP have provided foci for those who are studying the affective disorders and lithium. Miscellaneous effects of lithium have been reported and interactions between adenyl cyclase and cyclic AMP provide an attractive unifying hypothesis. In three recent reviews,[34-36] the literature has been evaluated.

The Effect of Lithium on Membranes.—It has been suggested that the concentration of lithium in the red blood cells (erythrocytes) during prophylactic treatment with lithium may have clinical value either in the prediction of concentrations of lithium in the brain[37] or for discrimination between unipolar and bipolar patients and responders and non-responders to the drug.[38, 39] These proposals have been widely disputed.[40, 41]

The kinetics of uptake of lithium by normal red cells have been studied in

[26] P. K. Kajda and N. J. Birch, *J. Inorg. Biochem.*, 1981, **14**, 275.

[27] P. K. Kajda and N. J. Birch, unpublished work.

[28] N. J. Birch, P. K. Kajda, and R. Wareing, *Br. J. Pharmacol.*, 1981, **72**, 576P.

[29] E. T. Mellerup, P. Plenge, and O. J. Rafaelsen, *Dan. Med. Bull.*, 1974, **21**, 88.

[30] E. T. Mellerup and O. J. Rafaelsen, in 'Lithium Research and Therapy', ed. F. N. Johnson, Academic Press, London, 1975, p. 381.

[31] P. B. Vendsborg and O. J. Rafaelsen, *Acta Psychiatr. Scand.*, 1973, **49**, 601.

[32] E. S. Haugaard, A. Frazer, J. Mendels, and N. Haugaard, *Biochem. Pharmacol.*, 1975, **24**, 1187.

[33] H. Rasmussen, D. P. Goodman, N. Friedman, J. E. Allen, and K. Kurokawa, in 'Handbook of Physiology' Sect. 7 (Vol. III), American Physiological Society, Washington, D.C., 1976, p. 225.

[34] R. P. Ebstein and R. H. Belmaker, in ref. 8, p. 703.

[35] J. Forn, in ref. 30, p. 485.

[36] E. Friedman, in 'Lithium: its Role in Psychiatric Research and Treatment', ed. S. Gershon and B. Shopsin, Plenum Press, New York, 1973, p. 75.

[37] A. Frazer, J. Mendels, S. K. Secunda, C. M. Cochrane, and C. P. Bianchi, *J. Psychiatr. Res.*, 1973, **10**, 1.

[38] A. Elizur, B. Shopsin, S. Gershon, and A. Ehlenberger, *Clin. Pharmacol. Ther.*, 1972, **13**, 947.

[39] J. Mendels and A. Frazer, *J. Psychiatr. Res.*, 1973, **10**, 9.

[40] C. R. Lee, S. M. Hill, M. Dimitrakoudi, F. A. Jenner, and R. J. Pollit, *Br. J. Psychiatry*, 1975, 596.

[41] J. Rybakowski, M. Chlopocka, Z. Kapelski, B. Hernacka, Z. Szajnerman, and K. Kasprzak, *Int. Pharmacopsychiatry*, 1974, **9**, 166.

detail.[42-45] Four different components of lithium transport occur in the red cell:
  (i) Ouabain-sensitive uptake of lithium by the sodium–potassium ATPase system
  (ii) 'Leak' diffusion of lithium
  (iii) Phloretin-sensitive uptake resulting from an electrogenic sodium–lithium counter-transport
  (iv) An anion exchange, where lithium is transported as a result of ion-pairing with bicarbonate.

Aspects of these transport systems have been confirmed by other workers.[46, 47]

A study also showed an irreversible inhibition of the uptake of choline by erythrocytes following the administration of lithium.[48, 49] Dorus *et al.*[45] have identified a genetically related effect on the lithium-transport mechanism that is present in the families of some manic-depressive psychotics. Canessa *et al.*[50] have shown an increased sodium–lithium counter-transport in essential hypertension: if this is confirmed, a major advance has been made in the detection of this disease, which accounts for over 90% of all cases of hypertension and for which there is no specific diagnostic procedure except the exclusion of other causes.

Two recent studies have shown decreased uptake of 5-hydroxytryptamine by blood platelets from manic-depressives; this defect was reversed by long-term treatment with lithium.[51, 52]

Lithium stimulates sodium–potassium ATPase in the human red cell[53, 54] to cause sodium efflux which is ouabain-sensitive and stimulated by sodium, potassium, and lithium. Glen[55] has confirmed that lithium has a potassium-like action on the postassium-sensitive side of the membrane, and concluded that there is substantial evidence to support the view that lithium acts to stimulate transport during depression and to inhibit it during mania. A recent report of the stimulation of magnesium-activated and calcium/magnesium-activated ATPase by lithium in the rat iris and visual cortex[56] suggests that the action of lithium is not limited to the sodium pump but may be seen in other ATPases.

Effects of Lithium on the Kidney.—The most contentious issue in the field of lithium research during recent years has been the possibility of long-term damage

[42] J. Duhm and B. Becker, *J. Membr. Biol.*, 1979, **51**, 263.
[43] W. Greil and F. Eisenried, in 'Lithium in Medical Practice', ed. F. N. Johnson and S. Johnson, M.T.P. Press, Lancaster, U.K., 1978, p. 415.
[44] G. N. Pandey, E. Dorus, J. M. Davis, and D. C. Tosteson, *Arch. Gen. Psychiatry*, 1979, **36**, 902.
[45] E. Dorus, G. N. Pandey, R. Shaughnessy, M. Gavaria, E. Val, S. Eriksen, and J. M. Davis, *Science*, 1979, **205**, 932.
[46] H. L. Meltzer, C. J. Rosoff, S. Kassir, and R. R. Fieve, *Life Sci.*, 1976, **19**, 371.
[47] A. Frazer, J. Mendels, and D. Brunswick, *Commun. Psychopharmacol.*, 1977, **1**, 255.
[48] K. Martin, in ref. 43 p. 167.
[49] C. Lingsch and K. Martin, *Br. J. Pharmacol.*, 1976, **57**, 323.
[50] M. Canessa, N. Adragna, H. S. Solomon, T. M. Connolly, and D. C. Tosteson, *N. Engl J. Med.*, 1980, **302**, 772.
[51] G. V. R. Born, G. Grignanai, and K. Martin, *Br. J. Clin. Pharmacol.*, 1980, **9**, 321.
[52] A. Coppen, C. Swade, and K. Wood, *Br. J. Psychiatry*, 1980, **136**, 235.
[53] A. I. M. Glen, M. W. B. Bradbury, and J. Wilson, *Nature (London)*, 1972, **239**, 399.
[54] L. A. Beaugé and E. del Campillo, *Biochim. Biophys. Acta*, 1976, **433**, 547.
[55] A. I. M. Glen, in ref. 8, p. 768.
[56] H. W. Reading and T. Isbir, *Biochem. Pharmacol.*, 1979, **28**, 3471.

to the kidney following treatment with lithium. A number of patients who had been treated with lithium for a long term developed serious damage that was detected on biopsy of the kidney; apparently characteristic histological lesions were seen in a high proportion of patients.[57-59] Lithium has been reported to be associated with polyuria; this has been considered by some to be an almost inevitable side-effect but which is well tolerated by the patients. Lithium inhibits the action of antidiuretic hormone,[60, 61] and it was assumed that the polyuria was a consequence of this.

The appearance of the histological lesions in patients who were demonstrating polyuria gave rise to the misconception that these two phenomena were linked. However, polyuria is completely non-specific, and it may be seen in a large number of patients, for various reasons. Patients with behavioural disturbances show excessive drinking, and hence polydipsia-induced polyuria. In the early days of lithium therapy, because the dangers of toxicity were appreciated, it was common to insist that patients should ensure an adequate water intake, and this instruction, together with the lack of a counter instruction, may have resulted in the resetting of the thirst mechanism by the persistent habit of drinking. As a result of the reported nephrotoxic effect of lithium, the renal physiology of lithium has been well documented,[60-62] and a special symposium was organised.[63]

The histological lesions certainly exist in some patients following lithium therapy, and it is clear that lithium does have effects on renal physiology.[60-62] However, reports of nephrotoxic effects also coincide with a high incidence of lithium toxicity. The histological changes that have been demonstrated suggest previous lithium intoxication. The renal concentrating ability, *i.e.* the renal function, of patients on lithium is not impaired permanently, and this is tested ultimately by the report by Glen *et al.*[64] that, in 784 patients receiving lithium, over a period of ten years, in Edinburgh, 33 died during the period of study, of which none could be attributed to renal damage. In studies of rats, Christiansen and Hansen[65] have demonstrated that the impaired renal concentrating ability that followed the administration of lithium was reversed on discontinuation of the drug.

2 New Anti-tumour Complexes: Titanium, Vanadium, and Molybdenum Metallocenes, and Dialkyltin Dihalides

All of these complexes, which are active in animal models, have been described as analogues of *cis*-[Pt(NH$_3$)$_2$Cl$_2$], since they all have *cis*-halides as ligands.

Dichlorobis-(η^5-cyclopentadienyl)titanium(IV) (1) and the analogous vanadium(IV) and molybdenum(IV) complexes exhibit activity towards the Ehrlich

[57] J. Hestbech, H. E. Hansen, A. Amdisen, and S. Olsen, *Kidney Int.*, 1977, **12**, 205.
[58] H. E. Hansen, J. Hestbech, J. L. Sorensen, K. Norgaard, J. Heilskov, and A. Amdisen, *Q. J. Med.*, 1979, **48**, 577.
[59] G. D. Burrows, B. Davis, and P. Kincaid-Smith, *Lancet*, 1978, **1**, 1310.
[60] F. A. Jenner and P. R. Eastwood, in ref. 43, p. 247.
[61] F. A. Jenner, *Arch. Gen. Psychiatry*, 1979, **36**, 888.
[62] P. Vestergaard, in ref. 6, p. 345.
[63] Editorial, *Lancet*, 1979, **2**, 1056.
[64] A. I. M. Glen, *Neuropsychobiology*, 1979, **5**, 167.
[65] S. Christensen and B. Hansen, *Renal Physiol.*, in the press.

ascites tumour system.[66-68] The application of 80 or 90 mg of vanadocene dichloride per kg of body weight to mice 24 hours after transplantation of the Ehrlich ascites tumour results in 100% inhibition of tumours until day 30.

The anti-tumour activity of a series of diorganotin dihalide and di-pseudo-halide complexes $[R_2SnX_2L_2]$ ($R = Me$, Et, Pr^n, Bu^n, or Ph; $X = F$, Cl, Br, I, or NCS; $L = O$- or N-donor ligands) has been investigated.[69] The most encouraging results were found for the 1,10-phenanthroline and 2,2'-bipyridyl adducts of diethyltin complexes (2), which exhibit anti-leukaemic activity against the

(1)   (2)

P338 tumour system. They are less active than platinum compounds, but, as yet, there is no evidence of high nephrotoxicity for organotin compounds. It has been suggested that the mode of action involves transportation of $[Et_2SnX_2]$ into the cell, followed by hydrolysis. The Cl–metal–Cl bond angle in diethyltin dichloride (96°) is very similar to that in *cis*-$[Pt(NH_3)_2Cl_2]$ and $[(\eta^5\text{-}C_5H_5)_2MCl_2]$ ($M = Ti$, V, or Mo) complexes.

3 Vanadium: a Natural Regulator of ATPase?

Macara has summarized[70] recent progress in the search for a biological role for vanadium. Much of the biological interest results from the close similarity of phosphate and vanadate. Almost all enzymes that are currently known to be inhibited by vanadate are phosphohydrolases which employ phosphorylated enzyme as intermediates (see Table 2). It is unclear if control of the sodium pump could be based on $V^V \rightleftharpoons V^{IV}$ and linked to the redox state of cells.

Analyses of various tissues[71] show vanadium contents ranging from 0.06 (cow brain) to 0.66 (dog kidney) nmol g^{-1}, but the amount appears to be inadequate to function as a physiological regulator of ATPase. In ischaemia, the physiological concentrations of ions and adenine nucleotides may be favourable for the inhibition of ATPase by vanadate. In isolated horse muscle ATP (*e.g.* equine ATP purchased from Sigma Chemical Company Ltd), vanadium is present as VO^{2+} that is complexed to ATP. The mystery remains as to why the blood cells of the tunicate *Ascidia nigra* should accumulate vanadium from sea-water (*ca* 5×10^{-8} mol l^{-1}) and store it at a high concentrattion (*ca* 1 mol l^{-1}), effectively in a dilute solution of sulphuric acid (pH < 2).

[66] H. Köpf and P. Köpf-Maier, *Nachr. Chem. Tech. Lab.*, 1979, **27**, 169.
[67] P. Köpf-Maier and H. Köpf, *Z. Naturforsch., Teil. B*, 1979, **34**, 805.
[68] P. Köpf-Maier, M. Leitner, R. Voitländer, and H. Köpf, *Z. Naturforsch., Teil. C*, 1979, **34**, 1174.
[69] A. J. Crowe and P. J. Smith, *Chem.-Biol. Interact.*, 1980, **32**, 171.
[70] I. G. Macara, *Trends Biochem. Sci.*, 1980, **5**, 92.
[71] R. L. Post. D. P. Hunt, M. O. Walderhaug, R. C. Perkins, J. H. Park, and A. H. Beth, in 'Na,K-ATPase: Structure and Kinetics', ed. J. C. Skou and J. G. Noerby, Academic Press, London, 1979, p. 389.

Table 2 *Enzymes that are inhibited or activated by vanadium[a]*

Enzyme	Postulated binding[b]	Apparent inhibition or dissociation constants/mol l^{-1}
Ribonuclease	E $\diagdown$ His^{12} ... V^V·U (His^{12}, Lys^{44})	10^{-5}
	E $\diagdown$ His^{12} ... V^{IV} (His^{12}, His^{119})	6×10^{-5}
Acid phosphatase	E–His–V^V	2×10^{-7}
Alkaline phosphatase	E–His–V^V	4×10^{-6}
	E–His–V^{IV}	4×10^{-7}
(Na$^+$-K$^+$)-ATPase	K·E$_2$–Asp–V^V·Mg	4×10^{-9}
Myosin ATPase	E$_2$·ADP·V^V	$\rightarrow 0$ (?)
Ca^{2+}-ATPase	E$_2$–V^V·Mg (?)	7×10^{-6}
Adenylate kinase[c]	E·V$_{20}$O$_{28}$H$_5$ (?)	$< 1 \times 10^{-6}$
Adenyl cyclase[c]	E$_2$·GDP·V^V (?)	activated

(a) Other references may be found in ref. 70, from which this has been adapted. (b) U = uridine; E–V indicates bonding between vanadium and the enzyme; (c) in crystalline adenylate kinase, oligovanadate sits across the phosphate sites of ATP and AMP (see E. F. Pai, W. Sachsenheimer, R. H. Schirmer, and G. E. Schulze, *J. Mol. Biol.*, 1977, **144**, 37).

4 Iron

Uptake and Removal.—Brown *et al.*[72] have discussed the use and design of Fe^{III} chelates in the oral treatment of iron-deficiency anaemia. The use of iron(III) acetohydroxamate is recommended, on the basis of results from animal studies. Acetohydroxamic acid[72] has a proton dissociation constant of 9.4 for the dissociation shown in reaction (1), but complexation with Fe^{III} occurs even at pH 2.

$$\text{Me—C=O} \quad \rightleftharpoons \quad \text{Me—C=O} + H^+ \qquad (1)$$
$$\text{HN—OH} \qquad\qquad \text{HN—O}^-$$

In solutions containing 3mM-Fe^{III} and a five-fold excess of acetohydroxamic acid, the only stable species between pH 5.5 and pH 9 is the 1:3 chelate Fe^{III}-(AHA)$_3$, for which $\log\beta_3 = 24.6$. This is a monomeric, high-spin species, although polymeric (hydroxo-bridged) complexes are formed even at pH 7 if less than a five-fold excess of ligand is present in solution.

Bergeron *et al.* have reported[73] the synthesis of N^1,N^8-bis-(2,3-dihydroxy-benzoyl)spermidine hydrochloride (3) and suggested that it may be a useful chelating agent for the treatment of iron-overload diseases such as haemochromatosis. Unlike desferrioxamine B, which is currently the most widely used

[72] D. A. Brown, M. V. Chidambaram, J. J. Clarke, and D. M. McAleese, *Bioinorg. Chem.*, 1978, **9**, 255.
[73] R. J. Bergeron, P. S. Burton, K. A. McGoven, E. J. St. George, and R. R. Streiff, *J. Med Chem.*, 1980, **23**, 1130.

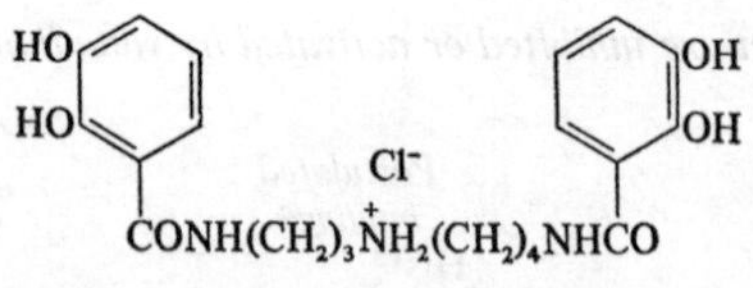

(3)

drug, there is a possibility that it may be taken orally. It is stable in acidic solutions, has a low toxicity, and is readily synthesized. Trials in rats are in progress.

Iron and Infection.—It has been known for several years that iron compounds can abolish the antibacterial effects of body fluids *in vitro* and enhance bacterial virulence *in vivo*. Although there is plenty of iron in the body fluids of vertebrates (20 μmol l^{-1} in human plasma), most of it is strongly bound to transferrin in blood and lymph, and to lactoferrin in secretions. The iron that is available in low-molecular-weight forms ('free', *ca* 10^{-18} mol l^{-1}) is far too low for bacterial growth. Pathogenic organisms which multiply successfully under these conditions have mechanisms for assimilating bound iron or acquiring it from liberated haem.[74] These often involve the secretion of phenolates or hydroxamates as (ferric) iron-chelating agents.

The genera *Salmonella*, *Escherichia*, and *Klebsiella*, for example, secrete enterochelin (4) (also called enterobactin) under conditions of iron restriction

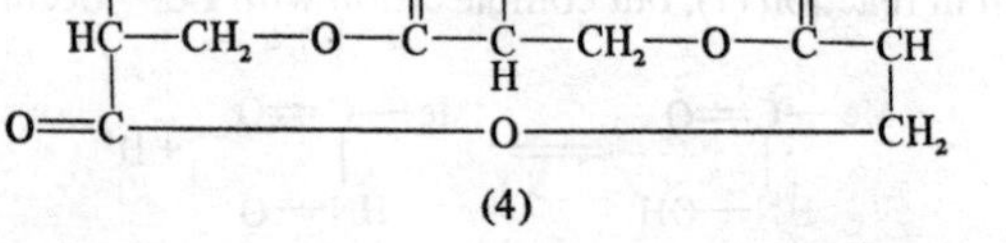

(4)

in vitro. It appears that serum and milk can contain antibodies which interfere with the production or secretion of enterochelin, so producing a second line of defence in combination with transferrin. It has been shown for *E. coli* 0111 [75] that the bacterial antigen against which the antibody is directed is colitose (3,6-dideoxy-L-galactose), which is the terminal monosaccharide of the *O*-specific side-chain of the lipopolysaccharide sequence GlcNAc-Glc-Gal-.

Col	Col

In addition to secreting enterochelin, *Escherichia coli* that grow under iron-restricted conditions produce new outer-membrane proteins which act as recep-

[74] E. Griffiths, H. J. Rogers, and J. J. Bullen. *Nature (London)*. 1980, **284**, 508.
[75] S. P. Fitzgerald and H. J. Rogers, *Infect. Immun.*, 1980, **27**, 302.

tors for ferric enterochelin[76] and which contain altered species of tRNA.[77] The tRNA's that are involved are those for Phe, Tyr, Trp, Ser, Cys, and Leu, all of which recognize codons with a 5'-terminal uridine residue. Iron specifically affects the synthesis of the hypermodified nucleoside 2-methylthio-N^6-(Δ^2-isopentenyl)adenosine. When iron is added back to the growth medium, the abnormal tRNA's return to normal. Such modifications have been demonstrated in *E. coli* 0111 that was recovered from peritoneal cavities of lethally infected guinea-pigs.[77]

In some cases, the iron-sequestering system may be plasmid-mediated. The virulent fish pathogen *Vibrio anguillarum* loses its ability to grow under conditions of iron restriction that are imposed by transferrin on losing the plasmid.[78]

It has recently been demonstrated[79] that siderophores are produced by phytopathogenic bacteria which infect plants. *Agrobacterium tumefaciens* produces a new catechol-type siderophore, agrobactin (5).

(5)

5 Copper

Copper and Inflammation.—Williams and co-workers[80, 81] have examined the binding of copper to various components of serum and to ligands of low molecular weight with a view to understanding the role of copper in rheumatoid arthritis. Their computer models (thermodynamic and based on stability constants) suggest that the circulating pool of copper species of low molecular weight is composed mainly of [Cu(His)(Cys)]⁻, [Cu(His)(CysH)], [Cu(His)$_2$], and [Cu(His)(Thr)]. The concentration of 'free' copper is very low (10^{-11} to 10^{-18} mol l⁻¹), and much is bound to the α_2-macroglobulin caeruloplasmin. Although Cu^{II}–aspirinate complexes are reported to produce less ulceration of the stomach and intestine than aspirin alone, calculations appear to suggest that Cu^{II}–acetylsalicylate complexes should not be present in the stomach or intestine, or in the plasma.

Brown *et al.* have prepared a range of copper complexes and evaluated their anti-inflammatory activity and irritancy after oral, subcutaneous, and local

[76] M. A. McIntosh and C. F. Earhart, *J. Bacteriol.*, 1977, **131**, 331.
[77] E. Griffiths, J. Humphreys, A. Leach, and L. Scanlon, *Infect. Immun.*, 1978, **22**, 312.
[78] J. H. Crosa, *Nature (London)*, 1980, **284**, 566.
[79] S. A. Ong, T. Peterson, and J. B. Neilands, *J. Biol. Chem.*, 1979, **254**, 1860.
[80] G. E. Jackson, P. M. May, and D. R. Williams, *J. Inorg. Nucl. Chem.*, 1978, **40**, 1189
[81] G. Arena, G. Kavu, and D. R. Williams, *J. Inorg. Nucl. Chem.*, 1978, **40**, 1221.

administration in rats and guinea-pigs.[82] The inflammation model that was used was the induction of paw oedema by administration of kaolin. They tested Cu^I and Cu^{II} complexes with ligands which are potentially active alone (aspirin, d-penicillamine, levamisole, flufenamic acid, L-histidine, and 2-amino-2-thiazoline) and with ligands which are themselves inactive (pyridine, EDTA, diethyl-dithiocarbamate, anthranilic acid, thiourea, Ph_3P, and sulphonamides). In general, more copper compounds produced an anti-inflammatory response after subcutaneous than after oral administration, and both Cu^I and Cu^{II} complexes had similar activity once absorbed. Copper complexes with ligands which were themselves anti-inflammmatory were often more active than the ligand alone, and less irritant. An unexpected toxicity of $[Cu^{II}(His)_2(NO_3)_2]\cdot 2H_2O$ was noted.

Therapy with Copper/Zinc Superoxide Dismutase.—Bovine liver superoxide dismutase has been marketed by Diagnostic Data Inc. (California) as an anti-inflammatory drug. The non-proprietary name is orgotein and the trade name Palosein. For human use, it is prepared under the trade name Ontosein.[83] There is a remarkable immunological tolerance to orgotein when it is administered to Man and animals (in mg quantities) by a variety of routes. This has been attributed to the conservation of its amino-acid composition and secondary structure from species to species, its good solubility, its compact size, and its favourable clearance kinetics.

Orgotein appears to remain extracellular after administration, and to be an anti-inflammatory agent without antipyretic, analgesic, or immunosuppressive actions. Initial clinical experience[84] has shown that the onset of effects due to orgotein can take several weeks in some patients, but, once established, these can last one month beyond the termination of therapy. Its effectiveness in rheumatoid arthritis has been compared to gold compounds and penicillamine.

It is possible that orgotein could remove O_2^- that is produced during respiratory bursts by phagocytes in their plasma membranes. This might prevent lysis of cells in areas of tissue destruction.

Leuthauser et al.[85] have reported that local injections of orgotein have a slight effect on tumour growth and on the survival of mice bearing a Sarcoma 180 tumour. Their rationale for using orgotein was that, in many tumour systems that have been examined, the Mn-superoxide dismutase levels are much reduced compared to normal cell types. In mammalian cells, two types of superoxide dismutase are found. The Cu/Zn enzyme is located in the cytosol whereas the Mn enzyme is present in the mitochondrion and in the cytosol. However, added enzyme will not penetrate cell membranes, and for this reason the low-molecular-weight lipid-soluble copper complex $[Cu^{II}(3,5\text{-di-isopropylsalicylate})_2]$ (6) was also tested. This was known to have superoxide dismutase activity and to be an anti-inflammatory agent,[86] and has now been shown to exhibit anti-

[82] D. H. Brown, W. E. Smith, J. W. Teape, and A. J. Lewis, *J. Med. Chem.*, 1980, **23**, 729.
[83] W. Huber and M. G. P. Saifer, in 'Superoxide and Superoxide Dismutases', ed. A. M. Michelson, J. M. McCord, and I. Fridovich, Academic Press, London, 1977, p. 517.
[84] K. B. Menander-Huber and W. Huber, in ref. 83, p. 537.
[85] S. W. C. Leuthauser, L. W. Oberley, T. D Oberley, and J. R. J. Sorenson, *Cancer Res.*, in the press.
[86] J. R. J. Sorenson, *J. Med. Chem.*, 1976, **19**, 135.

(6)

tumour activity against a solid form of Ehrlich carcinoma in mice, apparently by delaying the process of metastasis rather than by direct cytotoxicity. A group of mice receiving 0.5 mg of the complex for 10 doses almost doubled their survival time.

The effectiveness of complexes of this type seems to parallel their solubility in lipids,[87] *i.e.* $[Cu^{II}(3,5\text{-di-isopropylsalicylate})_2] > [Cu^{II}_2(aspirinate)_4] \cdot 4$ pyridine $> [Cu^{II}_2(aspirinate)_4] \cdot 4$ DMSO.

Copper Complexes as Anticonvulsant Drugs.—Sorenson and co-workers [88, 89] have examined the anticonvulsant activity of a range of copper salicylates, acylsalicylates, and amino-acid complexes. Copper levels are high in the normal human brain (*ca* 370 μg per g of tissue ash), and copper-dependent enzymes are known to be required for the normal development and functioning of the brain. These include tyrosinase (Tyr → dopa) and dopamine-β-hydroxylase (dopamine → norepinephrine). Levels of copper in serum are elevated in epileptic patients, but levels in brain are markedly reduced. Both quaking mice and mottled mice exhibit tremors as well as degeneration of the neural and central nervous systems as symptoms of their genetic deficiency of copper.

$[Cu^{II}(L\text{-Thr})(L\text{-Ser})]$, $[Cu^{II}(L\text{-Thr})(L\text{-Ala})]$, $[Cu^{II}(L\text{-Val})_2]$, $[Cu^{II}(L\text{-Thr})_2]$, $[Cu^{II}(Ser)_2]$, and $[Cu^{II}(L\text{-Ala})_2]$ are effective against metrazol-induced seizure in mice, 0.5 and 4 hours after administration. Anticonvulsant activity was also found for $[Cu^{II}_2(adamantylsalicylate)_4]$ and $[Cu^{II}_2(acetylsalicylate)_4]$, and $[Cu^{II}(3,5\text{-di-isopropylsalicylate})_2]$ is effective in preventing maximal electro-shock and metrazol-induced seizures. No activity was found for $[Cu^{II}_2(acetate)_4(H_2O)_2]$ or for $[Cu^{II}Cl_2(H_2O)_2]$. Many of the complexes have low neurological toxicities (rotating rod test), and, since some are known to have potent anti-ulcer activity and to lack gastrointestinal irritancy, they may have some advantage over existing organic anti-epileptic drugs.

6 Zinc

Zinc Citrate Complexes in Milk.—Lönnerdal *et al.*[90] have suggested that complexation of zinc with citric acid (7) may enhance the uptake of zinc by normal newborn children and by patients with acrodermatitis enteropathica. They have

---

[87] L. W. Oberley, S. W. H. C. Leuthauser, G. R. Buettner, J. R. J. Sorenson, T. D. Oberley, and I. B. Bize, in 'Active Oxygen in Medicine,' Raven Press, in the press.

[88] J. R. J. Sorenson, *J. Appl. Nutr.*, 1980, **32**, 4

[89] J. R. J. Sorenson, D. O. Rauls, K. Ramakrishna, R. E. Stull, and A. N. Voldeng, in 'Proceedings of the Conference on Trace Substances in Environmental Health, XIII', ed. D. D. Hemphill, University of Missouri Press, 1979.

[90] B. Lönnerdal, A. G. Stanislowski, and L. S. Hurley, *J. Inorg. Biochem.*, 1980, **12**, 71.

$$
\begin{array}{c}
CH_2CO_2H \\
| \\
HO-C-CO_2H \\
| \\
CH_2CO_2H
\end{array}
$$

(7)

purified a zinc-binding compound of low molecular weight from human milk by ultrafiltration, gel filtration, and ion-exchange chromatography. They suggest a formula close to $Zn_3(C_6H_5O_7)_2$. The concentration of citrate ion in human milk ultrafiltrates is *ca* 1 mmol l^{-1} and the concentration of zinc is *ca* 4 μmol l^{-1}. In bovine milk it appears that much of the zinc is bound to casein, and unavailable for absorption.

In the rat, it appears that the zinc content of rat milk increases with consecutive milking, and this result is reflected in the concentration of zinc in the liver of the pup.[91] This highlights the difficulties of studying the composition of 'normal' rat milk. A further study on zinc-deficient and normal rats showed that zinc deficiency during pregnancy severely impaired parturition and prevented the rise in production of RNA in the mammary gland that is normally seen in the initiation of lactogenesis. Shorter periods of deficiency have no obvious effect on the metabolism of the mammary gland, and it may be that the timing of the deficiency is critical. The mammary gland appears to have priority in obtaining any available zinc in the body.[92]

Studies of the zinc concentration in the plasma of human infants suggest that they approach those of adults if optimal zinc nutrition is available. The bioavailability of zinc from human milk is considerably greater than that of bovine milk, owing to the zinc-binding ligand that is discussed above, and this is reflected in the concentration of zinc in the plasma of human infants (six months old) that were breast fed and fed on a formula containing bovine milk.[93] Supplementation with zinc of the formula containing bovine milk, although increasing the resultant concentration of zinc in plasma in the infants, did not restore this concentration to that of the breast-fed group.

The clinical physiology of zinc has recently been reviewed,[94] with particular regard to the effects of parenteral nutrition following surgery or traumatic injury. It has been concluded that it is important to ensure repletion of trace metals following trauma, and that indeed recovery is consequently more rapid if this condition is attained.

Several years ago, Boursnell *et al.*[95] suggested that zinc–citric acid complexes are present in boar vesicular sections and in seminal plasma. Since citric acid is potentially a multidentate ligand, equilibria in solution may be very complicated. Shpilev[96] has described the formation of $Zn_3(C_6H_5O_7)_2$, but, in alkaline

[91] C. L. Kleen, B. L. Lönnerdal, M. V. Sloan, and L. S. Hurley, *Physiol. Behav.*, 1980, **24**, 613.
[92] P. B. Mutch and L. S. Hurley, *Am. J. Physiol.*, 1980, **238**, E26.
[93] K. M. Hambidge, P. A. Walravens, C. E. Casey, R. M. Brown, and C. Bender, *J. Pediatr.*, 1979, **94**, 607.
[94] G. S. Fell and R. R. Bruns, in 'Advances in Parenteral Nutrition', ed. I. D. A. Johnston, M.T.P. Press, Lancaster, U.K., 1978, p. 241.
[95] J. C. Boursnell, P. A. Briggs, V. Lavon, and E. J. Butler, *J. Reprod. Fertil.*, 1973, **34**, 73.
[96] F. S. Shpilev, *J. Gen. Chem. USSR*, 1939, **9**, 1286.

solution, the deprotonated hydroxyl group, as well as the three carboxyl groups, can co-ordinate to give $Na_3ZnC_6H_4O_7$. There do not seem to be any reported crystal structures of such complexes, but mixtures of Cu^{2+} and citric acid, for example, deposit the insoluble species $Cu_2(Cit)\cdot2H_2O$,[97] which contains a three-dimensional array of Cu^{2+} ions linked by alkoxo-bridges and three types of carboxylate bridge from the septidentate $C_6H_4O_7{}^{4-}$ (citrate) anions.

Zinc citrate complexes feature in the patent literature as dentifrices and mouthwashes for the prevention of plaque. The astringent taste and undesirable side-effects of oral pharmaceuticals containing water-soluble zinc salts appear to be alleviated when 2-hydroxy- or α-amino-acids are used as complexing agents.[98] Other metal complexes of citric acid have pharmaceutical uses, for example Bi (anti-ulcer), Ga (neoplasm uptake), Ni (bacteriocide), and Fe (anaemia). Complexation of citric acid by iron is probably involved in its conversion into isocitric acid by the iron-containing enzyme aconitase.

Zinc as an Anti-viral Agent. Herpes simplex type 2 virus (HSV-2) causes a common sexually transmitted disease. In 1967 Falke[99] observed that zinc chloride that was added to HSV-infected tissue culture cells blocked the formation of Herpes-induced giant cells. At concentrations of $mmol\,l^{-1}$, Zn^{2+} does not cause changes in cellular DNA or in the morphology of uninfected cells, but causes a marked reduction of the synthesis of HSV, perhaps due either to inhibition of enzymatic activity of a virus-induced DNA polymerase or to blocking of the HSV-induced synthesis of polypeptides. Chvapil[100, 101] has developed a collagen sponge, cross-linked by reaction with glutaraldehyde, which shows topical antiviral activity (in mice) after soaking in $0.1M$-$ZnSO_4$ at pH 5.5. Systemically administered $ZnSO_4$ (doubling the level of zinc in serum), on the other hand, exacerbated the vaginitis. It was suggested that an elevated level of zinc in serum inhibited the function of phagocytic cells.

7 The Platinum Metals

Rhodium(II) Carboxylates as Potential Anti-cancer Drugs.—Several (tetra-μ-carboxylato)dirhodium(II) complexes (8) have been shown to increase the life-

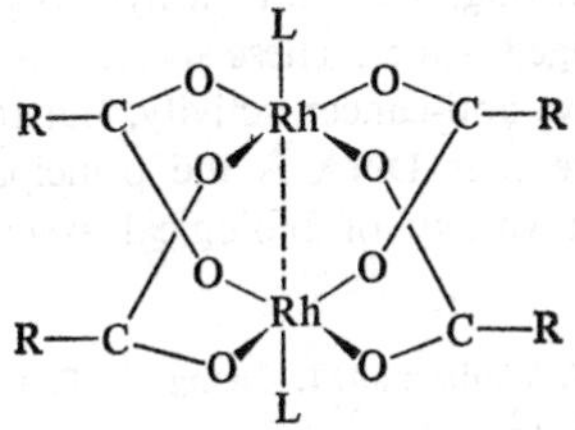

(8)

Rh–Rh distance is *ca* 2.45 Å

[97] D. Mastropaolo, D. A. Powers, J. A. Portenza, and H. J. Schugar, *Inorg. Chem.*, 1976, **15**, 1444.
[98] K. E. Jonsson and N. F. E. Moren, *Chem. Abstr.*, 1976, **84**, 65 262.
[99] D. Falke, *Z. Med. Mikrobiol. Immunol.*, 1967, **53**, 175.
[100] P. O. Tennican, G. Z. Carl, and M. Chvapil, *Curr. Chemother.*, 1978, 363.
[101] P. O. Tennican .G. Z. Carl, and M. Chvapil, *Life Sci.*, 1979, **24**, 1877.

span and to produce cures for mice bearing Ehrlich ascites, sarcoma 180 ascites, and leukaemia P388 tumours.[102-104] The concentrations which inhibit proliferation of exponentially growing L1210 cells parallel the observed toxicities of the compounds *in vivo*. They vary with carboxylate in the order methoxyacetate < acetate < propionate < butyrate. The butyrate complex is extremely cytotoxic at relatively low concentrations: on treatment at 0.8 μmol l^{-1} for six hours, the survival of HeLa and CHO cells is essentially zero.[105] The retardation of cell-cycle progression of HeLa cells appears to be due to inhibitory effects of the drug on the synthesis of DNA and of protein. Synthesis of RNA is unaffected.[106] Stationary-phase cells are the most sensitive. However, rhodium(II) carboxylates do not seem to bind to DNA, and their inhibition of the synthesis of DNA may be related to inhibition of the enzymes involved in the synthesis of DNA.

The axial ligands L (usually H_2O) are readily replaced by donor molecules such as histidine. When an *O*-donor is replaced by an *N*-donor, the colour of the complex changes from blue-green (λ_{max} *ca* 587 and 447 nm) to pink. Studies of the inhibition of enzymes *in vitro* have shown that rhodium(II) carboxylates irreversibly inhibit enzymes that have sulphydryl groups in or near their active sites; for example, glyceraldehyde-3-phosphate dehydrogenase. Studies with radioactively labelled rhodium(II) acetate have shown that the dimeric cage structure breaks down within a few hours after injection of the complex, and none of the complex is excreted unchanged.[107] This breakdown process may also involve thiols.

Anti-tumour Platinum Compounds.—Rosenberg has reviewed the history of 'cisplatin', *cis*-[Pt(NH$_3$)$_2$Cl$_2$], and its possible mechanisms of action.[108] Combination therapies with the platinum drug are now a safe and effective treatment for a number of different types of cancer. The major toxic side-effects are damage to the kidney, and nausea and vomiting. Hydration of the patient ameliorates the kidney toxicity so well that in some clinical reports it is no longer considered to be the dose-limiting side-effect. Toxic effects on normal cells, such as those in the kidney, may be related to the hydrolysis of the dichloride drug to produce the monoaquo-monohydroxy-form. This species has a half-life of only a few minutes at pH 7,[109] and it oligomerizes to a hydroxo-bridged centrosymmetric dimer and a hydroxo-bridged trimer. These species are more toxic than cisplatin itself. The monomer shows anti-cancer activity, but the oligomers do not.

The wealth of evidence that DNA is the principal target site for neutral platinum complexes in a variety of biological systems has been thoroughly

[102] A. Erck, L. Rainen, J. W. Whileyman, I. Chang, A. P. Kimball, and J. L. Bear, *Proc. Soc. Exp. Med.*, 1974, **145**, 1278.
[103] J. L. Bear, H. B. Gray, and L. Rainen, *Cancer Chemother. Rep.*, 1975, **59**, 611.
[104] K. M. Kadish, K. Das, R. A. Howard, A. M. Dennis, and J. L. Bear, *Bioelectrochem. Bioenerg.*, 1978, **5**, 741.
[105] R. A. Howard, A. P. Kimball, and J. L. Bear, *Cancer Res.*, 1979, **39**, 2568.
[106] P. N. Rao, M. L. Smith, S. Pathak, R. A. Howard, and J. L. Bear, *J. Nat. Cancer Inst.*, 1980, **64**, 905.
[107] R. A. Howard, T. G. Spring, and J. L. Bear, *J. Clin. Haematol. Oncol.*, 1977, **7**, 391.
[108] B. Rosenberg, 'Nucleic Acid and Metal Ion Interactions', ed. T. G. Spiro, J. Wiley and Sons, New York, 1980, p. 1.
[109] B. Rosenberg, *Cancer Treat. Rep.*, 1979, **63**, 1433.

discussed by Roberts and Thomson.[110] Reactions with DNA impair its function as a template for further DNA replication. Attack by platinum at N(7)–O(6) of guanine bases of DNA (chelation) may help to explain the selective cytotoxicity toward tumour cells in the presence of normal cells, which may have enzymes that are able to repair such lesions. This could also account for the inactivity of the *trans*-isomers. However, more chemical evidence of attack at O-6 is still required.

Lippard has found[111] that *cis*- and *trans*-$[Pt(NH_3)_2Cl_2]$ react differently with nucleosome core particles (DNA that is wrapped around an assembly of histone proteins as it is in the cell). The *cis*-isomer causes partial unwinding of the DNA helix, whereas the *trans*-isomer cross-links either protein to protein or DNA to protein. Interestingly, experiments on small, closed circular DNA[111] indicate that the specific binding site for the *cis*-isomer is in a region of four guanine bases that are hydrogen-bonded to four cytosine bases. Evidence for considering the primary binding sites for platinum to be N-7 of purines and N-3 of pyrimidines can be found in the first volume of a new series on 'Metal Ions in Biology'.[112] However, binding to the phosphate backbone and hydrogen-bonding interactions between DNA and co-ordinated platinum ligands cannot be ignored.

A book on cisplatin has recently been published.[113] It consists of papers presented at a meeting in Atlanta, Georgia in September 1979, and it aims to present new information on the clinical application of cisplatin in a wide variety of adult and pediatric malignancies, and to discuss the mechanism by which the drug appears to cause the death of tumour cells. Section I is devoted to pre-clinical studies, including chemistry, anti-tumour activity in animals, toxicity, and the development of analogues. Cleare *et al.*[114] have summarized current knowledge of structure–activity relationships for *cis*-$[PtA_2X_2]$ complexes, as shown in Table 3. They singled out six complexes as potential second-generation platinum-containing anti-tumour drugs. Three of these [(9), (10), and (11)] contain an aliphatic amine or ammonia and three more [(12), (13), and (14)] contain 1,2-diaminocyclohexane as a ligand.

Some interesting statistical data have been presented by Wolpert-DeFilippes.[115] The Division of Cancer Treatment at the National Cancer Institute, in Bethesda, USA, tested 11 745 compounds of 55 metals between 1955 and 1979. Of these, 898 (8%) met minimal standards for activity in at least one system (*in vivo* or *in vitro*) that is used by the Drug Evaluation Branch for screening. Eighteen per cent (185 out of 1055 tested) of the platinum compounds met the minimal standards of activity. It is apparent from Figure 1 that many classes of metal compounds have yet to be thoroughly investigated.

Section II of the book presents the current status of clinical applications of cisplatin.

[110] J. J. Roberts and A. J. Thomson, *Prog. Nucleic Acid Res.*, 1979, **22**, 71.
[111] S. J. Lippard and J. D. Hoeschele, *Proc. Natl. Acad. Sci. USA*, 1979, **76**, 6091 and *Chem. Eng. News*, 1980, Jan. 21, p. 35.
[112] J. K. Barton and S. J. Lippard, in ref. 108, p. 31.
[113] 'CISPLATIN: Current Status and New Developments', ed. A. W. Prestakyo, S. T. Crooke, and S. K. Carter, Academic Press, New York, 1980.
[114] M. J. Cleare, P. C. Hydes, D. R. Hepburn, and B. W. Malerbi, in ref. 113, p. 149.
[115] M. K. Wolpert-DeFilippes, in ref. 113, p. 183.

Table 3 *Structure–activity studies on cis-*[PtA^{I}X^I] *species*

Effective A groups	*Examples*
(n-Alkyl)amines	C$_2$—C$_4$
(Isoalkyl)amines	C$_3$—C$_5$[a]
Alicyclic amines	C$_3$—C$_8$[a]
Diaminoalkanes	Ethylenediamines (also alkyl-substituted), propylenediamine
Diaminocycloalkanes	1,2-Diaminocyclohexane
Heterocyclic amines	Ethyleneimines, pyrrolidine
Aromatic diamines	o-Phenylenediamine

Effective X groups	*Examples*
Halide	Chloride, bromide
Oxyanions	Sulphate, nitrate
Carboxylates	Halogeno-acetates
Dicarboxylates	Oxalate, malonate, substituted malonates, phthalates

(a) The solubility in water is very low if the carbon chain is long.

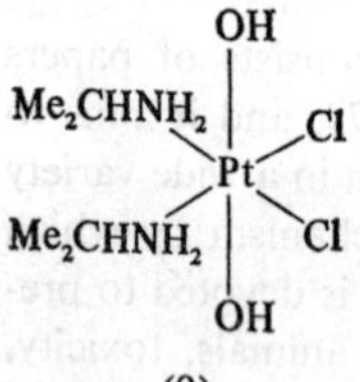

(9)

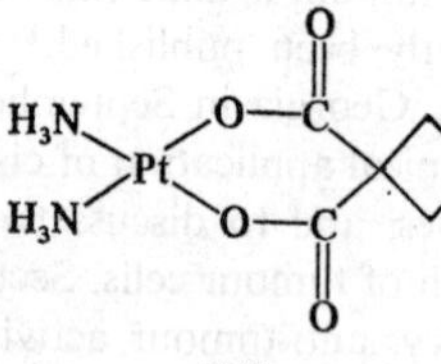

(10)

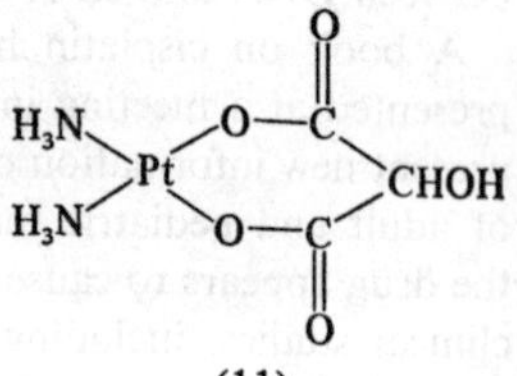

(11)

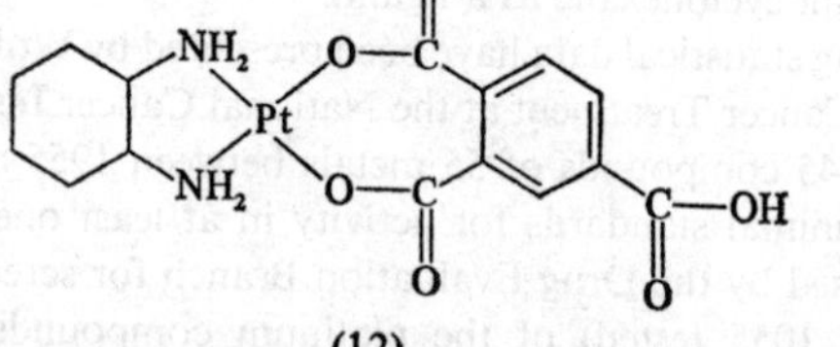

(12)

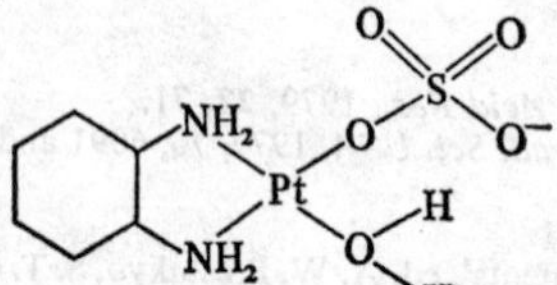

(13)

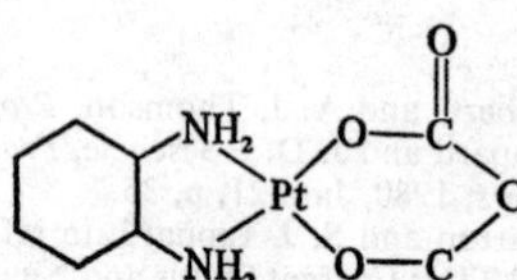

(14)

```
                                                              | - | - |
    | - | - |                               | - | - | - | - | - | - |
    | - | - |                               | Al | - | - | - | - | - |
    | - | - | Sc | Ti | V | Cr | Mn | Fe | Co | Ni | Cu | Zn | Ga | Ge | As | Se | - | - |
    | - | - | Y  | Zr | Nb | Mo | - | Ru | Rh | Pd | Ag | Cd | In | Sn | Sb | Te | - | - |
    |       | La* | Hf | Ta | W | Re | Os | Ir | Pt | Au | Hg | Tl | Pb | Bi | - | - | - |
    |       |
```

*lanthanides   Ce | Pr | Nd | Pm | Sm | Eu | Gd | Tb | Dy | Ho | Er | Tm | Yb | Lu |

Key (i) **bold type**, at least 30 different complexes were tested, of which 5% or more were active.
   (ii) *italic type*, more than 500 different complexes were tested.
   (iii) ***italic bold type***, conditions (i) and (ii) both apply.

Figure 1 *The distribution over the Periodic Table of metals whose compounds have been tested for anti-tumour activity by the National Cancer Institute, USA, between 1955 and 1979 (in all, 11 475 compounds of 55 metals; data are taken from ref. 115)*

8 Gold

Gold Compounds as Anti-tumour Agents?—The ion $[Au^{III}Cl_4]^-$ is known to react with a variety of nucleosides such as adenine, guanine, and xanthine to form insoluble complexes and with nucleotides such as AMP, GMP, ADP, ATP, and cyclic AMP to give soluble, but polymeric, complexes.[116] There is a report[117] that an Au^{III}-diazouracil complex, $[Au(5\text{-}du)_2Cl_2]Cl\cdot HCl$, shows anti-tumour activity in mice. Recent interest has focused on the possibility that Au^I complexes can also exhibit anti-tumour activity.

Agrawal *et al.*[118] have allowed triphenylphosphinegold acetate, $(Ph_3P)Au$-(OAc), to react with 2-thiouracil, 5-fluorouracil, 5-fluorodeoxyuridine, thymidine, and 6-mercaptopurine, presumably to give complexes in which an *N*-bonded or *S*-bonded (with a thio-group present) nucleoside replaces acetate. The resultant complexes have anti-neoplastic activity against P388 leukaemia in mice. Curiously, the thiouracil complex is inactive but the mercaptopurine complex is active. The thymidine complex was the most active, with a T/C value (the ratio of survival times of treated and control animals) of 171% when a dose of 30 mg per kg was administered on days 1, 3, and 5, beginning 24 hours after transplantation.

Lorber *et al.* have suggested[119] that the orally active gold(I) triethylphosphine thioglucose derivative auranofin may be useful for treatment of tumours such as metastatic colon cancer. Auranofin exerts irreversible dose-dependent inhibition of the synthesis of DNA and inhibits the uptake of [³H]thymidine and (to a lesser extent) of [³H]leucine and of [³H]uridine by HeLa cells at concentrations of gold of *ca* 4 μmol l⁻¹. HeLa cells have reduced viability and uptake of

116 D. W. Gibson, M. Beer, and R. J. Barrett, *Biochemistry*, 1971, **10**, 3669.
117 C. Dragulescu, J. Heller, A. Maurer, S. Policec, V. Topcui, M. Csalci, S. Kirschner, S. Kravitz, and R. Moraski, *Int. Coord. Chem., Conf. XVI*, 1974, abstr. 1.9.
118 K. C. Agrawal, K. B. Bears, D. Marcus, and H. B. Jonassen, *Cancer Abstr.*, 1978, abstract 110.
119 T. M. Simon, D. H. Kunishima, G. J. Vibert, and A. Lorber, *Cancer (Brussels)*, 1979, **44**, 1965.

oxygen after exposure to auranofin, and exhibit dose-dependent morphological changes, *i.e.* blebbing and pitting. Membrane blebbing has also been observed *in vivo* on lymphocytes from arthritic patients who had been treated with auranofin.[120] This may be related to cell-cycle sensitivity or may merely be a manifestation of cell death.

An outstanding feature of the clinical trials using auranofin for rheumatoid arthritis has been the lack of serious side-effects, although transient elevation of levels of enzymes in liver and diarrhoea have been noted.[121]

Chrysotherapy.—Shaw[122] has reviewed in detail the mammalian biochemistry of gold, focusing attention on the action of gold in the therapy of rheumatoid arthritis. There now appears to be some evidence from animal studies that a metallothionein-like protein (of mol. wt 10 000) is present in the cytosol of liver and kidney tissue following the administration (*in vivo*) of aurothiomalate, gold(I) bis(thioglucose), $(Et_3P)AuCl$, and $HAuCl_4$ (adjusted to pH 6). Mogilnicka and Piotrowski[123] also found gold complexes of low molecular weight in mitochondria and nuclei after administration of chloroauric acid (at pH 6). It is notable that the low-molecular-weight gold-containing proteins which they isolated also contained a high concentration of copper (Au:Cu molar ratio of *ca* 1:3). Copper has long been implicated in the aetiology of rheumatoid arthritis.

The proceedings of an international workshop and symposium on gold salts in the rheumatic diseases (held in Montreal, Canada on 5 February 1979) are available as a supplement of the Journal of Rheumatology.[124]

Gottlieb[125] has discussed the metabolism and distribution of aurothioglucose and aurothiomalate (see Table 4) in comparison with auranofin. The beneficial effects of the latter, orally active, compound appear sooner (after 5 weeks) than with the gold thiolates (after several months). In addition, adverse reactions such as stomatitis, dermatitis, and proteinuria are absent, although gastrointestinal symptoms can appear. However, cumulative doses of up to fifteen grams of autothiomalate have been given to patients without evidence of cumulative toxicity or adverse reactions.

Taylor *et al.*[126] have studied the metabolism of ^{14}C-labelled sodium aurothiomalate in rats. They found that 50% of the injected radioactivity is excreted in urine in the first day; 10% appears in the expired air in the first six hours and none thereafter. About 3% is present in the faeces. Radioactivity was found in muscle, bone, and skin, with residual activity in all other tissues examined. A similar distribution pattern was found for gold. They concluded that although some thiomalate is metabolized *via* the tricarboxylic acid cycle, the widespread distribution of ^{14}C and gold suggests that the complex itself, and not just Au^I, is relevant to metabolic studies of aurothiomalate.

[120] A. Lorber, T. M. Simon, J. Leeb, A. Peter, and S. A. Wilcox, *J. Rheumatol.*, 1979, 6 (Suppl. 5), 82.
[121] F. E. Berglof, K. Berglof, and D. T. Walz, *J. Rheumatol.*, 1978, **5**, 68.
[122] C. F. Shaw, *Inorg. Perspect. Biol. Med.*, 1979, **2**, 287.
[123] E. M. Mogilnicka and J. K. Piotrowski, *Biochem. Pharmacol.*, 1979, **28**, 2625.
[124] *J. Rheumatol.*, 1979, 6 (Suppl. 5), pp. 1—164.
[125] N. L. Gottlieb, in ref. 124, p. 2.
[126] A. Taylor, L. J. King, and V. Marks, *Ann. Clin. Biochem.* (*Suppl.*), 1979, **16**, 12.

Table 4 *Levels and distribution of gold after weekly administration of* 50 mg *of aurothiomalate (AuTM) or aurothioglucose (AuTG)*

Level of gold in serum rises gradually for 0—8 weeks: peak is *ca* 36 μmol l^{-1} within 2 h for AuTM, but AuTG is absorbed more slowly than AuTM, and its peak is *ca* 24 μmol l^{-1}, with a plateau of *ca* 15 μmol l^{-1}. The $T_{\frac{1}{2}}$ for Au*TM is *ca* 6 days.
40% of administered dose is excreted weekly (70% in urine, 30% in faeces).

Distribution is very wide, and is high in lymph nodes, adrenal gland, liver, kidney, bone marrow, spleen > other internal organs, endocrine glands, muscle, skin > eye lens, hair, nails, body fluids. Skin accumulates gold (mostly in dermis).
Storage of gold: bone marrow > liver > skin > bone (accounts for 85% of body total of gold).

Aurosomes.—Aurosomes are lysosomal bodies that contain gold. Administration of colloidal gold produces aurosomes that contain spherical electron-dense granules, whereas soluble gold complexes produce aurosomes containing lamellar filamentous, and rod-like profiles, studied with particles and granules. The appearance of aurosomes is not affected by the particular gold complex that is used (aurothiomalate, aurothioglucose, or aurothiosulphate) or by the route of its administration. Even orally administered (Et$_3$P)AuCl produces aurosomes of similar morphology in the synovial intimal cells and subsynovial macrophages. Electron microprobe analysis of the aurosomes gives an approximate composition Au:P:S = 1:1.17:3.07.[127]

It seems likely that gold enters pre-existing lysosomes and alters their behaviour so as to prevent the release of enzymes which would perpetuate the inflammatory process and destroy articular tissues. Lorber and Simon[128] have discussed possible ways in which gold may modify immune processes.

9 The Actinides: Bioinorganic Chemistry

Bulman[129] has reviewed the bioinorganic chemistry of the actinides. Most studies have been directed towards uranium and plutonium, but interest in americium and curium has grown, since they are produced by fast-breeder nuclear reactors which use plutonium-239 as the fissionable-fuel element. Plutonium, americium, and curium are regarded as major hazards to health, as they emit α-particles.

The most stable oxidation state of plutonium in biological fluids appears to be Pu^{IV}. Aquated ions rapidly polymerize to species such as [Pu(OH)$_4$]$_n$, which has a very low solubility product ($K_s \approx 7 \times 10^{-56}$). Polymerization can be prevented by carboxylate ligands such as citrate. In blood, a major binding site appears to be transferrin. Uptake of Pu^{4+}, like that of Fe^{3+}, requires the presence of bicarbonate.

Actinides are readily taken up into mineralizing tissue, either phosphate- or

127 F. N. Ghadially, in ref. 124, pp. 25, 45.
128 A. Lorber and T. M. Simon, *Gold Bull.*, 1979, 12, 149.
129 R. A. Bulman, *Coord. Chem. Rev.*, 1980, 31, 221.

carbonate-based. The ion $UO_2{}^{2+}$, for example, becomes associated with areas of active calcification, alveolar crests of the mandible, and the primary spongiosa of the long bones. The uptake of radioelements on to the surface of bone poses a major threat to health because of the carcinogenicity that is associated with the irradiation of osteogenic cells that are close to the endosteal surface. It seems likely that lipophilic chelating agents will be required to remove them.

Diethylenetriaminepenta-acetic acid, DTPA, is superior to EDTA for the mobilization of plutonium in experimental animals. However, it does not cross cell membranes. If administered soon after the uptake of Pu^{4+}, Am^{3+}, or Cm^{3+}, it is effective in removing them from the serum. Amongst the chelating agents which are ineffective in clearing plutonium from experimental animals are *N*-stearoyl-desferrioxamine, Pyridine-2,6-dihydroxamic acid, and a phosphatidylethanol-amido-EDTA derivative (all lipophilic), as well as rhodotorulic acid and 2,3-dihydroxybenzoyl-*N*-glycine.[130] Two encouraging recent developments have occurred at Berkeley[131] and at Harwell.[132] In the latter laboratories, dialkyl-amido-derivatives of DTPA have been synthesized (*e.g.* Puchel). Puchel is partially lipophilic, and it removes reduced plutonium contamination in hamsters which had inhaled mixed oxides of Pu and Na to 40% of that of control animals in 30 days. A report[133] that the simultaneous administration of DTPA and salicylic acid is effective in removing actinides from the body has not been verified in other laboratories.[134, 135]

10 The Cardiovascular System

In the first volume of these Reports[1] we discussed various aspects of the effects of magnesium and its deficiency on the cardiovascular system. This story has now developed somewhat, and, in addition, a number of studies with other elements have made an updating of the literature appropriate.

The possible role of vanadium in the regulation of Na-K ATPase was reviewed in Volume 2 of these Reports.[2] Since much of the activity of the heart and circulation is regulated by fluxes of sodium and potassium, for instance in initiation of the heartbeat, contraction of vascular smooth muscle, and regulation of the ionic constituents of the red cell, it is of interest that studies have now been carried out on the effects of vanadium in the intact animal. It has been reported that the infusion of sodium vanadate into dogs produces constriction in the vascular beds of skin and in the skeletal muscles. As a consequence of this, the arteriolar and venous blood pressures increased, owing to the decrease in peripheral circulation.[136, 137] The individual effects were also investigated, and the cardiac output was shown to the decreased in dog heart whilst the blood pressure

[130] R. A. Bulman, R. J. Griffin, and A. T. Russell, *Health Phys.*, 1979, **37**, 729.
[131] F. L. Weitl, K. N. Raymond, W. L. Smith, and T. R. Howard, *J. Am. Chem. Soc.*, 1978, **100**, 1170.
[132] R. A. Bulman, J. W. Stather, J. C. Strong, P. Rodwell, R. J. Griffin, and A. Hodgson, *Nat. Radiol. Protect. Board NRPB/R&D 3, 19780*, HMSO, London, 1979, p. 128.
[133] J. Schubert and S. K. Derr, *Nature (London)*, 1978, **275**, 311.
[134] R. A. Bulman, F. E. H. Gawley, and D. A. Geden, *Nature (London)*, 1979, **281**, 406.
[135] J. Shubert, *Nature (London)*, 1979, **281**, 406.
[136] D. J. Inciarte, R. P. Steffen, B. T. Swindall, and F. J. Haddy, *Physiologist*, 1978, **21**, 57.
[137] D. J. Inciarte, R. P. Steffen, B. T. Swindall, J. Johnston, D. E. Dobbins, and F. J. Haddy, *Fed. Proc.*, 1979, **38**, 1036.

increased.[138] Strain gauges were introduced into the heart and the blood pressure in the coronary arteries was measured. It was demonstrated that vanadate increased the resistance of the coronary blood vessels but had little effect on the contractile force of the heart.[139] Finally, a study on the peripheral blood vessels themselves, this time in a preparation *in vitro*, showed that vanadate caused an increase in tension of the isolated venous smooth muscle which is not prevented by α-adrenergic blocking agents. The effect does not appear to be due to the inhibition of the Na–K pump, as had originally been assumed, and the question of the mode of action of vanadium on the cardiovascular system now remains open.

Such effects on blood pressure are of course important from the theoretical point of view, since over 90% of cases of elevated blood pressure (hypertension) are due to causes which have not yet been identified. This essential hypertension can therefore be diagnosed only by excluding other causes of increased blood pressure, such as increased secretion of aldosterone or adrenaline, or the failure of the kidney. Since hypertension is increasing in incidence in the Western world, and commonly results in cerebrovascular accidents (stroke), it is important that the cause of the elevated blood pressure in essential hypertension should be identified. About 10% of the population are at risk, and there is a familial trait. Recent studies implicating changes in the regulation of sodium and potassium in red cells[140] therefore provide the exciting possibility that the genetic defect in essential hypertension may be in the regulation of these ions in the various tissues of the cardiovascular system. The vanadium story discussed above may well be related. It has been reported that the abnormality is in the sodium–sodium counter-exchange system rather than in the Na-K ATPase, and that this may be identified by the substitution of lithium on the internal surface of the red cell. A diagnostic test for essential hypertension, using lithium transport, has been proposed.[50] This remains to be confirmed. The role of increased salt intake in human hypertension has been known for many years, and is the subject of a recent review.[141]

In addition to the identification of the role of metals in hypertension, effects have been seen of chronic low levels of cadmium and lead on the function of the heart and on its metabolism.[142] In rats that were fed with 5 p.p.m. of cadmium and/or lead over a period of fifteen months there was no detectable difference in growth or in the intake of food or water, but myocardial contractility and response to β-adrenergic stimulation were decreased. These changes were associated with depressed phosphorylation of the cardiac myofibrillar proteins.

[138] R. P. Steffen, D. J. Inciarte, B. T. Swindall, J. Johnston, and F. J. Haddy, *Fed. Proc.*, 1979, **38**, 1440.
[139] S. Huot, S. Muldoon, M. Pamnani, D. Clough, and F. J. Haddy, *Fed. Proc.*, 1979, **38**, 1036.
[140] R. P. Garray, J. L. Elghozi, G. Dafher, and P. Meyer, *N. Engl. J. Med.*, 1980, **302**, 769.
[141] F. C. Bartter, T. Fujita, C. S. Delea, and T. Kawasaki, in 'Biological and Behavioral Aspects of Salt Intake', ed. M. R. Kare, M. J. Fregly, and R. A. Bernard, Academic Press, New York, 1980, p. 341.
[142] S. J. Kopp, M. Barany, M. Erlanger, E. F. Perry, and H. M. Perry, *Toxicol. Appl. Pharmacol.*, 1980, **54**, 48.

Further studies[143] indicated that cadmium selectively depressed the conduction in the Bundle of His near the atrioventricular node while lead and (lead + cadmium) caused an impairment at a site in the conducting tissue beyond the Bundle of His. These results demonstrate that significant changes may be seen in the performance of the heart despite the absence of overt heavy-metal toxicity. In contrast, a study of hypertensive and non-hypertensive subjects indicated no difference in the cadmium content of their kidneys, as determined by neutron activation.[144] However, both of these groups were selected from those who were known not to be especially exposed to cadmium, and this study does not contradict the known possibility of hypertensive effects during cadmium toxicity.[145] The levels of cadmium in the kidney were elevated in smokers of tobacco.

In a review of minerals, coronary heart disease, and sudden coronary death, a number of reports of hypertension resulting from the administration of cadmium are considered; it is reported that, not only is there danger of ingestion of cadmium from industrial sources, but the element may be dissolved from drinking-water pipes in areas in which the water is soft, or be present in populations which consume large quantities of fish that have become contaminated with the element from industrial effluents.[145] Chromium deficiency has also been reported to be associated with ischaemic heart disease.[145, 146]

The role of the hardness of water in the development of ischaemic heart disease has been reviewed.[146-148] There is strong evidence to implicate an increased incidence of ischaemic heart disease with a low intake of magnesium, and various experimental studies have been carried out on effects of magnesium deficiency on the cardiovascular system. Altura has carried out a wide range of studies on such systems, and has concluded that the magnesium concentration in plasma regulates the permeability of membranes to calcium and may also play an important role in regulating the muscular tone of vascular smooth muscle.[149] He has further proposed that sudden-death ischaemic heart disease is a result of a decrease in magnesium concentration in and around the coronary arterioles, and that hypomagnesaemia produces progressive vasoconstriction, vasospasm, and ischaemia, leading eventually to sudden death.[150] In contrast, a study from Finland indicated that no specific histopathological changes were found in the heart in pigs which were fed a low-magnesium diet.[151] This contrasts with the data which suggest that humans who succumb to heart attacks generally have a lower magnesium content in heart tissue.[147, 148, 152] This latter evidence has

[143] S. J. Kopp, H. M. Perry, T. Glonek, M. Erlanger, E. F. Perry, and M. Barany, *Am. J. Physiol.*, 1980, **239**, H22.

[144] P. E. Cummins, J. Dutton, C. J. Evans, W. D. Morgan, A. Sivyer, and P. C. Elwood, personal communication.

[145] H. Karppanen, R. Pennanen, and L. Passinen, *Adv. Cardiol.*, 1978, **25**, 9.

[146] J. Vobecky and D. Shapcott, *Can. Med. Assoc. J.*, 1975, **113**, 922.

[147] T. W. Anderson, L. C. Neri, G. B. Schreiber, F. D. F. Talbot, and A. Zdrojewski, *Can. Med. Assoc. J.*, 1975, **113**, 199.

[148] J. Marier, L. C. Neri, and T. W. Anderson, *Nat. Res. Counc. Can. Publ.*, No. 17 581, 1979.

[149] B. M. Altura and B. T. Altura, *Blood Vessels*, 1978, **15**, 5.

[150] B. M. Altura, *Med. Hypotheses*, 1979, **5**, 843.

[151] P. J. Muobanne, R. P. Raunio, P. Saukko, and H. Karpannen, *Br. J. Nutr.*, in the press.

[152] J. R. Marier, in 'Medecine et Nutrition', ed. H. Gounelle de Pontanel, La Simarre, Tours, 1980, Vol. XVI, p. 23.

been sufficiently provocative that the public health authorities in Finland, which has a particularly high incidence of coronary heart disease, have been persuaded to allow an intervention study in which selected geographical areas have their domestically consumed salt supplemented with magnesium (Karppanen, personal communication).

The relationship of the calcium ion to the development of myocardial damage has been investigated. The strength of myocardial contraction produced by the β-adrenargic stimulus of isoprenaline in low doses and the myocardial necrosis that is produced by large doses have been suggested to be due to alterations in calcium metabolism. Following a dose of isoprenaline, a rapid influx of calcium into the heart muscle occurs; this may be prevented by the β-adrenergic blocking agent propranolol. However, the blocking agent did not completely prevent the myocardial necrosis, which suggests that additional mechanisms are involved.[153, 154] Recent studies have suggested that the calcium influx occurs during the process of cell injury, and is the determinant of whether or not a cell will survive. Calcium prevents phosphorylation and inhibits the uptake of oxygen. The rapid influx of calcium may quickly inactivate the cell; this may serve a protective function by leaving oxygen available for adjacent cells.[155]

[153] S. Bloom and D. Davis, in 'Myocardial Biology', ed. N. S. Dhalla, University Park Press, Baltimore, 1974, p. 581.
[154] S. Bloom and D. L. Davis, *Am. J. Pathol.*, 1972, **69**, 459.
[155] J. Tsokos and S. Bloom, *J. Mol. Cell. Cardiol.*, 1977, **9**, 823.